D1431885

Conformational Analysis

Conformational Analysis

Ernest L. Eliel *University of Notre Dame*

Norman L. Allinger *Wayne State University*

Stephen J. Angyal *The University of New South Wales*

George A. Morrison *University of Leeds*

Washington, DC
1981

833066

Conformational analysis.
   Reprint. Originally published: New York: Interscience Publishers, 1965.

   Includes bibliographies and index.

   1. Conformational analysis.
   I. Eliel, Ernest Ludwig, 1921-

| QD481.C68 | 1981 | 547.1'223 | 81-10783 |
| ISBN 0-8412-0653-8 | | | AACR2 |

© 1965 by the American Chemical Society

Previously published by Interscience Publishers, a division of John Wiley & Sons, Inc.

Reprinted 1981 by the American Chemical Society

All rights reserved. No part of this book may be reproduced or transmitted in any form
or by any means—graphic, electronic, including photocopying, recording, taping, or information storage and retrieval systems—without written permission from the American
Chemical Society.

PRINTED IN THE UNITED STATES OF AMERICA

To Derek H. R. Barton

Preface

The history of conformational analysis parallels the development of many other scientific concepts and, indeed, the history of mankind itself. The subject originated with Sachse in 1890 but failed to develop for almost thirty years. Revived in 1918, it grew at a very slow rate for over thirty years more until 1950 when, after the appearance of a pioneering paper by the man to whom this book is dedicated, an explosive growth set in and as a result conformational analysis has pervaded nearly all of organic chemistry. Now, less than fifteen years later, it is almost platitudinous to say that a chemist who does not understand conformational analysis does not understand organic chemistry. Even the area of physical chemistry related to molecular structure and physical properties has fallen heavily under the sway of conformational concepts, owing primarily to the pioneering work of Pitzer, Mizushima, and Hassel. The appearance of a monograph on conformational analysis thus seems timely—if indeed the best moment for it has not already passed, for the subject is proliferating so rapidly that, between the writing of a monograph and its appearance, not only new facts but entire new avenues of thought in the field come to light. Thus, for example, the subject of protein structure, in which conformational concepts are now becoming increasingly important, is not included in this book because, at the time the book was conceived, the subject was not yet considered ripe for systematic treatment.

This is a co-authored book. As such, it represents the combined effort and the combined judgment of four authors on the subject matter included. Despite the wide scope of the book—from ethane to rather complex terpenes and steroids—which would have made it a most difficult monograph for an individual author to write, we have striven to preserve a sense of unity. This, we hope, will make it a readable book for the novice. At the same time, we hope that the expert will find some interesting morsels in the pages to follow, even though there may not be as much detail in each of the many conceivable subdivisions of the subject as could have been provided by a multi-authored, more voluminous, less coherent, and, alas, more expensive treatise.

A book of this size could not have been produced without the help of many previous monographs and reviews in the field to whose authors, collectively, we express our appreciation. We would also like to express our special thanks to Dr. Janet Allinger, to Dr. J. A. Mills who supplied a draft which aided us in the writing of Secs. 6-5a and 6-5b, to Eva Eliel who prepared the author index and helped with the subject index, and to a number of co-workers who assisted in the checking of proof.

ERNEST L. ELIEL
NORMAN L. ALLINGER
STEPHEN J. ANGYAL
GEORGE A. MORRISON

Contents

Introduction

Definition

By "conformations" are meant the non-identical arrangements of the atoms in a molecule obtainable by rotation about one or more single bonds. Obviously, only rotation about bonds between atoms carrying at least one other substituent can lead to differences in conformation. Thus the simplest molecule in which conformation can be defined is hydrogen peroxide, H—O—O—H, where rotation about the O—O bond may lead to an infinite number of different conformations. In contrast, water, H—O—H, cannot give rise to a family of conformations, for, although rotation about the O—H bonds is possible, such rotation does not lead to distinctive arrangements. When there are more than two substituted atoms (so-called "tops") which can rotate with respect to each other, like the methyl groups in propane, H_3C—CH_2—CH_3, differences in conformation will, of course, result from rotation about any or all of the salient bonds.

As will be shown in Sec. 1-2, rotation about single bonds is, in most instances, not free and some conformations are more stable than others. "Conformational analysis" is an analysis of the physical and chemical properties of a compound in terms of the conformation (or conformations) of the pertinent ground states, transition states, and (in the case of spectra) excited states.

It is clear that the different conformations of a molecule are stereoisomeric, since they differ in spatial arrangement, but not in the number and kind of atoms and the way in which these atoms are linked. The delineation of conformational isomerism from other types of stereoisomerism, especially geometrical isomerism and optical isomerism of the biphenyl type, is thus a matter of importance, lest "conformational analysis" become synonymous with an analysis of physical and chemical properties in terms of all types of stereochemical differences (which it is not). Since the required delineation depends on quantitative

1

considerations and cannot simply be defined in a qualitative way, it will be deferred to Sec. 1-2.

Historical

The theory of the tetrahedral geometry of the linkages to saturated carbon was first proposed by van't Hoff[1] in 1874. About a decade later, Baeyer proposed his famous strain theory for ring compounds.[2] According to this theory, the carbon skeleton of a cyclic compound is assumed to have the shape of a planar polygon, and the strain in such a cycle is then one-half the absolute value of the difference between the internal angle of the polygon and the normal tetrahedral angle of 109° 28'. This leads to the prediction that the strain in a cycloalkane decreases from that in cyclopropane to a minimum in cyclopentane and then increases again for larger rings, presumably indefinitely. Experimental data on heats of combustion of cycloalkanes now available[3] clearly show that the strain theory is incorrect for any rings larger than five-membered ones. Even at the time Baeyer wrote his original paper, he had some reservations on six-membered rings (which appeared more stable than predicted by his theory), and it was not long before Sachse pointed out[4] that six-membered rings may be constructed in puckered shapes such that all the valence angles are tetrahedral. Sachse recognized that there were two puckered forms of cyclohexane free of angle strain, one of which he called the flexible or unsymmetric form, the other the rigid or symmetric form. (These forms are described in more detail in Sec. 2-1 where the former is called "flexible" and the latter "chair.") Sachse also realized that there could be two monosubstitution products of the rigid or chair form (one having what we now call an equatorial substituent and the other an axial substituent) which could be interconverted by a process of chair inversion (cf. Sec. 2-1) and would thus be in dynamic equilibrium. By virtue of this insight, Sachse may well be considered the founder of conformational analysis. Unfortunately, Sachse eventually came to think that chair interconversion occurred only on heating; in particular, he ventured to suggest that there were two forms (the diequatorial and the diaxial) of

[1] J. H. van't Hoff, *Bull. Soc. Chim. France*, [2] **23**, 295 (1875); the original (Dutch) version appeared in 1874.

[2] A. Baeyer, *Ber.*, **18**, 2269 (1885).

[3] (a) S. Kaarsemaker and J. Coops, *Rec. Trav. Chim.*, **71**, 261 (1952); H. van Kamp, Dissertation, Free University of Amsterdam, Amsterdam, Netherlands, 1957; (b) see also Table 1 in Sec. 4-1.

[4] H. Sachse, *Ber.*, **23**, 1363 (1890); *Z. Physik. Chem.*, **10**, 203 (1892).

trans-hexahydroterephthalic acid. This suggestion, regrettably, became the touchstone of his theory, and, when it was later proved wrong, the theory fell into disrepute. The coup-de-grace was applied in a popular textbook[5] about a decade later in which it was stated that, even if a monosubstituted cyclohexane existed in two chair forms, obviously the forms were in rapid equilibrium and therefore it would be more desirable to consider cyclohexane in the average planar form. This point of view was cited in the literature for almost forty years thereafter, with the unfortunate result that organic chemists were discouraged from thinking of cyclohexane in its true chair-shaped geometry, and it was actually not until 1950 (see below) that this handicap was fully overcome.

For almost thirty years, Sachse's hypothesis lay completely dormant. In 1918 Mohr pointed out[6] that the two chair forms of a substituted cyclohexane should be readily interconvertible by rotation about single bonds* (assumed to be either free or only slightly hindered) and that therefore no isolable isomers were to be expected from this source (as, indeed, none had been found). Mohr furthermore predicted that decalin should be capable of existence in two geometrically isomeric forms (*cis* and *trans*) both of which should be strain-free. This prediction was contrary to Baeyer's theory of planar rings—which would lead to the expectation that *trans*-decalin should either be incapable of existence or highly strained—and when it was verified by Hückel[7] in 1925 that decalin did, in fact, exist in two isomeric forms of which the *trans* was the more stable, the Sachse-Mohr theory of the puckered rings became finally established. Confirmatory chemical evidence came at about the same time from Boëseken's elegant investigations[8] on the effect of cyclic 1,2-diols on the conductivity of boric acid and on their reaction with acetone. However, acceptance of the concepts remained sporadic for another twenty-five years,† despite the fact that physical evidence for the chair shape of cyclohexane was obtained, first tenuously from X-ray diffraction data of the benzene hexahalides[9] and then more firmly from Raman

---

[5] O. Aschan, *Chemie der Alicyclischen Verbindungen*, Vieweg Verlag, Brunswick, Germany, 1905.

[6] E. Mohr, *J. Prakt. Chem.*, [2] **98**, 315 (1918); *Ber.*, **55**, 230 (1922).

[7] W. Hückel, *Ann.*, **441**, 1 (1925).

[8] J. Boëseken and J. van Giffen, *Rec. Trav. Chim.*, **39**, 183 (1920); J. Boëseken, *ibid.*, **40**, 553 (1921); *Ber.*, **56**, 2409 (1923).

[9] R. G. Dickinson and C. Bilicke, *J. Am. Chem. Soc.*, **50**, 764 (1928).

* Some bond angle strain is involved in the interconversion also.

† Some of the salient early contributions to an understanding of conformational principles and to their application came in the carbohydrate field from chemists such as Sponsler and Dore, Haworth, Hudson, Isbell, Hassell and Ottar, and Reeves. These contributions are summarized in Sec. 6-3a.

spectra,[10] electron diffraction data[11] and calculations of thermodynamic properties.[12] That in the chair form of cyclohexane there are two types of bonds (now called equatorial and axial; cf. Sec. 2-1) was first experimentally demonstrated by Kohlrausch.[10] It was confirmed by Hassel[11] and by Pitzer;[12] but the real breakthrough in the conformational analysis of cyclohexane derivatives did not come until 1950 when D. H. R. Barton, in a pioneering paper,[13] pointed out the manifold chemical consequences of the difference between equatorial and axial substituents. Barton's ideas were accepted avidly both by workers interested in natural products and by those concerned with mechanistic studies, and it is the ensuing rapid development of the subject since 1950 which forms the major portion of the subject matter of this book.

[10] K. W. F. Kohlrausch, A. W. Reitz, and W. Stockmair, *Z. Physik. Chem.*, **B32,** 229 (1936).

[11] O. Hassel, *Tidsskr. Kjemi Bergvesen Met.*, **3,** 32 (1943); cf. *Quart. Rev. (London)*, **7,** 221 (1953).

[12] C. W. Beckett, K. S. Pitzer, and R. Spitzer, *J. Am. Chem. Soc.*, **69,** 2488 (1947).

[13] D. H. R. Barton, *Experientia*, **6,** 316 (1950).

Chapter 1

The Conformation of Acyclic Molecules

1-1. Introductory Remarks

This chapter deals with the application of conformational concepts to acyclic compounds. Aside from the existence of biphenyl isomerism recognized[1,2] in 1926, the first suggestion of the importance of conformation (i.e., rotational arrangement) on reactivity in acyclic compounds came, characteristically, in a paper by Boëseken[3] on the reaction of the diastereoisomeric 2,3-butanediols with acetone and on their effect on the conductivity of boric acid. The first suggestion that rotation about single bonds (other than biphenyl bonds) was not free came two years later in a paper[4] on the difference of the dipole moments of the stilbene dichlorides (*meso*, 1.27 D, *dl*, 2.75 D; the configurational assignment was made later). This conclusion was supported by several other investigations of physical properties in the early 1930's,[5] which in turn led to speculations about the energy barrier to rotation about carbon-carbon single bonds. The early work culminated in 1936 in the important suggestion by Kemp and Pitzer[6] that there is a barrier to rotation in ethane of approximately 3 kcal./mole, an assumption which made possible reconciliation of the

[1] F. Bell and J. Kenyon, *Chem. & Ind. (London)*, **4**, 864 (1926); W. H. Mills, *ibid.*, p. 884; cf. also E. E. Turner and R. J. W. LeFèvre, *ibid.*, pp. 831, 883.

[2] For reviews, see R. Adams and H. C. Yuan, *Chem. Rev.*, **12**, 261 (1933), and R. L. Shriner and R. Adams, "Optical Isomerism," in H. Gilman, *Organic Chemistry*, 2nd ed., John Wiley and Sons, New York, 1943, pp. 343–382.

[3] J. Boëseken and R. Cohen *Rec. Trav. Chim.*, **47**, 839 (1928).

[4] A. Weissberger and R. Sängewald, *Z. Physik. Chem.*, **B9**, 133 (1930). See also the almost simultaneous discussion by K. L. Wolf, *Trans. Faraday Soc.*, **26**, 315 (1930).

[5] Cf. S. Mizushima, *Structure of Molecules and Internal Rotation*, Academic Press, Inc., New York, 1954, Chaps. 1–3, for a more detailed discussion of the early work, e.g., S. Mizushima and K. Higasi, *J. Chem. Soc. Japan*, **54**, 226 (1933).

[6] J. D. Kemp and K. S. Pitzer, *J. Chem. Phys.*, **4**, 749 (1936); *J. Am. Chem. Soc.*, **59**, 276 (1937); K. S. Pitzer, *J. Chem. Phys.*, **5**, 473 (1937). See also J. B. Howard, *Phys. Rev.*, **51**, 53 (1937).

experimental data on the heat capacity and entropy of ethane[7] with calculations based on statistical mechanics. The establishment of the ethane barrier probably marks the highlight of the history of conformational analysis in acyclic systems, although there has been much subsequent important work, such as the spectral studies by Mizushima and his school[5] as well as others, studies on the conformation of polypeptides initiated by Pauling and Corey,[8] and the more detailed considerations of the influence of conformation on reactivity in acyclic systems initiated by Curtin.[9] This work will be discussed in the following sections.

1–2. Energy Considerations and Terminology. Ethane. Butane

Figure 1 represents the change in potential energy of ethane in the course of one complete rotation about the carbon-carbon bond. Plotted vertically is potential energy, whereas the abscissa represents the so-called "angle of torsion, τ." By "angle of torsion," sometimes called "dihedral angle" or "azimuthal angle" (Fig. 2), is meant the angle between two planes, one defined by the carbon-carbon bond and one of the carbon-X bonds on the front carbon, the other similarly defined by the C—C bond and a C—Y bond on the rear where X and Y are designated atoms. (In the case of

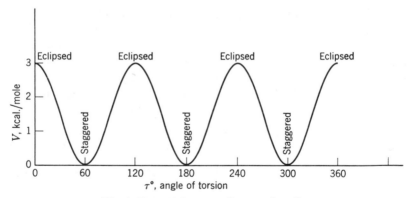

Fig. 1. Potential energy diagram for ethane.

[7] R. K. Witt and J. D. Kemp, *J. Am. Chem. Soc.*, **59**, 273 (1937).

[8] L. Pauling, R. B. Corey, and H. R. Branson, *Proc. Natl. Acad. Sci. U.S.*, **37**, 205 (1951). See also ref. 5, Chap. 6; S. J. Singer, *Advan. Protein Chem.*, **17**, 1 (1962); W. Kauzmann, *ibid.*, **14**, 1 (1959); E. R. Blout, in C. Djerassi, ed., *Optical Rotatory Dispersion*, McGraw-Hill Book Co., Inc., New York, 1960, Chap. 17.

[9] (a) P. I. Pollak and D. Y. Curtin, *J. Am. Chem. Soc.*, **72**, 961 (1950), and later papers; (b) cf. D. Y. Curtin, *Record Chem. Progr. Kresge-Hooker Sci. Lib.*,**15**, 111 (1954).

ethane, X would be a specified hydrogen on C_1 and Y a specified hydrogen on C_2.) In the case of ethane, there are six conformations of special interest: the so-called "opposed" or "eclipsed" conformations which correspond to energy maxima and for which $\tau = 0°$, $120°$, or $240°$; and the so-called "staggered" conformations which correspond to energy minima[10] and for which $\tau = 60°$, $180°$, or $300°$. Common representations for these conformations in terms of the so-called "saw-horse" formulas and "Newman projection" formulas[11] are shown in Fig. 3. (Other representations have been used but will not be employed in this book.) As mentioned in the Introduction, reconciliation of the experimental and calculated heat capacity and entropy of ethane requires a potential barrier (cf. Fig. 1) of approximately 3 kcal./mole. (At the time of writing, the best thermodynamic value is given[12] as 2875 ± 125 cal./mole and the best spectroscopic value[13] as 3030 ± 300 cal./mole. Unfortunately the height of the ethane barrier is not known with great accuracy.)

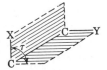

Fig. 2. Angle of torsion.

The variation of potential energy with dihedral angle (Fig. 1) is approximately sinusoidal and may be expressed quite closely by the expression

$$V(\tau) = \tfrac{1}{2} V_0 (1 + \cos 3\tau)$$

where V_0 is the height of the potential barrier. The source of the potential barrier is not clear at the present time, although there has been much speculation on this point.[14] van der Waals repulsion between hydrogen atoms in the eclipsed conformation is clearly inadequate, for these

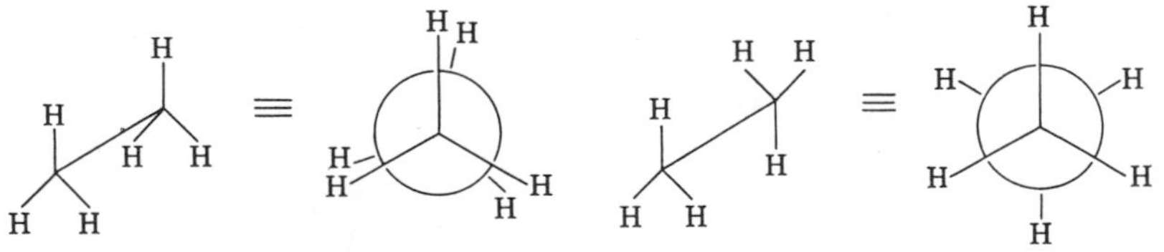

Eclipsed conformation*   Staggered conformation

Fig. 3. Saw-horse and Newman projection formulas for ethane.

[10] That the staggered conformation of ethane is, in fact, the stable one has been deduced from infrared spectral evidence: L. G. Smith, *J. Chem. Phys.*, **17**, 139 (1949).

[11] M. S. Newman, *J. Chem. Educ.*, **32**, 344 (1955).

[12] K. S. Pitzer, *Discussions Faraday Soc.*, **10**, 66 (1951).

[13] D. R. Lide, *J. Chem. Phys.*, **29**, 1426 (1958).

[14] E. B. Wilson, "The Problem of Barriers to Internal Rotation in Molecules," *Advan. Chem. Phys.*, **2**, 367–393 (1959).

* For the sake of clarity, eclipsed conformations in Newman projection formulas are drawn with a dihedral angle of about 10° instead of the required angle of 0°.

hydrogens are 2.3 Å apart and thus barely touch, since the van der Waals radius of hydrogen is 1.2 Å. Even considering that there are three pairs of interacting hydrogens, the total interaction is at most 360 cal./mole.[15] If this were indeed all the ethane barrier amounted to, rotation would be essentially free, for the thermal rotational energy of ethane is RT or approximately 600 cal./mole at room temperature. The suggestion has been made[16] that the barrier is due to quadrupole repulsion of the polarized C—H bonds, but this also appears unlikely, for it has been calculated[17] that an almost completely ionic bond, C^-H^+, would be required to account for the experimentally observed barriers in some common molecules. Other possibilities which have been considered are that the barrier is due to some property of the carbon-carbon bond (rather than of the attached bonds or atoms),[17] and that it is due to quantum mechanical exchange interactions of C—H bond electrons.[18] The latter hypothesis requires that the C—H bond be not a pure sp^3 hybridized bond, but that it have some d and f character (about 2% of each). It leads to the correct prediction that the barrier is about the same in all compounds of type

$$R_1R_2R_3C—CR_4R_5R_6$$

regardless of the nature of the R's (cf. Chap. 3, Table 1, p. 140),* and that the barriers in CH_3NH_2 and CH_3OH should be roughly two-thirds and one-third, respectively, of the barrier in CH_3CH_3. Finally, it has been suggested[19] that the barrier is due not to destabilization of the eclipsed form but to stabilization of the staggered form by some hyperconjugative effect of the C—H bond electrons, believed to be at a maximum when the H—C—C—H system has the *anti* geometry (see below):

This arrangement occurs only in the staggered, not in the eclipsed, form of ethane.

[15] H. Eyring, *J. Am. Chem. Soc.*, **54**, 3191 (1932).

[16] E. N. Lassettre and L. B. Dean, *J. Chem. Phys.*, **17**, 317 (1949).

[17] E. B. Wilson, *Proc. Natl. Acad. Sci. U.S.*, **43**, 816 (1957).

[18] L. Pauling, *Proc. Natl. Acad. Sci. U.S.*, **44**, 211 (1958).

[19] H. Eyring, G. H. Stewart, and R. P. Smith, *Proc. Natl. Acad. Sci. U.S.*, **44**, 259 (1958); G. H. Stewart and H. Eyring, *J. Chem. Educ.*, **35**, 550 (1958).

* See, however, J. G. Aston, P. E. Wills, and T. P. Zolki, *J. Am. Chem. Soc.*, **77**, 3939 (1955).

A considerable number of potential barriers have recently been measured by microwave spectroscopy.[14] The method is discussed in Sec. 3-2c and the results are tabulated in Table 1 on p. 140.

In ethane, all energy minima ("stable conformations") are identical. A different situation pertains in butane whose potential energy diagram is shown in Fig. 4. The angle of torsion plotted here is the angle between the C—C—CH_3 planes. It is immediately obvious that there are two different minima and two different barriers. The higher barrier corresponds to one CH_3-CH_3 and two H-H eclipsings and is not known with accuracy; estimates[20] range from 4.4 to 6.1 kcal./mole. The lower barrier, which involves two CH_3-H and one H-H interaction is assumed to be equal to the barrier in propane (3.3–3.4 kcal./mole).[21] The two energy minima differ by 0.8–0.9 kcal./mole;[22] a value of 0.8 kcal./mole will be used in this book.

It is clear from Fig. 4 that butane has two stable conformations, called "conformational isomers" or sometimes "conformers." The Newman projections of these forms are shown in Fig. 5. The less stable isomer is

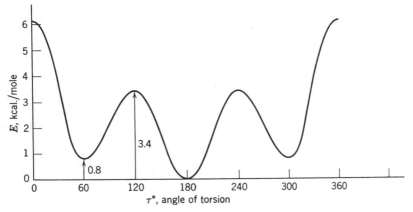

Fig. 4. Potential energy diagram for butane.

[20] (a) W. G. Dauben and K. S. Pitzer, in *Steric Effects in Organic Chemistry*, M. S. Newman, ed., John Wiley and Sons, New York, 1956, p. 8; (b) K. Ito, *J. Am. Chem. Soc.*, **75**, 2430 (1953).

[21] K. S. Pitzer, *J. Chem. Phys.*, **12**, 310 (1944), and earlier work there cited. Actually, propane has one CH_3-H and two H-H interactions. However, since the CH_3-H eclipsing leads to hardly more interaction energy than the H-H eclipsing—note the 3.3–3.4 kcal./mole barrier in propane compared to the 2.8–3.1 kcal./mole barrier in ethane—this is barely of significance.

[22] (a) K. S. Pitzer, *J. Chem. Phys.*, **8**, 711 (1940); (b) G. J. Szasz, N. Sheppard, and D. H. Rank, *ibid.*, **16**, 704 (1948); (c) R. A. Bonham and L. S. Bartell, *J. Am. Chem. Soc.*, **81**, 3491 (1959).

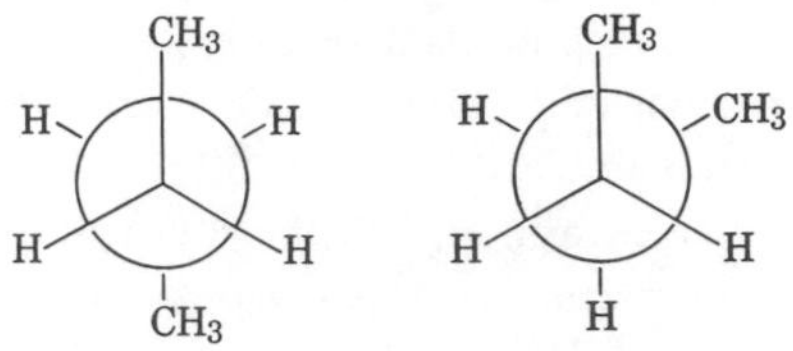

trans or anti   gauche or skew

Fig. 5. Conformational isomers of butane.

called the "*gauche*" or "*skew*" isomer. The more stable isomer will be called "*anti*" in this book; more commonly the term "*trans*" has been used, but, since this term is also used for configurational designation, it is confusing and will be avoided here. The two conformational isomers could also be described in terms of the angle of torsion, τ, between the methyl groups: $\tau = 60°$ for the *gauche* form and $180°$ for the *anti* form.[23] Since, in many cases, the exact angle of torsion is not known, the following nomenclature in terms of approximative dihedral angles has been proposed:[23]

Angle of Torsion	Designation	Symbol
$-30°$ to $+30°$	$\pm syn$-periplanar	$\pm$sp
$+30°$ to $+90°$	$+syn$-clinal	$+$sc
$+90°$ to $+150°$	$+anti$-clinal	$+$ac
$+150°$ to $210°$ (or $-150°$)	$\pm anti$-periplanar	$\pm$ap
$-30°$ to $-90°$	$-syn$-clinal	$-$sc
$-90°$ to $-150°$	$-anti$-clinal	$-$ac

It might be noted that *syn*-periplanar is synonymous with eclipsed, *syn*-clinal is synonymous with *gauche*, and *anti*-periplanar is synonymous with *anti*.

In butane, since the *gauche* and *anti* forms differ in potential energy by 0.8 kcal./mole (in favor of the *anti* form), there will be about two molecules in the *anti* form for each molecule in one of the *gauche* forms at room temperature. This estimate is arrived at as follows: the potential energy difference may be equated to an enthalpy difference, $\Delta H°$, favoring the *anti* form. However, since there are two *gauche* forms for each *anti*, the *gauche* form has a double probability which leads to its being favored,

[23] W. Klyne and V. Prelog, *Experientia*, **16**, 521 (1960). τ is taken to be positive if, in going from one substituent (methyl group) to the other, one describes a turn of a right-handed screw, negative if a left-handed screw turn is involved. The original article should be referred to for the procedure to follow when more than one substituent on each carbon differs from hydrogen.

with respect to entropy, by $R \ln 2$. The free energy difference between *anti* and *gauche* is $\Delta G° = \Delta H° - T \Delta S°$. Thus, at room temperature, $\Delta G° = -0.8 - (-RT \ln 2) = -0.8 + 0.42 = -0.38$ kcal./mole, corresponding to $K = e^{-\Delta G°/RT}$ of about 1.9. (Table 1 gives a useful relationship between $\Delta G°$, K, and the percent of stable isomer at room temperature.)

Table I. Equilibrium Constant-Free Energy Relationships

Stable Isomer, %	K	$\Delta G°_{25}$, cal./mole
50	1	0
55	1.22	119
60	1.50	240
65	1.86	367
70	2.33	502
75	3.00	651
80	4.00	821
85	5.67	1028
90	9.00	1302
95	19.00	1745
99	99.00	2723
99.9	999.00	4092

An alternative way of computing the free energy difference between the conformational isomers is to take into account that the *gauche* isomer is a *dl* pair and therefore favored by an entropy of mixing of $R \ln 2$.[24]

It is convenient to define a term "conformational free energy" as the excess free energy of a given conformation over that of the conformation of minimum free energy. According to this definition, the conformational free energy of the *gauche* form of butane is 0.4 kcal./mole. (We shall often round off such energy differences to the nearest 0.1 kcal./mole inasmuch as the second decimal is rarely experimentally significant.)

It should be noted that it may be incorrect to assume a dihedral angle of exactly 60° for the *gauche* form of butane, as was done above. In the *gauche* form there is van der Waals repulsion between the methyl groups which will tend to increase the dihedral angle above 60°. As the angle increases, torsional strain will, however, be set up, as the molecule moves

[24] E. L. Eliel, *Stereochemistry of Carbon Compounds*, McGraw-Hill Book Co., Inc., New York, 1962, p. 32. Both the *gauche* and *anti* forms of butane have a symmetry number of 2 (cf. *ibid.*, p. 215); hence there is no entropy difference due to difference in symmetry number.

toward an eclipsed conformation. Nevertheless it may be expected that, for a small turn beyond 60°, the relief of van der Waals repulsion will exceed the increase in torsional strain. Thus, using the equation for the torsional potential on p. 7 with $V_0 = 3.4$ kcal./mole gives $V(75°) = 0.5$ kcal./mole [as compared to $V(60°) = 0$]. If somewhat more than half the total van der Waals repulsion of 0.8 kcal./mole (at a torsional angle of 60°) were relieved by increasing the torsional angle to 75°, it is clear that the 75° position would be favored over the 60° position. In fact, it has been calculated[25] that the high energy conformer of butane corresponds to a dihedral angle of approximately 85° (rather than 60°). Experimental evidence[19c] based on electron diffraction data favors an angle of $63 \pm 8°$, not as large as the calculated angle, but probably in excess of 60°.

It is important to recognize that, although conformational isomers are distinct entities whose existence can be demonstrated by physical and chemical measurements (in contrast, for example, to resonance hybrids), the energy barriers between them, of the order of 3 kcal./mole, are so low that one cannot hope to isolate the isomers at room temperature or even much below. (It requires a barrier of about 16–20 kcal./mole between interconvertible isomers to permit their separation at room temperature; thus, isolation of isomers separated by a barrier of only 3 kcal./mole would be feasible only in the vicinity of 50°K.) There do exist, however, rotational isomers which are separated by barriers high enough to permit their individual isolation in some instances, namely the biphenyls.[2] Such isomers are called "atropisomers." Although they will not be discussed extensively in this book (good summaries being available elsewhere[2]), it is well to recognize that atropisomerism is a special case of conformational isomerism in which the barrier is high enough to permit isolation of the isomers. Clearly the delineation between ordinary conformational isomerism and atropisomerism is a question of temperature,* and there are many biphenyls (such as 2,2'-dicarboxybiphenyl) in which the enantiomers cannot be separated at room temperature because the barrier is too low; such enantiomers are conformational isomers but not atropisomers.

In principle there should also be a gradual transition between conformational isomerism and geometrical isomerism, inasmuch as geometrical isomers are rotational isomers separated by barriers of the order of 40 kcal./mole and sometimes become readily interconvertible at elevated temperature. Nevertheless, geometrical isomerism of olefins is *not* customarily considered a special case of conformational isomerism.

[25] N. W. Luft, *Trans. Faraday Soc.*, **49,** 118 (1953).

* In biphenyls there are only two stable conformations rather than the three normally found in ethanes, but this is not pertinent to the definition of atropisomerism.

1-3. Relative Stability of Acyclic Conformational Isomers

a. Saturated Molecules. n-Butane illustrates the first factor which determines the relative stability of conformational isomers, namely, steric repulsion. In the *gauche* form there are the following individual *gauche* interactions (cf. Fig. 5): Me-Me, 2 Me-H and 3 H-H. The corresponding interactions in the *anti* form are 4 Me-H and 2 H-H, the difference being an Me-Me and an H-H interaction in the *gauche* form versus 2 Me-H interactions in the *anti* form. Evidently the sum of the former two interactions exceeds the sum of the latter two.[*] In general, if two groups of different size, L (larger) and S (smaller) are disposed in a molecule abcC—Cdef in such a way that there will be two L-S *gauche* interactions, this disposition will be more favorable than one in which there is an L-L and an S-S interaction, if only steric factors are operative.

In more complex molecules, interactions other than steric ones need to be considered. At the same time, such molecules offer new tools of investigation beyond those mentioned in the preceding section (thermodynamic considerations, electron diffraction, and microwave spectroscopy). For example, the preferred conformation of 1,2-dichloroethane is the *anti* conformation, not so much because of the relatively mild steric interaction of the chlorine atoms[†] but mainly because of their dipole repulsion in the *gauche* conformation. Among the many methods[27] which have been used to study 1,2-dichloroethane, dipole moments and infrared and Raman spectra are prominent. The dipole moment of gaseous 1,2-dichloroethane varies from 1.12 D at 32°C. to 1.54 D at 270°C.[27a] The expected dipole moment of the *anti* form is, of course, zero, that of the *gauche* form is of the order of 3.2 D. Clearly the *anti* form predominates as predicted; the ΔH value of 1.1 kcal./mole[27b] is only partially offset by the entropy of mixing of the *gauche* form, which forms a *dl* pair. Since $K = e^{\Delta S/R}e^{-\Delta H/RT}$, the contribution of the *gauche* form increases with increasing temperature and thus the dipole moment increases. (See Sec. 3-5, p. 159, for pertinent calculations.) In the condensed phase, dipolar repulsion decreases (see

---

[26] Ref. 24, pp. 339, 341.

[27] (a) Ref. 5, pp. 7 ff.; (b) ref. 20a, p. 53; (c) N. Sheppard, *Advan. Spectr.*, **1**, 295–310 (1959).

[*] In this particular case it appears that the Me-H and H-H interactions are energetically negligible and that the instability of the *gauche* form of butane may be ascribed entirely to the Me-Me interaction taken to be 0.8 kcal./mole.

[†] *cis*-1,2-Dichloroethylene is more stable than the *trans* isomer by about 0.5 kcal./mole,[26] indicating that, perhaps, attractive London forces more than overcome the usual repulsive steric interaction: H. A. Stuart, *Phys. Z.*, **32**, 793 (1931); see also *Chem. Eng. News*, Oct. 7, 1963, p. 38.

Sec. 3-5) and therefore the contribution of the *gauche* form increases (i.e., the dipole moment increases) to the point where, in pure liquid 1,2-dichloroethane, ΔH between the two conformations is zero.[27b,c]

The infrared spectrum of gaseous 1,2-dichloroethane shows a considerable number of lines, some due to the *anti* form and others due to the *gauche*. When the substance is frozen, many of the lines disappear because crystallization occurs exclusively in the *anti* conformation; the infrared spectrum of this conformation is not rich because the structure has a center of symmetry and many vibrations are infrared-inactive. When the material is melted or vaporized, the lines due to the *anti* conformation (assigned from the spectrum of the solid) persist, but additional lines, assigned to the *gauche* isomer, now appear also.* From the temperature dependence of the two sets of bands (cf. Sec. 3-4a, p. 147) the enthalpy difference of the conformational isomers in the gaseous state was calculated to be 1.25 ± 0.05 kcal./mole.[28]

Infrared and Raman data have been used to assess the conformations of most of the 1,2-dihaloethanes; dipole moment data have been applied to molecules such as the hydrobenzoins, $C_6H_5CHOHCHOHC_6H_5$; the diethyl tartrates, $H_5C_2O_2CCHOHCHOHCO_2C_2H_5$, and the stilbene dichlorides, $C_6H_5CHClCHClC_6H_5$.[27a]

Nuclear magnetic resonance studies have been used[29,30] to assess the conformational population of the 2,3-dibromobutanes. The basis of the method, discussed in more detail in Sec. 3-4d, is that the coupling constant is greater for protons in the *anti* conformation than for protons in the *gauche* conformation; where both conformations contribute, intermediate coupling constants will be found whose magnitude increases as the contribution of the *anti* conformation becomes more important. In the case of the *meso* isomer (Fig. 6) it is found, surprisingly, that the *gauche* conformations contribute as much as 30% in the pure liquid; in dilute carbon disulfide solution (less polar) this value, not unexpectedly, decreases to 20% (see the earlier discussion on dichloroethane).[29] For the *dl* isomer (Fig. 7) the NMR spectrum suggests that of the order of 20% of the conformation has the hydrogen atoms *anti* (B), the remainder being in conformations A and C. The small contribution of conformation B is not surprising; in this conformation both the two methyl groups and the two bromine atoms are *gauche* to each other, and as explained earlier *gauche*

[28] H. J. Bernstein, *J. Chem. Phys.*, **17**, 258 (1949).

[29] A. A. Bothner-By and C. Naar-Colin, *J. Am. Chem. Soc.*, **84**, 743 (1962).

[30] F. A. L. Anet, *J. Am. Chem. Soc.*, **84**, 747 (1962).

* That the new set of bands is not due to the *cis* isomer has been established through an ingenious argument based on the infrared spectrum of 1,2-dibromoethane-1-*d*: J. T. Neu and W. D. Gwinn, *J. Chem. Phys.*, **18**, 1642 (1950).

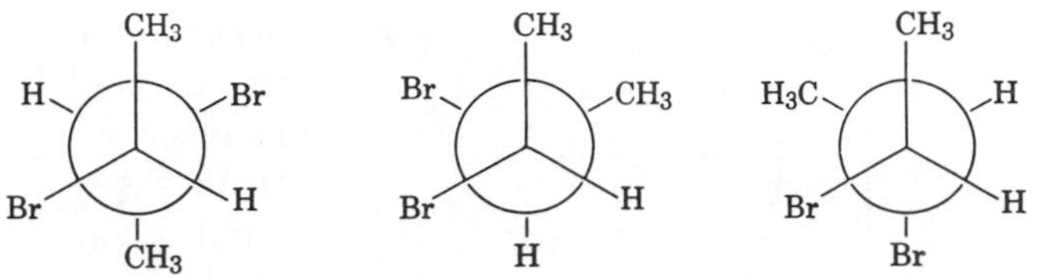

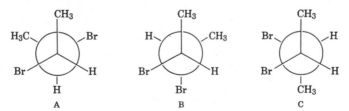

Fig. 6. Conformations of *meso*-2,3-dibromobutane.

methyl groups are less favored than *anti* methyl groups and, in a non-polar solvent, *gauche* bromine atoms are less good than *anti* bromine atoms. In conformation A the methyl groups are *gauche* but the bromine atoms are not, whereas in conformation C the bromine atoms are *gauche* though the methyl groups are not. It is true that both A and C have two methyl-bromine *gauche* interactions, but, as will be seen below, these may be attractive rather than repulsive.

NMR measurements have also been applied to the 2,3-dichlorobutanes, the 2,3-diacetoxybutanes, and the 2,3-diphenylbutanes.[27] The data for the *meso*-2,3-diacetoxybutanes are interesting in that they suggest a predominance of the *gauche* isomer, presumably[29] because of an electrostatic attraction of the appropriately oriented acetoxy dipoles. Infrared data on propyl bromide suggest that in this compound the *gauche* form is lower in internal energy than the *anti* form by 0.1–0.5 kcal./mole,[27c] reflecting an even greater (by RT ln 2) difference in free energy in favor of the *gauche* because of its existence as a *dl* isomer. In the case of propyl chloride, infrared,[27c] electron diffraction,[31] and microwave[32] studies agree in indicating that the *gauche* isomer is lower in internal energy by 0.0–0.6 kcal./mole; and, in the case of propyl fluoride, microwave data[33] indicate greater stability of the *gauche* isomer by about 0.5 kcal./mole. It is of some interest to speculate why the *gauche* forms of the propyl halides tend to be somewhat more stable than the *anti*. Steric interactions, as in the

Fig. 7. Conformations of *dl*-2,3-dibromobutane.

[31] Y. Morino and K. Kuchitsu, *J. Chem. Phys.*, **28**, 175 (1958).
[32] T. N. Sarachman, *J. Chem. Phys.*, **39**, 469 (1963).
[33] E. Hirota, *J. Chem. Phys.*, **37**, 283 (1962).

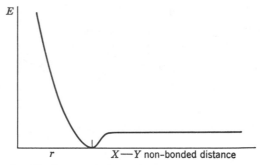

Fig. 8. van der Waals curve (schemàtic) (r = van der Waals distance).

gauche form of butane, would at first sight seem to lead to the opposite prediction. It has been suggested[34] that dipole attraction between the CH_3-C and C-X (X = halogen) dipoles stabilizes the *gauche* form. This is probably not the entire explanation, however, for, if it were, propyl fluoride should be the most stable of the propyl halides in the *gauche* form since fluorine is the smallest of the halogens. (The three C—X bond moments X = F, Cl, Br are not very different.) Also, on the basis of this explanation it is hard to see why the *gauche* forms of the 1,2-dihaloethanes are not less stable than, in fact, they are, since here both steric and dipole interactions appear unfavorable. It would seem, actually, that the steric repulsions between X and Y in the *gauche* form of XCH_2CH_2Y, where X is halogen or methyl and Y is halogen (i.e., in the dihaloethanes or 1-halopropanes), is not very severe at all because the X—Y distance is such that X and Y fall either into or very close to the attractive part of the van der Waals curve (Fig. 8). This is particularly likely in the case of halides because of the high polarizability of the halogen atoms, which leads to a rather large attraction in the vicinity of the van der Waals distance as a result of London forces, i.e., to a deep well in the curve in Fig. 8.* Calculations have been performed for both the 1,2-dihaloethanes and the 1-halopropanes.[35] It turns out that for the chlorides the van der Waals (including the London) energy is attractive, the electrostatic (dipole interaction) energy is repulsive for 1,2-dichloroethane and attractive for 1-chloropropane and the total calculated energy difference between the *anti* and *gauche* is in good agreement with the experimental. In the

---

[34] G. J. Szasz, *J. Chem. Phys.*, **23**, 2449 (1955).

[35] M. M. Kreevoy and E. A. Mason, *J. Am. Chem. Soc.*, **79**, 4851 (1957).

* The explanation is somewhat marred by the fact that in propyl mercaptan, $CH_3CH_2CH_2SH$, despite the high polarizability of sulfur, the *anti* form is more stable than the *gauche* by 0.4 kcal./mole: R. E. Pennington, D. W. Scott, H. L. Finke, J. P. McCullough, J. F. Messerly, I. A. Hossenlopp, and G. Waddington, *J. Am. Chem. Soc.*, **78**, 3266 (1956). See also Chap. 7.

bromides, however, the van der Waals energy is repulsive, and the calculation for 1-bromopropane predicts a strong predominance of the *anti* form, contrary to experiment. The situation is obviously not yet completely understood, and it is not even clear whether the theoretical or experimental data on propyl bromide are in error. In this connection it is of interest that electron diffraction data on 1-bromobutane[36] suggest that the all-*anti* (zig-zag) conformation is preferred. In contrast, in *n*-butyl chloride,[37] conformations in which chlorine is *gauche* to ethyl abound and it appears that, whereas the *anti* form is preferred about the CH_3CH_2—CH_2CH_2Cl bond by about 0.4 kcal./mole, the *gauche* form predominates about the $CH_3CH_2CH_2$—CH_2Cl bond by about 0.3 kcal./mole. Surprisingly, the *gauche-gauche* form in which the methyl and chlorine approach most closely (Fig. 9) contributes as much as 24%, suggesting that this form may be endowed with some special stability, possibly because of London attraction between methyl and chlorine.

Cl   CH₃

H_2C CH_2

C

H   H

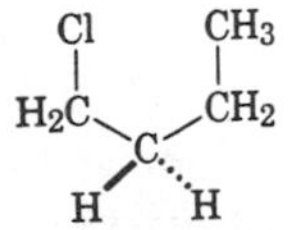

Fig. 9. gauche-gauche Form of 1-chlorobutane.

Another molecule of interest is isobutyl chloride (Fig. 10).[38] Although on the basis of the preceding discussion of propyl chloride, one might expect the *gauche-gauche* isomer (A) to predominate, the molecule in fact exists in the *gauche-anti* conformation (B) to the extent of 80%. It is significant that the angle of torsion between methyl and chlorine in this conformation was found to be 66°, suggesting, perhaps, that a favorable interaction between *gauche* methyl and chlorine requires a slight increase of the normal 60° angle. In conformation A (Fig. 10) it is, of course, not possible to increase the Me—Cl dihedral angle on one side without diminishing it on the other at the same time.*

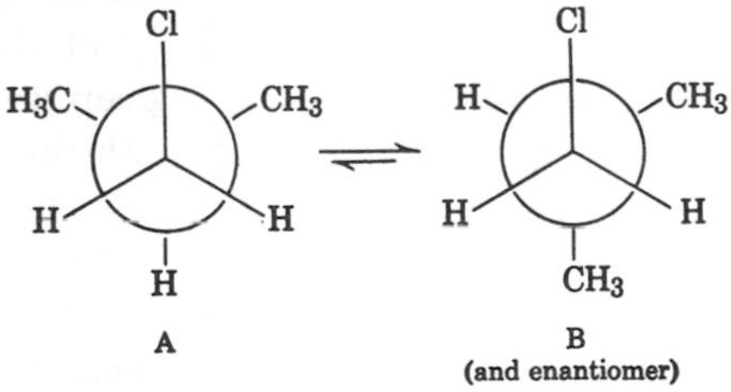

Fig. 10. Conformations of isobutyl chloride.

[36] F. A. Momany, R. A. Bonham, and W. H. McCoy, *J. Am. Chem. Soc.*, **85**, 3077 (1963).

[37] T. Ukaji and R. A. Bonham, *J. Am. Chem. Soc.*, **84**, 3631 (1962).

[38] G. H. Pauli, F. A. Momany, and R. A. Bonham, *J. Am. Chem. Soc.*, **86**, 1286 (1964).

* The situation with respect to this angle in propyl chloride is ambiguous. Microwave spectra (ref. 32) suggest it to be 70°, but electron diffraction (ref. 31) gives a value of 59°.

Another factor which may lead to enhanced stability of *gauche* conformations is hydrogen bonding. Clearly, hydrogen bonding requires that the acidic hydrogen atom in the —X—H donor* group be close to an acceptor* atom Y which has unshared pairs of electrons. Contrary to a common misconception, it does not, however, require that X, H, and Y be collinear. Thus *cis*-1,2-cyclopentanediol in which the OH groups are nearly eclipsed (*syn*-periplanar) forms a strong intramolecular hydrogen bond,[39] *cis*- and *trans*-1,2-cyclohexanediol in which the OH groups are skew (*syn*-clinal) form weak intramolecular hydrogen bonds,[39] and *trans*-1,2-cyclopentanediol in which the angle of torsion between the hydroxyl groups is about 120° (the OH groups are *anti*-clinal) shows no evidence of intramolecular hydrogen bonding.[39] Since ethylene glycol, CH_2OHCH_2OH, shows evidence of hydrogen bonding,[39,40] it is clear that an appreciable number of molecules must be in the *gauche* form since no hydrogen bonding can occur in the *anti* (or *anti*-periplanar) form.† For ethylene chlorohydrin, CH_2OHCH_2Cl, an estimate of the potential energy difference between the two conformations has been obtained[41] from the temperature dependence of the intensity of the free and bonded OH-stretching bands (cf. Sec. 3-4a), and it appears that the *gauche* conformation is more stable by 0.95 kcal./mole. The enhanced stability of the *gauche* conformer appears to be due to the extra energy of the intramolecular hydrogen bond.‡

Another interesting conformational situation is found in 1,1,2,2-tetrabromoethane, $CHBr_2CHBr_2$. In the liquid state this molecule exists predominantly in the *gauche* conformation which is favored by about 0.8 kcal./mole[27c] over the *anti* isomer. (A similar situation is found in 1,1,2,2-tetrachloroethane.) This could not have been predicted on the basis of the nearly equal population of the *anti* and *gauche* conformations in liquid 1,2-dichloroethane and the predominance of the *anti* conformation in 1,2-dibromoethane. It has been pointed out[42] that, because of the

[39] L. P. Kuhn, *J. Am. Chem. Soc.*, **74**, 2493 (1952).

[40] M. Kuhn, W. Lüttke, and R. Mecke, *Z. Anal. Chem.*, **170**, 106 (1959).

[41] S. Mizushima, T. Shimanouchi, T. Miyazawa, K. Abe, and M. Yasumi, *J. Chem. Phys.*, **19**, 1477 (1951).

[42] I. Miyagawa, T. Chiba, S. Ikeda, and Y. Morino, *Bull. Chem. Soc. Japan*, **30**, 218 (1957).

* The terms donor and acceptor refer to donation and acceptance of the hydrogen bridge, *not* to donation or acceptance of an electron pair.

† In ref. 40, an attempt is made to estimate the relative population of the *gauche* and *anti* forms of various molecules of type CH_2OHCH_2Y from the relative intensity of the unbonded and bonded OH stretching frequency. This does not seem permissible, however, since the extinction coefficients for the two frequencies may well be unequal and may, moreover, differ relative to each other for different Y's.

‡ This explanation has been argued against in ref. 40.

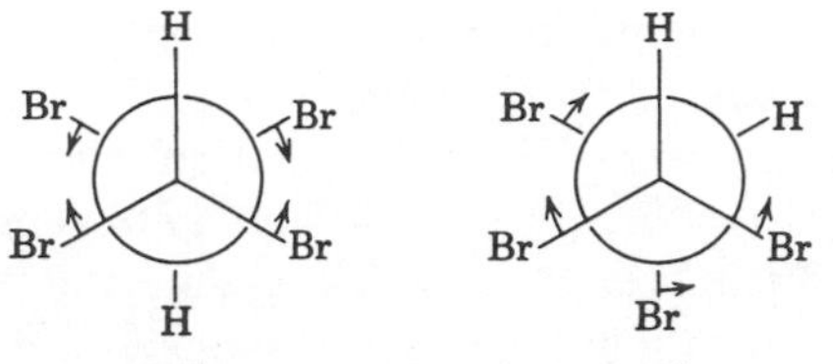

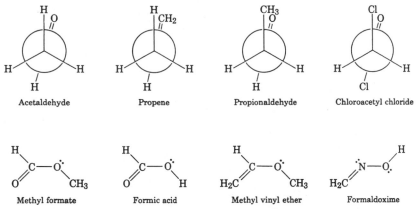

Fig. 11. Angle deformation in the two conformations of 1,1,2,2-tetrabromoethane.

spreading out of the Br—C—Br bond angle, the *gauche* bromines are forced to approach each other in the *anti* isomer but recede from each other in the *gauche* isomer (Fig. 11), thus explaining the enhanced stability of the latter. Other peculiarities of the $CHBr_2CHBr_2$ molecule are its low potential barrier[43] (1.8 kcal./mole) and the fact that it may crystallize in either the *gauche* or the *anti* conformation.[43] (Most other molecules crystallize in a single conformation, usually the more stable one.)

b. Unsaturated Molecules. The simplest unsaturated molecules in which conformational considerations play a part are acetaldehyde and propene (Fig. 12). Surprisingly (at least at first sight) both molecules exist with the double bond (C=O or C=C) eclipsed with a hydrogen (Fig. 12). For acetaldehyde this has been demonstrated by microwave spectroscopy[44] as well as electron diffraction;[45] in the case of propene the preferred conformation follows from microwave study.[46] An analogous conformation had

Acetaldehyde	Propene	Propionaldehyde	Chloroacetyl chloride

Methyl formate	Formic acid	Methyl vinyl ether	Formaldoxime

Fig. 12. Some simple unsaturated molecules in their preferred conformations.

[43] R. E. Kagarise, *J. Chem. Phys.*, **24**, 300 (1956), and earlier papers there cited; K. Krebs and J. Lamb, *Proc. Roy. Soc. (London)*, **A244**, 558 (1958).

[44] R. W. Kilb, C. C. Lin, and E. B. Wilson, *J. Chem. Phys.*, **26**, 1695 (1957).

[45] R. H. Schwendeman and L. O. Brockway, unpublished observations cited in ref. 44.

[46] D. R. Herschbach and L. C. Krischer, *J. Chem. Phys.*, **28**, 728 (1958).

been previously postulated, on thermochemical grounds, to be the preferred one for *cis*-2-butene.[47] Eclipsing of C=X with methyl C—H and the concomitant staggering of the aldehyde hydrogen in CH_3CHO and the 2-hydrogen in $CH_3CH=CH_2$ with the methyl hydrogens appear reasonable in an old-fashioned ball-spring-and-stick model; the "banana character" of the double bond finds some theoretical support in modern theory, and, since, as mentioned earlier, the source of bond eclipsing barriers is apparently found at or near the atom from which the bond emanates, the

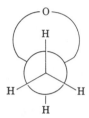

Fig. 13. Banana-type double bond.

banana-type double bond would be staggered rather than eclipsed with respect to the carbon-hydrogen bond on the adjacent atom (Fig. 13). Acetyl fluoride[48] and acetyl chloride[49] have stable conformations similar to that of acetaldehyde.

Even more surprising is the finding, based on NMR coupling constants, that propionaldehyde[50] (Fig. 12) exists mainly in the conformation in which a *methyl group* (rather than a hydrogen atom) is eclipsed with the double bond. In 1-butene (Fig. 14) the analogous conformation is as populated as are the conformations with eclipsed hydrogen.[51] Earlier Raman and electron diffraction studies[52] had already indicated an analogous conformation in chloroacetyl chloride, $ClCH_2COCl$ (Fig. 12), bromoacetyl chloride, and bromoacetyl bromide. For the latter two compounds, the conformations analogous to that indicated in Fig. 12 were found more stable, by 1.0 and 1.9 kcal./mole, respectively, than the next most stable conformation (probably the one in which

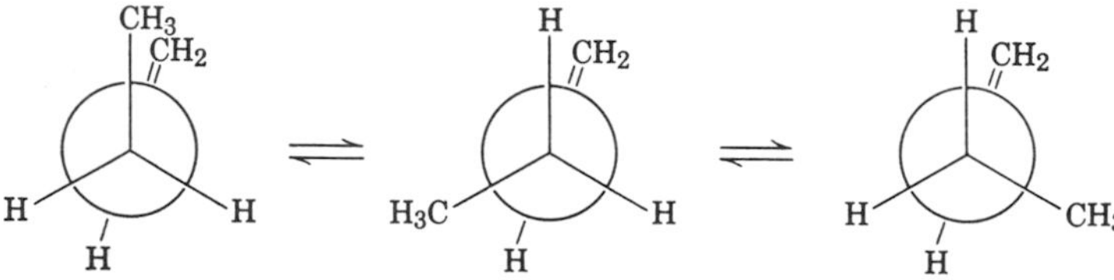

Fig. 14. Conformation of 1-butene.

[47] Ref. 20a, p. 59.

[48] L. Pierce, *Bull. Am. Phys. Soc.*, **1**, 198 (1956).

[49] K. M. Sinnott, *Bull. Am. Phys. Soc.*, **1**, 198 (1956); cf. E. B. Wilson, *Proc. Natl. Acad. Sci. U.S.*, **43**, 816 (1957).

[50] R. J. Abraham and J. A. Pople, *Mol. Phys.*, **3**, 609 (1960).

[51] A. A. Bothner-By, C. Naar-Colin, and H. Gunther, *J. Am. Chem. Soc.*, **84**, 2748 (1962).

[52] I. Nakagawa, I. Ichisima, K. Kuratani, T. Miyazawa, T. Shimanouchi, and S. Mizushima, *J. Chem. Phys.*, **20**, 1720 (1952); Y. Morito, unpublished observation there cited.

carbonyl oxygen is eclipsed with hydrogen). The *cis* conformation, in which methyl is eclipsed with the double bond (Fig. 12), is also preferred in methyl formate, as indicated by microwave spectrum[53] and electron diffraction studies[54] as well as infrared investigations.[55] Methyl acetate has the analogous conformation (methyl and C═O *cis*, methyl and methyl *anti*).[55] Formic acid (Fig. 12) also exists in the *cis* conformation, which is 2.0 kcal./mole more stable than the *anti*.[56] The *cis* conformation similarly predominates in methyl vinyl ether[57] (Fig. 12), but the *anti* conformation predominates in formaldoxime (Fig. 12).[57a] The latter results have been rationalized in terms of repulsion between *p*- and pi-electron clouds *cis* to each other in compounds such as methyl formate, formic acid, methyl vinyl ether and *p*- and *p*-electrons in formaldoxime, but such an explanation clearly does not cover propionaldehyde and 1-butene. For propionaldehyde the attraction of the C—CH$_3$ and C═O dipoles has been invoked to account for the preferred conformation.[51] The case of 1-butene (Fig. 14) may be dismissed by saying that to eclipse methyl with methylene requires no more energy than to eclipse methyl with hydrogen. This is perhaps not unexpected in view of the fact that, in models, there appears to be little steric repulsion between methyl and methylene in the *cis* conformation of 1-butene, and the eclipsing potential as such is known to be relatively independent of what is at the end of the bond. As one goes from 1-butene, CH$_3$CH$_2$CH═CH$_2$, to neopentylethylene

$$(CH_3)_3CCH_2CH═CH_2$$

models now indicate a strong steric repulsion between the *t*-butyl group and the eclipsed methylene group in the *cis* conformation of the latter molecule, and, accordingly, neopentylethylene is found largely in the conformation in which the methylene is eclipsed with hydrogen rather than with *t*-butyl.[51]

A few remarks about conjugated diunsaturated molecules will conclude this section. Such molecules, 1,3-butadiene and *trans*-3-penten-2-one for example (Fig. 15), may exist in s-*cis* and s-*trans* conformations, the prefix s before *cis* and *trans* indicating that conformation about the central

[53] R. F. Curl, *J. Chem. Phys.*, **30**, 1529 (1959).

[54] J. M. O'Gorman, W. Shand, and V. Schomaker, *J. Am. Chem. Soc.*, **72**, 4222 (1950).

[55] J. K. Wilmhurst, *J. Mol. Spectr.*, **1**, 201 (1957); T. Miyazawa, *Bull. Chem. Soc. Japan*, **34**, 691 (1961).

[56] T. Miyazawa and K. S. Pitzer, *J. Chem. Phys.*, **30**, 1076 (1959).

[57] N. L. Owen and N. Sheppard, *Proc. Chem. Soc.*, **1963**, 264.

[57a] I. N. Levine, *J. Mol. Spectr.*, **8**, 276 (1962).

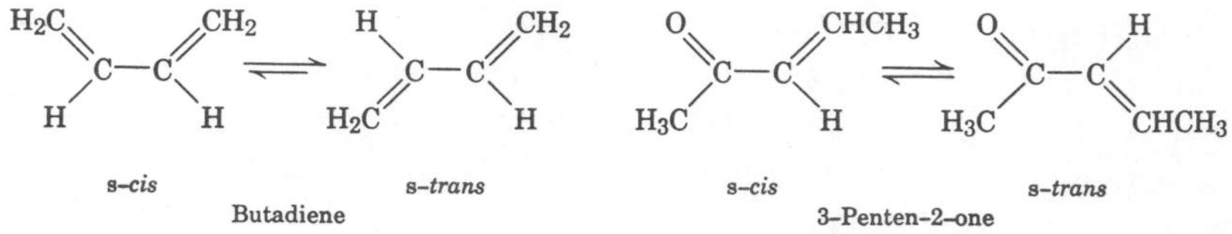

s-cis   s-trans   s-cis   s-trans
        Butadiene   3-Penten-2-one

Fig. 15. Conformation of conjugated diunsaturated molecules.

single bond is referred to.[58] In butadiene, a number of different methods, such as calculation from thermodynamic properties,[59] quantum mechanical calculations,[60] electron diffraction measurements,[61] infrared measurements,[62] and microwave investigation,[63] all lead to the conclusion that the s-*trans* form is more stable than the s-*cis* by at least 2 kcal./mole, and an indirect chemical argument[64] is in general agreement with this finding. In *trans*-3-penten-2-one, on the other hand, it appears from infrared and Raman spectra that both the s-*cis* and the s-*trans* conformations are appreciably populated.[65] Model considerations suggest steric interference of hydrogen atoms in the s-*cis* (but not the s-*trans*) form of butadiene;[58] in the enone, on the other hand, hydrogen-hydrogen interference appears between methyl and hydrogen at C-3 in the s-*trans* form, but this is not very severe.* It might be noted, finally, that the ready participation of butadiene in Diels-Alder reactions indicates that the molecule can quite readily assume the s-*cis* conformation when necessary; in fact, the barrier for rotation about the central single bond is estimated[59] to be about 4.9 kcal./mole.

Space considerations do not permit discussion of a number of other simple molecules whose conformations have been elucidated; leading references to recent work in this area may be found elsewhere.[66]

---

[58] Cf. ref. 24, pp. 331–333.

[59] J. G. Aston, G. Szasz, H. W. Wolley, and F. G. Brickwedde, *J. Chem. Phys.*, **14,** 67 (1946).

[60] R. G. Parr and R. A. Mulliken, *J. Chem. Phys.*, **18,** 1338 (1950).

[61] A. Almenningen, O. Bastiansen, and M. Tratteberg, *Acta Chem. Scand.*, **12,** 1221 (1958).

[62] D. J. Marias, N. Sheppard, and B. P. Stoicheff, *Tetrahedron*, **17,** 163 (1962).

[63] D. R. Lide, *J. Chem. Phys.*, **37,** 2074 (1962).

[64] W. B. Smith and J. L. Massingill, *J. Am. Chem. Soc.*, **83,** 4301 (1961).

[65] K. Noack and R. N. Jones, *Can. J. Chem.*, **39,** 2225 (1961).

[66] D. J. Millen and R. F. M. White, *Ann. Rept. Progr. Chem. (Chem. Soc. London)*, **59,** 190 (1962); D. J. Millen, in *Progress in Stereochemistry*, Vol. 3, P. B. D. de La Mare and W. Klyne, eds., Butterworths, London, 1962, Chap. 4.

* However, the quantum mechanical calculations in ref. 60 suggest that the major source of instability of the s-*cis* form of butadiene is unfavorable arrangement of the pi orbitals, not hydrogen-hydrogen interaction.

I-4. Equilibrium Studies in Acyclic Systems

Because of the low energy barrier between conformational isomers, conformational equilibria cannot be studied by conventional chemical means. They can sometimes be studied indirectly by chemical methods (e.g. Sec. 2-2c), but, of more interest, from the chemical point of view, is the influence of conformation on position of configurational equilibria and on reactivity. Equilibria depend only on ground states and their conformation and will be dealt with in this section; reaction rates require consideration of transition states as well and will be considered in the next section.

One of the simplest chemical equilibria in an acyclic system is that between diastereoisomers, such as *meso* and *dl* forms or *erythro* and *threo* forms. When there are two adjacent asymmetric carbon atoms we have the situation depicted in Fig. 16: each diastereoisomer exists in three staggered conformations. Because, from first principles, we might expect to assess the stability of an individual conformer but not that of a configurational isomer as such, it is important to be able to compute the enthalpy and entropy of each isomer from those of its contributing conformations. Once this is done, the difference in enthalpy and entropy between the configurational isomers can be immediately deduced. Fortunately, the task is relatively simple. If N_1, N_2, and N_3 are the mole fractions of one of the isomers in its three possible stable conformations and H_1, H_2, and H_3 are the enthalpies of the three conformations, then the total enthalpy is $H = N_1H_1 + N_2H_2 + N_3H_3$. For the general case where there are i conformations, $H = \sum_i N_iH_i$. The mole fractions can be

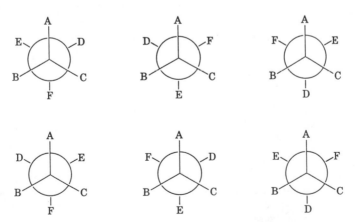

Fig. 16. Configurational isomers and their conformations.

assessed from the relationship $N_1/N_2 = e^{-H_1/RT}/e^{-H_2/RT}$, etc.,* on the assumption that the various conformations are equal in entropy—which is often justified. Because of the appearance of H_i as an exponential term in the computation of N_i, it is permissible to disregard conformations whose enthalpy is very much higher (say about 2.5 kcal./mole) than that of the most stable conformation; in fact, for qualitative purposes it is frequently permissible to restrict oneself exclusively to a consideration of the most stable conformation. The entropy of a species is computed from the entropy-of-mixing formula:

$$\Delta S = -2.3R(N_1 \log N_1 + N_2 \log N_2 + N_3 \log N_3)$$

or, in the general case, $\Delta S = -2.3R \sum_i N_i \log N_i$. Here, again, the entropy of mixing tends toward zero as one of the mole fractions tends toward unity and the others tend toward zero, i.e., when one conformation is very much more stable than the others.

We may illustrate the case by means of the 2,3-dibromobutanes (Figs. 6 and 7). We shall assume the following *gauche* interactions: CH_3-CH_3, 0.8 kcal./mole; Br-Br, 0.7 kcal./mole (liquid phase value); CH_3-Br, 0.2 kcal./mole. First we compute the interactions in the *meso* isomer (Fig. 6). The *anti* form has only two CH_3-Br interactions totaling 0.4 kcal./mole, but the *gauche* forms have a CH_3-CH_3 interaction, a Br-Br interaction, and a CH_3-Br interaction totaling 1.7 kcal./mole. The difference of 1.3 kcal./mole corresponds to a population ratio of *anti* to individual *gauche* of $e^{-1300/RT}$ or 9.0 at 25°C. Hence $N_A/N_B = 9.0$, $N_A/N_C = 9.0$, and $N_A + N_B + N_C = 1$, whence $N_A = 0.82$; i.e., 82% of the *meso*-2,3-dibromobutane molecules should be in the *anti* form. This is in reasonable agreement with the experimental value deduced from NMR spectroscopy[26,27] (see previous section) indicating that the parameters chosen in the calculation above are reasonable. A similar calculation can be made for the active isomer; it turns out that the interactions in forms A, B, and C (Fig. 7) are 1.2, 1.5, and 1.1 kcal./mole, respectively, and that the corresponding populations at 25°C are 36% A, 22% B, and 42% C. The figure of 22% B is consistent with the NMR studies,[29,30] which, however, give no information about the relative populations of A and C.†

[67] S. Winstein and R. E. Wood, *J. Am. Chem. Soc.*, **62**, 548 (1940).

[68] H. G. Trieschmann, *Z. Physik. Chem.*, **B33**, 283 (1936).

* Since $N_1/N_2 = K = e^{-\Delta G/RT} = e^{\Delta S/R}e^{-\Delta H/RT}$; so, if $\Delta S = 0$, $N_1/N_2 = e^{-\Delta H/RT} = e^{-(H_1-H_2)/RT} = e^{-H_1/RT}/e^{-H_2/RT}$.

† For reasons which are not clear to us, it is assumed in both ref. 29 and ref. 30 that A is greatly preferred over C. There appears to be no experimental basis for this assumption except possibly some early dipole moment data (refs. 67, 68) which indicate that the moments of the two 2,3-dibromobutanes are sizable and nearly equal to each other.

We are now in a position to compute the enthalpies due to *gauche* interactions of the two isomers. For the *meso* isomer this will be $0.82 \times 0.4 + 0.18 \times 1.7$ or 0.63 kcal./mole; and, for the active isomer, $0.36 \times 1.2 + 0.22 \times 1.5 + 0.42 \times 1.1$ or 1.22 kcal./mole. Also, we may calculate the entropy of mixing of the *meso* isomer to be $-2.3R$ $(0.82 \log 0.82 + 0.09 \log 0.09 + 0.09 \log 0.09)$ or 1.18 cal. mole^{-1} deg.$^{-1}$. (It is necessary to consider the enantiomeric *gauche* forms separately in the computation of the entropy of mixing.) Similarly, for the active isomer, the entropy of mixing is calculated to be 2.12 cal. mole^{-1} deg.$^{-1}$.

Thus the free energy due to group interaction at room temperature (298°K), or $G = H - TS$, is $0.63 - 0.35$ or 0.28 kcal./mole for the *meso* isomer and $1.22 - 0.63$ or 0.59 kcal./mole for the active isomer. (It might be noted, parenthetically, that the free energy of the *meso* isomer is not very different from the enthalpy of the *anti* form A, because the *gauche* forms contribute little; but the free energy of the active form is appreciably lower than the enthalpy of its most stable conformation because of the admixture of another, nearly equally stable, conformation as well as of a third conformation of only little more energy.) In computing the free-energy difference between the *dl* and the *meso* isomer (which is the quantity usually desired) two further factors affecting entropy must be taken into account: the entropy of mixing of the *dl*-pair and the symmetry contribution to entropy. The former favors the *dl*-pair by R ln 2, but the latter lowers the entropy of the *dl*-pair (symmetry number 2) *vis-a-vis* the *meso* form (symmetry number 1) by an equal amount because of a loss in rotational entropy caused by the symmetry axis. In the overall aggregate the two entropy effects cancel and the difference in free energy between *meso* and *dl* is 0.59-0.28 = 0.31 kcal./mole in favor of the *meso*. The only data available to test this prediction are quite old, but do suggest[68] that the *meso* isomer is slightly more stable than the *dl* pair. It is only fair to point out that the calculation is quite sensitive to the choice of interaction parameters, that it should be correct only in polar solvents (because of the variation of the Br-Br *gauche* interaction with solvent polarity), and that it is based on the assumption that, apart from group interaction, the two diastereoisomers have the same enthalpy and entropy.

It would be desirable if a number of predictions of equilibrium such as the one above could be tested experimentally; unfortunately, however, the data are largely lacking. In the case of the stilbene dibromides, $C_6H_5CHBrCHBrC_6H_5$, the *meso* isomer appears to be more stable than the *dl* by at least 1.4 kcal./mole.[69] [This is perhaps more than one would have expected after comparing the isomers labeled A (but with phenyl instead

    [69] R. E. Buckles, W. E. Steinmetz, and N. G. Wheeler, *J. Am. Chem. Soc.*, **72**, 2496 (1950).

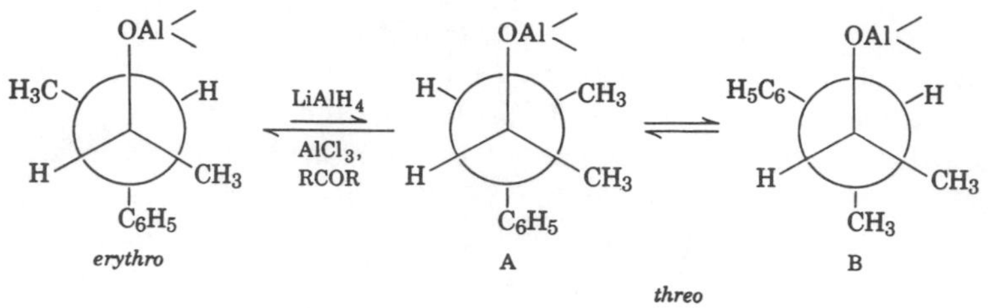

Fig. 17. Equilibration of 3-phenyl-2-butanol complexes.

of methyl) in Figs. 6 and 7.] It must be noted that the equilibration in this case was carried out in the solid state, so that intermolecular interactions may have been very important. For the 2,3-dimethylsuccinic acids, $HO_2CCH(CH_3)CH(CH_3)CO_2H$, it has been claimed[70] that the *meso* isomer and its diamide are more stable than the corresponding *dl* isomers, but later work has led to the opposite conclusion.[70a] In the case of the tartaric acids, both the *meso* and the *dl* form are present at equilibrium, the *dl* apparently predominating,[71] but, since only about half of the total acid could be recovered, it is not certain that the ratio of acids isolated corresponds to the ratio in solution. The greater stability of the active isomer is, however, supported by the heat of combustion of the gaseous dimethyl tartrates:[72] active isomer, 640.2 kcal./mole, *meso* isomer, 641.6 kcal./mole, giving a difference in enthalpy (not free energy) of 1.4 kcal./mole in favor of the active form. Here the situation is undoubtedly influenced by the possibility of intramolecular hydrogen bonding between OH and CO_2CH_3 and between OH and OH; the latter kind of bonding is more easily attained in the active isomer than in the *meso*.

One of the few equilibria in acyclic systems checked by modern methods is that of the 3-phenyl-2-butanols using lithium aluminum hydride-aluminum chloride.[73] The equilibrium shown in Fig. 17 corresponds to 68% *erythro* and 32% *threo* isomer. The result is qualitatively reasonable on the basis of the most populous conformations shown in Fig. 17, since the aluminum-complexed oxygen is a very large group.[74] The *threo* isomer A has an extra methyl-methyl *gauche* interaction; that it is as stable as it is is probably due to the presence of an appreciable amount of the alternative conformation B, leading to an enhancement in entropy due to entropy of mixing.

[70] R. P. Linstead and M. Whalley, *J. Chem. Soc.*, **1954**, 3722.
[70a] L. Eberson, *Acta Chem. Scand.*, **13**, 203 (1959).
[71] A. F. Holleman, *Org. Syn.*, Coll. Vol. **1**, 497 (1941).
[72] H. C. Blanck and K. L. Wolf, *Z. Physik. Chem.*, **B32**, 139 (1936).
[73] B. P. Thill, Ph.D. thesis, University of Notre Dame, Notre Dame, Ind., 1963.
[74] E. L. Eliel and M. N. Rerick, *J. Am. Chem. Soc.*, **82**, 1367 (1960).

I-5. Reactivity in Acyclic Systems

Conformational effects on reactivity may be understood in terms of two now well-recognized relationships. Both relationships are based on transition state theory, i.e., the interpretation of rate in terms of the difference in free energy of the ground state and the transition state for a given reaction. The first relationship is $k = \sum_i N_i k_i$, where k is the rate of a given reaction of a conformationally heterogeneous system, N_i is the mole fraction of the material in the ith conformation, and k_i is the specific rate of the reaction studied for that conformation. This equation means that the reaction is considered as the sum of the reactions of all the contributing conformations. Because the equation is most useful in cyclohexanoid systems, it will be derived and considered in detail in Sec. 2-2c. The relationship does, however, apply in acyclic systems as well.[9b]

An example is the reaction of *meso-* and *dl*-2,3-dibromobutane with potassium iodide in methanol; the rate of formation of *trans*-2-butene from the *meso* isomer is about twice the rate of formation of *cis*-2-butene from the *dl* isomer.[75] To understand this we must consider first the mechanism of debromination which is, itself, conformationally controlled in the sense that the bromine atoms to be eliminated should be *anti* for facile reaction.[76] As Fig. 18 shows, this leads to formation of *trans*-2-butene from the *meso*-dibromide and *cis*-2-butene from the *dl*-dibromide.*

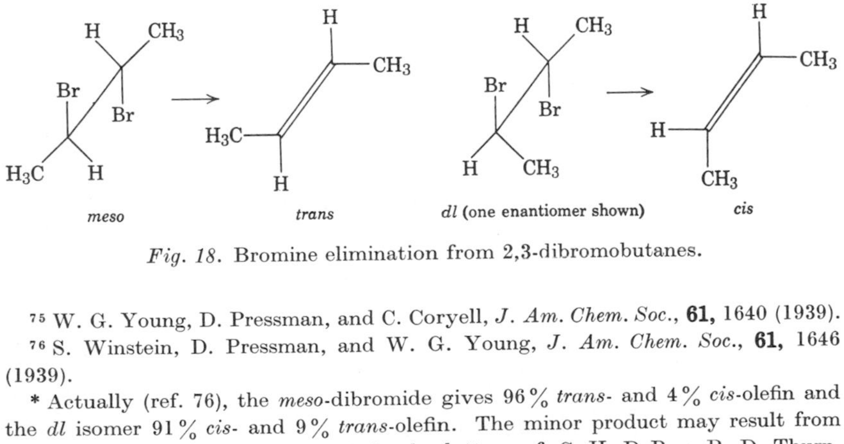

Fig. 18. Bromine elimination from 2,3-dibromobutanes.

[75] W. G. Young, D. Pressman, and C. Coryell, *J. Am. Chem. Soc.*, **61**, 1640 (1939).

[76] S. Winstein, D. Pressman, and W. G. Young, *J. Am. Chem. Soc.*, **61**, 1646 (1939).

* Actually (ref. 76), the *meso*-dibromide gives 96% *trans-* and 4% *cis*-olefin and the *dl* isomer 91% *cis-* and 9% *trans*-olefin. The minor product may result from *gauche* or *cis* elimination—more likely the latter: cf. C. H. DePuy, R. D. Thurn, and G. F. Morris, *J. Am. Chem. Soc.*, **84**, 1314 (1962)—or, perhaps more plausibly, may be the product of iodide displacement followed by anti elimination: cf. H. L. Goering and H. H. Espy, *J. Am. Chem. Soc.*, **77**, 5023 (1955), and W. M. Schubert, H. Steadly, and B. S. Rabinovitch, *ibid.*, **77**, 5755 (1955).

The reason, presumably, is that the orbitals involved in the sigma bonds to the bromines are in the best location for forming the pi bond of the olefin when the bromines are *anti*.

The over-all rate of elimination from each dibromide may be expressed as $k = k_{anti} \cdot N_{anti} + k_{gauche} \cdot N_{gauche}$, but, as we have just seen, k_{gauche} is small and so the second term is probably quite negligible. It follows that, if we compare the rates of elimination from the *meso* isomer (k_{meso}) and from the *dl* isomer (k_{dl}), we have $k_{meso}/k_{dl} = k_{anti}^{meso} \cdot N_{anti}^{meso}/k_{anti}^{dl} \cdot N_{anti}^{dl}$. If we now make the further assumption that the specific rates of elimination for the *anti* conformations of the *meso* and the *dl* isomers are the same,* i.e., $k_{anti}^{meso} = k_{anti}^{dl}$, it follows that $k_{meso}/k_{dl} = N_{anti}^{meso}/N_{anti}^{dl}$. These mole fractions were computed earlier (p. 24) and found to be 0.82 and 0.36, respectively, so that the calculated value for k_{meso}/k_{dl} is 0.82/0.36 or 2.3, in perhaps fortuitously excellent agreement with the experimental value.

The second principle which governs the effect of conformation on reactivity is the Curtin-Hammett principle.[9b] The principle has been derived mathematically elsewhere;[77] it states that, if two or more isomeric forms of a compound which are in rapid equilibrium (such as two different conformations of a compound) undergo a reaction in which each isomeric form gives rise to its own characteristic product, the ratio of the products so formed is independent of the relative energy levels of the various starting forms and depends only on the relative energy level of the transition states by which the products are formed—provided that the activation energy for product formation is large compared to the activation energy for the interconversion of the isomeric starting materials.

An example of the operation of the Curtin-Hammett principle is the elimination of dimethylhydroxylamine from dimethyl-*sec*-butylamine oxide to give, besides a major amount of 1-butene, *trans*- and *cis*-2-butene in a ratio of about 2:1.[78] As shown in Fig. 19, since the dimethylamine oxide grouping is clearly much more bulky than the methyl group, the preferred conformation of the starting material must be A; nevertheless

[77] Ref. 24, pp. 237–239.

[78] A. C. Cope, N. A. LeBel, H. H. Lee, and W. R. Moore, *J. Am. Chem. Soc.*, **79**, 4720 (1957).

* At first sight, this assumption may appear unreasonable, since elimination toward *cis*-2-butene leads to eclipsing of the methyl groups and elimination toward *trans*-2-butene leads only to eclipsing of methyl with hydrogen. Certainly, the assumption made may not hold in all cases. However, for 2-butene, the interaction energy of the methyl groups in the *cis* isomer is only about 1 kcal./mole, very similar to the energy of the *gauche* methyl groups in the starting state, and thus the assumption of equal activation energy for the *anti* conformations of the *meso* and *dl* forms going to their respective transition states may not be out of order.

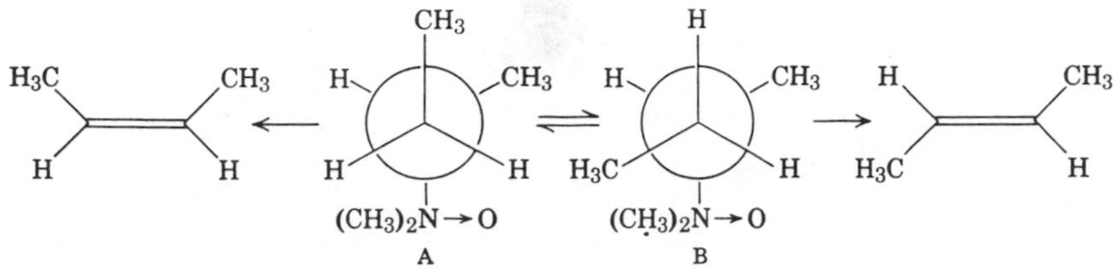

Fig. 19. Elimination of dimethylhydroxylamine from dimethyl-*sec*-butylamine oxide.

the product corresponds to reaction from the less favored conformation B, the reason being that the conformational population of A and B is in fact immaterial as far as product is concerned, and *trans*-2-butene is the favored product because in the transition state leading to it there is not the methyl-methyl eclipsing found in the transition state leading to the *cis* isomer.* Another example (Fig. 20) is the reaction of phenylbenzylcarbinyl triethylbenzoate with potassium *t*-butoxide which gives rise to

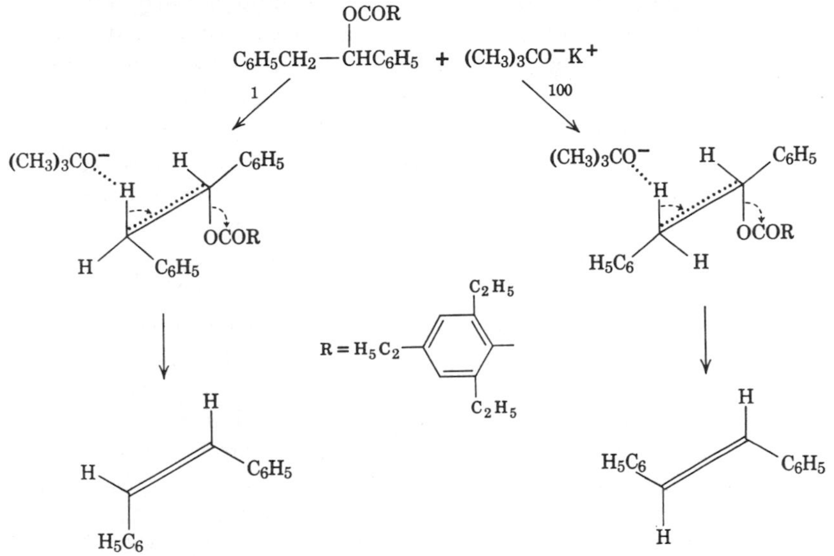

Fig. 20. Elimination from phenylbenzylcarbinyl triethylbenzoate. From E. Eliel, *Stereochemistry of Carbon Compounds*, copyright 1962, McGraw-Hill Book Company, Inc., by permission of the publishers.

* The example is not so elegant as one would wish, as conformation A also can give rise to *trans*-2-butene, and it might be argued that the predominance of the *trans*-olefin is merely due to the fact that it can be derived from both conformations of the starting material whereas the *cis* isomer can only be obtained from one of them.

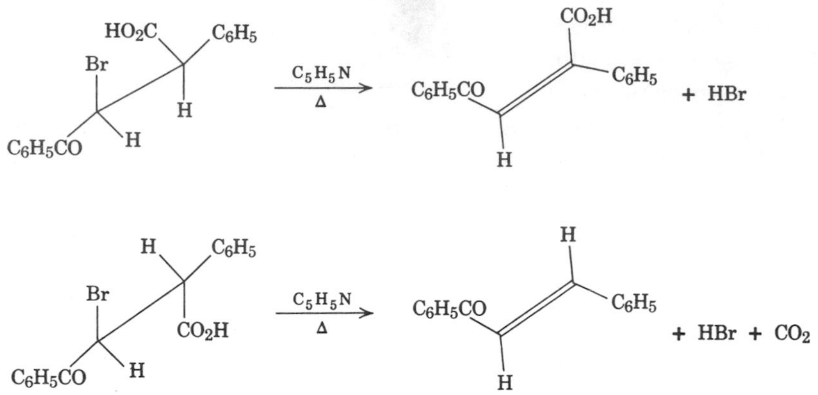

Fig. 21. Reaction of 3-benzoyl-3-bromo-2-phenylpropionic acids with pyridine.

trans-stilbene and *cis*-stilbene in a ratio of 100:1, the transition state leading to the *cis*-olefin being much less favorable because of the incipient eclipsing of the phenyl groups.* An interesting example in which two diastereoisomerically different starting materials give rise to different products is the reaction of stilbene dibromide, $C_6H_5CHBrCHBrC_6H_5$, with pyridine.[79] The *dl*-dibromide undergoes dehydrohalogenation to give bromo-*trans*-stilbene, *trans*-$C_6H_5CBr{=}CHC_6H_5$, but the *meso*-dibromide does not undergo dehydrobromination with hot pyridine to give the corresponding *cis* isomer (in a reaction which would involve serious phenyl eclipsing in the transition state) but, instead, loses bromide in *anti* fashion to give *trans*-stilbene. A similar example is presented by the reaction of the 3-benzoyl-3-bromo-2-phenylpropionic acids with hot pyridine (Fig. 21);[80] one diastereoisomer readily undergoes the expected dehydrobromination to a keto acid, whereas the other—for which dehydrohalogenation would involve a crowded transition state with eclipsed phenyl and benzoyl groups—undergoes decarboxylative debromination to benzalacetophenone. One of the cases in which the Curtin-Hammett principle does not apply is the rearrangement of 1,1-diaryl-2-aminopropanols, $ArAr'COHCH(NH_2)CH_3$, with nitrous acid to the ketones, $ArCOCHAr'CH_3$ or $Ar'COCHArCH_3$.[9b, 81] It is generally found that whether Ar or Ar' migrates depends on which diastereoisomer of the starting material is subjected to deamination; and it was believed at first[9b] that the

[79] P. Pfeiffer, *Ber.*, **45**, 1810 (1912).

[80] E. P. Kohler, W. D. Peterson, and C. L. Bickel, *J. Am. Chem. Soc.*, **56**, 2000 (1934).

[81] For example, D. Y. Curtin and M. C. Crew, *J. Am. Chem. Soc.*, **77**, 354 (1955).

* This example, unfortunately, lacks elegance too because the major product happens to correspond also to the more populated starting conformation.

relative stability of the two transition states dictated the result. However, the amine-nitrous acid reaction is one in which the rate-determining step (loss of nitrogen from the intermediate diazonium ion, $R—N_2^+$) has a very low activation energy, and therefore one of the premises of the validity of the Curtin-Hammett principle, namely, that the activation energy of the reaction under consideration is large compared to the barrier to rotation, is not fulfilled. In the extreme opposite, when the activation energy of the reaction is small compared to the barrier to internal rotation, the product ratio would be dictated by the ratio of reacting conformations in the starting material. In actual fact it has been shown[82] that the deamination of amino alcohols falls into neither category, but that it is an intermediate case in which the activation barrier and the barrier to rotation are of comparable magnitude.

Besides the already mentioned rearrangement[83a] and ionic[83b] and molecular[83c] elimination reactions, addition reactions, especially those proceeding through molecular,[83d] ionic,[83e] and free radical[83f] mechanisms are subject to important conformational restraints. Because the conformational aspects of these reactions are discussed in the standard works on stereochemistry[83] or reaction mechanisms, they will not be developed here in detail. One aspect of ionic elimination, namely, the conformational dependence of isotope effects recently studied by Shiner and co-workers,[84] needs, however, to be discussed. This study is concerned with the over-all secondary isotope effects (k_H/k_D) in the solvolysis, in 60% ethanol, of the mono-, di-, and trideutero-t-butyl chlorides shown in Fig. 22. (The hexa- and nonadeutero compounds were also studied.) The results may be summarized by saying that only a deuterium atom *anti* to the leaving chloride has a substantial isotope effect (1.30), whereas the secondary isotope effect of a *gauche*-situated deuterium is nearly negligible (1.01).* The best explanation of these findings is expressed in terms of hyperconjugation, i.e., overlap of C—H sigma bonds with the vacant p-orbital of the incipient carbonium ion; this overlap is good only for the hydrogen in position *anti* to the leaving group (C—H bond parallel to p-orbital) but

[82] B. M. Benjamin, H. J. Schaeffer, and C. J. Collins, *J. Am. Chem. Soc.*, **79**, 6160 (1957). See also J. C. Martin and W. G. Bentrude, *J. Org. Chem.*, **24**, 1902 (1959), and ref. 24, pp. 153–156, for a discussion of this work.

[83] For example, ref. 24: (a) pp. 142–149; (b) pp. 140–142; (c) p. 149; (d) pp. 350–355, 357–361; (e) pp. 355–357; (f) pp. 361–365.

[84] V. J. Shiner, B. L. Murr, and G. Heinemann, *J. Am. Chem. Soc.*, **85**, 2413 (1963).

* The original article should be consulted for a quantitative demonstration of the difference between three deuterium atoms exercising equal, cumulative isotope effects and three deuterium atoms enhancing the isotope effect merely through being favorably located conformationally as explained above.

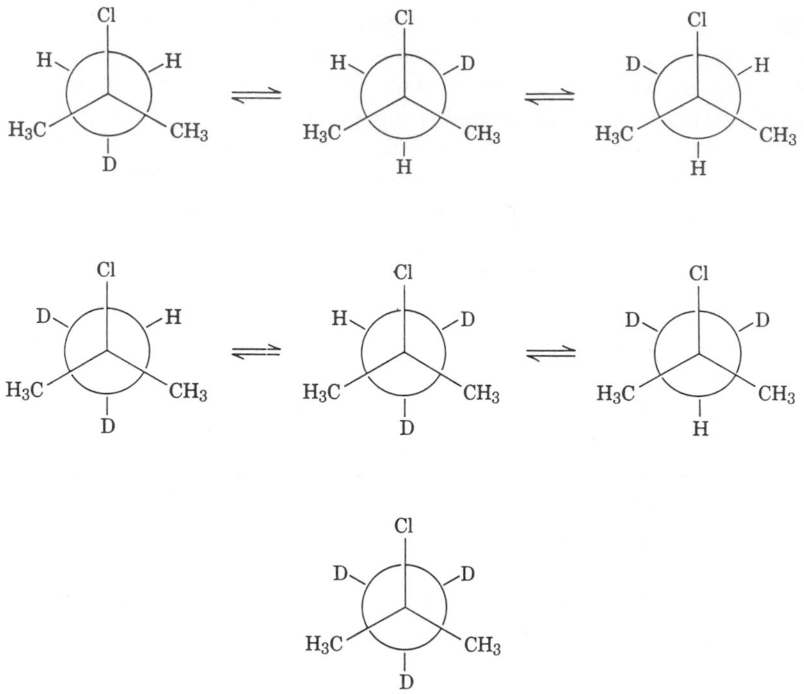

Fig. 22. Deuterated *t*-butyl chlorides used in solvolysis study.

quite unfavorable for the two *gauche* hydrogens (which are at a 30° angle of torsion to the *p*-orbital) as shown in Fig. 23.[85]

We shall turn next to conformational factors in the reactivity of unsaturated systems. Pertinent here are two rules predicting the stereochemical outcome of certain carbonyl addition reactions, namely, Cram's rule[86a,b] and Prelog's rule.[87a] Cram's rule states that in kinetically controlled addition reactions (other than catalytic hydrogenation) of a carbonyl compound, *LMSCCOR* (e.g., Grignard addition, hydride reduction), the predominantly formed stereoisomer may be predicted as follows: The asymmetric carbon is so rotated that the carbonyl group is flanked by the two smaller groups, *M* (medium) and *S* (small), attached to the

[85] V. J. Shiner, *J. Am. Chem. Soc.*, **78**, 2653 (1956).

[86] (a) D. J. Cram and F. A. Abd Elhafez, *J. Am. Chem. Soc.*, **74**, 5828 (1952). (b) D. J. Cram and K. R. Kopecky, *ibid.*, **81**, 2748 (1959); J. H. Stocker, P. Sidisunthorn, B. M. Benjamin, and C. J. Collins, *ibid.*, **82**, 3913 (1960). (c) For an application of similar principles to a different reaction, see G. Zweifel, N. R. Ayyangar, and H. C. Brown, *ibid.*, **84**, 4342 (1962).

[87] (a) V. Prelog, *Helv. Chim. Acta*, **36**, 308 (1953). (b) For an application to a different system, see H. M. Walborsky, L. Barash, A. E. Young, and F. J. Impastato, *J. Am. Chem. Soc.*, **83**, 2517 (1961).

asymmetric carbon, with the large group L eclipsing R. The reagent will then approach from the side of the smaller group S, as shown in Fig. 24. The model implied in Fig. 24 does not apply when one of the groups S, M, or L attached to the asymmetric carbon is capable of complexing with the organometallic (e.g., OH, NH_2); then a different model is to be used.[86b] The question may well be raised whether even when the three groups do not complex (alkyl, aryl, hydrogen) the rule does more than give a formal prediction; i.e., if the model implied in Fig. 24 has any mechanistic significance or whether its predictive success is purely coincidental. It might be noted, first of all, that the stable ground state conformation of a carbonyl compound is *not* that implied in Fig. 24, but one in which the carbonyl group is eclipsed and the R group staggered. For, as already mentioned, in acetaldehyde, a hydrogen of the methyl group eclipses the oxygen (with the other two hydrogens staggering the aldehyde hydrogen);[44] in diethyl ketone, oxygen may be eclipsed by hydrogen;[88] and, in propionalde-

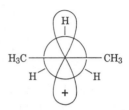

Fig. 23. Hyperconjugation in incipient *t*-butyl carbonium ion.

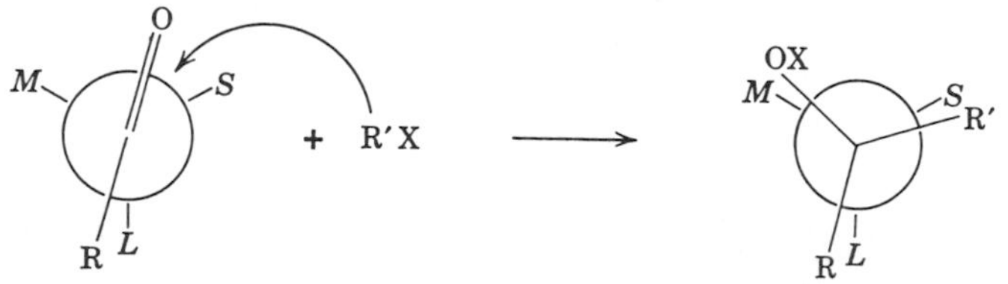

hyde, methyl eclipses oxygen.[50] This, however, does not necessarily militate against a *transition state* for carbonyl addition which corresponds to the picture shown in Fig. 24, inasmuch as, according to the Curtin-Hammett principle, ground state conformation is not decisive in determining product. Moreover, it has been shown[73] that increasing the steric requirements of the approaching reagent in the reduction of 3-phenyl-2-butanone from $LiAlH_4$ in ether to $LiAlH(O\text{-}t\text{-}Bu)_3$ in ether to $LiAlH_4$ in tetrahydrofuran to $LiAlH(O\text{-}t\text{-}Bu)_3$ in tetrahydrofuran to $LiAlH_4\text{-}AlCl_3$ brings about an increase in stereoselectivity from a minimum of 72.5% in favor of the *threo*-3-phenyl-2-butanol to a maximum of 82% of the same isomer. Since the reagents are in ascending steric requirement, as indicated by their stereoselectivity in the reduction of 3,3,5-trimethylcyclohexanone,[89] it is at least reasonable to assume that steric interference in the

Fig. 24. Cram's rule.

[88] R. N. Jones and K. Noack, *Can. J. Chem.*, **39**, 2214 (1961). See, however, C. Romers and J. E. G. Creutzberg, *Rec. Trav. Chim.*, **75**, 331 (1956).

[89] H. Haubenstock and E. L. Eliel, *J. Am. Chem. Soc.*, **84**, 2363 (1962).

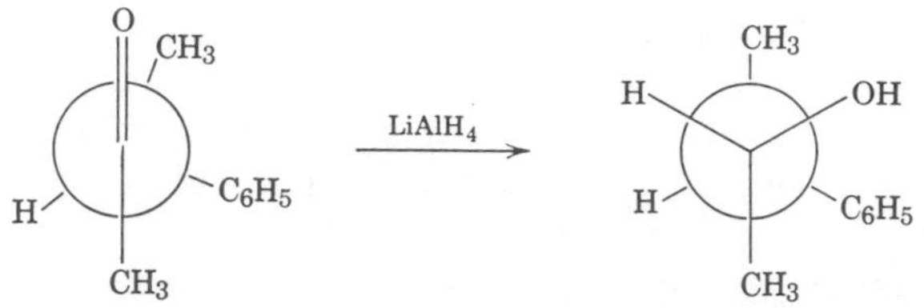

Fig. 25. Cram's rule: alternative reactive conformation.

transition state, as implied in Fig. 24, dictates the predominating isomer, increasing steric interference leading to increasing stereoselectivity. The same point is suggested by the reaction of 3-phenyl-2-butanone with the methyl, ethyl, and phenyl Grignard reagent, the expected *erythro* product predominating[86a] by a ratio of 2:1, 3:1, and 5:1,[73] respectively, in the three cases. It must be conceded, however, that an alternative conformation (Fig. 25) in which methyl (not hydrogen) eclipses oxygen and the approaching reagent comes from the side of the (smaller) hydrogen can also explain the results.

Prelog's rule[87a] deals with the approach of a reagent to the ketone carbonyl of an α-keto ester of an asymmetric alcohol, $RCOCOOCSML$, and is symbolized in Fig. 26. That the alcohol part of the ester is *cis* to the ester carbonyl (and therefore *trans* to the ketone carbonyl) is reasonable on the basis of the earlier-cited spectroscopic studies of such simple molecules as methyl formate and methyl acetate,[55] but the conformation of the three groups L, M, and S about the alkyl oxygen bond is arbitrary, and, in fact, it is clear that two of the three possible staggered conformations lead to a reasonable answer if one assumes that the approaching reagent comes from the side of the smaller of the two groups which protrude sideways (M in preference to L, or S in preference to M). With so much conformational freedom in the molecule, it would appear nearly hopeless to obtain a true model of the transition state.*

In conclusion, it might be stressed that, in the application of both Cram's and Prelog's rules, stereoselectivity is not high—perhaps a maximum of 10:1 (but often as little as 2–3:1) in Cram's rule and less than

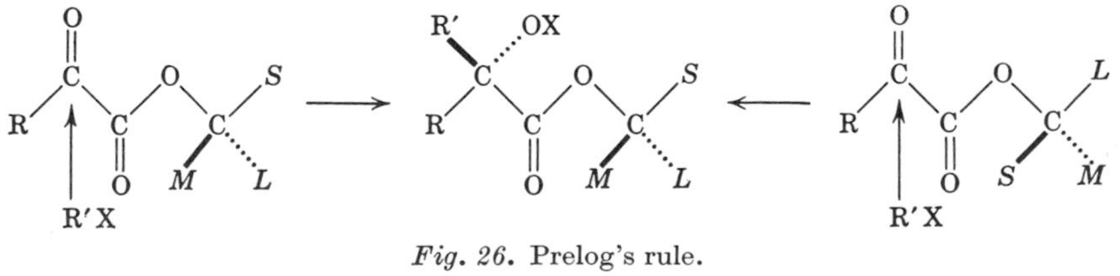

Fig. 26. Prelog's rule.

* For an exception to Prelog's rule in hydride reduction, see J. A. Berson and M. A. Greenbaum, *J. Am. Chem. Soc.*, **81**, 6456 (1959).

that in Prelog's rule (where the asymmetric center is farther from the reaction site and there is more conformational ambiguity). This illustrates a general principle according to which stereoselectivity is, in general, moderate in acyclic systems because the systems can, with relatively little energetic effort, adjust themselves to the stereoelectronic demands of the reaction. Greater stereoselectivity is to be expected in the more rigid cyclic systems to be considered in the following chapter.

General References

D. Y. Curtin, "Stereochemical Control of Organic Reactions," *Record Chem. Progr.* (*Kresge-Hooker Sci. Lib.*), **15,** 111 (1954).

W. G. Dauben and K. S. Pitzer, "Conformational Analysis," in M. S. Newman, ed., *Steric Effects in Organic Chemistry*, John Wiley & Sons, Inc., New York, 1956, Chap. 1.

H. H. Lau, "Prinzipien der Konformationsanalyse," *Angew. Chem.*, **73,** 423 (1961).

D. J. Millen, "Restricted Rotation about Single Bonds," in P. B. D. de la Mare and W. Klyne, eds., *Progress in Stereochemistry*, Vol. 3, Academic Press, Inc., New York, 1962, Chap. 4.

S. Mizushima, *The Structure of Molecules and Internal Rotation*, Academic Press, Inc., New York, 1954.

S. Mizushima, "Japanese Researches on Internal Rotation," *Pure Appl. Chem.*, **7,** 1 (1963).

Chapter 2

Basic Principles of Conformational Analysis—Cyclohexane

2–1. Chair and Flexible Forms

It was said in the Introduction that Sachse recognized the existence of two forms of cyclohexane free of angle strain, the rigid or chair form and the flexible form, later sometimes misnamed the boat form (see below). The chair form is shown in Fig. 1. There are two types of bonds in the chair form, those pointing up and down, called "axial bonds" (a in Fig. 1), and those pointing outward, called[1] "equatorial bonds" (e in Fig. 1).* This chair form is not only free of angle strain but also free of torsional (Pitzer) strain (Sec. 1-2) and free of van der Waals strain, since no two non-bonded atoms approach each other within their van der Waals radii. Since the same is not true of the flexible form (see below), it may be expected that cyclohexane and most of its derivatives will exist in the chair form. That this is in fact so has been established in a number of instances by physical measurements, some of which were mentioned in Sec. 1-2 (see also Chap. 3). Perhaps the most salient characteristic of the chair form is that, because of the staggering of the bonds, the *cis* substituents (e_1 and a_2 in Fig. 1) and one pair of *trans* substituents (e_1 and e_2 in Fig. 1) are equidistant.† (The same is not true

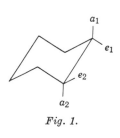

a_1
e_1
e_2
a_2

Fig. 1.

[1] D. H. R. Barton, O. Hassel, K. S. Pitzer, and V. Prelog, *Science*, **119**, 49 (1953); *Nature*, **172**, 1096 (1953).

* Before 1953, the axial bonds were called "polar." The designations "ϵ" (first letter of Greek word for upright) and "κ" (for Greek for prostrate) were also used for axial and equatorial bonds, respectively.

† Instead of "equidistant" the correct wording would read "nearly equidistant." As will be seen later, (Sec. 2-7), substituted cyclohexanes, and even cyclohexane itself, may be slightly deformed chairs, and as a result the distances in question become slightly unequal.

36

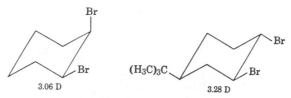

Fig. 2. Dipole moments of 1,2-dibromocyclohexanes.

for all adjacent substitutents in the flexible form.) As a result, the dipole moments of such compounds as *cis*-1,2-dibromocyclohexane[2] and *cis*-3-bromo-*trans*-4-bromo-*t*-butylcyclohexane[3] (Fig. 2) are nearly the same (3.06 D vs. 3.28 D in carbon tetrachloride solution). Similarly, the strength of the internal hydrogen bond in *cis*-1,2-cyclohexanediol is about the same as that in *trans*-1,2-cyclohexanediol, as evidenced by a similar shift of the infrared stretching frequency of the bonded with respect to the unbonded hydroxyl group[4] (*cis*, 39 cm^{-1}; *trans*, 32 cm.$^{-1}$).* Other physical evidence for the chair shape of cyclohexane and a number of its derivatives comes from X-ray diffraction,[5] electron diffraction,[6] infrared[7] and Raman[8] spectroscopy, and thermodynamic data.[9] Some of this evidence will be discussed in greater detail in Chap. 3.

Direct chemical support for the chair form of cyclohexane is more difficult to pinpoint, but the combined chemical evidence to be discussed in this chapter is probably uniquely consistent with the assumption that the great majority of cyclohexane derivatives are chair-shaped.

The flexible form of cyclohexane has often been erroneously equated with a "boat form." Models of this form (Fig. 3, A) indicate that, although the boat is free of angle strain, there is bond eclipsing leading to torsional strain on the side of the boat and there is also van der Waals interaction between the two hydrogen atoms occupying the "bowsprit" and "flagpole" positions. (These hydrogens are within 1.83 Å of each other, whereas the sum of their van der Waals radii is 2.4 Å.) As a result,

[2] P. Bender, D. L. Flowers, and H. L. Goering, *J. Am. Chem. Soc.*, **77,** 3463 (1955).

[3] E. C. Wessels, Ph.D. Dissertation, University of Leiden, Holland, 1960.

[4] L. P. Kuhn, *J. Am. Chem. Soc.*, **74,** 2492 (1952).

[5] S. B. Hendricks and C. Bilicke, *J. Am. Chem. Soc.*, **48,** 3007 (1926); R. G. Dickinson and C. Bilicke, *ibid.*, **50,** 764 (1928); O. Hassel and H. Kringstad, *Tiddskr. Kjemi Bergvesen Met.*, **10,** 128 (1930).

[6] O. Hassel and H. Viervoll, *Acta Chem. Scand.*, **1,** 149 (1947).

[7] R. S. Rasmussen, *J. Chem. Phys.*, **11,** 249 (1943).

[8] K. W. F. Kohlrausch and H. Wittek, *Z. Phys. Chem.*, **48B,** 177 (1941); H. Gerding, E. Smit, and R. Westrik, *Rec. Trav. Chim.*, **61,** 561 (1942).

[9] J. G. Aston, S. C. Schumann, H. L. Fink, and P. M. Doty, *J. Am. Chem. Soc.*, **63,** 2029 (1941).

* The reason for the small but real difference will be returned to on pp. 77–78.

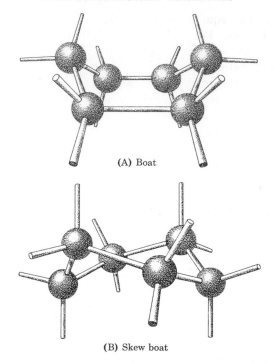

(A) Boat

(B) Skew boat

Fig. 3. Boat and skew-boat forms of cyclohexane. From E. Eliel, *Stereochemistry of Carbon Compounds,* copyright 1962, McGraw-Hill Book Company, Inc., by permission of the publishers.

the boat is calculated[10a] to be 6.9 kcal./mole less stable than the chair. In manipulating a model of the flexible form one may readily convince oneself, however, that the boat does not represent a conformation of minimum potential energy. In rotating bonds so as to pass from one boat to another, one obtains a conformation in which the bowsprit-flagpole interaction is less severe than in the boat and the eclipsing of C—H bonds is also somewhat alleviated. This conformation represents the true energy minimum of the flexible form; it has been called[12a] a "skew boat," a "twist form," or a "stretched form" (Fig. 3, B). The term skew boat will be used here in referring to this particular conformation of the flexible form. Its energy has been calculated[10a] to be 1.6 kcal./mole less than that of the true boat or 5.3 kcal./mole more than that of the chair form.

An experimental determination of the enthalpy difference between the chair and the flexible forms comes from a study[11] of the temperature

[10] (a) J. B. Hendrickson, *J. Am. Chem. Soc.,* **83,** 4537 (1961); see also (b) N. L. Allinger, *ibid.,* **81,** 5727 (1959); (c) R. Pauncz and D. Ginsburg, *Tetrahedron,* **9,** 40 (1960); (d) K. E. Howlett, *J. Chem. Soc.,* **1957,** 4353; (e) P. Hazebroek and L. J. Oosterhoff, *Discussions Faraday Soc.,* **10,** 87 (1951).

[11] N. L. Allinger and L. A. Freiberg, *J. Am. Chem. Soc.,* **82,** 2393 (1960).

dependence of the equilibrium between *cis-* and *trans-*1,3-di-*t*-butyl-cyclohexane (Fig. 4). For the equilibrium as shown, $\Delta H° = 5.9 \pm 0.6$ kcal./mole and $\Delta S° = 4.9 \pm 1.0$ cal./deg. mole. The large positive entropy change militates against the *trans* isomer being in the chair form with an axial *t*-butyl group but suggests that the large axial *t*-butyl-hydrogen-hydrogen interaction (see below) in this conformation forces the *trans* isomer into the flexible form, which has a high entropy.* If this argument is accepted, and if it is granted that the extra interaction energy terms caused by the presence of the *t*-butyl groups are about the same in the *cis* chair form and in the *trans* flexible form,[12] then the measured $\Delta H°$ value of 5.9 kcal./mole represents the difference between the chair and the flexible forms. The good agreement with the calculated value of 5.3 kcal./mole (see above) is encouraging.

Another experimental estimate[12a] of the enthalpy difference between the chair and flexible forms comes from the heats of combustion of the lactones of the 3-hydroxy-*trans*-decalin-2-acetic acids shown in Fig. 5. The diequatorial *trans* isomer (A) can form a lactone in the chair form, but the diaxial *trans* isomer (B) must be converted to a flexible form before it can be lactonized. The difference in heats of combustion of the two lactones is 4.1 ± 0.4 kcal./mole. After correction for eclipsing interactions occurring in the flexible but not the chair form, and allowing for the fact that the five-membered ring attached to the rigid chair in (A) is more strained than the same ring attached to the flexible form in (B), the enthalpy difference between the chair and flexible form is calculated[12] to be 5.5 kcal./mole, in agreement with the values cited earlier.

A similar value, 4.79 ± 0.94 kcal./mole, has been derived from the heats of combustion[12b] of *trans-syn-trans-* and *trans-anti-trans-*perhydro-anthracene; as will be shown in Chap. 4, the central ring in the latter isomer is in the twist form.

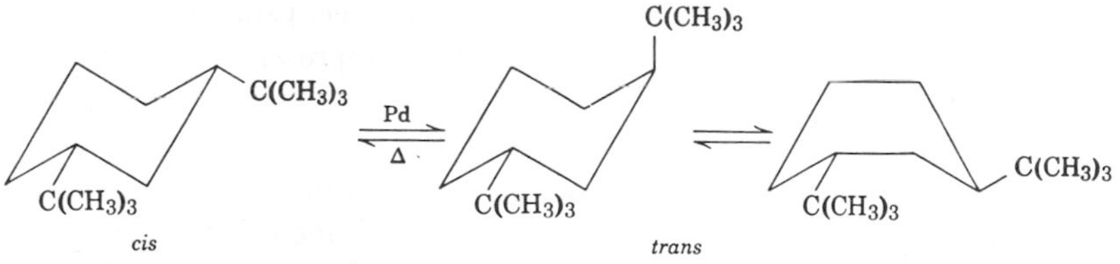

Fig. 4.

[12] (a) W. S. Johnson, V. J. Bauer, J. L. Margrave, M. A. Frisch, L. H. Dreger, and W. N. Hubbard, *J. Am. Chem. Soc.*, **83**, 606 (1961); (b) J. L. Margrave, M. A. Frisch, R. G. Bautista, R. L. Clarke, and W. S. Johnson, *ibid.*, **85**, 546 (1963); (c) see also B. J. Armitage, G. W. Kenner, and M. J. T. Robinson, *Tetrahedron*, **20**, 747 (1964).

* A rigid boat form is also incompatible with the large increase in entropy of the reaction shown in Fig. 4.

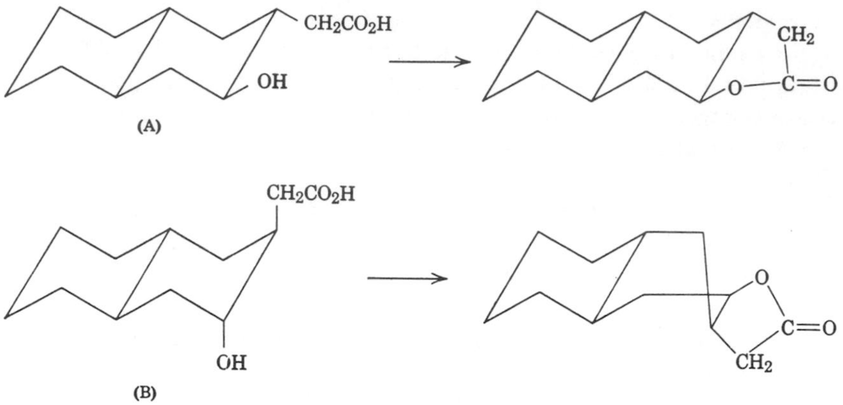

(A)

(B)

Fig. 5.

trans-1,3-Di-*t*-butycyclohexane and some *cis*-1,4-di-*t*-alkylcyclohexane-2,5-diols[13] are the only known monocarbocyclic saturated cyclohexane derivatives which exist largely in the flexible form. (For the diols[13] this has been clearly shown through infrared study of intramolecular hydrogen bonding which, on geometrical grounds, can occur only in the flexible form and not in the chair.) Normally, the chair form of a cyclohexane derivative is so much more stable, at room temperature, than the flexible form that the population of the latter is negligible. For example, in cyclohexane itself, for the transformation chair $\rightleftharpoons$ flexible form, with $\Delta H° = 5.3$ kcal./mole (the value calculated by Hendrickson[10a]) and $\Delta S° = 4.9$ cal./deg. mole (the value found by Allinger for the *trans*-1,3-di-*t*-butyl compound[11]),* $\Delta G°$ at 25° is about 3.8 kcal./mole; this means that only one or two molecules in a thousand will be in the flexible form at room temperature. Exceptions occur among substituted cyclohexanones and cyclohexadiones (Sec. 2-6) as well as in a number of polycyclic cyclohexane derivatives and will be considered elsewhere in this book.[14,15]

Just as ethane can exist in three equal but distinct staggered conformations which are interconvertible by internal rotation (Chap. 1), cyclohexane can exist in two distinct chair conformations. In passing from one of these to the other, the chair is first converted to the flexible form which,

[13] R. D. Stolow, *J. Am. Chem. Soc.*, **83**, 2592 (1961); R. D. Stolow and M. M. Bonaventura, *ibid.*, **85**, 3636 (1963).

[14] J. Levisalles, *Bull. Soc. Chim. France*, **1960**, 551. See also Sec. 7-4.

[15] See also M. Balasubramanian, *Chem. Rev.*, **62**, 591 (1962).

* This value is in excess of that calculated for the chair $\rightleftharpoons$ boat equilibrium ($R \ln 6 - R \ln 2$ or 2.2 cal./deg. mole) assuming that the entire entropy difference is due to a difference in symmetry numbers (see p. 57) between the chair (6) and the boat (2).

in turn, is converted to the other chair, as shown in Fig. 6. In the process of "flipping" or chair inversion, all equatorial bonds become axial and all axial bonds become equatorial. Chair inversion occurs rapidly. This may be inferred from the fact that the nuclear magnetic resonance spectrum of cyclohexane shows only a single sharp line. Since, in principle, axial and equatorial protons in cyclohexane resonate at different fields,[16] one might have expected that two lines (probably further divided by spin-spin coupling) would appear in the spectrum. The fact that only one line is seen implies that the two chair forms are in rapid equilibrium.[17] Under these circumstances a spectrum of an "average" of the two chair forms is seen, and, inasmuch as each hydrogen atom is then equatorial 50% of the time and axial 50% of the time, they all have the same average position and thus give the same signal.

When cyclohexane is cooled, the rate of chair interconversion is reduced, and at low temperatures ($-90°$ to $-100°$), one does see two families of bands in the NMR spectrum, corresponding to the equatorial and axial protons.[18-22] As cyclohexane is warmed from $-100°$ to room temperature these two bands gradually coalesce. From the temperature of coalescence ($-66.7°$) a rate of chair inversion of 105 sec^{-1} at $-66.7°$ corresponding to a free energy barrier of 10.1 kcal./mole has been calculated;[18] an alternative calculation[19] gives a free energy barrier of 10.3 kcal./mole associated with values of 9 kcal./mole for $\Delta H^{\ddagger}$ and $+2.9$ cal./deg. mole for $\Delta S^{\ddagger}$. The complete potential energy diagram for cyclohexane is thus as shown in Fig. 7.* The energy maximum has been calculated[10a]

Fig. 6. Chair inversion.

[16] For example, R. U. Lemieux, R. K. Kullnig, H. J. Bernstein, and W. G. Schneider, *J. Am. Chem. Soc.*, **80**, 6098 (1958). See also Sec. 3-4d.

[17] J. D. Roberts, *Nuclear Magnetic Resonance*, McGraw-Hill Book Co., New York, 1959, p. 63.

[18] F. R. Jensen, D. S. Noyce, C. H. Sederholm, and A. J. Berlin, *J. Am. Chem. Soc.* **84**, 386 (1962); see also preliminary communication, *ibid.*, **82**, 1256 (1960).

[19] F. A. L. Anet, M. Ahmad, and L. D. Hall, *Proc. Chem. Soc.*, **1964**, 145; F. A. Bovey, F. P. Hood, E. W. Anderson, and R. L. Kornegay, *ibid.*, **1964**, 146; R. K. Harris and N. Sheppard, *ibid.*, **1961**, 418.

[20] L. W. Reeves and K. O. Strømme, *Can. J. Chem.*, **38**, 1241 (1960).

[21] G. Van Dyke Tiers, *Proc. Chem. Soc.*, **1960**, 389.

[22] L. W. Reeves and K. O. Strømme, *Trans. Faraday Soc.*, **57**, 390 (1961).

* This figure actually corresponds to the more general case of a substituted cyclohexane in which the two chair forms are of unequal stability.

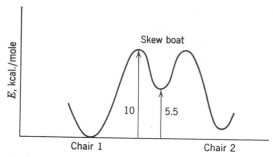

Fig. 7. Energy diagram for cyclohexane chair inversion. From E. Eliel, *Stereochemistry of Carbon Compounds*, copyright 1962, McGraw-Hill Book Company, Inc., by permission of the publishers.

to correspond to a "half-chair" form of cyclohexane (see the discussion on cyclohexene in Sec. 2-6) in which four of the ring atoms are in one plane. The height of the energy barrier has also been estimated by ultrasonic relaxation measurements (cf. Sec. 3-10), and values of $\Delta H^{\ddagger}$ of 8.3 kcal./mole[23] and 10.8 kcal./mole[24] and of $\Delta S^{\ddagger}$ of -11 cal./deg. mole[23] for methylcyclohexane and of $\Delta H^{\ddagger} = 12.0$ kcal./mole for chloro- and bromocyclohexane[24] have been obtained. NMR barrier values on record for substituted cyclohexanes are $\Delta G^{\ddagger} = 10.9$ kcal./mole for bromocyclohexane,[20] 11.9 kcal./mole for 1,2 dibromo- and 1,2-dichlorocyclohexane,[22] and $\Delta H^{\ddagger} = 7.5$ kcal./mole and $\Delta S^{\ddagger} = 10.7$ cal./deg. mole for perfluorocyclohexane.[21]

2–2. Monosubstituted Cyclohexanes. Conformational Equilibrium

a. Introductory Remarks. A monosubstituted cyclohexane may exist either in the conformation with an equatorial substituent or in that with an axial substituent (Fig. 8). The barrier for interconversion of the two chair forms is of the same order of magnitude as in cyclohexane itself.[20,25] The two conformational isomers are thus in rapid

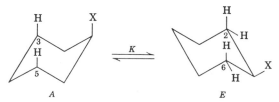

Fig. 8. Monosubstituted cyclohexane.

[23] M. E. Pedinoff, *J. Chem. Phys.*, **36**, 777 (1962).
[24] J. E. Piercy, *J. Acoust. Soc. Am.*, **33**, 198 (1961).
[25] A. J. Berlin and F. R. Jensen, *Chem. Ind. (London)*, **1960**, 998.

equilibrium with each other, and no case has yet been found among substituted cyclohexanes where they can be isolated. There is, however, much physical evidence for their separate existence (cf. Chap. 3).

The distance between hydrogen atoms on adjacent carbons in cyclohexane (equatorial-equatorial or equatorial-axial) is 2.5 Å, the same as the distance between axial hydrogens on the same side of the ring (1-3).* However, for substituents larger than hydrogen, the distance from the axial position to the axial hydrogens at C_3 and C_5 is less than the distance from an equatorial position to the adjacent hydrogens at C_2 and C_6. For example, for carbon the former distance is 2.55 Å and the latter is 2.8 Å. As a result, an axial substituent will encounter more van der Waals repulsion from the axial hydrogens at C_3 and C_5 than an equatorial substituent does from the two extra hydrogens at C_2 and C_6;† that is, the monosubstituted cyclohexane will normally be more stable in the equatorial conformation (Fig. 8, E) than in the axial conformation (Fig. 8, A), and thus the "conformational equilibrium constant" K (Fig. 8) will exceed unity.

The point may be illustrated further by the example of methylcyclohexane. In a Dreiding model the distance of closest approach of the hydrogens of the equatorial methyl group (staggered with respect to the ring) and the ring hydrogens is 2.5 Å, just outside the combined van der Waals radii of two hydrogens (2.4 Å). On the other hand, when the methyl group is axial one of its hydrogens (the one pointing inside the ring) approaches the syn-axial ring hydrogens to within ca. 1.8 Å. Therefore the axial methyl group is subject to van der Waals repulsion, the equatorial is not. The extent of the repulsion may be estimated by noting that the interaction between the axial methyl group and the syn-axial hydrogen is of the same type as the interaction encountered in the $gauche$ form of butane. Since there are two such interactions in the axial methylcyclohexane, the enthalpy difference between the axial and the equatorial forms of methylcyclohexane should be twice the $gauche$-$anti$ enthalpy difference in butane (0.8–0.9 kcal./mole) or 1.6–1.8 kcal./mole. This is in good agreement with the experimental findings shown in Table 1.‡ From this table it may be inferred that at room temperature about 95% of all

[26] E. J. Prosen, W. H. Johnson, and F. D. Rossini, *J. Res. Natl. Bur. Std.*, **39**, 173 (1947).

[27] C. W. Beckett, K. S. Pitzer, and R. Spitzer, *J. Am. Chem. Soc.*, **69**, 2488 (1947).

* Such hydrogens will sometimes be called syn-axial in this book.

† For a more detailed and sophisticated treatment of this problem, other repulsions as well as, in some cases, attractions need be considered. See Sec. 7-2.

‡ The best experimental value is 1.6–1.7 kcal./mole in the liquid phase[26] and 1.9 kcal./mole in the gas phase.[26] A grand over-all value of 1.8 kcal./mole[27] has often been used. We prefer to use 1.7 kcal./mole for the liquid and 1.9 kcal./mole for the vapor. See also p. 54.

methylcyclohexane molecules are in the equatorial form and only about 5% in the axial form. (For a more detailed discussion see Sec. 2-3.)

b. Tabulation of Conformational Equilibria. The negative of the free energy difference ΔG_X° corresponding to the conformational equilibrium constant K in Fig. 8 (called the conformational free energy difference or, sometimes, simply the conformational energy) for a variety of substituents X is listed in Table 1. The term "A-value" has also been used[28]

Table I. Conformational Free Energy Differences

Group	$-\Delta G_X^\circ$, kcal./mole	Group	$-\Delta G_X^\circ$, kcal./mole
F	0.25	NH(CH$_3$)$_2^+$	2.4
Cl	0.0–0.5; 0.4	CH$_3$	1.5–2.1; 1.7
Br	0.2–0.7; 0.5	C$_2$H$_5$	1.65–2.25; 1.8
I	0.40–0.45; 0.4	i-C$_3$H$_7$	1.8–2.5; 2.1
OH	0.25–1.25; 0.7	n-C$_3$H$_7$	2.1[c]
OCH$_3$	0.6–0.75; 0.7	n-C$_4$H$_9$	2.1[c]
OC$_2$H$_5$	0.9–1.0; 0.9	(CH$_3$)$_3$C	~ 5[c]
OCOCH$_3$	0.4–1.5; 0.7	(CH$_3$)$_3$CCH$_2$	2.0[c]
OSO$_2$C$_6$H$_4$CH$_3$-p	0.6–1.7; 0.7	C$_6$H$_5$	2.0–3.1; 3.1
OCOC$_6$H$_4$NO$_2$-p	1.0	C$_6$H$_{11}$	2.15
OCOC$_6$H$_4$CO$_2^-$-o	1.2	CO$_2$H	0.7–1.7; 1.2
SH	−0.4–0.9; 0.9	CO$_2^-$	2.1–2.3; 2.3
SC$_6$H$_5$	0.8	CO$_2$CH$_3$	1.1
NH$_2$	1.1–1.8; 1.2,[a] 1.8[b]	CO$_2$C$_2$H$_5$	1.0–1.2; 1.1
NH$_3^+$	1.8–2.0; 1.9	CN	0.15–0.25; 0.2
N(CH$_3$)$_2$	2.1	HgBr	ca. 0

[a] In aprotic solvent.
[b] In hydrogen donor solvent.
[c] Crude estimate.

for $-\Delta G_X^\circ$, but since the term is not well founded on rational nomenclature it will be avoided here.

Conditions of temperature and solvent for the data in Table 1 will be detailed in Sec. 7-1. Most data refer to room temperature (25°). Changes with temperature are to be expected when $\Delta S_X^\circ \neq 0$; such conformational entropy differences are probably unimportant for spherically symmetrical groups such as halogens or methyl (the matter has not been extensively investigated) but will surely have to be considered for unsymmetrical groups such as isopropyl for which, then, ΔG_X° is appreciably temperature-dependent. This point will be considered in Sec. 2-4. The effect of solvent has not been extensively investigated as yet; in

[28] S. Winstein and N. J. Holness, *J. Am. Chem. Soc.*, **77**, 5562 (1955).

some groups, such as hydroxyl, it is apparently unimportant;[29],[30] in others, such as amino, it is clearly important;[31] and in yet others, such as bromo, there is conflicting evidence.[32],[33] It now appears that the importance or unimportance of solvent will have to be established for many more groups, although in some cases a rational prediction may be possible (Sec. 2-5). Methods of obtaining the individual values summarized in Table 1 are detailed, with references, in Sec. 7-1. Among the physical methods are comparison of calculated with experimental thermodynamic properties,[34] electron diffraction methods, infrared methods, and methods employing NMR spectroscopy. These and other methods will be discussed in Chap. 3. Chemical methods available for determining K are equilibrium methods and a kinetic method, to be discussed later in this chapter.[35]

The values for ΔG_X° in Table 1 are listed as an experimental range with the most recommended value for general use indicated in italic type.

Several features of the data in Table 1 merit discussion. The values for monovalent atoms (F, Cl, Br, I) are remarkably small—less than 0.5 kcal./mole—even for atoms as large as iodine. The values for bivalent atoms (O, S, Hg) are also remarkably small, in the 0–1 kcal./mole category. The second substitutent makes relatively little difference in these cases (compare —OH with —OAc, —OTs, —OCH$_3$, and —OC$_2$H$_5$ and —SH with —SC$_6$H$_5$). The largest groups are those with two or three substituents (other than the ring), such as CO$_2$C$_2$H$_5$, CO$_2$H, CH$_3$, C$_2$H$_5$, and (CH$_3$)$_2$CH.

The fact that the number rather than the type of substituents on the atom next to the ring governs the magnitude of $-\Delta G^\circ$ is perhaps not surprising. Model considerations suggest that the large substituents can usually be rotated into positions where they cause no extra interference with the axial ring hydrogens. Notable exceptions are the methyls of the t-butyl group, one of which must necessarily point into the ring when the group is axial; accordingly, this group has a compulsion to be equatorial by about 5 kcal./mole.

More striking are the relatively small $-\Delta G^\circ$ values for large substituents on the ring, such as bromine, iodine, mercury, and sulfur. Two principal reasons have been suggested[35],[36] for these small values. One is

[29] G. Chiurdoglu and W. Masschelein, *Bull. Soc. Chim. Belges*, **69,** 154 (1960).

[30] G. Chiurdoglu and W. Masschelein, *Bull. Soc. Chim. Belges*, **70,** 767 (1961).

[31] E. L. Eliel, E. W. Della, and T. H. Williams, *Tetrahedron Letters*, **1963,** 831.

[32] G. Chiurdoglu, L. Kleiner, W. Masschelein, and J. Reisse, *Bull. Soc. Chim. Belges*, **69,** 143 (1960).

[33] E. L. Eliel, *Chem. Ind. (London)*, **1959,** 568.

[34] C. W. Beckett, K. S. Pitzer, and R. Spitzer, *J. Am. Chem. Soc.*, **69,** 2488 (1947).

[35] See also E. L. Eliel, *J. Chem. Educ.*, **37,** 126 (1960).

[36] E. L. Eliel and R. G. Haber, *J. Am. Chem. Soc.*, **81,** 1249 (1959).

that, as the group X becomes larger, although its van der Waals radius increases, its bonding radius increases also. Since the C—H distance of the interfering axial hydrogens remains constant at 1.08 Å, increasing the C—X distance (Fig. 8) will increase the axial-axial X—H distance. For example, this distance is 2.55 Å for a methyl group (C—CH_3 bond length 1.54 Å) but 2.67 Å for a bromine atom (C—Br atom distance 1.94 Å). Since bromine and methyl have about the same van der Waals radius, the Br-H-H axial repulsion should be less than CH_3-H-H. A second important contributing factor has to do with the polarizability of the atom X. The axial H—X distance is generally just a few tenths of an angstrom unit less than the combined van der Waals radii. As a result, van der Waals repulsion is moderate and is offset, to some extent, by London attraction due to the polarizability interaction of the electron clouds of H and X.* This attraction will be particularly important for highly polarizable atoms, such as bromine, iodine, sulfur, mercury. As a result, these atoms do not interact as much in the axial position as their size might suggest.

Because of van der Waals repulsion between an axial substituent and *syn*-axial hydrogens, it is to be expected that the axial substituent will bend outward—either by itself or, more likely, by a deformation (flattening) of the ring as a whole.[37] Such deformation will continue until the total potential energy,—i.e., the residual van der Waals repulsive energy (which normally includes any offsetting London attraction) plus any energy due to angle strain and Pitzer strain set up in the deformation— is at a minimum. In principle, if one knew accurately all interatomic distances as well as the shape of the van der Waals, angle deformation, and torsional potential curves, the shape of the molecule at the energy minimum could be calculated by means of a high speed computer.[37] So far, the calculation has been carried out only for methyl, and even then the result is only approximative.[37] At the same time, there are only a few meager experimental data to support the idea of a deformation and to

[37] J. B. Hendrickson, *J. Am. Chem. Soc.*, **84**, 3355 (1962).

* Since London attraction varies inversely with the sixth power of the distance whereas van der Waals repulsion varies with the inverse twelfth power, London energy is of particular significance just within and just outside the van der Waals distance. In fact, it is responsible for the familiar dip of the van der Waals potential curve at the van der Waals distance. Much beyond the van der Waals distance both the van der Waals repulsion and the London attraction are negligible; much within the van der Waals distance the repulsive force completely outweighs the attractive force. The magnitude of the London energy is not fairly reflected in the mere measurement of the van der Waals radius; it does, however, affect the shape of the van der Waals potential function. The larger the London energy, the more gentle will be the slope of the repulsive part of the curve just within the van der Waals distance. (See also Sec. 7-2.)

estimate its magnitude. In particular, electron diffraction measurements[38],[39] of *trans*-1,4-dichloro- and *trans*-1,4-dibromocyclohexane suggest a deviation of the C—X bond in the diaxial isomers of 6.3° in the dichloride and 7.7° in the dibromide from the direction of the other axial bonds. Analogous deformations undoubtedly occur in other axially substituted cyclohexanes (see Sec. 2-7). As a result the magnitude of the C—C—X bending force constant may also affect the magnitude of the conformational free energy difference, as will the C—X distance by virtue of the fact that a given angle deformation will give rise to a deflection which is the larger the farther X is from C.

c. Computation of Conformational Equilibria by the Kinetic Method. Since a monosubstituted cyclohexane usually exists in two discrete conformations, it is convenient to consider the reactions of such a compound in terms of separate reactions of the equatorial and axial conformers.[28],[40],[41] Referring to Fig. 8, the rate of reaction of the axial isomer may be put equal to $k_a[A][P]$, where k_a is the specific rate of reaction of the axial isomer, [A] is its concentration, and [P] is the product of the concentrations of any other species which enter into the kinetic equation. (For a first-order process, [P] = 1; for a second-order process, [P] would be the concentration of the appropriate species—e.g., for a saponification reaction, [P] = [OH$^-$].) Similarly, the rate of the equatorial isomer is $k_e[E][P]$ so that the total rate becomes

$$\text{Rate} = k_e[E][P] + k_a[A][P] \qquad (1)$$

But, empirically, one has

$$\text{Rate} = k[C][P] \qquad (2)$$

where k is the empirical over-all rate constant and [C] is the stoichiometric concentration. Equating equations (1) and (2), one obtains

$$k[C] = k_e[E] + k_a[A] \qquad (3)$$

But [C] = [E] + [A] and K = [E]/[A]. Dividing equation (3) by [A] and substituting appropriately, one obtains

$$k([E] + [A])/[A] = k_e K + k_a$$

or

$$k(K + 1) = k_e K + k_a$$

[38] V. A. Atkinson and O. Hassel, *Acta Chem. Scand.*, **13,** 1737 (1959).

[39] O. Hassel and E. Wang Lund, *Acta Cryst.*, **2,** 309 (1949).

[40] E. L. Eliel and C. A. Lukach, *J. Am. Chem. Soc.*, **79,** 5986 (1957).

[41] E. L. Eliel and R. S. Ro, *J. Am. Chem. Soc.*, **79,** 5995 (1957).

whence

$$k = (k_e K + k_a)/(K + 1) \tag{4}$$

Equation (4), derived by Eliel and Ro[42] and Eliel and Lukach,[40] is equivalent to

$$k = N_e k_e + N_a k_a \tag{5}$$

derived by Winstein and Holness,[28] which can be obtained most easily from equation (3) by dividing by [C] and noting that $[E]/[C] = N_e$, the mole fraction of material in the equatorial conformation, and $[A]/[C] = N_a$, the mole fraction of material in the axial conformation.

Equations (4) and (5) as given are mainly of heuristic interest—they allow one to think qualitatively and quantitatively of a reaction of a monosubstituted cyclohexane in terms of the discrete reactions of its equatorial and axial conformers. For example, knowing that only axial but not equatorial tosylates undergo bimolecular elimination reactions with sodium ethoxide in ethanol[35] and knowing that cyclohexyl tosylate exists about 75% in the equatorial form at room temperature,* one may put $k_e = 0$ and $N_a = 0.25$, $N_e = 0.75$ in equation (5), whence $k = 0.25 k_a$; i.e., cyclohexyl tosylate will be expected to undergo bimolecular elimination with sodium ethoxide in ethanol quite readily, despite the fact that the unreactive equatorial isomer predominates. In fact, the rate should be about one-fourth that of a purely axial tosylate. The experimentally found value for this ratio[28] is 1:3.44 at 75°. (See below for the method of ascertaining k_a.)

Equation (4) may be transposed into

$$K = (k_a - k)/(k - k_e) \tag{6}$$

and in this form it serves, in principle, to calculate the equilibrium constant K in Fig. 8. In practice, of course, there is the difficulty that, although k can be readily ascertained for appropriate reactions (by measuring the specific rate of reaction of the substituted cyclohexane in question), k_e and k_a cannot be so ascertained, because the individual conformations E and A are not stable enough to be studied separately. A way out of this difficulty was suggested by Winstein and Holness,[28] who use 4-(or 3-)t-butyl-substituted cyclohexanes to evaluate k_a and k_e.†

[42] E. L. Eliel and R. S. Ro, *Chem. Ind.* (*London*), **1956,** 251.

* Using a $-\Delta G°$ value of 0.7 from Table 1.

† A historical note may be pertinent here. Following earlier speculations on the conformational behavior of mobile systems [E. L. Eliel, *Experientia*, **9,** 91 (1953)], E. L. E., in the fall of 1954, developed the quantitative expression embodied in equation (4) as shown above and communicated some of the results to Professors

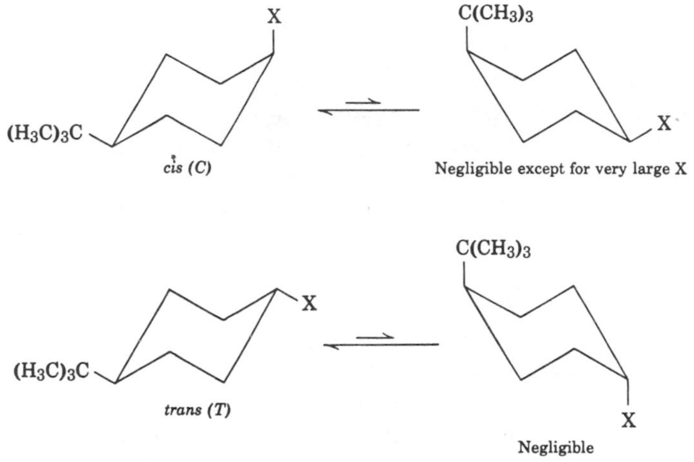

Fig. 9.

As shown in Fig. 9, because of the previously mentioned reluctance of the *t*-butyl group to occupy the axial position, a *cis*-4-*t*-butyl-substituted cyclohexyl-X(*C*) will have the substituent X exclusively in the axial position (provided X does not approach the *t*-butyl group in size; it normally does not) and the corresponding *trans*-4-*t*-butyl-substituted cyclohexyl-X(T) will have the group X exclusively in the equatorial position. On this basis, Winstein and Holness suggest that the specific rate of reaction of (*C*) may be taken to be equal to k_a and the corresponding rate for (*T*) may be taken to be equal to k_e. Using these data, then, equation (6) may be used to evaluate K. Several of the equilibrium constants given in Table 1 were obtained in this fashion.

Since k_e and k_a in equation (6) actually stand for the (experimentally inaccessible) rate constants of the equatorial and axial conformers of cyclohexyl-X, the use of rate constants obtained from *trans*- and *cis*-4-*t*-butylcyclohexyl-X for k_e and k_a requires justification. The assumption that (*T*) and (*C*) are exclusively in the conformations shown in Fig. 9 is

---

W. G. Dauben and D. Y. Curtin (private communications, dated December 20 and December 22, 1954). In January 1955, Professor S. Winstein kindly sent to E. L. Eliel the manuscript of ref. 28 in which equation (5) was derived by a different method. Thus equations (4) and (5) were developed independently in the two laboratories. On the other hand, the suggestion to use the *t*-butyl group as a holding group came exclusively from Winstein and Holness (ref. 28). Eliel's plan had been to utilize data from methyl-substituted cyclohexyl compounds to evaluate K (see below); after learning from Winstein and Holness about the utility of the *t*-butyl group as a holding group, he decided to include *t*-butyl-substituted compounds in investigations subsequently published in 1956 and 1957 (refs. 40–43).

[43] E. L. Eliel and R. S. Ro, *J. Am. Chem. Soc.*, **79**, 5992 (1957).

undoubtedly sound, for the reasons already stated.* In addition, however, it must be assumed that the *t*-butyl group has no effect on the reactivity of the substituent X. In detail, this means:[35] (1) the *t*-butyl group must not exert a polar effect at the 4-position; (2) it must not exert a steric effect at the 4-position; and (3) it must not distort the ring in either the ground state or the transition state. Assumption 1 is supported by a comparison of the dissociation constants of *trans*-4-*t*-butylcyclohexanecarboxylic acid, $pK_a = 7.79$ in 66% aqueous dimethylformamide,[44] and of cyclohexanecarboxylic acid, $pK_a = 7.82$. (See also Chap. 3.) If the 4-*t*-butyl group exerted an appreciable polar influence, the two acids would differ more extensively in acidity.† The absence of steric effects of the *t*-butyl group across the ring can be deduced from scale models. The weakest assumption of the three, and the one which will require further study, is that the *t*-butyl group produces no distortion of the ring. This point will be returned to in Sec. 2-7. Applications of the kinetic method for determining K will be discussed in Sec. 2-5. It should be mentioned that the method has recently been subjected to criticism.[45a]

2–3. Di- and Polysubstituted Cyclohexanes

a. Introductory Remarks. Like cyclohexane and monosubstituted cyclohexanes, disubstituted cyclohexanes may exist in two alternative chair

[44] R. D. Stolow, *J. Am. Chem. Soc.*, **81**, 5806 (1959).

[45] H. van Bekkum, P. E. Verkade, and B. M. Wepster, *Koninkl. Ned. Akad. Wetenschap. Proc.*, **64B**, 161 (1961).

[45a] H. Kwart and T. Takeshita, *J. Am. Chem. Soc.*, **86**, 1161 (1964). See also W. Hückel and M. Hanack, *Ann.*, **616**, 18 (1958).

* Some authors have stated that compounds (*T*) and (*C*) cannot flip into the alternative conformations shown in Fig. 9 (right side). This statement may be misinterpreted to mean that there is an unusually high barrier to ring inversion in these compounds. However, there is no indication that this is so. The barrier to ring inversion is approximately 10 kcal./mole, and ring inversion thus occurs rapidly. However, since the free energy difference between the conformational isomers is probably of the order of 5 kcal./mole, only a few molecules in 10,000 will be in the conformation with the axial *t*-butyl group at room temperature, this being the real reason why 4-*t*-butyl-substituted cyclohexyl compounds may be considered "conformationally homogeneous." The confusion is caused by the fact that an energy difference of 5 kcal./mole is "large" and indicates that almost all molecules will be in the more stable state at room temperature.
energy difference of 5.5 kcal./mole is "large" and indicates that almost all molecules will be in the more stable state at room temperature.

† What little difference there is appears to be due to the presence of a small amount of the axial conformer (of higher pK; cf. Sec. 3-11) in the unsubstituted acid. See, however, ref. 45.

Fig. 10. cis-1,2-Disubstituted cyclohexane (identical substituents).

conformations. In a 1,1-disubstituted cyclohexane, chair inversion interchanges the equatorial and axial substituents, and the molecule will exist predominantly in the conformation in which the larger group is equatorial. In fact, the free energy difference between the two conformational isomers might be expected to be the difference in $\Delta G°$ values (Table 1) between the two substituents. No experimental test of this prediction seems yet to be available.* When the 1,1-substituents are the same, as in 1,1-dimethylcyclohexane, the two possible conformations will be superimposable.

A 1,2-disubstituted cyclohexane may exist in *cis* and *trans* configurations. The *trans* isomer (*dl* pair) will be either diequatorial (*e,e*) or diaxial (*a,a*), with the diequatorial conformation predominating by far. The *cis* isomer will be either equatorial-axial or axial-equatorial. When the two substituents are different, the conformation in which the larger substituent is equatorial will predominate, the situation being similar to that expected for the 1,1-isomer, except that one now

Fig. 11. 1,3-Diaxial interaction.

has a resolvable *dl* pair. When the two substituents are the same (as in *cis*-1,2-dimethylcyclohexane), the two conformational isomers are mirror images of each other. In this case one is dealing with a rapidly interconverting, non-resolvable *dl* pair (Fig. 10).[46] The fact that the two conformational isomers are not superimposable but are mirror images of each other affects the entropy of the *cis*-1,2-isomer (see below).

In 1,3-disubstituted cyclohexanes it is the *cis* isomer which will be diequatorial, whereas the *trans* isomer will be equatorial-axial. The diaxial conformation of the *cis* isomer is particularly unfavorable here because in this conformation the substituents crowd each other strongly, as shown in Fig. 11. This crowding gives rise to repulsive energy terms

---

[46] Cf. H. G. Derx, *Rec. Trav. Chim.*, **41**, 312 (1922).

* What little information there is rather seems to contradict the assumption. Thus D. H. R. Barton, A. da S. Campos-Neves, and R. C. Cookson, *J. Chem. Soc.*, **1956**, 3500, have assigned to 1-methylcyclohexyl chloride the configuration with equatorial chlorine and axial methyl, on the basis of the apparently equatorial C—Cl stretching band in the infrared (772 cm⁻¹). This is contrary to what one would have thought on the basis of the conformational free energies of methyl (1.7 kcal./mole) and chlorine (0.4 kcal./mole).

listed in Table 2. When the two substituents are identical, the *cis* isomer has a plane of symmetry and is *meso*; otherwise it is a *dl* pair. The *trans* isomer is always a *dl* pair. When the two substituents are the same, ring inversion converts one *e,a* conformation of the *trans* isomer into a superimposable *a,e* conformation rather than into a mirror image conformation, as in the case of the 1,2-isomer.

Table 2. 1,3-Diaxial Interactions[a]

Interaction	OH-OH	OAc-OAc	OH-CH$_3$	CH$_3$-CH$_3$
Energy, kcal./mole	1.9	2.0	1.9–2.4[b]	3.7
Ref.	47	48	40, 49, 50	51

[a] See also ref. 12c. [b] See last footnote on p. 66.

For the 1,4-isomer both *cis* and *trans* configurations have a plane of symmetry and are therefore optically inactive. The *trans* isomer will exist predominantly in the *e,e* conformation, the *a,a* conformation contributing little. The *cis* isomer will exist in either an *e,a* or an *a,e* conformation. When the two substituents are different, these conformations are distinct and the one with the larger substituent in the equatorial position will predominate, as in the corresponding case of the 1,2- and 1,3-isomers. When the two substituents are the same, the *e,a* and *a,e* conformations will be superimposable.

It might be noted that, whereas for 1,2- and 1,4-disubstituted cyclohexanes the *trans* isomer (*e,e*) will be more stable than the *cis* (*e,a*) because of the greater number of equatorial substituents, the reverse is true for the 1,3-disubstituted cyclohexanes where the *cis* isomer (*e,e*) is more stable than the *trans* (*e,a*). Experimentally, this fact was first confirmed in the case of the 1,3-dimethylcyclohexanes, for which the thermodynamically less stable isomer[52] was shown to be capable of optical activity.[53] Thus this isomer must have the *trans* configuration. Another notable case is that of the 3-methylcyclohexanols, for which it was shown[54] that the

[47] S. J. Angyal and D. J. McHugh, *Chem. Ind. (London)*, 947 (1955). Cf. Chap. 6.
[48] R. U. Lemieux and P. Chu, Abstracts, San Francisco Meeting, *Am. Chem. Soc.*, 31N (1958).
[49] E. L. Eliel and H. Haubenstock, *J. Org. Chem.*, **26**, 3504 (1961).
[50] G. Chiurdoglu and W. Masschelein, *Bull. Soc. Chim. Belges*, **70**, 782 (1961).
[51] N. L. Allinger and M. A. Miller, *J. Am. Chem. Soc.*, **83**, 2145 (1961). Cf. Sec. 2–4b.
[52] J. E. Kilpatrick, H. G. Werner, C. W. Beckett, K. S. Pitzer, and F. D. Rossini, *J. Res. Natl. Bur. Std.*, **39**, 523 (1947).
[53] M. Mousseron and R. Granger, *Bull. Soc. Chim. France*, [5] **5**, 1618 (1938).
[54] H. L. Goering and C. Serres, *J. Am. Chem. Soc.*, **74**, 5908 (1952); D. S. Noyce and D. B. Denney, *ibid.*, **74**, 5912 (1952); S. Siegel, *ibid.*, **75**, 1317 (1953).

Table 3. Conformations and Enthalpies of the Dimethylcyclohexanes

Isomer	Conformation	No. gauche[a]	Interaction, kcal./mole	ΔH, calcd., kcal./mole	ΔH, exptl.,[b] kcal./mole
cis-1,2	(e,a) / (a,e)	3	2.7		
trans-1,2	e,e	1	0.9	1.8	1.87
trans-1,2	a,a	4	3.6		
cis-1,3	a,a	4	5.5[c]		
cis-1,3	e,e	0	0		
trans-1,3	(e,a) / (a,e)	2	1.8	−1.8	−1.96
cis-1,4	(e,a) / (a,e)	2	1.8		
trans-1,4	e,e	0	0	1.8	1.90
trans-1,4	a,a	4	3.6		

[a] Number of butane-*gauche* interactions.

[b] Value for *cis* isomer minus value for *trans* isomer in the gas phase at 25° (refs. 52, 56). See also text below.

[c] Includes diaxial methyl-methyl interaction.

isomer with the lower heat of combustion[55] and the higher thermodynamic stability[43] is the *cis* isomer.

b. Dialkylcyclohexanes. As the thermodynamic properties of the dimethylcyclohexanes have been studied very carefully, these compounds will be used to illustrate the enthalpy, entropy, and free energy situation in disubstituted cyclohexanes. Table 3 summarizes the possible chair conformations of the 1,2-, 1,3-, and 1,4-dimethylcyclohexanes, the number of *gauche* interactions in each, the calculated interaction energy (computed

[55] A. Skita and W. Faust, *Ber.*, **64,** 2878 (1931).

[56] E. J. Prosen, W. H. Johnson, and F. D. Rossini, *J. Res. Natl. Bur. Std.*, **39,** 173 (1947).

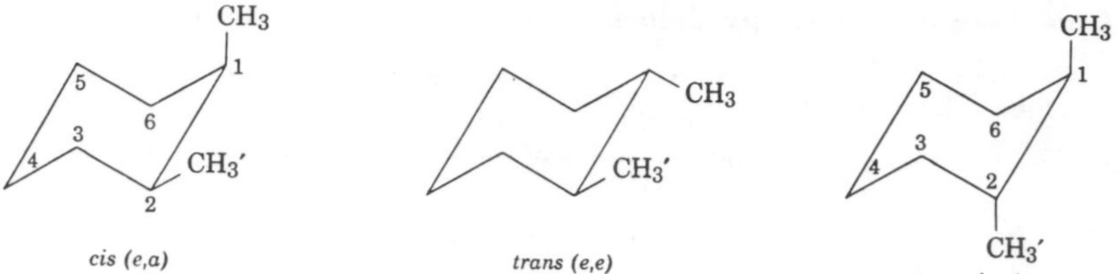

Fig. 12. *Gauche* interaction in 1,2-dimethylcyclohexanes.

at 0.9 kcal/mole per *gauche* interaction; cf. p. 43), the calculated enthalpy difference, and the observed enthalpy difference from heat of combustion data.[52,56]

The number of *gauche* interactions may be readily counted in a model, as shown in Fig. 12 for the 1,2-dimethylcyclohexanes.

As may be seen, the experimental enthalpy differences between pairs of *cis* and *trans* isomers are in good agreement with the calculated ones; in fact, it may be inferred from the data that the enthalpy difference between equatorial and axial methyl is approximately 1.9 kcal./mole at 25° in the gas phase; this is one of the values entered in Table 1. Corresponding combustion data for the liquid phase[56] support a value of 1.6–1.7 kcal./mole;* the value 1.7 kcal./mole will be used in the sequel instead of the more customary value of 1.8 kcal./mole.

It must be pointed out here that it is not, strictly speaking, accurate to equate the enthalpy difference of, for example, *cis*- and *trans*-1,4-dimethylcyclohexane with the conformational enthalpy difference for methyl, indicated in Table 1. *cis*-1,4-Dimethylcyclohexane exists as two conformational isomers whose enthalpies are, however, equal. The *trans* isomer, on the other hand, exists in two conformationally isomeric forms of unequal enthalpy, and its enthalpy is therefore given by the equation given in Chap. 1, viz.,

$$H = \sum_i n_i H_i$$

In the particular case of *trans*-1,4-dimethylcyclohexane, setting $H_{e,e}$ arbitrarily as zero (Fig. 13), $H_{a,a} = 3.6$ kcal./mole (Table 3). Assuming

---

* The reason why the liquid phase value is less than the gas phase value is discussed in Sec. 3-7. It might be noted that the 0.9 kcal./mole parameter for the methyl *gauche* interaction is an over-all average value; better agreement with the experimental data is obtained by using 0.85 kcal./mole in the liquid phase and 0.95 kcal./mole in the gas phase.

that there is no entropy difference between the e,e and the a,a conformations, this gives $N_{a,a} = 0.003$, corresponding to 99.7% of the e,e isomer. Therefore

$$H_{trans} = 0.003 \times H_{a,a} + 0.997 \times 0 = 0.003H_{a,a}$$

Similarly,

$H_{cis} = H_{e,a}$, the conformational enthalpy difference for methyl

Hence

$$H_{cis} - H_{trans} = H_{e,a} - 0.003H_{a,a}$$

Making the reasonable assumption (Fig. 13) that $H_{a,a} = 2H_{e,a}$, one then has

$$H_{cis} - H_{trans} = 0.994H_{e,a}$$

But, experimentally, $H_{cis} - H_{trans} = 1.90$ kcal./mole (gas phase). Therefore $H_{e,a} = 1.90/0.994$ or 1.91 kcal./mole. Fortunately it turns out that the correction is only 0.6 part per hundred and therefore of the same order of magnitude as the inherent experimental error of the measurements. Nevertheless, it is well to recognize that there is, in principle, a small difference between $H_{cis} - H_{trans}$ and $H_{e,a}$ (or ΔH°_{Me}); it is also of interest that, because of the slightly differing contribution of the diaxial isomer in the 1,2, 1,3, and 1,4 series (Table 3), the correction is not quite the same in the three cases. Calculations similar to those above indicate

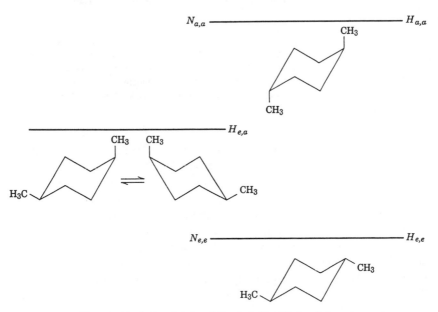

Fig. 13. Enthalpy levels of the 1,4-dimethylcyclohexanes.

that the experimental enthalpy difference of the epimers in the 1,3 series (1.96 kcal./mole) is directly equal to $-\Delta H_{\text{Me}}^{\circ}$, whereas in the 1,2 series there is a correction of 0.04 kcal./mole which leads to a calculated $-\Delta H_{\text{Me}}^{\circ}$ of 1.91 kcal./mole from the experimental enthalpy difference (Table 3) of 1.87 kcal./mole. It is noteworthy that, although these corrections bring the values from the three series into better alignment, they still leave a

Table 4. Entropy Differences in Dimethylcyclohexanes[a]

Isomer	Symmetry No. σ	$-R \ln \sigma$	Entropy Due to dl Forms ($R \ln 2$)	Due to e,e-a,a Equilibrium[b]	Total Calcd.	Exptl. Entropy Difference[d]
cis-1,2	1	0	1.38	0	1.38[c]	
						0.72
trans-1,2	2	−1.38	1.38	0.11	0.11	
cis-1,3	1	0	0	ca. 0	0	
						1.24
trans-1,3	1	0	1.38	0	1.38	
cis-1.4	1	0	0	0	0	
						1.19
trans-1,4	2	−1.38	0	0.03	−1.35	

[a] Entropies in cal./deg. mole.

[b] $-R(N_1 \ln N_1 + N_2 \ln N_2)$, where N_1 and N_2 are the mole fractions of e,e and a,a isomers, respectively, computed from $N_1 + N_2 = 1$ and $N_1/N_2 = e^{-\Delta G/RT}$. ΔG is set equal to the calculated ΔH of the two conformational isomers (cf. Table 3).

[c] This value is considerably in excess of the experimental value because interference with methyl rotation is not taken into account. If this interference is taken into account, a value of 0.86 may be calculated (ref. 34).

[d] Ref. 57.

noticeable difference between the 1,3 and the 1,4 value which may have its origin in deformation of the chair (Sec. 2-7).

The entropies of the dimethylcyclohexanes have been measured by classical methods[57] (Sec. 3-3), and the data are shown in Table 4. Most of the entropy differences between isomers are of "trivial" origin, being made up of entropy-of-mixing and symmetry-number terms as explained in Chap. 1. The other factors contributing to the total entropy (vibrational, translational, and rotational entropy terms) may reasonably be considered the same for a pair of epimers. (See, however, footnote c in Table 4 above.)

[57] H. M. Huffman, S. S. Todd and G. D. Oliver, J. Am. Chem. Soc., 71, 584 (1949).

A small entropy of mixing must be taken into account for the *trans*-1,2- and *trans*-1,4-dimethylcyclohexanes (which exist as *e,e-a,a* mixtures), whereas for the *cis*-1,3 isomer the entropy of mixing is negligible because the *a,a* isomer (1,3-diaxial methyl) does not contribute appreciably (cf. Table 2). An entropy-of-mixing term also enters for *cis*-1,2-dimethylcyclohexane (which is a *dl* pair, even though non-resolvable, the conformational isomers being mirror images rather than identical) but not for *trans*-1,3- or *cis*-1,4-dimethylcyclohexane, where ring inversion leads to an equivalent structure.

Entropy due to symmetry is $\Delta S_{sym} = -R \ln \sigma$, where σ is the "symmetry number" or the number of ways in which the molecule may be rotated to give an equivalent structure.[*,58] Thus the symmetry number for *trans*-1,2- and *trans*-1,4-dimethylcyclohexane is 2 because the molecule may

Fig. 14.

be rotated around a two-fold axis of symmetry so that one methyl group takes the place of the other. None of the other isomers can be rotated into superimposable positions other than the original one, and their symmetry number is therefore 1. From the experimental $\Delta H°$ and $\Delta S°$ values for the dimethylcyclohexane pairs, $\Delta G°$ may be calculated at any temperature. Comparison with experimental equilibrium values will be made in the next section (2-4). At 25°, the calculated $\Delta G°$ values (*trans* → *cis* isomer) for the dimethylcyclohexanes (gas phase) are: 1,2-isomer, 1.66 kcal./mole; 1,3-isomer, -1.59 kcal./mole; 1,4-isomer, 1.55 kcal./mole.

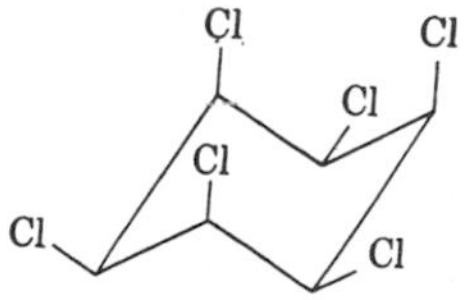

No reliable thermochemical data appear to be available for disubstituted cyclohexanes other than the dimethyl compounds. Older data available for the methylcyclohexanols[55] are not sufficiently accurate for practical purposes. Thermodynamic data for the 1,3,5-trimethylcyclohexanes have been derived from equilibrium measurements; they will be discussed in the next section. No other thermodynamic data on polysubstituted cyclohexanes appear to be known other than those on the inositols (cyclohexanehexols) discussed in Chap. 6.

The conformational situation in some polycarboxy- and polyhalocyclohexanes has been elucidated by physical measurements. It is of interest that the triaxial (all-*cis*) hexachlorocyclohexane (so-called *ι* isomer, Fig. 14) is not known, although the corresponding inositol (OH instead

---

[58] Cf. S. W. Benson, *J. Am. Chem. Soc.*, **80**, 5151 (1958).

[*] The more symmetrical a molecule, the fewer different rotational states it will have, as certain rotational arrangements which differ in a non-symmetrical molecule will be identical in a symmetrical one. As a result, the rotational entropy decreases with symmetry.

of Cl) has been prepared.[59] The conformation of a number of other polysubstituted cyclohexanes has been discussed in a review.[60]

2–4. Chemical Equilibria

a. Introductory Remarks. Conformational consideration of chemical equilibria in substituted cyclohexanes is of great interest from two entirely different points of view. On the one hand, equilibria between configurational isomers may be evaluated to yield the conformational preference of atoms or groups as between equatorial and axial positions. A number of the data listed in Tables 1 and 2 were obtained in this way. On the other hand, conformational considerations may be called on to predict positions of equilibrium between epimers, and experimentally established equilibria between epimers of unknown configuration may be used to assign configuration to such compounds. The most useful applications of the latter technique are in the field of natural products, and some instances will be presented in Chap. 5.

Qualitatively speaking, equatorial isomers are generally more stable than axial isomers, as pointed out on p. 43. An exception may occur in 1,2,2,6,6-pentasubstituted compounds (attention being focused on the substituent in position 1) or for that matter in a 1,2,6-trisubstituted compound in which the substituents at positions 2 and 6 are axial.* Here the two extra *gauche* interactions of the equatorial C_1 substituent with the axial substituents at C_2 and C_6 may outweigh the interaction of the axial C_1 substituent with the axial hydrogens at C_3 and C_5, as implied in Fig. 15, A.[61] An example has supposedly been provided[61] in the degradation acid derived from the sesquiterpenoid eudesmol (Fig. 15, B). It should be noted, however, that the observed epimerization of the degradation acid from the *trans* (relative to the acetic acid side chain) to the *cis* configuration is better explained as an axial → equatorial change in the alternative conformations shown in Fig. 15, C, which avoid the unfavorable 1,3-diaxial methyl-methyl interactions.

b. Alkylcyclohexanes. In turning to quantitative aspects of equilibration, one may consider first the dimethylcyclohexanes whose equilibration

[59] S. J. Angyal and D. J. McHugh, *J. Chem. Soc.*, **1957**, 3682.

[60] H. D. Orloff, *Chem. Rev.*, **54**, 347 (1954).

[61] D. H. R. Barton, *Chem. Ind. (London)*, **1953**, 664.

* If only two axial substituents at C_2 and C_6 are present, the point has to be illustrated in a rigid system (*trans*-decalin, steroid, etc.), as in a mobile system the molecule would invert to the conformation in which the 2- and 6-substitutents are both equatorial.

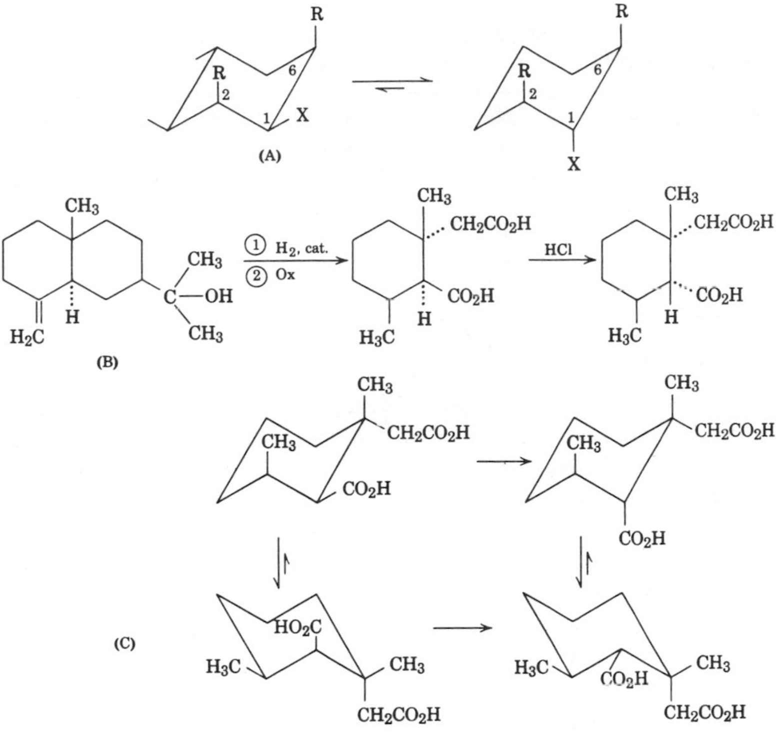

Fig. 15.

has been studied by a number of investigators[62,63] and for which equilibrium data may be compared with the thermodynamic data cited in the preceding section. The earlier data[63a,b] were obtained by equilibration with Lewis acids, such as concentrated sulfuric acid or aluminum chloride which leads to rather complex mixtures of the 1,2-, 1,3-, and 1,4-dimethyl-cyclohexanes which could not be analyzed with very high precision. Later results were obtained by the use of hydrogenation catalysts at elevated temperature; these catalysts produce clean *cis-trans* isomerization. Typical results are shown in Table 5 which also contains a comparison with the free energy difference calculated from thermodynamic

[62] N. L. Allinger and Shih-En Hu, *J. Org. Chem.*, **27**, 3417 (1962).

[63] (a) G. Chiurdoglu, P. J. C. Fierens, and C. Henkart, *Bull. Soc. Chem. Belges*, **59**, 140 (1950); G. Chiurdoglu, J. Versluys-Evrard, and J. Decot, *ibid.*, **66**, 192 (1957); (b) A. K. Roebuck and B. L. Evering, *J. Am. Chem. Soc.*, **75**, 1631 (1953); (c) C. E. Boord and K. W. Greenlee, unpublished results, cited in ref. 56; (d) C. Boelhouwer, G. A. M. Diepen, J. van Elk, P. Th. van Raaij, and H. I. Waterman, *Brennstoff-Chem.*, **39**, 299 (1958).

Table 5. Equilibration of Dimethylcyclohexanes

Isomer	Temp., °C.	% Stable Isomer	$\Delta G°$ Found, kcal./mole	$\Delta G°$, Calcd., kcal./mole[a]	Ref.
1,2	175	82	1.35	1.55	63d
1,3	175	84	1.48	1.40	63d
	225	78	1.25	1.34	63c
	280	73.6	1.12	1.27	62
	327	70.4	1.03	1.22	62
1,4	175	80	1.23	1.37	63d
	280	73.1	1.03	1.19	62

[a] From Tables 3 and 4.

data (Sec. 2-3b). From the data of Allinger and Hu[62] one may calculate $\Delta H° = 1.97$ kcal./mole, $\Delta S° = 1.57$–1.58 cal./deg. mole for the 1,3- and 1,4-dimethylcyclohexane isomerizations. The enthalpy value is in excellent agreement with the thermochemical one given in Table 3, whereas the entropy value is slightly greater than that given in Table 4.

A hydrogenation-dehydrogenation catalyst was also employed to establish the equilibrium of the 1,3,5-trimethylcyclohexanes (Fig. 16).[64] From the change of equilibrium with temperature, $\Delta H°_{25°}$ was computed to be 2.11 kcal./mole, slightly greater than the analogous value for the dimethylcyclohexanes. The value for $\Delta S°_{25°}$ of 2.3 cal./deg. mole is in good agreement with the entropy difference of $R \ln 3$ or 2.2 cal./deg. mole to be expected on the basis that the *cis* isomer has a symmetry number of 3, whereas the *trans* isomer has $\sigma = 1$. $\Delta H°$ varies little with temperature (e.g., $\Delta H°_{323°} = 1.96$ kcal./mole), but $\Delta G°$ does, of course, decrease with temperature (because of the entropy difference) from 1.41 kcal./mole at 25°C. to 0.75 kcal./mole at 323°.

In the case of the methyl group it was assumed that there is no inherent entropy difference between the equatorial and the axial position. This justifies taking the $\Delta H°$ values given above and in the preceding section and entering them as $-\Delta G°$ in Table 1.* The assumption is probably valid for a spherically symmetrical group, such as methyl, but it clearly cannot be extended to an unsymmetrical substituent such as ethyl or

[64] C. J. Egan and W. C. Buss, *J. Phys. Chem.*, **63**, 1887 (1959).

* The reader should not be confused by the signs attached to the thermodynamic parameters. By convention, we shall always write the (more stable) equatorial isomer on the right in the conformational equilibrium of a monosubstituted cyclohexane (Fig. 8); hence $\Delta G°$ is negative in this case. This is why the conformational free energy difference is defined throughout as $-\Delta G°$. For other equilibria, such as the one in Fig. 16, however, the more stable isomer may be written on either the right or the left, and so $\Delta G°$ may be negative or positive.

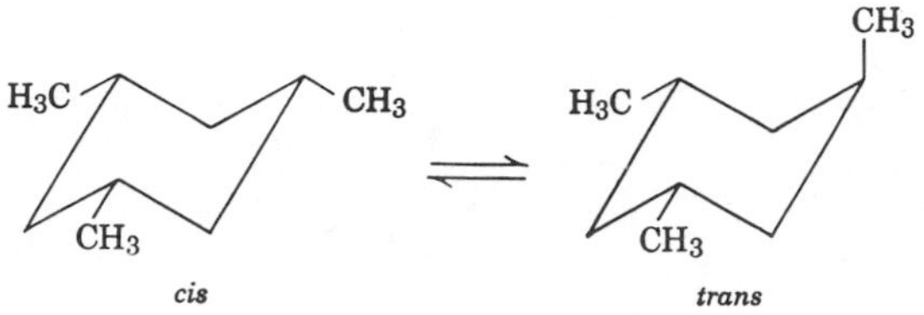

Fig. 16.

isopropyl. Model considerations show that in the equatorial position there are two equally favored and one less favored staggered conformations of the ethyl group (by virtue of rotation about the ring —C_2H_5 bond), whereas in the axial position there are only two admissible conformations (of equal energy). (The third staggered conformation, in which the methyl end of the ethyl group points into the ring, would involve very severe steric compression and is therefore excluded.) As a result, the equatorial isomer has a greater entropy of mixing; it has been estimated that it is favored by 0.51 cal./deg. mole over the axial isomer.[65] It has also been possible to estimate the enthalpy difference between an equatorial and an axial ethyl group at 1.71 kcal./mole;* thus the calculated free energy difference at 25° is 1.86 kcal./mole, only slightly more than the 1.8 kcal./mole assumed for methyl.† Using the same type of argument, one obtains excellent agreement between the calculated equilibrium of the 1,3- and 1,4-diethylcyclohexanes at 554°K. (76% diequatorial isomer) and the found value (75.5–77.3% at temperatures varying from 532° to 565°K.); this agreement, of course, supports the basis of the theoretical calculation. Similar arguments have been applied to the isopropyl group[62]

[65] N. L. Allinger and Shih-En Hu, *J. Am. Chem. Soc.*, **84**, 370 (1962).

* The reason why this is somewhat less than the 1.8 kcal./mole value assumed for methyl is that outlined on p. 23; in assessing the enthalpy of the axial ethylcyclohexane one considers two conformations each having one methyl *gauche* interaction (0.9 kcal./mole) more than axial methyl; but in the equatorial isomer, in addition to two analogous conformations, one must consider a third which has the equivalent of an axial methyl and is therefore 1.8 kcal./mole higher in enthalpy than an equatorial methyl group. Since, in computing the over-all enthalpy of the equatorial ethyl group, one has to take into account all three conformations in proportion to their population, it is clear that the high enthalpy conformation contributing to the equatorial (but not axial) ethylcyclohexane will serve to diminish the enthalpy difference between the two. In the over-all calculation of free energy at 25° the slightly enhanced enthalpy of the equatorial isomer resulting from the contribution of a less favorable conformation is, of course, more than offset by the gain in entropy of mixing.

† Clearly the exact conformational free energy difference of ethyl will depend on that for methyl in this argument, but the conclusion that the ethyl value exceeds the value for methyl by 0.06 kcal./mole is not dependent on the exact value chosen for $-\Delta G^{\circ}_{Me}$.

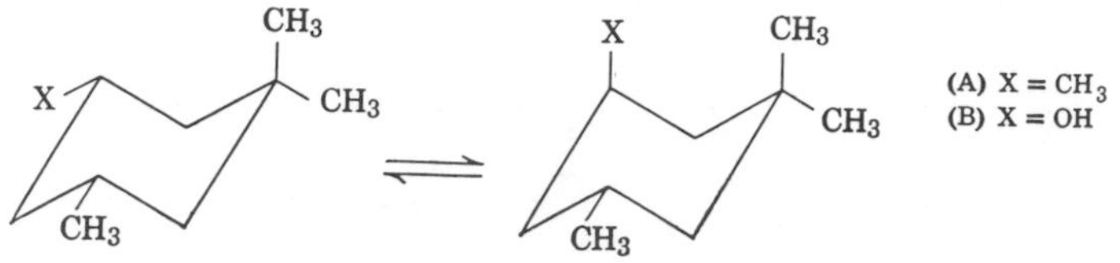

(A) X = CH$_3$
(B) X = OH

Fig. 17.

and have led to the following calculated differences between equatorial and axial isopropyl: $\Delta H^\circ_{25^\circ}$, -1.63 kcal./mole, $\Delta S^\circ_{25^\circ}$, -1.59 cal./deg. mole, $\Delta G^\circ_{25^\circ}$, -2.10 kcal./mole. In this case the agreement of calculation with experiment in the equilibration of the 1,3-diisopropylcyclohexanes was not so good as in the case of the diethylcyclohexanes, possibly because of intervention of appreciable amounts of the flexible forms (rather than only well-defined chair forms) at the temperatures of the experiment.

The equilibration of the 1,1,3,5-tetramethylcyclohexanes (Fig. 17, A) over a palladium catalyst in the range from 247° to 358° has also been studied.[51] From the temperature dependence of equilibrium the values $\Delta H^\circ_{300^\circ C.} = 3.70$ kcal./mole and $\Delta S^\circ_{300^\circ C.} = 1.65$ cal./deg. mole were arrived at. The good agreement of the observed entropy difference with the difference (1.38 cal./deg. mole) calculated on the basis that the *cis* isomer is a *meso* form but the *trans* is a *dl* pair suggests[50] that, even at the relatively high temperatures of the experiment, the *trans* isomer exists very largely in a chair form (though presumably deformed, as there would be a large repulsion of the two axial methyl groups in an undeformed chair) rather than in the flexible form, which should have a considerably higher entropy. The ΔH° value of 3.7 kcal./mole may thus be taken to represent the extra interaction of the two axial methyl groups with each other and is thus entered in Table 2. (Both isomers shown in Fig. 17, A have two other CH$_3$-H axial interactions which may be assumed to cancel out when their energy difference is considered.)

c. Other Substituted Cyclohexanes.

Unlike the thermochemical method, the equilibration method has been applied to several substituents other than alkyl. Thus equilibration of the epimeric ethyl 4-*t*-butylcyclohexanecarboxylates[66] (Fig. 18, A) with sodium ethoxide in ethanol at 78° leads to an equilibrium mixture containing 84.7 ± 0.6% of the *trans* (equatorial) isomer, corresponding to a ΔG° of 1.2 kcal./mole. If, as assumed earlier in the discussion of the kinetic method (p. 47) the *t*-butyl groups have no perturbing effect, this free energy difference may be

[66] E. L. Eliel, H. Haubenstock, and R. V. Acharya, *J. Am. Chem. Soc.*, **83**, 2351 (1961); see also N. L. Allinger, L. A. Freiberg, and Shih-En Hu, *ibid.*, **84**, 2836 (1962).

assumed to be equal to the free-energy difference between equatorial and axial carbethoxyl in the conformational equilibrium of ethyl cyclohexanecarboxylate (Table 1). Fortunately, for the carbethoxyl group, the value for the conformational free energy difference obtained by equilibrium is in excellent agreement with the values obtained by the kinetic method (to be discussed in the next section) as well as with a value derived from NMR spectroscopy (Sec. 3-4d). Finally, independent support comes from the position of equilibrium in the diethyl cyclohexane-1,3-dicarboxylates[67] which corresponds to 71% *cis* (*e,e*) and 29% *trans* (*e,a*) isomer at 78°, corresponding to a $\Delta G°$ of 0.63 kcal./mole. Since the *trans* isomer is a *dl* pair, its entropy is presumably higher than that of the *cis* isomer by $R \ln 2$ or 1.38 cal./deg. mole. If one corrects for this, one finds that the free energy difference between the *meso-cis* (*e,e*) and active *trans* (*e,a*) is $0.63 + (351 \times 1.38)/1000$ or $0.63 + 0.48 = 1.1$ kcal./mole at 78°C, in excellent agreement with the value of $\Delta G°_{CO_2Et}$ cited above.

cis- and *trans*-4-*t*-Butylcyclohexyl cyanides have been equilibrated in similar fashion,[68,69] equilibrium corresponding to $-\Delta G°_{66°C.} = 0.15$–$0.25$ kcal./mole. The equilibrium of the 3-cyanocholestanes[70]—ca. 60% of β (*e*) to 40% α (*a*) at 245° is in agreement with this free energy difference.

The 3-carboxycholestanes have also been epimerized,[70] but the exact position of equilibrium is not known. Prolonged heating of *cis*-4-phenylcyclohexanecarboxylic acid at 195° gives a mixture containing 89% *trans* epimer[71] corresponding to $-\Delta G° = 1.95$ kcal./mole on the questionable assumption that the acid is conformationally homogeneous; on the other hand, heating either 4-*t*-butylcyclohexanecarboxylic acid at 230° gives a mixture containing only 76% *trans* corresponding to 1.15 kcal./mole;[72] these findings bracket the value of 1.5–1.7 kcal./mole for $-\Delta G°_{CO_2H}$

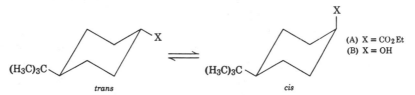

Fig. 18. Epimerization equilibria.

[67] N. L. Allinger and R. J. Curby, *J. Org. Chem.*, **26**, 933 (1961).

[68] B. Rickborn and F. R. Jensen, *J. Org. Chem.*, **27**, 4606 (1962).

[69] N. L. Allinger and W. Szkrybalo, *J. Org. Chem.*, **27**, 4601 (1962).

[70] G. Roberts, C. W. Shoppee, and R. J. Stephenson, *J. Chem. Soc.*, **1954**, 2705.

[71] H. E. Zimmerman and H. J. Giallombardo, *J. Am. Chem. Soc.*, **78**, 6259 (1956).

[72] E. L. Eliel and R. P. Gerber, unpublished observations. See also ref. 12c.

found from acidity measurements (see the next section). The thermal equilibrium of the 4-*t*-butylcyclohexylcarbonyl chlorides[72] corresponds to a $-\Delta G°$ value of the COCl group of 1.2 kcal./mole at 135°.

For the hydroxyl group (Fig. 18, B) equilibration has been achieved by reversible oxidation-reduction over Raney nickel,[30] equilibrium corresponding to 67–69% *trans* (equatorial) isomer at 90°C., corresponding to a conformational free energy difference of 0.54 kcal./mole, in good agreement with the kinetic value of 0.5 kcal./mole.[40] A slightly lower value (0.41–0.50 kcal./mole) was found[30] for $-\Delta G°$ in cyclohexane at 110–170°, possibly owing to the difference in solvent or (less likely) temperature.* A higher value had previously been indicated on the basis of the equilibration of 4-*t*-butylcyclohexanol with aluminum isopropoxide in isopropyl alcohol-acetone[43] (79-21 equilibrium at 82° corresponding to $-\Delta G° = 0.94$ kcal./mole), but it is possible that this value is distorted because of the persistence of some of the 4-*t*-butylcyclohexanols in the form of their aluminate complexes whose equilibrium is different from that of the free alcohols (see below). This possibility is supported by a study of the equilibrium as produced by sodium alcoholate as a function of alcoholate concentration which indicates that the amount of *trans* (equatorial) isomer increases with increasing amount of alcoholate up to a maximum of about 85% in benzene at 80° (this may correspond to the equilibrium of the sodium 4-*t*-butylcyclohexoxides) but decreases to an extrapolated value of about 66% *trans* at zero alcoholate concentration.[73]

It is of interest to consider an analogous equilibrium situation for two epimers which are not conformationally homogeneous, such as the 4-methylcyclohexanols (Fig. 19).[43,74] It is clear that here

$$K_{epi} = \frac{A + A'}{B + B'} = \frac{A}{B}\frac{1 + A'/A}{1 + B'/B} = K_{OH}\frac{1 + K_{trans}^{-1}}{1 + K_{cis}}$$

where K_{epi} is the observed epimerization equilibrium constant, and the other equilibrium constants have the meaning indicated in Fig. 19. If one assumes that conformational enthalpies are additive (i.e., that the enthalpy difference between A and A' is the sum of the enthalpy differences between equatorial and axial methyl and between equatorial and axial hydroxyl, and, moreover, that there is no entropy difference between A and A' so that the free energy difference between A and A' is

[73] G. Chiurdoglu, H. Gonze, and W. Masschelein, *Bull. Soc. Chim. Belges*, **71**, 484 (1962); see also O. R. Rodig and L. C. Ellis, *J. Org. Chem.*, **26**, 2197 (1961).

[74] See also W. G. Dauben and R. E. Bozak, *J. Org. Chem.* **24**, 1596 (1959); E. L. Eliel, *Record Chem. Progr.* (*Kresge-Hooker Sci. Lib.*), **22**, 129 (1961); and ref. 107.

* Since entropy considerations should favor the equatorial isomer, if any, one might have expected greater preponderance of the equatorial rather than the axial isomer at higher temperature.

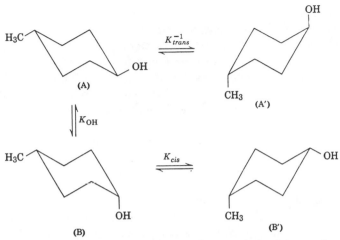

Fig. 19. Epimerization equilibrium for the 4-methylcyclohexanols.

equal to their enthalpy difference), then K_{trans}^{-1} and K_{cis} may be computed from the known conformational free energies of CH_3 and OH. Choosing 1.7 and 0.5 kcal./mole, respectively, for these quantities, one calculates: $K_{epi}^{1,2} = 0.90K_{OH}$; $K_{epi}^{1,3} = 0.85K_{OH}$; $K_{epi}^{1,4} = 0.89K_{OH}$. This way of estimating the epimerization equilibrium constant is simpler than a separate calculation of the expected $\Delta H°$ (taking into account weighted enthalpies of all pertinent states, as explained earlier) and $\Delta S°$ (from entropies of mixing). It might be noted that the predicted equilibrium constant is smallest for the 1,3-isomer for which the *e,e* (*cis*) epimer lacks an accessible alternative conformation and is therefore relatively disfavored by entropy-of-mixing considerations. Unfortunately, the experimental results[43,50] so far do not support the *a priori* considerations; K_{epi} for methylcyclohexanol pairs appears to be greatest (instead of least) for the 1,3-isomer, and in one study[50] K_{epi} is actually greater than the assumed K_{OH} instead of smaller for all three isomers. Some but not all of the difficulty may arise because the assumption of additivity of $\Delta H°$ is not a good one, possibly because of deformation of the ring (see Sec. 2-7). The situation here is similar to that in the dimethylcyclohexanes (p. 56).

The 3,3,5-trimethylcyclohexanols (Fig. 17, B) have been equilibrated by means of aluminum isopropoxide also;[49] application of arguments similar to those given for the 1,1,3,5-tetramethylcyclohexanes led to a computation of the CH_3-OH axial-axial interaction energy of 2.4 kcal./mole (Table 2). The observed $\Delta G_{85°}°$ for the process shown in Fig. 17, B is 1.96 kcal./mole, which corresponds to the difference of two CH_3-H axial interactions in the *cis* isomer and one CH_3-H, OH-H, and CH_3-OH

interaction each in the *trans* isomer.* Taking OH-H as 0.45 kcal./mole†
and CH_3-H as 0.85 kcal./mole one gets $0.85 + 1.96 = X + 0.45$ or $X =$
2.4 kcal./mole for the CH_3-OH interaction. A very much lower value,
$\Delta G^\circ_{125^\circ} = 0.75$ kcal./mole, has been found for the same equilibrium (Fig.
17, B) by means of Raney nickel equilibration;[50] but this value (which
leads to 'an Me-OH interaction of 1.4 kcal./mole) is a composite of a
rather high ΔH° value of 3.13 kcal./mole and a rather high ΔS value of
ca. 6 cal./deg. mole,‡ so that the calculated $\Delta G^\circ_{85^\circ}$ is 0.88 kcal./mole,
corresponding to a CH_3-OH interaction of only 1.5 kcal./mole, taking
OH-H now as 0.25 kcal./mole as determined by Raney nickel equilibration. (At 25° the corresponding values would be 1.34 and 1.9 kcal./mole.)

d. Indirect Equilibration. There are many groups, such as —SH or —SR
to which the equilibration method does not seem to be applicable
directly, as the groups cannot be readily epimerized chemically. Sometimes indirect equilibration methods may be applied to such groups.
In general, the principle of these methods is to pit the group in question
against some other group which can be epimerized by conventional
methods. A few examples will illustrate the point.

When the 4-*t*-butylcyclohexanols are equilibrated with a lithium
aluminum hydride-aluminum chloride reagent (1:4), equilibrium lies
almost entirely (>99.5% in ether at room temperature) on the side of
the *trans* isomer.[75,76] The alcohols in this case are completely in the
form of complexes (possibly of the type $ROAlCl_2$), and this probably
accounts for the difference in position of equilibrium from that shown in
Fig. 18, B, since the complexing of the oxygen and the attendant solvation greatly enhance the size of $ROAlX_2$ over ROH. For the —$OAlX_2$
group, $-\Delta G^\circ \geq 3.1$ kcal./mole. On the basis of this finding, a method

[75] E. L. Eliel and M. N. Rerick, *J. Am. Chem. Soc.*, **82**, 1367 (1960).

[76] E. L. Eliel and T. J. Brett, *J. Am. Chem. Soc.*, **87**, 5039 (1965).

* Since the diaxial methyl-methyl interaction is so large (3.7 kcal./mole; Table 2),
the alternative conformation of the *trans* isomer may be disregarded. By going
through an argument analogous to that embodied in Fig. 19 one may convince
oneself that this disregard does not introduce a measurable error.

† One-half the observed value for the axial OH interaction measured by the
earlier mentioned equilibration of *cis*- and *trans*-4-*t*-butylcyclohexanol.

‡ The cause of the high value of ΔS° is puzzling unless one assumes that the *trans*
isomer is in the flexible form. On the basis of the results cited earlier for *trans*-
1,1,3,5-tetramethylcyclohexane, this does not appear likely. The found values for
ΔH° and ΔS° (ref. 50) cannot be considered very reliable because they were determined over a temperature interval of only 15 degrees. Recent experiments on
Raney nickel equilibration of the 3,3,5-trimethylcyclohexanols in boiling benzene
suggest $\Delta G^\circ_{80^\circ} = 1.8$ kcal./mole: E. L. Eliel and S. H. Schroeter, *J. Am. Chem.
Soc.*, **87**, 5031 (1965).

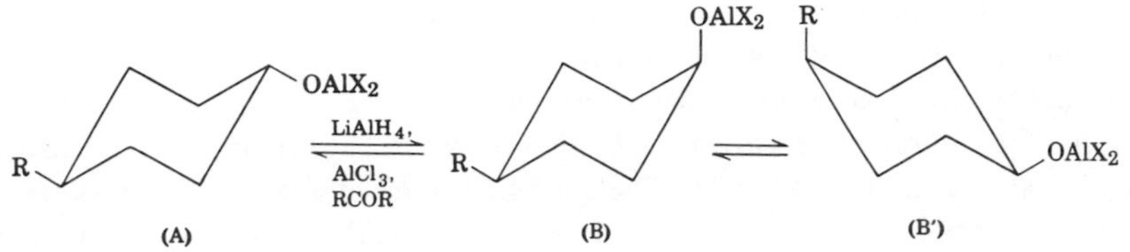

Fig. 20.

has been devised[75] to determine the conformational free energy of alkyl groups as outlined in Fig. 20. Clearly the observed equilibrium constant $K_{epi} = A/(B + B')$ or

$$\frac{1}{K_{epi}} = \frac{B}{A} + \frac{B'}{A} = \frac{1}{K_{OAlX_2}} + \frac{1}{K_R}$$

whence

$$\frac{1}{K_R} = \frac{1}{K_{epi}} - \frac{1}{K_{OAlX_2}}$$

For $K_{epi} \leq 49$ (less than 98% more stable isomer) the last term may certainly be neglected so that $K_R = K_{epi}$. If $K_{epi} > 49$, the last term may not be negligible, and a fairly exact knowledge of the magnitude of this term is required. The accompanying data have been obtained by this method.[75,76] These values generally agree with values obtained by

Group	$-\Delta G_R^\circ(1,3)$[a]	$-\Delta G_R^\circ(1,4)$[a]
CH_3	1.70	1.53
C_2H_5	1.82	1.62
$(CH_3)_2CH$	2.20	1.84

[a] In kcal./mole.

other methods (Table 1). The slight discrepancy between the data derived from 3- and 4-alkylcyclohexanols may again reflect deformations of the ring (Sec. 2-7), especially with the more mobile groups such as methyl. Unfortunately, the equilibration with $LiAlH_4$-$AlCl_3$ has not been successful so far with cyclohexanols containing polar groups such as chlorine or alkoxyl. In the case of 4-chlorocyclohexanol, the *cis* isomer preponderates at equilibrium[76] (to the extent of 56:44), although other data (Table 1) suggest that chlorine should be more stable in the equatorial position. The difficulty may be caused by some sort of dipole interaction.

For the 3- and 4-methoxycyclohexanols, equilibration is frustrated by the formation of insoluble complexes.[*, 77]

Another type of equilibration has been used to assess the conformational free energy for sulfur through an equilibration of the monothioketals shown in Fig. 21.[78,79] The configurations of the two stereoisomers were assigned tentatively on the basis of NMR spectra.[†] With this assignment, equilibrium corresponds to 58% of the isomer with axial sulfur when the hemithioketal ring is five-membered ($n = 2$) and to 55% of the isomer with equatorial sulfur in the six-membered ring hemithioketal ($n = 3$). Although these results are perhaps not completely consistent, they indicate definitely that sulfur and oxygen in the compounds shown in Fig. 21 are of comparable size.

A third indirect method for establishing conformational equilibria is based on the hydroxyacid-lactone equilibrium[80] shown in Fig. 22. The measured equilibrium constant between the total hydroxyacid and the lactone may be expressed as[‡]

$$K_{obs} = \frac{B}{A + A'} = \frac{B/A'}{A/A' + 1} = \frac{K_1}{K_2 + 1}$$

If K_2 is large compared with unity (as it usually is, unless the substituent R is very large), one may use the approximation $K_{obs} = K_1/K_2$. Comparing the unsubstituted lactone (R = H) with a substituted one, one has $K_{obs}^H = K_1/K_2^H$ and $K_{obs}^R = K_1/K_2^R$, the equilibrium constant K_1

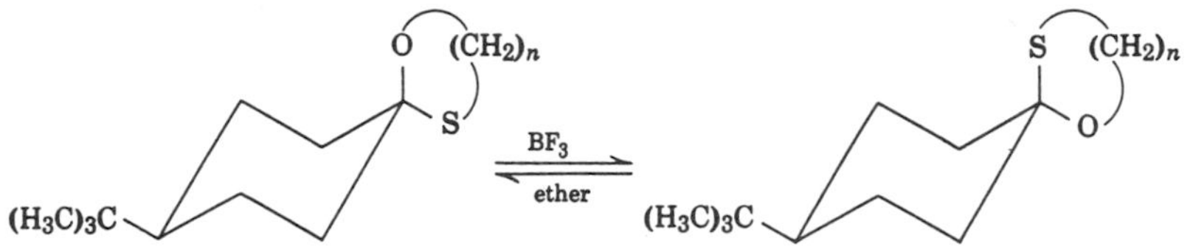

Fig. 21.

[77] E. L. Eliel and T. J. Brett, *J. Org. Chem.*, **28**, 1923 (1963).

[78] E. L. Eliel and L. A. Pilato, *Tetrahedron Letters*, **1962**, 103. See also M. P. Mertes, *J. Org. Chem.*, **28**, 2320 (1963).

[79] E. L. Eliel, E. W. Della and M. Rogić, *J. Org. Chem.*, **30**, 855 (1965).

[80] D. S. Noyce and L. J. Dolby, *J. Org. Chem.*, **26**, 3619 (1961).

[*] It is interesting[77] that the insoluble complexes, upon decomposition with acid, return nearly pure *cis*-3- and *cis*-4-methoxycyclohexanol, whereas the clear supernatant solution upon decomposition gives the nearly pure *trans* isomer. Apparently the solid precipitate is some form of cyclic aluminum chelate.

[†] This assignment has recently been confirmed by the relative oxidation rates of the corresponding sulfoxides; *cf*. Ref. 79.

[‡] The constant concentration of the water is absorbed in the constants K_{obs} and K_1.

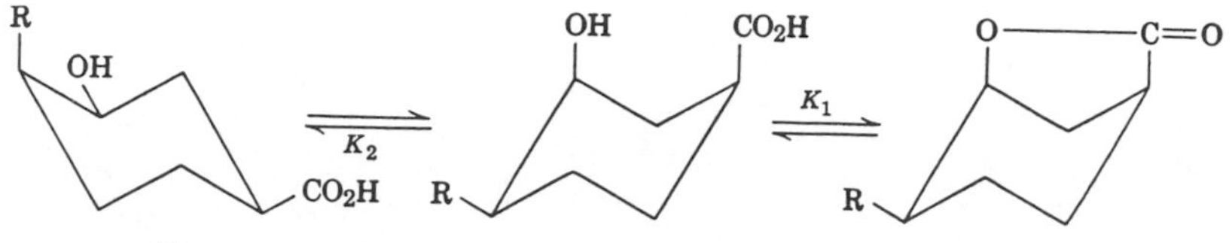

Fig. 22.

being assumed to be constant. Hence $K_{obs}^{H}/K_{obs}^{R} = K_2^{R}/K_2^{H}$. Converting to free energies ($\Delta G^{\circ} = -RT \ln K$) this becomes

$$\Delta G_{obs}^{H} - \Delta G_{obs}^{R} = \Delta G_2^{R} - \Delta G_2^{H}$$

But the only difference between ΔG_2^{R} and ΔG_2^{H} is the free energy expended in shifting the R group from the equatorial to the axial position, i.e., ΔG_R°. Hence $\Delta G_R^{\circ} = \Delta G_{obs}^{H} - \Delta G_{obs}^{R}$, the quantities on the right side being readily evaluated from experiment. The method assumes that the free energies of group translocation are additive and also that ΔG_R° is substantially smaller than the *syn*-axial interaction energies in the axial form of the hydroxyacid. Values obtained by the method are listed in Table 1 and include the value of 0.74 kcal./mole for methoxyl and 0.98 kcal./mole for ethoxyl. The values obtained for methyl (1.54 or 1.94 kcal./mole) depend on whether the substituent is in the 4- or 5-position and suggest the importance of ring deformation. The value for ethyl (2.27 kcal./mole) is somewhat higher than values obtained by other methods. Values for groups larger than ethyl are probably not reliable because the approximations made above that $K_2 + 1 \cong K_2$ is no longer valid.

The last type of equilibration to be considered here is that of diaxial and diequatorial dihalides. When *trans*-3-*cis*-4-dibromo-*t*-butylcyclo-hexane (diaxial dibromide, Fig. 23, top) is heated, it is partly equilibrated to the diequatorial dibromide *cis*-3-*trans*-4-dibromo-*t*-butylcyclohexane (Fig. 23).[81] Equilibrium corresponds to a nearly 50–50 composition, since the four relatively small 1,3-diaxial Br-H interactions in the diaxial isomer are offset by the steric and polar Br-Br repulsion in the diequatorial epimer. In contrast, in 2,3-[82] or 5,6-dihalogenosteroids,[83] equilibrium lies almost entirely on the side of the diequatorial isomer, since the diaxial

[81] E. L. Eliel and R. G. Haber, *J. Org. Chem.*, **24**, 143 (1959). Regarding the mechanism of this transformation, cf. C. A. Grob and S. Winstein, *Helv. Chim. Acta.* **35**, 782 (1952).

[82] G. H. Alt and D. H. R. Barton, *J. Chem. Soc.*, **1954**, 4284.

[83] D. H. R. Barton and A. J. Head, *J. Chem. Soc.*, **1956**, 932.

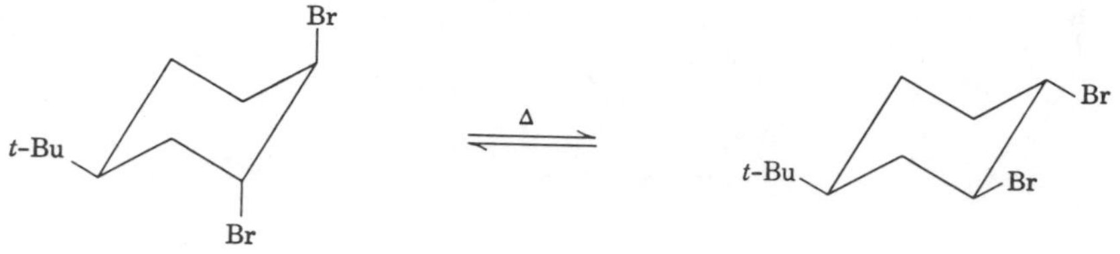

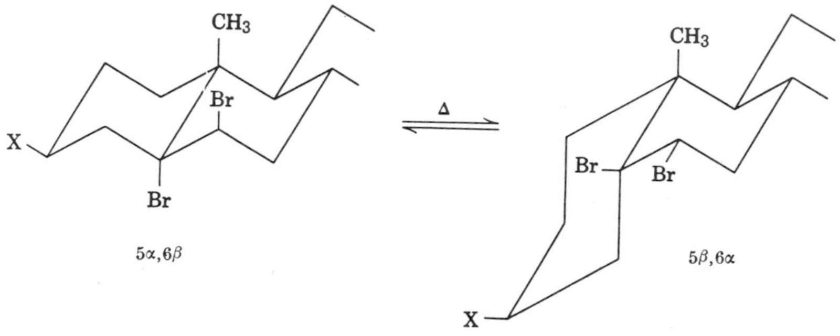

Fig. 23.

isomer has a CH$_3$-Br diaxial repulsion (Fig. 23, bottom). The equilibrium is less favorable to the diequatorial isomer when there is an equatorial substitutent at C$_3$ in the diaxial isomer, since this substituent will, in turn, become *syn*-axial with one of the bromines upon epimerization. Data shown in Table 6 indicate that the position of equilibrium depends

Table 6. Equilibria in 3β-Substituted 5,6-Dibromo-steroids

3β-Substituent	Dibromide at Equilibrium, %	
	5α,6β	5β,6α
H	<1	>99
OH	14.5	85.5
OCOC$_6$H$_5$	20	80
Cl	87	13
Br	90	10

strongly on the size of the group at C$_3$ and suggests that the 1,3-diaxial interactions are in the order[83] Br-H < Br-OH < Br-OBz ≪ Br-Cl < Br-Br. This sequence may come about as a result of several factors: steric crowding, dipolar repulsion, and intramolecular hydrogen bonding, of which the first two disfavor a 1,3-diaxial situation and the last favors it.

2–5. Conformation and Reactivity

a. Introductory Remarks. Perhaps the most fruitful application of conformational analysis has been in the prediction of relative reactivities of stereoisomers as well as, in some instances, positional isomers. With a few exceptions,[84] the extensive use of the method dates from a pioneering paper by D. H. R. Barton[85] in which the relationship between conformation and reactivity in several different reactions was first clearly pointed out. Before entering upon this subject matter, a few general remarks are in order.

Any cyclohexanoid system may be defined as either "mobile" or "rigid." A mobile system is of the type shown in Fig. 8. The substituent may be either equatorial or axial, and as a result the reactivity is complex, being governed by the equations developed on pp. 47 and 48. Such a system is obviously not ideal for demonstrating the relative reactivity of equatorial and axial substituents. A rigid system is one structurally incapable of undergoing chair inversion. An example is *trans*-decalin (Fig. 24, X = Y = H); since the chairs cannot be fused by two axial linkages, it is impossible to invert them, although it is possible (though generally energetically unattractive) to convert either or both of the chairs to a flexible form. In a derivative of *trans*-decalin the substituent can unequivocally be defined as equatorial or axial (see Fig. 24). Many of the fused and bridged ring systems found in steroids, terpenoids, and alkaloids (Chap. 5) are of this type. Even a mobile system, however, may illustrate the behavior of equatorial or axial substituents, provided the conformational equilibrium is nearly all on one side, as in the 4-*t*-butylcyclohexyl substituted compounds shown in Fig. 9. (There the *cis* isomer has the group X axial, the *trans* isomer equatorial; see also fourth footnote, p. 50.) Such systems may be called "biased." Menthol (Fig. 25) and neomenthol (Fig. 26) and their derivatives are examples often cited in the early

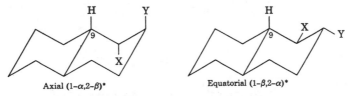

Axial (1–α,2–β)*   Equatorial (1–β,2–α)*

Fig. 24. Derivatives of *trans*-decalin.

[84] See, for example, J. Boëseken, *Rec. Trav. Chim.*, **40**, 553 (1921); H. S. Isbell, *J. Res. Nat. Bur. Std.*, **18**, 505 (1937).

[85] D. H. R. Barton, *Experientia*, **6**, 316 (1950).

* Regarding nomenclature, see p. 89 (seventh footnote).

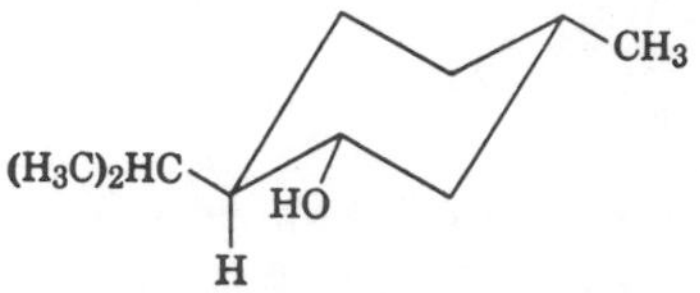

Menthol

Fig. 25.

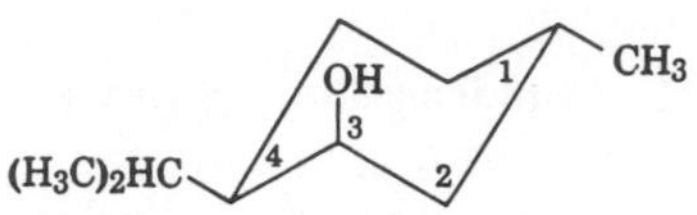

Neomenthol

Fig. 26.

papers. In menthol the hydroxyl is equatorial, in neomenthol, axial; the alternative conformational isomers in which both alkyl groups are axial (giving rise to an unfavorable energy term of about 4 kcal./mole) will obviously contribute very little, except in neomenthyl derivatives if the substituent at C_3 is very much larger than hydroxyl.

In considering the effect of conformation on reactivity it is convenient to distinguish reactions at exocyclic atoms and reactions involving positions in the ring. In the former, only steric factors* need usually be considered, and it will generally be found that the axial position is more hindered and therefore reacts more slowly than the equatorial position. In reactions affecting substituents on the ring itself the situation is more complex. Since the geometric orientation of the reacting groups is generally quite rigid in such cases, stereoelectronic factors* may play a major role. Steric factors also often play an important part, but, since their effect may be predominating in either the ground state or the transition state, they may result in either steric hindrance or steric assistance. The examples given below will elucidate these points.

b. Reactions at Exocyclic Positions. The best-known examples are saponification and esterification reactions. The cyclohexane moiety may be either in the acid or the alcohol part of the ester under consideration. The former case is illustrated by the saponification rates of the ethyl 4-*t*-butylcyclohexanecarboxylates in 70% ethanolic sodium hydroxide at 25° (Fig. 27).[66] The equatorial isomer reacts about 20 times as fast as the

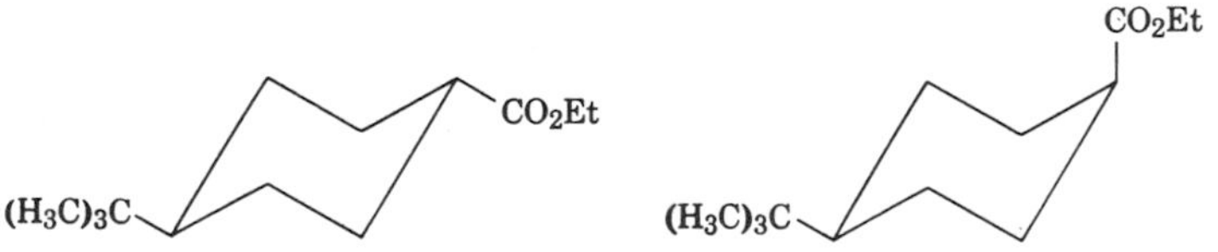

k_{trans} 8.50×10^{-4} l. mole^{-1} sec.$^{-1}$

k_{cis} 0.428×10^{-4} l. mole^{-1} sec.$^{-1}$

Fig. 27. Saponification rate of ethyl 4-*t*-butylcyclohexanecarboxylates.

[66] E. L. Eliel, *Stereochemistry of Carbon Compounds*, McGraw-Hill Book Co., New York, 1962.

* For a definition of these terms see ref. 86, p. 139.

axial isomer.[80,87] This fact has been put to use, for example, in the assignment of configuration of the alkaloids yohimbine and corynanthine.[88] Both alkaloids, on vigorous basic hydrolysis, yield the same acid, yohimbic acid, and are therefore epimeric only about C_{16} (the carbon α to the carboxyl group). As the fusion of rings D and E is *trans*, the conformations of yohimbine and corynanthine are as represented in Fig. 28. Experimentally, it was found that yohimbine is saponified considerably more rapidly than corynanthine;[89] therefore the carbomethoxy group is equatorial in yohimbine (formula 1) and axial in corynanthine (formula 2). This is in agreement with the finding that the common product of basic hydrolysis upon esterfication gives yohimbine, in which the carboxyl function thus occupies the more stable equatorial position.

When the alcohol function is part of the ring, the differences between equatorial and axial esterification and saponification rates are not so great. Thus the ratio of saponification rates of *trans*- and *cis*-4-t-butylcyclohexyl acetates[90] (Fig. 9, X = OAc) in 50% aqueous dioxane at 40°C. is 6.65, the ratio for the corresponding p-nitrobenzoates[91] in 80% acetone at 25°C. is 2.5, and the ratio for the corresponding acid phthalates in 20% aqueous ethanol at 70°C. is 5.70.[92] Similarly, the ratio of acetylation

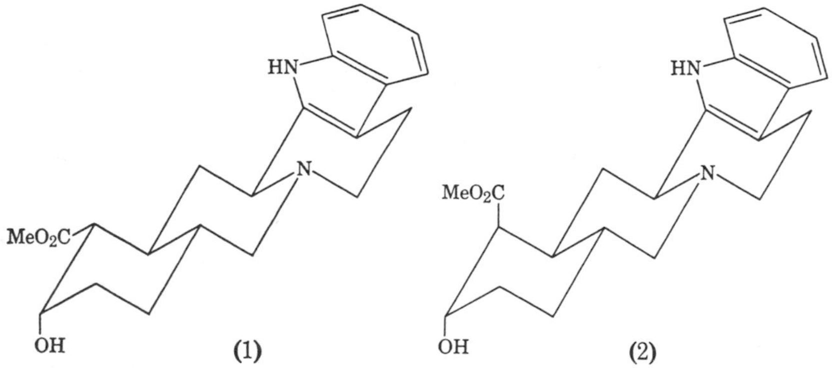

Fig. 28.

[87] E. A. S. Cavell, N. B. Chapman, and M. D. Johnson, *J. Chem. Soc.*, **1960**, 1413, found a ratio of 17 in 50% aqueous dioxane at 29.4°C for the methyl ester.

[88] R. C. Cookson, *Chem. Ind. (London)*, **1953**, 337.

[89] M.-M. Janot, R. Goutarel, A. Le Hir, M. Amin, and V. Prelog, *Bull. Soc. Chim. France*, **1952**, 1085.

[90] N. B. Chapman, R. E. Parker, and P. J. A. Smith, *J. Chem. Soc.*, **1960**, 3634.

[91] G. F. Hennion and F. X. O'Shea, *J. Am. Chem. Soc.*, **80**, 614 (1958).

[92] D. Capon, R. Cornubert, Y. Fagnoni, and G. Ivanowsky, *Bull. Soc. Chim. France*, **1961**, 240.

rates of *trans-* and *cis-*4-*t*-butylcyclohexanol (Fig. 9, X = OH) with acetic anhydride in pyridine[40] is 3.8, the corresponding ratio for propionylation is 3.3 and for isobutyrylation 2.5.* A large body of evidence regarding relative rates of acylation as well as rates of saponification of various axial and equatorial alcohols and their esters is found in the steroid field.[93]

A notable exception to the rule that equatorial esters are hydrolyzed faster than their axial epimers occurs when there is an axial hydroxyl group on the same side of the ring as the ester group to be saponified (i.e., in the 3- or 5-position, counting the site of the ester as 1).[94] Thus the saponification of 3-α-acetoxy-5-α-hydroxycholestane (partial formula A, Fig. 29) proceeds to the extent of 70% with potassium carbonate in benzene-methanol-water after 65 hours at 20°, whereas the corresponding 3-β compound (B, Fig. 29) is hydrolyzed only to the extent of 18% under the same conditions. Evidently the hydroxyl group at C_5 accelerates the reaction, presumably through neighboring group assistance by hydrogen bonding (Fig. 29, C), although there is also a non-steric component to the acceleration, inasmuch as 3-β-cholestanol itself (no hydroxyl at C_5) is not hydrolyzed at all under the mild conditions described above.[94] The fact that an axial acetyl group *syn*-axial† with an axial hydroxyl group is more readily solvolyzed than its equatorial epimer has led to a revision of the configuration of the steroidal alkaloid cevine.[95]

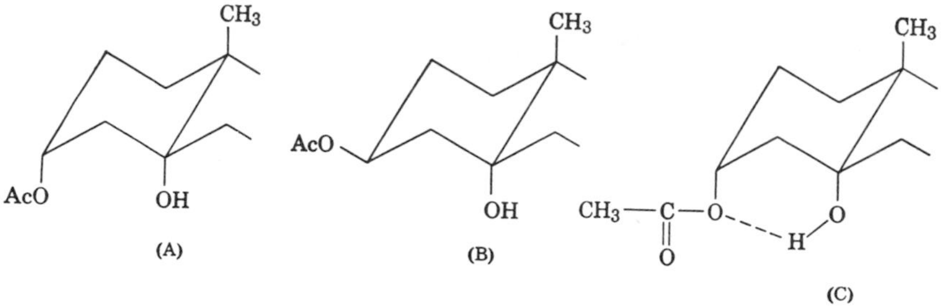

Fig. 29.

[93] Cf. L. F. Fieser and M. Fieser, *Steroids*, Reinhold Publishing Corp., New York, 1959, pp. 216–225. See also Chap. 5.

[94] H. B. Henbest and B. J. Lovell, *J. Chem. Soc.*, **1957**, 1965.

[95] S. M. Kupchan and W. S. Johnson, *J. Am. Chem. Soc.*, **78**, 3864 (1956).

* The decrease in reactivity ratio with increasing size of the anhydride is puzzling. It is possible that, with acetic anhydride, acetylation involves the (relatively large) acylpyridinium ion whereas, with the higher homologs, it may increasingly involve the free anhydride; cf. A. R. Butler and V. Gold, *J. Chem. Soc.*, **1962**, 976. Alternatively, interference with the *gauche* hydrogens at C_2 and C_6 may be more serious for the larger anhydrides than interference with the axial hydrogens at C_3 and C_5.

† The term *syn*-axial will sometimes be used to denote a second axial group on the same side of the ring as a functional axial substituent.

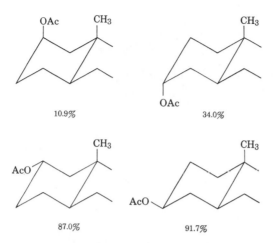

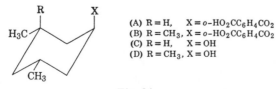

Fig. 30. Extent of saponification of cholestanyl acetates under standard conditions.

These data raise the question as to what effect a *syn*-axial non-participating group will have on rates of esterification and saponification. In the steroid series, saponification of the four isomeric 2- and 3-cholestanyl acetates has been studied[96] under standard conditions (0.01N ethanolic KOH at reflux for ½ hour) with the results shown in Fig. 30. Clearly the *syn*-axial 19-methyl group brings about a substantial further retardation of the already relatively slow rate of hydrolysis of the axial ester. On the other hand, comparison of the rate of saponification of the *trans*-3,*trans*-5-dimethylcyclohexyl acid phthalates (Fig. 31, A) with the corresponding *trans*-3,3,5-trimethyl compound (B) indicates only a minor effect of the axial substituent, the rates of saponification of B and A being in the ratio of 0.59:1.[92] The ratio of acetylation rates of the corresponding alcohols with acetic anhydride in pyridine[97] (Fig. 31, D and C) is 0.73. Possibly the greater flexibility of the monocyclic system leads to substantial relief of potential steric strain in the transition state by deformation of the chair.

(A) R = H,   X = o-$HO_2CC_6H_4CO_2$
(B) R = CH_3, X = o-$HO_2CC_6H_4CO_2$
(C) R = H,   X = OH
(D) R = CH_3, X = OH

Fig. 31.

[96] A. Fürst and P. A. Plattner, *Helv. Chim. Acta.*, **32**, 275 (1949).
[97] E. L. Eliel and F. Biros, unpublished observations.

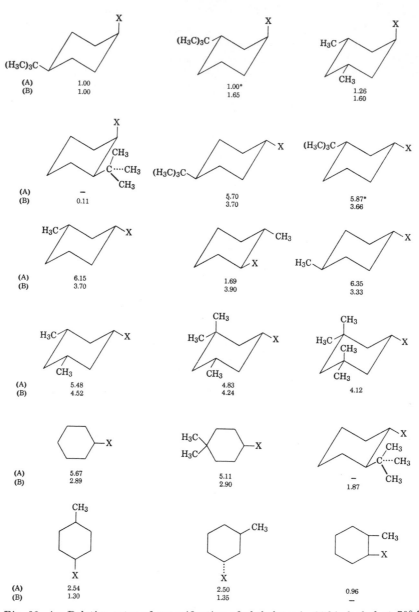

Fig. 32. A: Relative rates of saponification of phthalates in 20% alcohol at 70°;[97] starred data in water at 50°. B: Relative rates of esterification with acetic anhydride in pyridine at 25°.[30,97]

A comparison of rate data for conformationally analogous biased systems is given in Fig. 32. Unfortunately, the data for analogously placed groups are not so constant as one might hope. Careful inspection of the data for the 3- and 4-substituted systems suggests that neither inductive[98] nor direct steric effects can be blamed for the discrepancies, so that one is left with only ring deformation effects to fall back on (Sec. 2-7). The nature of these effects is, however, by no means clear. The situation in the acetylation of the 2-substituted cyclohexanols is somewhat more transparent. Whereas the *trans*-2-methyl substitutent produces a small acceleration, possibly inductive in origin, the *cis*-2-methyl and the *cis*- and *trans*-2-*t*-butyl substituents produce a steric retardation. It is significant that the retardation produced by the *cis*-2-*t*-butyl group is much greater than that by the *trans*-2-*t*-butyl group. This may best be shown by comparing the ratio 2-*t*-butyl/4-*t*-butyl, which enables one to compensate for the difference between equatorial and axial hydroxyl already discussed. The ratio is 0.51 for the *trans* (equatorial) isomers but 0.11 for the *cis* (axial) isomer, indicating a much more substantial reduction of rate by the *cis*-2-*t*-butyl group. This may be explained by postulating that, in the transition state, the hydroxyl group and the *t*-butyl group tend to bend away from each other (so as to relieve steric compression); inspection of models indicates that this "bending away" is much more facile in the *trans*-1,2- than in the *cis*-1,2-disubstituted compound.[99] The reason is that movement apart of the *trans*-1,2-placed groups (or movement together of the *cis*-1,2-placed groups) leads to a flattening of the chair toward the flexible form against a relatively soft potential barrier, whereas movement apart of the *cis*-1,2-placed groups (or movement together of the *trans*-1,2-placed groups) would require increased puckering of the chair against a potential barrier which is undoubtedly quite steep. The situation is depicted in Fig. 33. The results of this effect have already been mentioned in several instances. Thus it was mentioned (p. 37) that the hydrogen bond in *cis*-1,2-cyclohexanediol is somewhat stronger than in the corresponding *trans* isomer. This is not reasonable on the basis of a perfect chair model, but it becomes plausible when it is remembered that the *cis* groups can bend together more readily than the *trans* groups. Similarly, the low entropy difference between *cis*- and *trans*-1,2-dimethylcyclohexane (Table 4) was explained by hindered rotation of the methyl groups in the *cis* isomer. In the *trans* isomer, rotation is probably not so hindered because the methyl groups can readily spread apart.

We may now consider saponification and esterification rates in unbiased

[98] See, however, H. Kwart and T. Takeshita, *J. Am. Chem. Soc.*, **84**, 2833 (1962).
[99] E. L. Eliel, L. A. Pilato, and J. C. Richer, *Chem. Ind. (London)*, **1961**, 2007.

E

e,e bonds approaching
e,a bonds spreading

e,e bonds spreading
e,a bonds approaching

Dihedral angle
between *e,e* bonds

60°

Fig. 33.

mobile systems. The rate of saponification of ethyl cyclohexanecarboxylate in 70% ethanolic sodium hydroxide at 25°C. is 7.25×10^{-4} l. mole^{-1} sec.$^{-1}$.[66] Comparison with the data for the corresponding 4-*t*-butyl-substituted esters (Fig. 27) and application of equation (6) (p. 48) give $K_{\mathrm{CO_2Et}} = (0.428 - 7.25)/(7.25 - 8.50) = 5.46$ at 25°C. or $-\Delta G^{\circ}_{\mathrm{CO_2Et}} =$ 1.0 kcal./mole (Table 1), in good agreement with the equilibrium value given on p. 62.* A corresponding computation for ethyl *cis*-4-methylcyclohexanecarboxylate (Fig. 34), for which the measured saponification rate constant[61] is 2.65×10^{-4} l. mole^{-1} sec.$^{-1}$, gives $K = 0.38$ and $\Delta G^{\circ} = +0.57$ kcal./mole. Assuming additivity of the conformational energies (Fig. 34) and taking $\Delta G^{\circ}_{\mathrm{CH_3}} = -1.7$ kcal./mole in the liquid phase (p. 54), one obtains $\Delta G^{\circ}_{\mathrm{CO_2Et}} = \Delta G^{\circ}_{\mathrm{CH_3}} + \Delta G^{\circ} = -1.7 + 0.57 =$ -1.13 kcal./mole, in excellent agreement with the other values. In this case, at least, the assumption of additivity of the energies appears to be a good one.

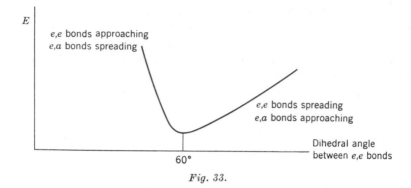

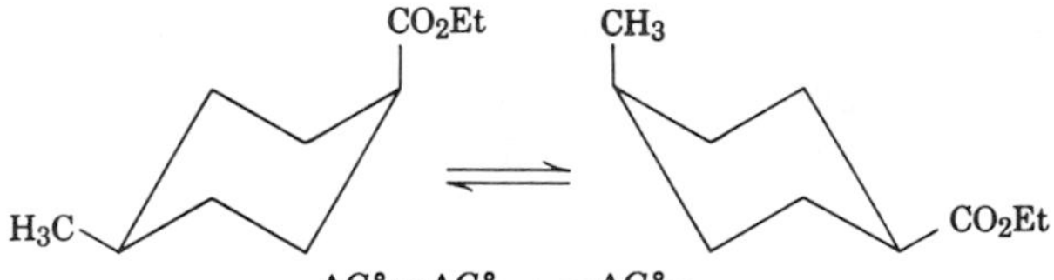

CO₂Et　　CH₃

H₃C

CO₂Et

$\Delta G^{\circ} = \Delta G^{\circ}_{\mathrm{CO_2Et}} - \Delta G^{\circ}_{\mathrm{CH_3}}$

Fig. 34.

* No value for $-\Delta G^{\circ}_{\mathrm{CO_2Me}}$ can be calculated from the data of ref. 87, since the reported specific rate of saponification of methyl cyclohexanecarboxylate is greater than the rate constant for the *trans*-4-*t*-butyl compound. Dr. N. B. Chapman (personal communication) has indicated that this is a peculiarity of the aqueous-dioxane solvent system; in aqueous methanol his data for the methyl esters are in general agreement with those of ref. 66 for the ethyl esters. The nature of the peculiarity of the water-dioxane system is not yet clear.

Corresponding calculations may be carried out for the acetylation[40] of cyclohexanol to yield $\Delta G^\circ_{OH} = -0.52$ kcal./mole and of *cis*-4-methyl-cyclohexanol to yield $\Delta G^\circ = 1.31$ kcal./mole, whence $\Delta G^\circ_{OH} = -0.39$ kcal./mole. The most accurate values of ΔG°_X will normally be obtained from a rate constant falling near the middle of the range defined by k_e and k_a.* Thus, as Table 7 indicates, the best value for ΔG°_{OH} is probably

Table 7

Group	k_e	k_a	k_H	$k_{cis\text{-}4\text{-}Me}$	Units	Ref.
—OH	10.65	2.89	8.37	3.76	$\times 10^{-5}$	40
—CO$_2$Et	8.50	0.428	7.25	2.65	$\times 10^{-4}$	66
o-HO$_2$CC$_6$H$_4$COO—	20.6	3.61	20.5	9.72	$\times 10^{-4}$	92
—OAc	24.0	3.61	22.3	8.77	$\times 10^{-3}$	90

Rate constants in l. mole^{-1} sec.$^{-1}$. k_e and k_a are rate constants for the *trans*- and *cis*-4-*t*-butyl derivatives, respectively.

that derived from cyclohexanol itself, but the best value for $\Delta G^\circ_{CO_2Et}$ is more likely that derived from the *cis*-4-methyl derivative. In the case of acid phthalate the value for the 4-H compound is too close to that for the *trans*-4-*t*-butyl compound for calculation, but from the value for *cis*-4-methylcyclohexyl acid phthalate one can calculate $\Delta G^\circ = 0.49$ kcal./mole, whence $\Delta G^\circ_{Phth} = -1.2$ kcal./mole. Finally, the value of $\Delta G^\circ_{OAc} = -1.5$ kcal./mole derived from the 4-H compound[90] is probably in error because of the close proximity of k_H and k_e; from the value for k_{Me} one obtains $\Delta G^\circ = +0.75$ kcal./mole, whence a more reasonable value of $-\Delta G^\circ_{OAc}$ of 0.95 kcal./mole is computed.†, ‡

From the rate constant of acetylation of 3,3-dimethylcyclohexanol,[40] 9.88×10^{-5} l. mole^{-1} sec.$^{-1}$, the conformational equilibrium constant for this compound (cf. Fig. 35) may be computed to be

$$K = (k_a - k)/(k - k_e) = 9.53,$$

for k_a the value for *trans*-3,3,5-trimethylcyclohexanol (Fig. 35) (2.45×10^{-5} l. mole^{-1} sec.$^{-1}$) being used and for k_e the value for *cis*-3-methylcyclohexanol (10.71×10^{-5} l. mole^{-1} sec.$^{-1}$) being used. Thus ΔG°

* Since $K = (k_a - k)/(k - k_e)$, it is clear that the probable error in K will be large when k is close to either k_e or k_a, since then one has to deal with small differences of large numbers.

† Quite possibly, the inconsistency in this case also may be due to the choice of aqueous dioxane as the solvent.

‡ Some contemplation will show that, taking $\Delta G^\circ_{CH_3} = -1.7$ kcal./mole, k_H will give the more accurate result for $-\Delta G^\circ_X$ if $-\Delta G^\circ_X < 0.85$ kcal./mole. For $-\Delta G^\circ_X > 0.85$ kcal./mole, the more accurate determination is based on $k_{cis\text{-}4\text{-}Me}$.

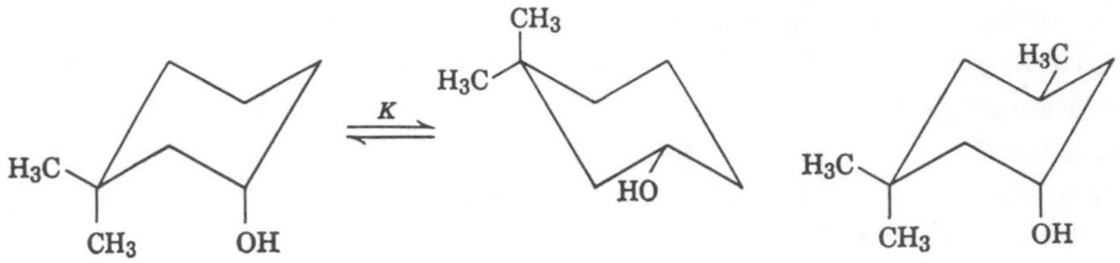

Fig. 35.

for the equilibrium shown in Fig. 35 is 1.34 kcal./mole whence, by a computation completely analogous to that given on p. 65, the interaction energy of axial CH_3 and OH is 1.9 kcal./mole. This value is somewhat lower than the equilibrium value given on p. 66 (which may be too high because it involves aluminum isopropoxide as catalyst) but agrees with the uncertain value at 25° derived from Raney nickel equilibrations (p. 66). Unfortunately, the kinetic value is also uncertain, both because of the proximity of k and k_e and because of the uncertainty of the proper choice of the model compounds used to find k_e and k_a.

Relatively little information is available on reaction rates at exocyclic positions other than esterification and saponification. The quaternization reaction of *N,N*-dimethylcyclohexylamines with methyl iodide (Menshutkin reaction) has been studied for the *cis*- and *trans*-3- and 4-methyl-substituted compounds,[100] but the rates vary only little, from 2.2 to 3.2×10^{-1} l. mole^{-1} min.$^{-1}$. This could be an indication that the *N,N*-dimethylamino group is rather bulky so that it occupies largely the equatorial position in either the *cis* or the *trans* series, relegating the methyl group to the axial position in the *cis*-1,4- and *trans*-1,3-isomers.[101] More surprising is the finding that the latter isomers—which must contain an appreciable mole fraction of conformers with axial—NMe_2 groups—react faster than their all-equatorial epimers; one is tempted to suspect that the preliminary assignment of configuration in the series, based on physical properties, may not be correct.

The reverse reaction, displacement of 4-*t*-butylcyclohexyltrimethyl-ammonium chlorides with potassium *t*-butoxide in *t*-butyl alcohol to give 4-*t*-butylcyclohexyldimethylamines and methyl *t*-butyl ether, has also been studied.[102] The reaction is complicated by concomitant Hofmann elimination in the case of the *cis*-4-*t*-butylcyclohexyl compound, but it was possible to estimate a ratio of 20:1 for the specific rates of displacement for the axial and equatorial quaternary salts, respectively.

[100] E. R. A. Peeling and B. D. Stone, *Chem. Ind.* (*London*), **1959**, 1625.

[101] M. Tichý, J. Jonáš, and J. Sicher, *Collection Czech. Chem. Commun.*, **24**, 3434 (1959).

[102] D. Y. Curtin, R. D. Stolow, and W. Maya, *J. Am. Chem. Soc.*, **81**, 3330 (1959).

The finding that the axial quaternary ammonium function undergoes displacement more readily suggests that steric hindrance in the transition state is not of major importance, but that the overriding factor is ground state strain of the axial $N(CH_3)_3^+$ group (comparable in size to t-butyl!) which is relieved in the transition state by the breaking away of one of the methyl substituents.

The reaction of cis- and $trans$-4-t-butylcyclohexylamine with 2,4-dinitrochlorobenzene[31] is noteworthy in that the cis (axial) isomer reacts faster than the $trans$ (equatorial) one. It has been suggested[31] that this most unusual finding is occasioned by excess solvation of the equatorial amine in the ground state by the alcoholic solvent. Comparison of the rate of reaction of cis-4-methylcyclohexylamine with that of the t-butyl compounds (cf. p. 79) gave[31] $-\Delta G^\circ_{NH_2} = 1.8$ kcal./mole; a value of 1.7 kcal./mole was obtained[103] by pK measurements of the same amines in aqueous methyl Cellosolve. NMR measurements[31] gave a slightly smaller value (1.4–1.5 kcal./mole) in alcoholic or chloroform solvents and showed that in solvents which are not hydrogen donors, such as cyclohexane, acetonitrile, and pyridine, $-\Delta G^\circ_{NH_2}$ is appreciably smaller (1.1–1.2 kcal./mole) than in hydrogen donors. This marked solvent effect might be explained by postulating that the preferred rotational conformation of the axial amino group is that in which the pair confronts the ring and is therefore not available for hydrogen bonding and the resulting gain in energy, but definite evidence on this point must await measurement of conformational entropy and enthalpy differences for NH_2.

c. Reactions at Ring Positions. Steric Control. The situation just described for displacements in cyclohexyltrimethylammonium salts where the axial isomer reacts faster because of steric strain in the ground state finds its parallel in a number of reactions affecting the ring carbons themselves. An acceleration of this type is often called[104] steric assistance and is likely to manifest itself whenever the difference in energy between an equatorial and an axial substituent in the starting state is diminished or nullified in the transition state. In that case the axial isomer, because of its higher ground state energy level, will have the lower activation energy.

A case in point is the oxidation of cyclohexanols with chromic acid. As a result of experiments originally carried out in the steroid series[105] and to be discussed in Chap. 5, it was suggested that the long-known[106]

[103] J. Sicher, J. Jonáš, and M. Tichý, *Tetrahedron Letters*, **1963**, 825.
[104] H. C. Brown, *Science*, **103**, 385 (1946); cf. ref. 86, p. 224.
[105] J. Schreiber and A..Eschenmoser, *Helv. Chim. Acta*, **38**, 1529 (1955).
[106] G. Vavon and C. Zaremba, *Bull. Soc. Chim. France*, **1931**, 1853.

higher rate of oxidation of axial cyclohexanols compared to their equatorial epimers was due not to greater accessibility of the equatorial carbinol hydrogen in the axial alcohol[85] but to steric acceleration.[105]* Data are now available[99] which illustrate the same point in monocyclic cyclohexanols; they are shown in Table 8. The most notable points are:

Table 8. Oxidation Rates of Monocyclic Cyclohexanols[a]

Entry	Compound	Rate[a]	Rel. Rate[b]
I	Cyclohexanol	5.53	1.38
2	*trans*-2-Methylcyclohexanol	5.8	1.44
3	*cis*-2-Methylcyclohexanol	23.2	5.78
4	*trans*-3-Methylcyclohexanol	18.51	4.60
5	*cis*-3-Methylcyclohexanol	4.74	1.18
6	*trans*-4-Methylcyclohexanol	4.21	1.05
7	*cis*-4-Methylcyclohexanol	12.15	3.00
8	4,4-Dimethylcyclohexanol	7.00	1.74
9	*trans*-3,3,5-Trimethylcyclohexanol	252.5	62.7
10	*cis*-3,3,5-Trimethylcyclohexanol	7.52	1.87
11	*trans*-2-*t*-Butylcyclohexanol	54.0	13.4
12	*cis*-2-*t*-Butylcyclohexanol	254.5	63.2
13	*trans*-3-*t*-Butylcyclohexanol	30.25	7.52
14	*cis*-3-*t*-Butylcyclohexanol	4.97	1.24
15	*trans*-4-*t*-Butylcyclohexanol	4.02	1.00
16	*cis*-4-*t*-Butylcyclohexanol	13.0	3.23

[a] In l. mole^{-1} sec.$^{-1}$ × 10^3 in 75% (by volume) acetic acid at 25.0°.
[b] *trans*-4-*t*-Butylcyclohexanol = 1.

(1) axial alcohols are oxidized 3–6 times more rapidly than their equatorial epimers; compare entries 12 and 11, 13 and 14, and 16 and 15; also 3 and 2, 4 and 5, and 7 and 6 (although in the latter three pairs the *e,a* isomers are not conformationally homogeneous, cf. p. 65). (2) Maximum acceleration is produced where there is a methyl group *syn*-axial with an axial hydroxyl group; see entries 9, 12, and 11 (regarding the situation with the 2-*t*-butyl groups, compare the diagrams in Fig. 32). (3) A *gauche* substituent (on the ring carbon next to that bearing the alcohol group) produces acceleration: compare entry 2 with 5 and 6. *A priori*, this acceleration might be either steric or polar in origin (an alkyl substituent

* An alternative explanation based on freezing of favorable rotational conformations of the intermediate chromate esters advanced by H. Kwart and P. S. Francis, *J. Am. Chem. Soc.*, **81**, 2116 (1959), and by H. Kwart, *Suomen Kemistilehti*, **A34**, 173 (1961), has been shown to be untenable: cf. J. Roček, F. H. Westheimer, A. Eschenmoser, L. Moldványi, and J. Schreiber, *Helv. Chim. Acta*, **45**, 2554 (1962).

in the 2-position does produce a palpable electron-donating inductive effect); the additional data[107] in Table 9 suggest that both polar and steric effects come into play. Thus the greater oxidation rate of II compared to I (factor of 3.2) may be either steric or polar in origin, but, since a substantial part of the acceleration persists in the pair IV-V

Table 9. Oxidation Rates with Chromium Trioxide[a]

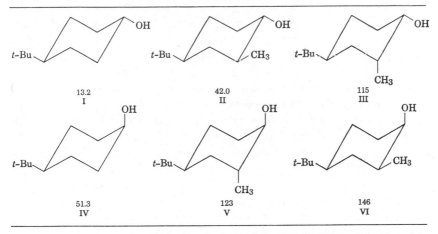

| 13.2 | 42.0 | 115 |
| I | II | III |

| 51.3 | 123 | 146 |
| IV | V | VI |

[a] In l. mole⁻¹ sec.⁻¹ at 25° in 89.7% acetic acid, 0.95M with respect to sulfuric acid, determined polarographically.[107]

(factor of 2.4) where there is no apparent extra steric compression (see, however, ref. 99), the effect may be largely polar. The further acceleration from II to III and from V to VI is, however, clearly steric in origin since polar effects would appear to be unchanged. The substantial difference between II and III illustrates once more the greater compression between adjacent e,a substituents (which cannot readily bend away from each other) as compared to adjacent e,e substituents (which can bend apart), as previously explained in the discussion of Fig. 33. It might also be noted that steric hindrance to hydrogen abstraction, originally suggested[85] as an explanation for the order of oxidation rates of equatorial and axial alcohols, cannot, in fact, be the cause. A clear example is provided by compound 10, Table 8 (see also Fig. 32), in which the axial carbinol hydrogen is screened by a syn-axial methyl group; nevertheless, the compound is oxidized somewhat faster than other equatorial alcohols (e.g., 15), probably because of the 3-alkylketone effect (Sec. 2-6c).

The generally accepted explanation for the observed order of rates is

[107] F. Šipoš, J. Krupička, M. Tichý, and J. Sicher, Collection Czech. Chem. Commun. 27, 2079 (1962).

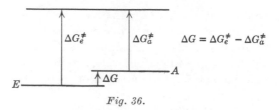

Fig. 36.

based on strain relief of the sterically crowded carbinols upon transformation to the corresponding ketones; this relief reflects itself in the transition state. In fact, it has been suggested[107] that the ratio of reaction rates (k_{ax}/k_{eq}) is a direct measure of the free energy differences between the alcohols being compared: $-\Delta G° = RT \ln k_{ax}/k_{eq}$. Although this assumption is clearly attended with practical success—for example, a conformational equilibrium constant of 3.23 corresponding to $-\Delta G°_{OH} =$ 0.70 kcal./mole which is in line with other values (Table 1) may be calculated from entries 15 and 16 in Table 8—it must be accepted with some reserve. In terms of energetics, it implies the situation depicted in Fig. 36, differences in ground state energies being wholly responsible for the observed differences in oxidation rates, the transition states for epimeric alcohols being, in fact, the same. This assumption may not be warranted in all cases. However, the ratio of oxidation rates of the epimeric 3,3,5-trimethylcyclohexanols (entries 9 and 10, Table 8) of 38.9 corresponds to a $-\Delta G°$ of 2.16 kcal./mole, in fair agreement with the equilibrium value of 1.96 kcal./mole (p. 65 and Fig. 18).

The solvolysis (S_N1) reaction of alkylcyclohexyl p-toluenesulfonates is another reaction series largely governed by considerations of strain relief. The interaction of an axial tosylate group with the *syn*-axial hydrogens is relieved when one goes to the carbonium ion intermediate or to the corresponding transition state which resembles the intermediate.[108] This will be true only if the intermediate is, in fact, a relatively bare carbonium ion. For the formolysis, acetolysis, and ethanolysis of the 4-t-butylcyclohexyl tosylates[28] at temperatures varying between 25° and 100°, ratios of k_{cis}/k_{trans} varying from 4.0 to 2.3 corresponding to a $-\Delta G°_{OTs}$ of between 0.62 and 0.82 kcal./mole have been found. These values of $-\Delta G°_{OTs}$ are in excellent agreement with two of the three values listed in Table 1 and obtained by entirely different methods: thus the situation in tosylate solvolysis seems to approach that implied by Fig. 36. The argument is further strengthened by the observation[109] that *trans*-3,3,5-trimethylcyclohexyl tosylate (cf. Fig. 31), in which there is a methyl *syn*-axial with the tosylate, reacts 12.6 times

[108] G. S. Hammond, *J. Am. Chem. Soc.*, **77**, 334 (1955).
[109] H. Fischer, C. A. Grob, and W. Schwarz, *Tetrahedron Letters*, **1962**, 25.

as fast as its equatorial isomer; if the transition states are comparable, this implies an extra compression energy of 1.75 kcal./mole in the axial isomer. If, as implied earlier, $-\Delta G^{\circ}_{\mathrm{OTs}}$ is comparable to or slightly larger than $-\Delta G^{\circ}_{\mathrm{OH}}$, this value is reasonable since, by arguments analogous to those given earlier (cf. Fig. 17), it would lead to an axial CH_3-OTs interaction of 1.9 kcal./mole, about the same as CH_3—OH.

In connection with the steric assistance argument, it is interesting that the solvolysis rates of *cis*- and *trans*-1,2-dimethylcyclohexyl bromide are virtually identical.[110] In one of these isomers there are two equatorial methyls and an axial bromine; in the other there is an equatorial methyl, an axial methyl, and an equatorial bromine. The ground state energies would be the same if the compression of axial methyl and axial bromine were the same. On the basis of $-\Delta G^{\circ}$ values (Table 1) it would appear, however, that bromine should prefer the axial position, although the opposite argument has also been made in the assertion, on the basis of infrared spectra, that the C—X stretching frequency in a 1-methylcyclohexyl halide indicates the halogen to be equatorial.

Despite the attractiveness of the argument that the transition states (and presumably the intermediates) for the solvolysis of the equatorial and axial tosylates are similar in energy, it cannot be reasoned that the intermediates are, in fact, structurally the same (e.g., a common carbonium ion), for the products from *cis* (axial) and *trans* (equatorial) 4-*t*-butylcyclohexyl tosylate are not identical. The *cis* isomer gives more olefin (83 vs. 67% with aqueous acetone), and the alcohol formed from either starting material has the inverted configuration: *trans*-alcohol from *cis*-tosylate, *cis*-alcohol from *trans*-tosylate.

Application of the equation $K = (k_a - k)/(k - k_e)$ (p. 48) to cyclohexyl tosylate and *cis*- and *trans*-4-*t*-butylcyclohexyl tosylate leads to a value of K_{OTs} of 9.0 at 50° corresponding to $-\Delta G^{\circ}_{\mathrm{OTs}}$ of 1.5 kcal./mole.[28] (The corresponding value at 75° is 1.8 kcal./mole.) These values appear unreasonably high and are out of line with the values of about 0.7 kcal./mole found by other methods (see also the discussion on p. 45), suggesting that, in this instance, the 4-*t*-butyl-substituted compounds do not provide good models for evaluating k_e and k_a, owing either to a deformation effect or, less likely, to excessive sensitivity of the reaction to the small inductive effect of the alkyl group. That the nature of the alkyl substituent does affect the solvolysis rate in ways other than by affecting the conformational equilibrium of cyclohexyl tosylates is further demonstrated by the 3-methylcyclohexyl tosylates.[111] Thus the specific ethanolysis rate of the purely equatorial *cis*-3-methylcyclohexyl tosylate

[110] T. D. Nevitt and G. S. Hammond, *J. Amer. Chem. Soc.*, **76**, 4124 (1954).
[111] W. Hückel and co-workers, *Ann.*, **624**, 142 (1959).

(0.61 × 10⁻⁶ sec.⁻¹ at 50°) is considerably less than that of the equally equatorial *trans*-4-*t*-butylcyclohexyl tosylate (1.07 × 10⁻⁶), and the rate constant for *trans*-3-methylcyclohexyl tosylate (2.24 × 10⁻⁶) when compared to that of *cis*-4-*t*-butylcyclohexyl tosylate (4.18 × 10⁻⁶) is not in line with the expected predominantly axial conformation of the tosylate group in the *trans*-3-methyl compound (which should react at a rate of about 3.7×10^{-6} sec.⁻¹, $-\Delta G_{OTs}^{\circ} = 0.7$ kcal./mole and $-\Delta G_{CH_3}^{\circ} = 1.7$ kcal./mole* being assumed).

Table 10. Relative Bimolecular Substitution Rates with Thiophenolate in 87% Ethanol

Entry	R	ROTs	Ref.	RBr	Ref.
1	*n*-Butyl	23	41	364	36
2	*sec*-Butyl	4.3	41	14	36
3	*cis*-4-*t*-Butylcyclohexyl	1.00[a]	41	1.00[b]	36
4	*trans*-4-*t*-Butylcyclohexyl	0.032	41	ca. 0.017	36
5	Cyclohexyl	0.28	41	0.25	36
6	*cis*-4-Methylcyclohexyl	0.50	41	0.70	36
7	*trans*-3,3,5-Trimethylcyclohexyl	0.78	116	—	

[a] 3.61×10^{-4} l. mole⁻¹ sec.⁻¹.
[b] 4.81×10^{-5} l. mole⁻¹ sec.⁻¹.

The S_N2 reaction of cyclohexyl tosylates[41] and bromides[36] with the nucleophiles thiophenolate and halide[112] has also been studied. In general, conformational difficulties cause the reactions of cyclohexyl halides to be slower than those of secondary aliphatic halides;[113] in part, these difficulties have been ascribed[114,115] to bond eclipsing in the transition state ("I-strain"), the bonds evidently being perfectly staggered in the ground state of cyclohexane derivatives. The data in Table 10 (entries 1–4) give a more detailed picture of the conformational effects in this reaction series.

Several conclusions may be drawn from the data in Table 10:

(1) The values for $-\Delta G_{OTs}^{\circ} = 0.64$ kcal./mole and $-\Delta G_{Br}^{\circ} = 0.69$ kcal./mole may be derived by using the equation $K = (k_a - k)/(k - k_e)$.

[112] S. Winstein, D. Darwish, and N. J. Holness, *J. Am. Chem. Soc.*, **78**, 2915 (1956).
[113] P. D. Bartlett and L. F. Rosen, *J. Am. Chem. Soc.*, **64**, 543 (1942).
[114] H. C. Brown, R. S. Fletcher, and R. B. Johannesen, *J. Am. Chem. Soc.*, **73**, 212 (1951).
[115] H. C. Brown and G. Ham, *J. Am. Chem. Soc.*, **78**, 2735 (1956).
[116] E. L. Eliel and R. P. Gerber, *Tetrahedron Letters*, **1961**, 473.

* Possibly the alkyl groups in the 3-position unlike those in the 4-position exert a polar effect; cf. ref. 45.

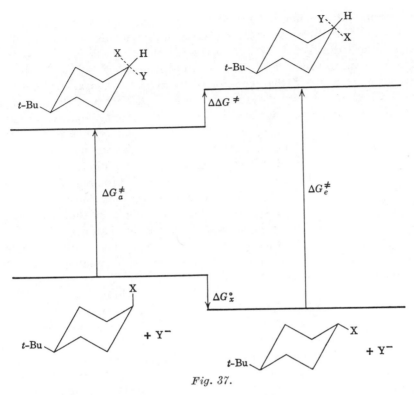

Fig. 37.

(2) The type of additivity relationship depicted in Fig. 34 works very poorly for the methylcyclohexyl tosylates and bromides. (This is also true in the chromic acid oxidation series shown in Table 8.) Perhaps this is not surprising, inasmuch as the change from sp^3 hybridization of the reaction site in the ground state of the solvolysis or oxidation to sp^2 in the transition state must involve substantial deformation of the cyclohexane ring, and the ease of such deformation is likely to be affected appreciably by the nature and position of the alkyl group in the ring.

(3) The ratio k_{cis}/k_{trans} (31 for the tosylates, 58 for the bromides) is much greater than the value of about 3 to be expected on the basis of ground state energy differences alone. Thus, clearly, the transition states for displacement of axial and equatorial substituents are not the same in energy. The actual situation is depicted in Fig. 37.

According to Fig. 37, $\Delta G_e^{\ddagger} = \Delta G_a^{\ddagger} + \Delta\Delta G^{\ddagger} - \Delta G_x^{\circ}$ or $\Delta\Delta G^{\ddagger} = \Delta G_e^{\ddagger} - \Delta G_a^{\ddagger} + \Delta G_x^{\circ}$. Now $\Delta G_e^{\ddagger} - \Delta G_a^{\ddagger} = RT \ln (k_e/k_a)$ amounts to 2.39 kcal./mole for bromide and 2.03 kcal./mole for tosylate. Deducting the $-\Delta G_x^{\circ}$ values given above leaves $\Delta\Delta G_{Br}^{\ddagger} = 1.7$ kcal./mole, $\Delta\Delta G_{OTs}^{\ddagger} = 1.4$ kcal./mole. Clearly, the transition state for the displacement of an equatorial group (involving an incoming axial substituent) lies higher than the transition state for the displacement of an axial group (involving an incoming equatorial substituent). This is perhaps not surprising in view of the fact that axial approach of the incoming nucleophile should be appreciably more hindered than equatorial approach. When axial approach is further hindered, e.g., by a methyl group *syn*-axial with the direction of approach of the nucleophile as in *cis*-3,3,5-trimethylcyclohexyl tosylate

Fig. 38.

(Fig. 38), the displacement with thiophenolate does, indeed, become extremely slow, as predicted.[72] It is of interest that hindrance to axial approach of the nucleophile is more pronounced in the cyclohexyl bromides than in the corresponding tosylates, despite the fact that the nucleophile (thiophenolate) is the same. This may be a reflection of a tighter transition state in the bromides (poorer leaving group!) which results in greater encumbrance of the incoming nucleophile.

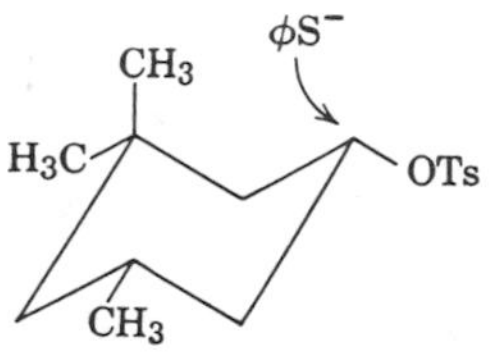

(4) So far we have considered the effect of the incoming group Y on the transition state (Fig. 37), but there is also interest in the steric situation with respect to the leaving group X. Focusing attention on the axial isomer,* one sees, *a priori*, that the leaving group X may encounter increased steric hindrance in the transition state (due to compression against the *syn*-axial hydrogens) but that this may be more than outweighed by the lesser hindrance to approach by the incoming group Y. Less likely is the possibility that X has completely receded from the area of compression, in which case $\Delta\Delta G^{\ddagger}$ would reflect entirely the greater hindrance of Y in the transition state of axial approach. Finally it is possible that the compression of X in the transition state is nearly the same as in the ground state; then the transition-state energy levels would reflect the ground-state energy levels were it not for the lesser ease of axial approach of Y. In that case, the compression energy of Y is represented not by $\Delta\Delta G^{\ddagger}$ but by $-\Delta G_x^{\circ} + \Delta\Delta G^{\ddagger}$. Intermediate situations may also be envisaged. To elucidate this matter requires comparison of two molecules, such as *cis*-4-*t*-butylcyclohexyl tosylate and *trans*-3,3,5-trimethylcyclohexyl tosylate (Fig. 39), where approach of Y to the equatorial side is equally easy but recession of X from the axial side might be seriously affected by the *syn*-axial methyl group in the 3,3,5 compound. In actual fact, the two rates (Table 10) differ by only 20%.[116] The surprising conclusion is thus that the leaving group X

---

* In the equatorial isomer the leaving group X is presumably unencumbered all along the reaction path.

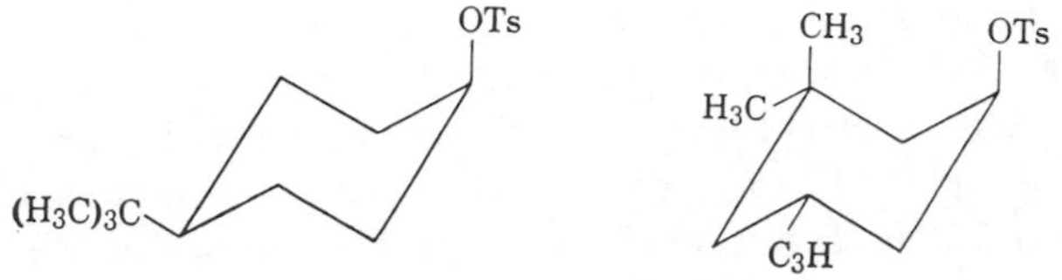

Fig. 39.

is about equally encumbered in the ground state and in the transition
state. This may be a fortuitous cancelling of factors (e.g., X may be
farther away in the transition state but more precariously inclined
toward the *syn*-axial hydrogens), or it may reflect the operation of a
special mechanism (see p. 97).

In addition to the rather well-known S_N1 and S_N2 reactions there is
another group of nucleophilic displacement reactions of less well-defined
mechanism. Among these are the reaction of amines with nitrous acid
to give alcohols and the reaction of alcohols with thionyl chloride,
phosphorus halides, or hydrogen halides to give alkyl halides. These
reactions sometimes involve retention of configuration, in which case
they are properly termed "S_Ni" reactions,[117] but in other instances they
involve inversion or extensive racemization.

The amine-nitrous acid reaction has been studied quite extensively in
the cyclohexane series, and it has been pointed out[118,119] that equatorial
amines generally give the corresponding equatorial alcohols in high yield.
Thus menthylamine appears to give mostly menthol,[120] and carvomenthyl-
amine gives largely carvomenthol.[121] 1β-Amino-*trans*-decalin* (equatorial
amine) gives only *trans*-1β-decalol,† and the equatorial 2α-amino-*trans*-
decalin gives only *trans*-2α-decalol.[122] Even the 9-amino-*cis*-decalin, in

---

[117] Cf. E. L. Eliel, "Substitution at Saturated Carbon Atoms," in M. S. Newman,
ed., *Steric Effects in Organic Chemistry*, John Wiley and Sons, New York, 1956,
pp. 128–130.

[118] J. A. Mills, *J. Chem. Soc.*, **1953**, 260.

[119] A. K. Bose, *Experientia*, **9**, 256 (1953).

[120] J. Read, A. M. Cook, and M. I. Shannon, *J. Chem. Soc.*, **1926**, 2223; see also
ref. 122a.

[121] A. K. Bose, *Experientia*, **8**, 458 (1952).

[122] (a) W. Hückel, *Ann.*, **533**, 1 (1937). (b) Regarding configuration see also
W. G. Dauben, R. C. Tweit, and C. Mannerskantz, *J. Am. Chem. Soc.*, **76**, 4420 (1954).

* A good deal of confusion has arisen in the use of the prefixes *cis* and *trans* as
applied to substituents in 1- and 2-substituted decalins, as the point of reference is
often not clear. We shall use the prefixes β and α in a fashion similar to the well-
known convention in the steroid series (e.g., ref. 86, p. 288), β meaning "on the
same side as the proximate hydrogen at C_9" and α meaning on the side opposite
from hydrogen, as indicated in Fig. 24.

† The prefix *trans* here as in similar cases will refer to the fusion of the decalin
ring system, not to the position of the substituent.

which the amino group is equatorial with respect to one ring and axial with respect to the other, gives the corresponding alcohol 9-hydroxy-*cis*-decalin with but a trace of its epimer, though along with a major amount of olefin (see below).[123] In the steroid series the equatorial 2α-, 3β-, 4α-, 6α-, and 7β-5α-cholestanylamines and the equatorial 3α,5β-cholestanylamine all give the corresponding alcohols in over 90% yield.[124]

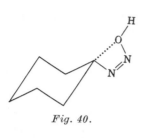

Fig. 40.

However, *trans*-4-*t*-butylcyclohexylamine gives some *cis*-4-*t*-butylcyclohexanol (13%) as well as olefin (10%) along with a major amount of *trans*-4-*t*-butylcyclohexanol.[125]

Several explanations have been suggested. The possibility that a relatively bare carbonium ion is formed and is attacked by solvent from the more accessible equatorial side must be discarded, because the epimeric axial amines (which should form the same carbonium ions) in some cases do not give the same products (see below). Mills has proposed a pyramidal transition state involving front-side attack of water in the case of the equatorial amines (or the corresponding diazonium salts) where attack from the rear is not favored (see the earlier discussion on the S_N2 reaction). However, there are no precedents for such transition states in displacement on carbon; moreover, the hydrolysis of *trans*-4-*t*-butylcyclohexyl tosylate which, being S_N1 (the corresponding ethanolysis is not accelerated by ethoxide) should proceed via a carbonium ion, gives *cis*-4-*t*-butylcyclohexanol (inversion, rearward attack) as the sole alcohol product.[28],* A plausible modification of Mills' mechanism[126] involves a cyclic transition state of a covalent diazonium hydroxide (Fig. 40). This transition state can be accommodated readily on the equatorial side of the ring, but not so readily on the axial side. A third suggestion[122b] involves front-side collapse of an unsymmetrically solvated carbonium ion, the solvation being on the equatorial side. A modification of this mechanism would involve front-side collapse of an ion pair R^+OH^- formed by extrusion of nitrogen from $R-N_2^+OH^-$.

The picture regarding the reaction of axial amines with nitrous acid is not so clear cut. Axial amines (in contrast to their equatorial epimers)

[123] W. G. Dauben, R. C. Tweit, and R. L. MacLean, *J. Am. Chem. Soc.*, **77**, 48 (1955).

[124] C. W. Shoppee, D. E. Evans, and G. H. R. Summers, *J. Chem. Soc.*, **1957**, 97; C. W. Shoppee, R. J. W. Cremlyn, D. E. Evans, and G. H. R. Summers, *ibid.*, **1957**, 4364.

[125] W. Hückel and K. Heyder, *Chem. Ber.*, **96**, 220 (1963).

[126] Cf. ref. 117, p. 83.

* Tosylate solvolyses and amine-nitrous acid reactions often give different products; for a detailed discussion see A. Streitwieser, *J. Org. Chem.*, **22**, 861 (1957).

yield substantial amounts of olefins. The reason for this will be explained in the next section. In some cases (e.g., 4β- and 6β-cholestanylamine,[124] $3a$-aminofriedelane[127]) the olefin is the exclusive product. In other cases, rearrangement products result; thus neomenthylamine (cf. Fig. 25) gives largely 4-menthanol. When unrearranged alcohols are obtained, configuration may be either retained or inverted, or mixtures of the two alcohols may result. Thus 1α-amino-*trans*-decalin is reported to give *trans*-1β-decalol as the only alcohol product, and 2β-amino-*trans*-decalin gives only *trans*-2α-decalol,[122a] complete inversion from the axial amine to the equatorial alcohol occurring in these cases. On the other hand, a number of steroidal amines—the 2β-, 3α-, and 7α-amino-5α-cholestanes as well as 3β-amino-5β-cholestane—give alcohols of retained configuration only (besides olefin).[124] *cis*-4-*t*-Butylcyclohexylamine gives a mixture of the corresponding alcohols.[125]

A special place is occupied by the 2- and 3-methylcyclohexylamines.[128] The equatorial isomers (*trans*-2 and *cis*-3) give alcohols with 85% or more retention, accompanied by little (3–4%) olefin. The epimeric amines (*cis*-2 and *trans*-3), though giving much more olefin (31% and 54%, respectively) also give alcohols of predominantly retained configuration (90% retention for *cis*-2, and 67% for *trans*-3). The interpretation may well lie in the conformational heterogeneity of this particular set of amines: since the conformational energy difference for methyl and amino (in aqueous solvent) is nearly the same (1.7–1.8 kcal./mole), the *cis*-2- and *trans*-3-methylcyclohexylamines will each exist in two nearly equally populated conformations, one with equatorial, the other with axial, NH_2.

The stereochemistry of conversion of cyclohexanols to cyclohexyl halides also does not seem to follow a uniform stereochemical path. For the phosphorus pentahalides the general picture is one of inversion. Thus the treatment of 5α-cholestan-3β-ol (equatorial OH) with phosphorus pentachloride gives 3α-chlorocholestane in high yield, whereas 5α-cholestan-3α-ol gives the 3β-chloride.[129] Similar results are obtained in the 5β-cholestan-3-ol series[130] and in the 5α-cholestan-7-ol series.[131] Phosphorus pentabromide also reacts with steroidal alcohols[126,130] as well as with 4-*t*-butylcyclohexanols[132] to give halides of inverted configuration. In general, the reaction proceeds better when one starts with the equatorial alcohol rather than with the axial alcohol.[130,131] The products, in some

[127] G. Drefahl and S. Huneck, *Chem. Ber.*, **93**, 1961 (1960).
[128] W. Hückel and K.-D. Thomas, *Ann.*, **645**, 177 (1961).
[129] C. W. Shoppee, *J. Chem. Soc.*, **1946**, 1138, and references there cited.
[130] R. J. Bridgewater and C. W. Shoppee, *J. Chem. Soc.*, **1953**, 1709.
[131] R. J. Cremlyn and C. W. Shoppee, *J. Chem. Soc.*, **1954**, 3794.
[132] E. L. Eliel and R. G. Haber, *J. Org. Chem.*, **24**, 143 (1959).

cases, are contaminated with vicinal dihalides.[130,132,133] In the case of menthol, however, phosphorus pentachloride gives rise to a mixture of menthyl chloride and neomenthyl chloride.[134] The amount of neomenthyl chloride (inversion) may be enhanced by the addition of pyridine,[134] whereas the amount of menthyl chloride may be enhanced by the addition of halides of aluminum, iron, or zinc.[134,135] Pyridine presumably promotes the formation of free chloride ions, $ROH + PCl_5 + C_5H_5N \rightarrow ROPCl_4 + C_5H_5NH^+ + Cl^-$, which then attack the $ROPCl_4$ species from the rear, whereas metal halides tie up the chloride in a complex, e.g., $ROH + PCl_5 + AlCl_3 \rightarrow ROPCl_4 + HAlCl_4$, and thus allow an $S_N i$ reaction within the complex: $ROPCl_4 \rightarrow [R^+ + OPCl_4^-] \rightarrow [R^+ + Cl^-] + POCl_3 \rightarrow RCl + POCl_3$, the species in brackets being paired ions. Such paired ions do not seem to play an appreciable part in the reaction of the previously mentioned cyclohexanols or steroidal alcohols with phosphorus penta-halides, but they do intervene in the corresponding reactions with thionyl chloride. With thionyl chloride, 5α-cholestan-7β-ol gives 7β-chloro-5α-cholestane[131] and 5α-cholestan-3β-ol and -3α-ol give the corresponding 3β- and 3α- halides with retention of configuration.[129,130] The reaction does not give pure halides in the 5β-cholestan-3-ol series, however. In general, it offers a plausible alternative synthesis of equatorial halides from equatorial alcohols where reaction of the corresponding axial alcohol with PCl_5 (which should also give equatorial alcohols) does not go well.

Hydrogen chloride is reported to react with menthol to give predominantly menthyl chloride;[134] the reaction may be improved by the addition of zinc chloride.[135] The reaction of the 4-t-butylcyclohexanols with hydrogen bromide, on the other hand, does not give pure products.[132]

d. Reactions at Ring Positions. Stereoelectronic Factors. So far, reactions have been discussed in which reactivity is affected by steric crowding in the vicinity of the reaction site. This crowding may give rise to either steric hindrance (if crowding is more serious in the transition state) or to steric assistance (if crowding is more prominent in the ground state). The rate factors caused by such effects are of the order of 2–60. In this section will be discussed rate differences caused by the more or less favorable arrangement of the bonding electrons toward the preferred geometry of the transition state. Such rate differences, due to "stereoelectronic causes," are generally much greater than those mentioned earlier, factors of several powers of ten not being uncommon.

Stereoelectronic factors have already been found to affect reaction rate in acyclic systems (Chap. 1). However, acyclic systems can usually adjust

[133] H. L. Goering and F. H. McCarron, *J. Am. Chem. Soc.*, **78**, 2270 (1956).

[134] W. Hückel and H. Pietrzok, *Ann.*, **540**, 250 (1939).

[135] J. G. Smith and G. F. Wright, *J. Org. Chem.*, **17**, 1116 (1952).

themselves to the stereoelectronic demands of the transition state, though sometimes only at the expense of 1–2 kcal./mole in energy. In contrast, cyclic systems—especially cyclopentanoid and cyclohexanoid ones—because of their relative rigidity often find it quite difficult to adjust in similar fashion. Either severe distortions, corresponding to an adverse energy of several kilocalories, may be incurred or the reaction may be forced to proceed via a transition state which is electronically unfavorable; in either event, activation energies will increase substantially.

A well-studied reaction of known stereoelectronic requirement is the bimolecular elimination (E_2 reaction) of a variety of compounds to give olefins as well as its converse, the addition reaction to olefins. Both reactions involve an *anti* pathway, as pointed out in Chap. 1. Thus elimination will proceed best when the groups or atoms to be removed are *trans* and diaxial (i.e., *anti*) but not when they are *trans* and diequatorial or when they are *cis* (i.e., *gauche* in both cases). Addition, in turn, will usually lead to the *trans* diaxial product.

A number of examples illustrating greater ease of elimination of diaxial as compared to diequatorial dihalides with iodide ion (to give olefins, bromide ion, and iodine) come from the steroid series.[82,136,137] Thus $3\alpha,4\beta$-dibromocholestane (diaxial) eliminates 91% of its bromine when treated with potassium iodide in acetone at 40°C. for 14 days, whereas the epimeric $3\beta,4\alpha$-dibromide (diequatorial) loses only 1% of its bromine under the same conditions.[82] Similar findings have been made for the 2,3-,[82] 5,6-,[137] and 11,12-[136] dihalides. Comparison of a diaxial (*trans*) and an equatorial-axial (*cis*) dibromide has been effected in the case of the 1,2-dibromocyclohexanes.[138] The *trans* isomer (which is predominantly diaxial; cf. Sec. 3-5) undergoes debromination 11.5 times as fast as the *cis* isomer (Fig. 41). This rate difference is, of course, quite small compared to that observed between *e,e* and *a,a* isomers in the steroid series, and it has been demonstrated that the reaction path of the *cis* (*a,e*) isomer is slow substitution by iodide ion to give the *trans*-1-bromo-2-iodocyclohexane, which will then rapidly eliminate IBr to give olefin (Fig. 41). Thus the observed rate factor of 11.5 is not, in fact, the ratio of the elimination rate of the *a,a* to that of the *e,a* isomer; it is the ratio of the elimination rate of the *a,a* to the substitution rate of the *e,a* isomer containing, in addition, a correction factor for the fact that the *trans*-dibromide is not, in fact, all *a,a*.* It might be noted in this connection

136 D. H. R. Barton and W. J. Rosenfelder, *J. Chem. Soc.*, **1951**, 1048.
137 D. H. R. Barton and E. Miller, *J. Am. Chem. Soc.*, **72**, 1066 (1950).
138 H. L. Goering and H. H. Espy, *J. Am. Chem. Soc.*, **77**, 5023 (1955).

* Since the mole fraction of the *a,a* species in the *trans* isomer is about 0.7, the true $k_e(a,a)/k_s(cis)$ is $11.5 \times 10/7$ or 16.4, if it is assumed that the diequatorial dibromide is inert.

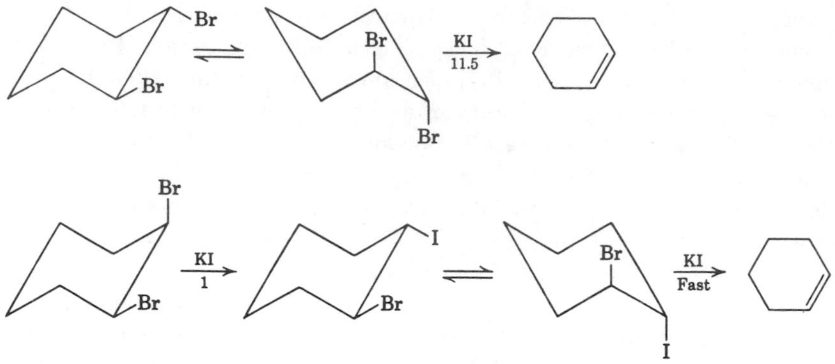

Fig. 41.

that even the 1% elimination observed in the diequatorial $3\beta,4\alpha$-dibromocholestane may not, in fact, represent true elimination but may represent a double substitution with inversion to the $3\alpha,4\beta$-diiodocholestane which would then suffer elimination rapidly. Alternatively, it may represent thermal conversion[137] of the *e,e* dibromide to the *a,a* epimer which is known to occur readily at higher temperature. Thus attempts to purify *cis*-3,*trans*-4-dibromo-*t*-butylcyclohexane (*e,e*) from contaminating *trans*-3,*cis*-4-epimer (*a,a*) by preferential destruction of the latter with iodide[132] failed because the two isomers were equilibrated rapidly under the conditions of the reaction.

A puzzling picture is presented by the iodide-induced elimination of the 2,3-dimethanesulfonates of the four epimeric (at C_2 and C_3) 22a,5α-spirostane-12-one-2,3-diols.[139] The diequatorial *trans*-dimesylate (2α,3β) and one of the equatorial-axial *cis*-dimesylates (2α,3α) undergo elimination with sodium iodide in acetone at 100°C., but the diaxial *trans*-dimesylate (2β,3α) and the other *cis*-dimesylate (2β,3β) do not. This seems in immediate contradiction to the principle of diaxial elimination, as is the observation[140] in the acyclic series that esters of α,β-dimesyloxy acids undergo apparent *cis* elimination with sodium iodide in acetone. The latter observation is now quite well understood, since it has been found[141] that *trans*-1,2-ditosyloxycyclohexane reacts with iodide first by displacement with inversion to give *cis*-1-iodo-2-tosyloxycyclohexane which then, presumably by further displacement to the *trans*-diiodide followed by elimination, gives cyclohexene and iodine. *cis*-1,2-Ditosyloxycyclohexane apparently reacts similarly by displacement to give the *trans*-1-iodo-2-tosyloxy compound which is not isolated, however, since it undergoes

[139] H. L. Slates and N. L. Wendler, *J. Am. Chem. Soc.*, **78**, 3749 (1956).

[140] R. P. Linstead, L. N. Owen, and R. F. Webb, *J. Chem. Soc.*, **1953**, 1218.

[141] S. J. Angyal and R. J. Young, *Australian J. Chem.*, **14**, 8 (1961).

rapid diaxial elimination to the olefin. Unfortunately, this does not entirely explain the results in the spirostane series; in particular, it fails to account for the facile reaction of the 2α compounds as compared to the 2β epimers, for, in the former, displacement requires approach of iodide on the hindered 2β side over the methyl group at C_{10} and it has already been indicated that such displacement is slow. This problem clearly deserves further study.

Base-catalyzed dehydrohalogenations and dehydrotosylations also proceed best when the hydrogen and the X group to be eliminated are *trans* diaxial. Thus menthyl chloride (Fig. 42) undergoes dehydrochlorination with sodium ethoxide in dry ethanol at 125° at a rate $\frac{1}{193}$ that of neomenthyl chloride.[142] In neomenthyl chloride (Fig. 42) the chlorine is axial, and there are two adjacent axial hydrogens; thus diaxial elimination can readily occur toward either side. In fact, a mixture of 3-menthene (78%) and 2-menthene (22%) results, the more stable 3-isomer predominating. In menthyl chloride the chlorine is normally equatorial and thus not prone to bimolecular elimination. Only in the alternative conformation of this molecule (in which all substituents are axial, and therefore its population is quite low) can elimination occur readily, and since, in this conformation the only adjacent axial hydrogen is at C_2, the exclusive product is 2-menthene. The over-all rate factor of 193 at 125° corresponds to a factor of 42.5 for the formation of 2-menthene alone; if this were to be attributed exclusively to the low population N_a of the axial conformation in menthyl chloride (k_a being assumed to be the same for the formation of 2-menthene from either neomenthyl or axial menthyl

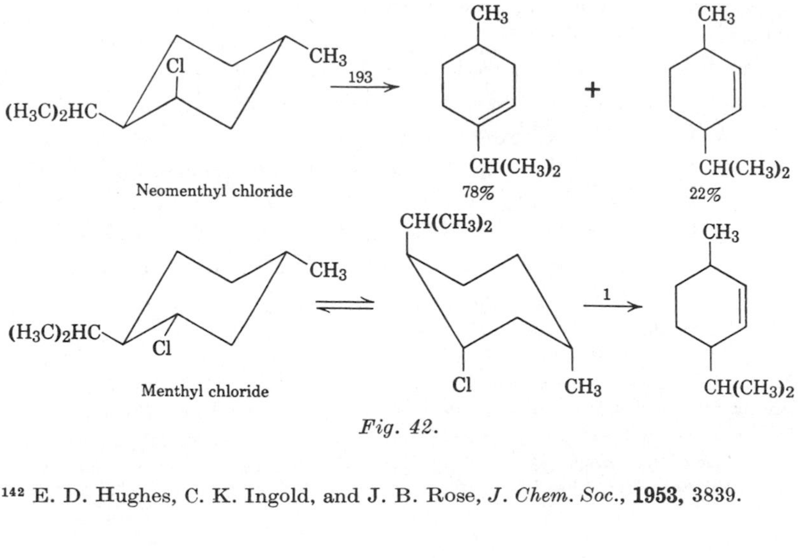

Fig. 42.

[142] E. D. Hughes, C. K. Ingold, and J. B. Rose, *J. Chem. Soc.*, **1953**, 3839.

chloride and k_e being assumed to be zero for stereoelectronic reasons), then, since $k = N_a k_a$ (see p. 48), $N_a = \frac{1}{42.5}$ corresponding to a conformational free energy difference of 3 kcal./mole. This seems somewhat small for moving three groups (Cl, Me, i-Pr, combined $-\Delta G^\circ$ values ca. 4.2 kcal./mole + syn-axial CH_3-Cl interaction) into the axial position, even though one must take into account also the unknown i-Pr-Cl *gauche* interaction in the equatorial isomer.* Other clear-cut examples demonstrating the stereoelectronic requirement of diaxial disposition of the leaving groups are *cis*- and *trans*-4-*t*-butylcyclohexyl tosylate[28] (cf. Fig. 18) and β-benzene hexachloride (Fig. 43).[143] In the 4-*t*-butylcyclohexyl tosylates the *cis* isomer (axial tosylate next to axial hydrogen) undergoes bimolecular elimination with sodium ethoxide readily, but in the *trans* isomer (equatorial tosylate) the bimolecular elimination is so slow as to be virtually undetectable in the presence of the concomitant E_1 reaction. The minimum rate factor between axial and equatorial tosylate has been estimated to be 70.[28]

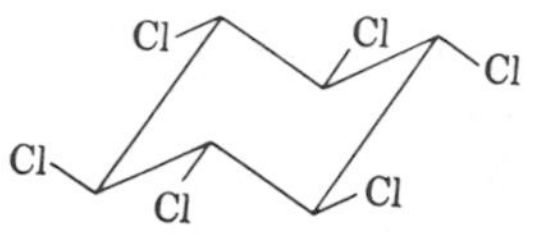

Fig. 43.

β-Benzene hexachloride (Fig. 43) is the only one of the known isomers of benzene hexachloride which undergoes bimolecular elimination with alkali in aqueous ethanol at an extremely slow rate, the activation energy of the process being 31–32 kcal./mole as compared to 18–22 kcal./mole for the other isomers. The β isomer, of course, is the only one which cannot have an axial hydrogen next to an axial chlorine in either conformation. The acid-catalyzed elimination of water from 4-alkylcyclohexanols to give olefins also proceeds easier in the *cis* isomers (hydroxyl largely axial) than in the *trans* isomers (hydroxyl equatorial).[144] The extensive formation of olefins in the treatment of axial cyclohexylamines with nitrous acid (as compared to their substantial absence in the corresponding reaction of the equatorial amines) mentioned previously may now also be explained on stereoelectronic grounds. Such an explanation rules out, once again, the formation of a common carbonium ion (which should produce the same degree of elimination for either epimeric starting amine) and suggests that there is some concerted (E_2) character to the reaction. A similar point had previously been made for the tosylate

---

[143] S. J. Cristol, *J. Am. Chem. Soc.*, **69**, 338 (1947); E. D. Hughes, C. K. Ingold, and R. Pasternak, *J. Chem. Soc.*, **1953**, 3832.

[144] G. Vavon, *Bull. Soc. Chim. France*, [4] **49**, 937 (1931).

* Inspection of models indicates that the situation is quite complex. What is normally the most stable conformation of an equatorial isopropyl group becomes unfavorable when there is an adjacent equatorial substituent. Thus the normal conformational free energy of isopropyl cannot be used in compounds such as menthol and menthyl chloride.

solvolysis (p. 85). The alternative possibility that the axial isomer reacts via a carbonium ion which gives substantial amounts of olefin whereas the equatorial isomer reacts by an $S_N i$ mechanism is not, however, ruled out. Preferred diaxial elimination from quaternary ammonium salts is documented[104] by the behavior of *cis*- and *trans*-4-*t*-butylcyclohexyltrimethyl ammonium chlorides with potassium *t*-butoxide in *t*-butyl alcohol at 75°; the *cis* (axial)* isomer suffers largely elimination of trimethylamine (90%) along with the adjacent axial hydrogen to give 4-*t*-butylcyclohexene; only 10% of the product is *N,N*-dimethyl-*cis*-4-*t*-butylcyclohexylamine formed by substitution of *t*-butoxyl on methyl (the substitution product being methyl *t*-butyl ether). In contrast, the equatorial *trans* isomer suffers no elimination at all, the exclusive product being *N,N*-dimethyl-*trans*-4-*t*-butylcyclohexylamine and (presumably) methyl *t*-butyl ether formed by substitution. Cyclohexyltrimethylammonium chloride itself has the functional group largely in the equatorial position and thus gives mainly *N,N*-dimethylcyclohexylamine by substitution (93%) and only a little trimethylamine (7%) by elimination.

Somewhat surprising results with regard to bimolecular elimination have been obtained in the case of the reaction of the 4-*t*-butylcyclohexyl tosylates or bromides with such feebly basic but relatively highly nucleophilic reagents as chloride,[145] bromide,[145] and thiophenolate.[36,41] These nucleophiles produced a much greater fraction of elimination (along with bimolecular substitution) than might have been expected on the basis of their feeble basicity. More surprisingly, substantial elimination was observed with the *trans* isomer (equatorial tosylate) as well as with the *cis* isomer. The fraction of elimination (as percentage of the total bimolecular reaction) varied between 40% and 60% with the axial tosylate and between 10% and 40% with the equatorial tosylate. The relative over-all elimination rates are reported only for the thiophenolate reaction and are $k_{el}^{cis}/k_{el}^{trans} = 40$ for the tosylate and 65 for the bromide. Winstein, Darwish, and Holness[145] were of the opinion that, in view of the stereoelectronic obstacles, this ratio was too small to be interpreted in terms of an E_2 reaction on the part of the equatorial *trans* isomer. Instead, they postulated a novel form of elimination-substitution which they called the "merged mechanism" (Fig. 44) according to which an intermediate is formed from the substrate and nucleophile in the rate-determining step; the intermediate can then partition itself between substitution and elimination products in a subsequent step not strongly

---

[145] S. Winstein, D. Darwish, and N. J. Holness, *J. Am. Chem. Soc.*, **78**, 2915 (1956).

* The compound may exist to a considerable extent in the flexible form which is not, however, well disposed toward diaxial elimination. Thus it is probably the portion in the chair form with axial NMe_3^+ which undergoes elimination.

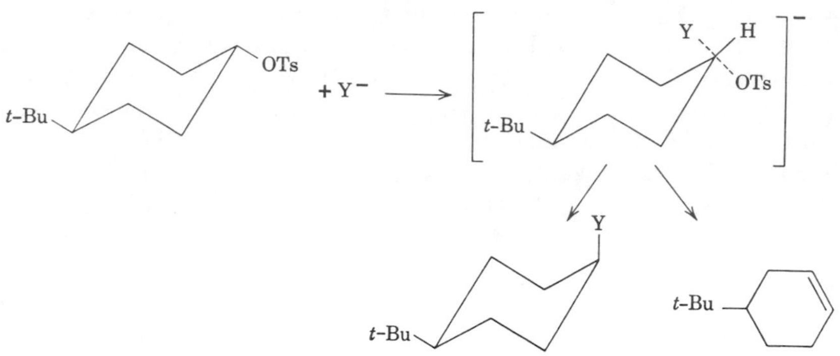

Fig. 44. Merged mechanism.

affected by stereoelectronic considerations. This mechanism has found some subsequent support,[36,146] but it has also been objected to.[147] The principal argument in the objection is that a ratio of k_{el}^{ax}/k_{el}^{eq} of 40–65 is not unreasonable for an E_2 reaction involving a nucleophile which turns out to be a remarkably good eliminating agent (thiophenoxide is about ten times as fast as ethoxide in producing elimination with *cis*-4-*t*-butyl-cyclohexyl tosylate[41]) and which may be less demanding, stereoelectron-ically, than a more basic but less effective nucleophile. The outcome of this controversy must await further experimentation.

Other instances in which the principle of diaxial elimination is broken concern 2-phenylcyclohexyltrimethylammonium salts[148] (Fig. 45, A) and 2-*p*-toluenesulfonylcyclohexyl tosylate (Fig. 45, B).[149] In both cases, reaction of the *trans* isomer with base gives directly the 1-substituted

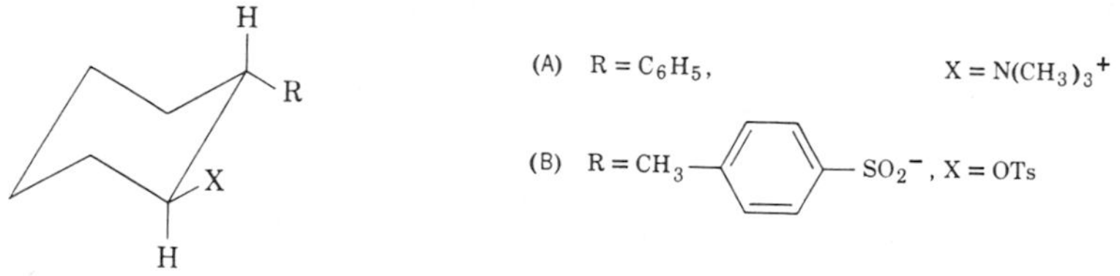

(A) $R = C_6H_5$, $X = N(CH_3)_3{}^+$

(B) $R = CH_3$—⟨benzene⟩—$SO_2{}^-$, $X = OTs$

Fig. 45.

[146] E. L. Eliel and R. S. Ro, *Tetrahedron*, **2**, 353 (1958).

[147] J. F. Bunnett, *Angew. Chem. Intern. Ed. Engl.*, **1**, 225 (1962).

[148] R. T. Arnold and P. N. Richardson, *J. Am. Chem. Soc.*, **76**, 3649 (1954);. J. Weinstock and F. G. Bordwell, *ibid.*, **77**, 6706 (1955); S. J. Cristol and F. R. Stermitz, *ibid.*, **82**, 4692 (1960); A. C. Cope, G. A. Berchtold, and D. L. Ross, *ibid.* **83**, 3859 (1961).

[149] (a) J. Weinstock, R. G. Pearson, and F. G. Bordwell, *J. Am. Chem. Soc.*, **76**, 4748 (1954); (b) J. Hine and O. B. Ramsay, *ibid.*, **84**, 973 (1962).

cyclohexene by *cis* elimination rather than the 3-substituted cyclohexene by *trans* elimination. This may be due to a combination of two factors: one is the large conformational energy of the trimethylammonio and *p*-toluenesulfonyl groups, which reduces the contribution of the diaxial conformation of the molecules shown in Fig. 45 to a very small fraction (thus discouraging the normal diaxial elimination into the 3-position*); and the other is the high acidity of the proton at C_1 which will encourage attack by the base at this position, compensating for the unfavorable stereoelectronic situation. In fact, it has been suggested[149b] that compound B, Fig. 45, may react via a carbanion intermediate formed by abstraction of the proton next to the *p*-toluenesulfonyl group (R). Such a carbanion would not be expected to be subject to stereoelectronic constraints in the further loss of the tosylate group (X) to form olefin.

Addition reactions to double bonds, whether electrophilic, free radical, or nucleophilic also proceed, in most cases, by a diaxial pathway. Thus (electrophilic) addition of bromine to cholest-2-ene gives 88–91% of the diaxial $2\beta,3\alpha$-dibromide.[82] Addition to 3-cholestene is even more stereoselective in giving 97% of the diaxial $3\alpha,4\beta$-dibromide.[82] Addition of bromide to 4-*t*-butylcyclohexene gives largely the diaxial *trans*-3,*cis*-4-dibromo-4-*t*-butylcyclohexane,[150] bromination of 3-methyl-2-cholestene gives the $2\beta,3\alpha$-dibromo-3β-methylcholestane,[151] and addition of chlorine to cholest-2-ene gives 72% of the diaxial $2\beta,3\alpha$-dichloride along with 28% of the diequatorial $2\alpha,3\beta$-dichloride.[82]. It is not clear whether the lack of complete stereoselectivity in some of these cases is inherent in the reaction, or whether it reflects a partial equilibration (cf. p. 94) of the initially formed diaxial dihalide to its diequatorial epimer, which proceeds via a transition state bearing resemblance to that of the electrophilic addition itself.[152] Hypohalous acids (HOCl and HOBr) also add predominantly diaxially.[154] In the addition of reagents of the type HX to cyclohexene the stereoelectronic requirements appear to be less stringent. Hydrogen bromide adds diaxially to 1,2-dimethylcyclohexene to give 1-bromo-*trans*-1,2-dimethylcyclohexane[153a] (equatorial methyls, axial H

---

[150] E. L. Eliel and R. G. Haber, *J. Org. Chem.*, **24**, 143 (1959).

[151] D. H. R. Barton, A. da S. Campos-Neves, and R. C. Cookson, *J. Chem. Soc.*, **1956**, 3500.

[152] C. A. Grob and S. Winstein, *Helv. Chim. Acta*, **35**, 782 (1952).

[153] (a) G. S. Hammond and T. D. Nevitt, *J. Am. Chem. Soc.*, **76**, 4121 (1954).

* It is of interest that *trans*-2-phenylcyclohexyl tosylate on treatment with base gives 3- and 1-phenylcyclohexene in a ratio of 2.65:1, whereas from the quaternary ammonium salt (Fig. 45, A) the ratio is 0.03:1. This variation may reflect the greater tendency of the tosylate to change to the diaxial conformation (OTs being much smaller than NMe_3^+) as well as the greater facility of carbanion formation in the ammonium salt (ref. 147).

and Br), but the nitric acid-catalyzed addition of water to the same olefin gives a mixture of the two epimeric 1,2-dimethylcyclohexanols.[153b] Finally, the addition of DCl to 3-methylcholest-2-ene is reported[151,154] to proceed diequatorially to give 3α-methyl-3β-chlorocholestane-2α-d. The configuration of the product was assigned on the basis of infrared spectroscopy (equatorial C—D and C—Cl stretching frequencies) and, as far as the halogen is concerned, on the basis of the fact that dehydrohalogenation with collidine gave 3-methylenecholestane (*anti* elimination) rather than the 3-methylcholest-2-ene to be expected from diaxial Saytzeff-type elimination of the (axial) 3α-chloride.

Another electrophilic addition reaction whose conformational course has been elucidated is the Prins reaction.* Addition of formaldehyde in the presence of acid to Δ²-*trans*-octalin gives 2β-hydroxymethyl-3α-hydroxy-*trans*-decalin (Fig. 46).[155]

Diaxial radical addition has been demonstrated in the reaction of 2-chloro-4-*t*-butylcyclohexene with hydrogen bromide in the presence of peroxide and with thiophenol (Fig. 47).[156] Diaxial anionic (nucleophilic) addition presumably occurs in the addition of *p*-toluenethiol to 1-*p*-toluenesulfonylcyclohexene in the presence of base (Fig. 47).[157] Because the product is the less stable of the two possible geometric isomers, the observed result is one of kinetic, rather than thermodynamic, control; however, diequatorial addition is not excluded by the facts observed.

The formation and opening of epoxides follow the same stereoelectronic principles as olefin-forming eliminations and olefin additions. Epoxides are formed readily from *trans* diaxial halohydrins but only very

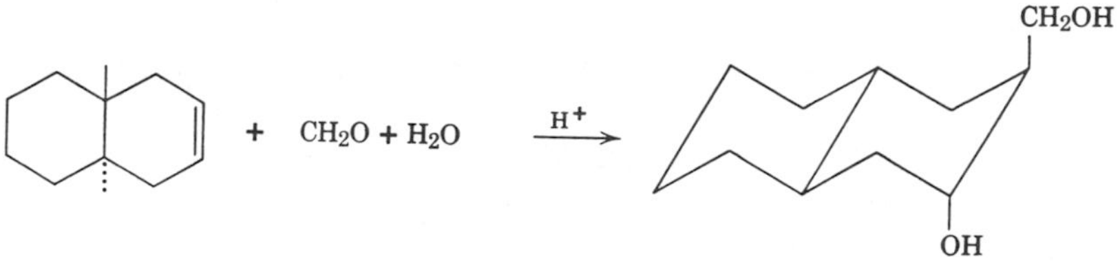

Fig. 46.

[153] (b) C. H. Collins and G. S. Hammond, *J. Org. Chem.*, **25**, 911 (1960).

[154] D. H. R. Barton, *Experientia Suppl.*, **II**, 121 (1955).

[155] E. E. Smissman and D. T. Witiak, *J. Org. Chem.*, **25**, 471 (1960).

[156] N. A. LeBel, Wayne State University, private communication.

[157] W. E. Truce and A. J. Levy, *J. Am. Chem. Soc.*, **83**, 4641 (1961).

* A number of electrophilic addition reactions which give *trans* products starting with cyclohexene but where the conformational course (diaxial as distinct from diequatorial addition) is not known are excluded from consideration here. See ref. 86, p. 355.

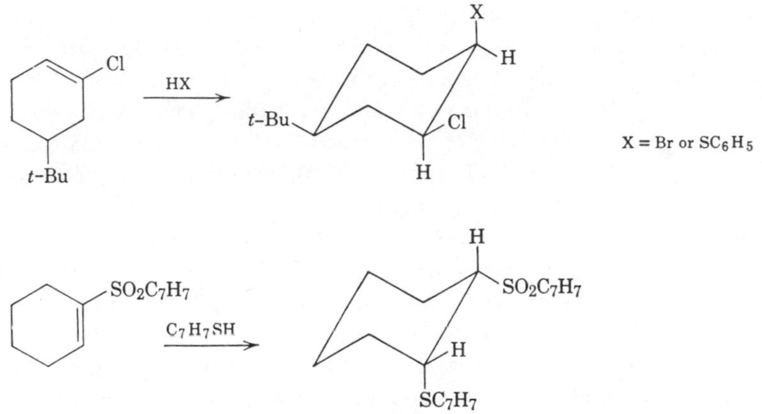

X = Br or SC$_6$H$_5$

Fig. 47.

slowly from *trans* diequatorial halohydrins; some pertinent data[154] are shown in Fig. 48. Under standard conditions (dilute sodium hydroxide in aqueous dioxane) ring closure of the diaxial bromohydrins is over 70% complete in less than a minute, whereas the diequatorial bromohydrin requires 76 hours for the same extent of reaction.[157a] The fact that the diequatorial bromohydrin forms an epoxide at all (albeit very slowly), whereas equatorial-axial *cis* halohydrins instead give rise to ketones or ring-contracted products[158] (see below) suggests that epoxide formation from diequatorial halohydrins may proceed via a flexible or boat form in which the halogen and hydroxyl are diaxial, or nearly diaxial, at least in the transition state.

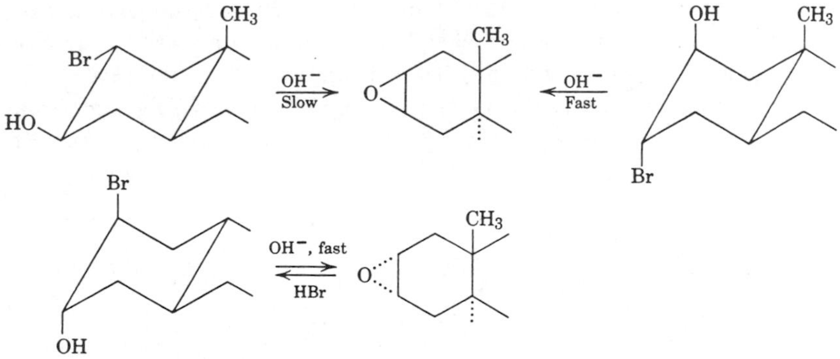

Fig. 48.

[157a] See also N. A. LeBel and R. F. Czaja, *J. Org. Chem.*, **26**, 4768 (1961).
[158] D. Y. Curtin and R. J. Harder, *J. Am. Chem. Soc.*, **82**, 2357 (1960).

Diaxial opening of epoxides is illustrated by the reaction of $2\alpha,3\alpha$-epoxycholestane by hydrogen bromide to give 2β-bromo-3α-hydroxycholestane (Fig. 48) by axial attack of bromide at C_2 (with inversion) on the protonated epoxide. A number of examples of this type have been pointed out by Fürst and Plattner.[159] Exceptions do occur occasionally; thus reduction of $5\beta,6\beta$-epoxycholestane with lithium aluminum hydride gives predominantly the 5β-ol by equatorial attack of hydride at C_6 and only a minor amount of the 6β-ol by axial attack at C_5.[160] Here the stereoelectronic conformational factors are offset by the tendency of hydride to attack the secondary rather than the tertiary carbon. Another exception occurs in the addition of hydrogen bromide to $2\beta,3\beta$-epoxylanost-8-ene to give 2α-bromo-lanost-8-ene-3β-ol.[161] This has been rationalized on the basis of the assumption that the epoxide is not in the usual distorted chair form (see Sec. 2-6b) but in a distorted boat because of the potential interaction of the epoxide function with the axial methyl groups at C_4 and C_{10} in the chair form.

In contrast to the addition reactions involving ionic or radical intermediates and the corresponding elimination reactions which generally seem to prefer *anti* coplanar geometry, there is another set of addition and elimination reactions in which groups simultaneously add to or depart from the same side of the molecule. Such reactions are called molecular addition (or elimination) reactions, and they generally prefer *cis* (equatorial-axial) geometry of the affected groups. Among the pertinent elimination reactions are the pyrolysis of acetates to give olefins and acetic acid, the pyrolysis of methyl xanthates ($ROCSSCH_3$) to give olefins, carbon oxysulfide, and methyl mercaptan, and the pyrolysis of alkyldimethylamine oxides to give olefins and N,N-dimethylhydroxylamine. The transition states for the three types are shown in Fig. 49. The reason for the preferred planar or near-planar geometry of the transition state in the case of the amine oxide is the five-membered ring, whereas in the case of the acetates and xanthates it is the benzene-like six-membered ring containing several incipient double bonds. It was mentioned previously (cf. Fig. 33)

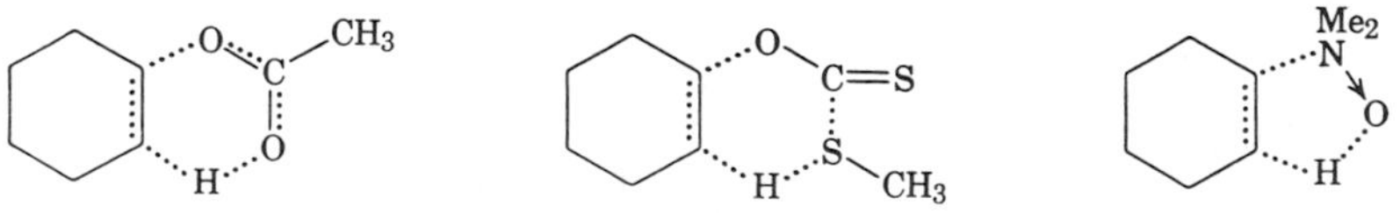

Fig. 49.

[159] A. Fürst and P. A. Plattner, Abstr. Papers 12th Intern. Congress Pure and Appl. Chem., New York, 1951, p. 409.

[160] A. S. Hallsworth and H. B. Henbest, *J. Chem. Soc.*, **1957**, 4604.

[161] D. H. R. Barton, D. A. Lewis, and J. F. McGhie, *J. Chem. Soc.*, **1957**, 2907.

that, although e,e and e,a bonds are equidistant in a perfect chair, the e,a (*cis*) bonds can bend into a plane much more readily than the e,e (*trans*) bonds. Examples are provided by the 2-phenylcyclohexyl acetates, xanthates, and dimethylamine oxides (Table 11). The data indicate that in the *trans* isomer, where *cis* elimination may occur either toward the phenyl side or away from it, the former mode leading to the conjugated. olefin 1-phenylcyclohexene is always much preferred. However, in the

Table 11. Stereochemistry of Elimination in the 2-Phenylcyclohexyl System[a]

	Stereochemistry	Products, %	
Substituent at 1		1-Phenyl-cyclohexene	3-Phenyl-cyclohexene
Acetate	cis	7	93
	trans	86.5	13.5
Xanthate[b]	cis	0–4	96–100
	trans	88	12
Dimethylamine oxide	cis	9	91
	trans	85	15

[a] Data cited in E. L. Eliel, J. W. McCoy, and C. C. Price, *J. Org. Chem.* **22,** 1533 (1957).

[b] Cf. H. R. Nace, *Org. Reactions*, **12,** 57 (1962).

cis isomer (phenyl and eliminating substituent *cis*) the only *cis* hydrogen is that away from the phenyl at C_3 and, accordingly, the major product by far is 3-phenylcyclohexene. An exception to *cis* elimination in the Chugaev reaction occurs in *trans*-2-*p*-toluenesulfonylcyclohexyl xanthate (Fig. 45 with X = OCSSCH$_3$ and R = C$_7$H$_7$SO$_2$) which gives 1-*p*-toluenesulfonyl-cyclohexene by *trans* elimination rather than the 3-isomer (by *cis* elimination), even though the 3-isomer does not rearrange to the 1-isomer under the conditions of the reaction.[162] It appears that in this case (as in the similar case of the base-induced elimination discussed earlier) the high acidity of the hydrogen next to the toluenesulfonyl group makes it so prone to elimination as to overcome the adverse stereoelectronic situation. This explanation is supported by the finding[162] that *trans*-2-*p*-tolylthio-cyclohexyl xanthate (sulfide instead of sulfone function) gives 3-tolylthio-cyclohexene by the normal *cis* elimination, the sulfide function not imparting any unusual acidity to the adjacent proton.

[162] F. G. Bordwell and P. S. Landis, *J. Am. Chem. Soc.*, **80,** 2450 (1958).

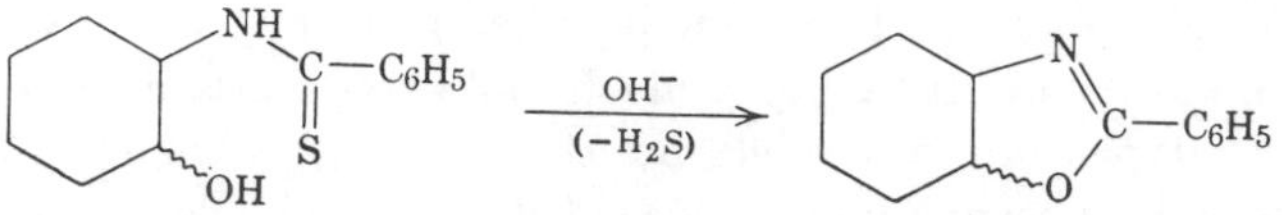

Fig. 50.

A number of addition reactions which are molecular in nature and involve *cis* geometry are also known. Catalytic hydrogenation of 1,2-dimethylcyclohexene over platinum gives mainly (up to 95%) *cis*-1,2-dimethylcyclohexane.[163a] Stereoselectivity is highest at high hydrogen pressure. With palladium catalysts, predominant *cis* hydrogenation does not occur.[163b]

Among the numerous occurrences of *cis* hydroxylation that of 5α, 22a-spirost-2-ene with osmium tetroxide to give the 2α,3α-diol[164] is of particular interest, for in addition to forming a *cis* diol this oxidation (and the related oxidation with potassium permanganate) produce the diol on the less hindered α side,* because they involve five-membered cyclic intermediates (osmate or manganate esters). Other steroids behave similarly.[165]

Any reaction involving formation of a five-membered ring fused to a six-membered ring—in the product, in an intermediate, or in a transition state—might be expected to prefer *cis* (*e,a*) geometry (see also p. 77); the amine oxide pyrolysis discussed above is a case in point. A somewhat different type of reaction which has been studied quantitatively is the cyclization of the 2-*N*-thiobenzamidocyclohexanols to the corresponding oxazolidines (Fig. 50).[166] The *cis* isomer reacts over three times as fast as the *trans* isomer. Since this rate factor is small, it is generally not important from the preparative point of view; thus both *cis*- and *trans*-2-benzamido-cyclohexanols form *N*-benzoyloxazolidines on treatment with benzaldehyde,[167] and both undergo N to O migration of the benzoyl group on treatment with hydrochloric acid, although the *cis* isomer reacts faster.[168] Both the resulting 2-aminocyclohexyl benzoates rearrange back to their

[163] (a) S. Siegel and G. V. Smith, *J. Am. Chem. Soc.*, **82**, 6082 (1960); (b) 6087 (1960).

[164] C. Djerassi, L. B. High, T. T. Grossnickle, R. Ehrlich, J. A. Morre, and R. B. Scott, *Chem. Ind. (London)*, **1955**, 474.

[165] For example, S. J. Angyal and R. J. Young, *J. Am. Chem. Soc.*, **81**, 5251 (1959).

[166] J. Sicher, J. Jonáš, M. Svoboda, and O. Knessl, *Collection Czech. Chem. Commun.*, **23**, 2141 (1958).

[167] G. E. McCasland and E. C. Horswill, *J. Am. Chem. Soc.*, **73**, 3923, (1951).

[168] G. Fodor and J. Kiss, *Acta Chim. Acad. Sci. Hung.*, **1**, 130 (1951).

* The β face of the steroid is screened by the angular methyl groups. For an ingenious indirect method of synthesizing the 2β,3β-*cis*-diol, cf. ref. 164.

N-benzoyl precursors when treated with base.[169] A difference in ease of e,e and e,a ring closure is also found in the reaction of the 1,2-cyclohexanediols with acetone: the *cis* isomer forms a cyclic ketal under certain conditions when the *trans* isomer does not.[170]

Formation of a cyclic intermediate[171] may also account for the faster rate of oxidation of *cis*- as compared to *trans*-1,2-cyclohexanediol with lead tetraacetate[172] and with periodate.[173]

Another large group of reactions of definite conformational stereoelectronic requirements are those involving rearrangements or neighboring group participation or both. Rearrangements may be of the Nametkin type (1,2-shift of substituent on the ring) or of the Wagner-Meerwein type (contraction from a six-membered to a five-membered ring) and may or may not be attended by neighboring group participation in the transition state resulting in rate enhancement (anchimeric assistance).* In turn, neighboring group participation may occur even when no rearrangement results. In all cases the conformational requirement is that the rearranging or participating groups be *anti*. In a Nametkin rearrangement this requires that the leaving and migrating groups be 1,2-diaxial. In ring contraction, on the other hand, the leaving group must be equatorial because the migrating entity is now one of the ring carbon-carbon bonds and model considerations make clear that these bonds are *anti* with respect to an equatorial substituent. Finally, the conformational requirements for anchimeric assistance are the same as those for rearrangement.

An example of both migration and ring contraction is afforded by the 2-aminocyclohexanols (Fig. 51).[174] The *trans* (e,e) isomer is disposed only toward ring contraction and yields exclusively cyclopentanecarboxaldehyde upon nitrous acid deamination. The *cis* (e,a) isomer, in contrast, may be disposed either to hydrogen migration (when the amino group is axial) or to ring contraction (when the amino group is equatorial); in fact, a mixture of the two predicted products, cyclopentanecarboxaldehyde and

[169] G. Fodor and J. Kiss, *J. Am. Chem. Soc.*, **72**, 3495 (1950).

[170] J. Boëseken and H. G. Derx, *Rec. trav. Chim.*, **40**, 519 (1921); see also W. R. Christian, C. J. Gogek, and C. B. Purves, *Can. J. Chem.*, **29**, 911 (1951); S. J. Angyal and C. G. Macdonald, *J. Chem. Soc.*, **1952**, 686; S. W. Fenton and L. L. Salcedo, Abstr. 130th Meeting, *Am. Chem. Soc.*, 7–O (1956); S. J. Angyal and R. M. Hoskinson, *J. Chem. Soc.*, **1962**, 2985.

[171] Cf. R. Criegee, E. Hoerger, G. Huber, P. Kruck, F. Marktscheffel, and H. Schellenberger, *Ann.*, **599**, 81 (1956); V. C. Bulgrin and G. Dahlgren, *J. Am. Chem. Soc.*, **80**, 3883 (1958).

[172] R. Criegee, L. Kraft, and B. Rank, *Ann.*, **507**, 159 (1933).

[173] C. C. Price and M. Knell, *J. Am. Chem. Soc.*, **64**, 552 (1942).

[174] G. E. McCasland, *J. Am. Chem. Soc.*, **73**, 2293 (1951).

* For terminology, see ref. 86, p. 146.

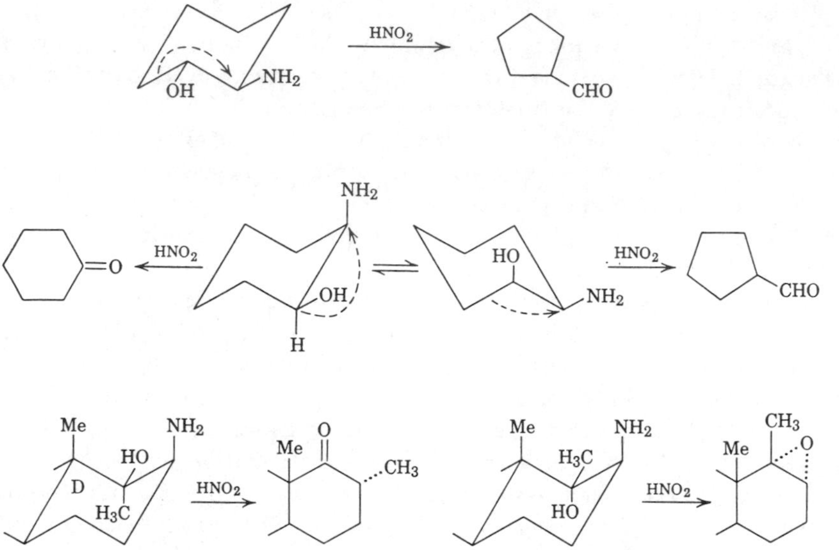

Fig. 51. Rearrangement of 2-aminocyclohexanols.

cyclohexanone, is obtained. It might be noted that only the diequatorial *trans*-2-aminoalcohol is disposed to ring contraction; its diaxial conformer would not be conformationally disposed to either ring contraction or hydrogen shift but would give an epoxide. An example from the D-homosteroid series is shown in Fig. 51: the 17aα-methyl-17aβ-hydroxy-17β-amine undergoes the expected methyl shift, whereas the 17a-epimer gives rise to the 17,17a-epoxide.[175] Examples of ring contractions are also found in a variety of steroid alcohols.[154]

Another elegant example of the effect of conformation on rearrangement behavior is afforded by the 2-bromo-4-phenylcyclohexanols (Fig. 52).[158] The *trans* isomers, whether *e,e* or *a,a*, afford epoxides when treated with base, although the *e,e* isomer undergoes ring contraction when treated with silver oxide. The *cis* isomers, on the other hand, suffer hydrogen shift* to 4-phenylcyclohexanone. For one of these isomers this entails reaction in an unfavorable conformation; the corresponding tertiary

---

[175] Cf. R. J. W. Cremlyn, D. L. Garmaise, and C. W. Shoppee, *J. Chem. Soc.*, **1953**, 1847. Surprisingly, only one of the 17α-amino compounds (equatorial NH$_2$) undergoes the expected ring contraction; the other gives epoxide by a diequatorial *trans* ring closure, in contrast to the monocyclic analog.

* The alternative of hydrogen bromide elimination to an enol followed by ketonization mentioned in ref. 158 appears unlikely on the basis of other results in the literature; cf. K. Mislow and M. Siegel, *J. Am. Chem. Soc.*, **74**, 1060 (1952); C. J. Collins, W. T. Rainey, W. B. Smith, and I. A. Kaye, *ibid.*, **81**, 460 (1959); J. B. Ley and C. A. Vernon, *J. Chem. Soc.*, **1957**, 2987.

methyl carbinol instead reacts in the more favorable conformation to give ring contraction.

Neighboring group participation has been called on to account for the large difference in acetolysis rate between neomenthyl tosylate (170) and menthyl tosylate (1);[176] as indicated in Fig. 42, only the neomenthyl compound has an axial tertiary hydrogen located in a position favorable for participation in the departure of the axial tosylate group. Surprisingly, no participation seems to occur in the acetolysis of the 2-*t*-butylcyclohexyl tosylates,[177] the rate factor of 2.1 at 50° being about what might be expected on the basis of the conformational difference between equatorial and axial tosylate (cf. p. 84). Anchimeric assistance seems to account for

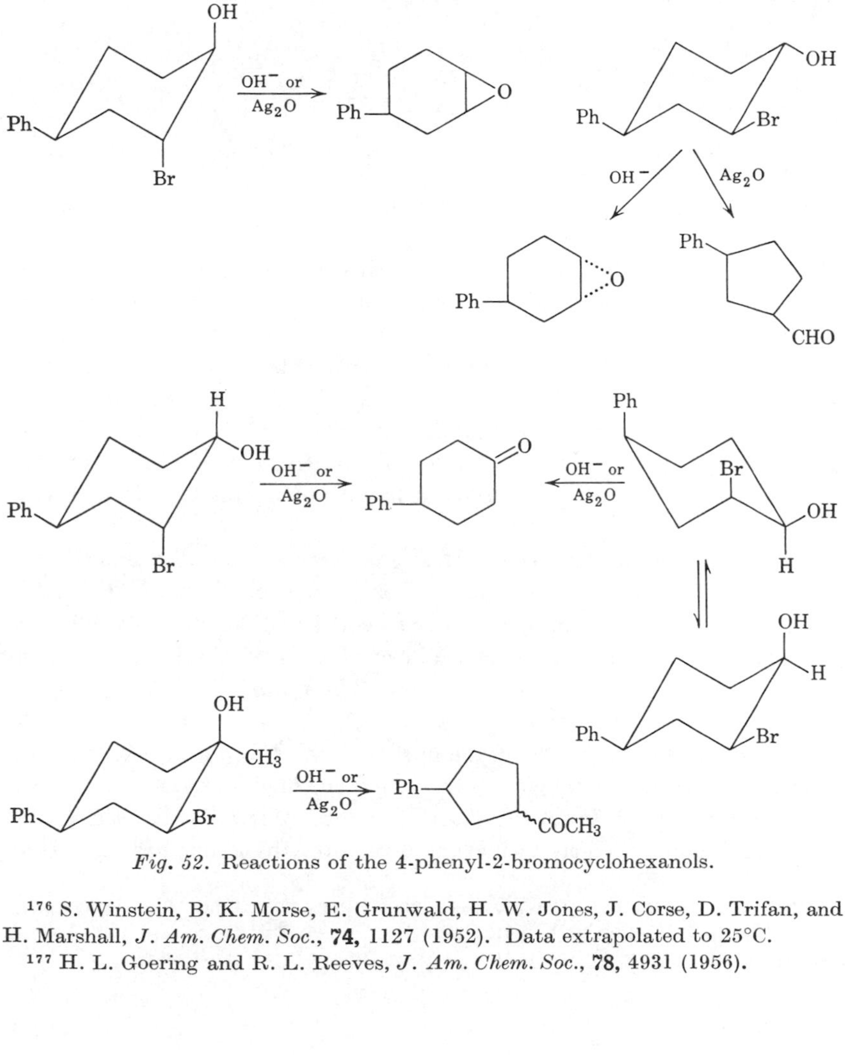

Fig. 52. Reactions of the 4-phenyl-2-bromocyclohexanols.

[176] S. Winstein, B. K. Morse, E. Grunwald, H. W. Jones, J. Corse, D. Trifan, and H. Marshall, *J. Am. Chem. Soc.*, **74**, 1127 (1952). Data extrapolated to 25°C.
[177] H. L. Goering and R. L. Reeves, *J. Am. Chem. Soc.*, **78**, 4931 (1956).

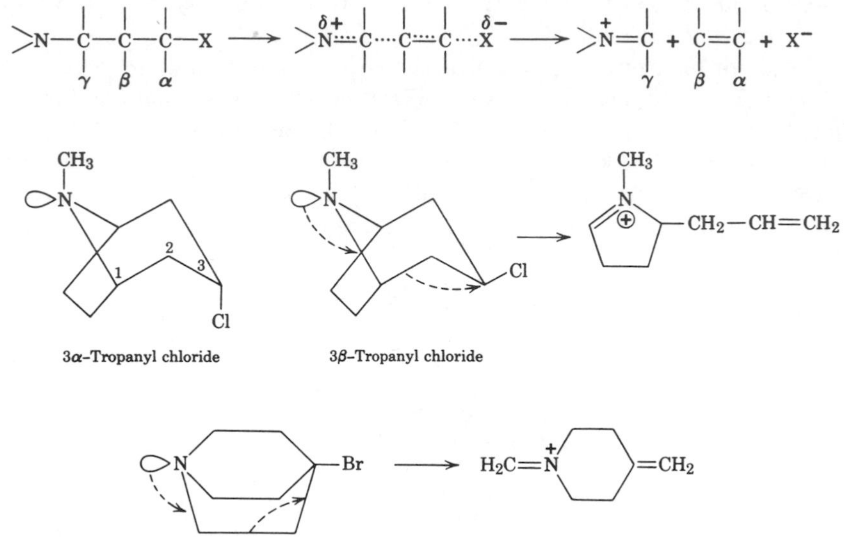

$$>N-C-C-C-X \longrightarrow >N\overset{\delta+}{\cdots}C\cdots C\cdots\overset{\delta-}{X} \longrightarrow >\overset{+}{N}=C + C=C + X^-$$

3α–Tropanyl chloride   3β–Tropanyl chloride

Fig. 53. Fragmentation.

the difference in behavior toward hydrogen bromide-acetic acid and toward thionyl chloride of 2β-bromocholestan-3α-ol (*trans*, diaxial) and 2α-bromocholestan-3β-ol (*trans*, diequatorial).[82] The former is converted, in good yield, to a diaxial dihalide (i.e., with retention of configuration): the 2β,3α-dibromide with HBr and a mixture of the 2β-bromo-3α-chloro and 2β-chloro-3α-bromo compounds with SOCl₂ (the fact that both isomers result supports the hypothesis of participation). The latter gives only an acetate with HBr-AcOH and is recovered unchanged after treatment with SOCl₂.

Another reaction of well-defined stereoelectronic requirements is fragmentation.[178] The general scheme of this reaction, shown in Fig. 53,[178] suggests that the reaction may occur in concerted fashion only when the β,γ bond to be broken is *anti* with respect to both the C—X bond and the orbital of the lone pair on nitrogen. The required geometry is found for the 2,3 bond in 4-bromoquinuclidine and for the 1,2 bond in 3β-tropanyl chloride, which do fragment readily. In contrast, 3α-tropanyl chloride, which lacks the required geometry, does not fragment. *cis*-3-Dimethylaminocyclohexyl tosylate fragments, albeit slowly; the stereoelectronically required conformation of the dimethylamino group in which

[178] C. A. Grob, *Bull. Soc. Chim. France*, **1960,** 1360; *Theoretical Organic Chemistry* (Kekulé Symposium), Butterworth and Co., Ltd., London, 1959, pp. 114–126; see also R. B. Clayton, H. B. Henbest, and M. Smith, *J. Chem. Soc.*, **1957,** 1982.

the lone pair is *anti* to the C_2—C_3 bond is not the preferred ground state conformation of the molecule, as may be seen readily in Dreiding models.

2–6. Cyclohexane Systems Containing Trigonal Ring Carbons

a. Cyclohexene.   The conformation of cyclohexene is that of a flattened chair or half-chair as shown in Fig. 54,[179] since carbons 1, 2, 3, and 6 are in a plane. A model which is only slightly distorted[180] can be constructed for this conformation as well as for the corresponding half-boat, but, because of Pitzer strain, the half-boat is less stable than the half-chair, the difference in energy, as computed by thermodynamic calculations,[181] being 2.7 kcal./mole. The half-chair has been shown, by X-ray crystallographic measurements, to be the conformation of the B ring in crystalline cholesteryl iodide[182] and in other substituted cyclohexenes.[183]

It might be noted that, in the half-chair form, cyclohexene is a dissymmetric molecule, but chair inversion readily converts one enantiomer to the other, the situation being similar to that in *cis*-1,2-dimethylcyclohexane (Sec. 2-3). The barrier to inversion is not known at the present time.

It has been pointed out[184] that, in cyclohexene, true axial (*a*) and equatorial (*e*) bonds are found only at C_4 and C_5. The bonds at C_3 and C_6 are somewhat differently disposed than in cyclohexane and are called "pseudoaxial" and "pseudoequatorial" (Fig. 54, *a'* and *e'*). To illustrate

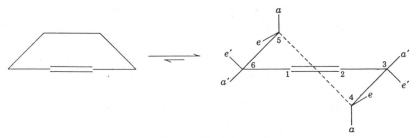

Fig. 54. Cyclohexene.

[179] J. Boëseken and J. Sturman, *Proc. Acad. Sci. Amsterdam*, **39,** 1 (1936); J. Boëseken and W. J. F. Rijck van der Gracht, *Rec. Trav. Chim.*, **56,** 1203 (1937). See also R. A. Raphael and J. B. Stenlake, *Chem. Ind. (London)*, **1953,** 1286; A. M. Mathieson, *Tetrahedron Letters*, **1963,** 81.

[180] E. J. Corey and R. A. Sneen, *J. Am. Chem. Soc.*, **77,** 2505 (1955).

[181] C. W. Beckett, N. K. Freeman, and K. S. Pitzer, *J. Am. Chem. Soc.*, **70,** 4227 (1948).

[182] C. H. Carlisle and D. Crowfoot, *Proc. Roy. Soc. (London)*, **A184,** 64 (1945).

[183] For example, R. A. Pasternak, *Acta. Cryst.*, **4,** 316 (1951).

[184] D. H. R. Barton, R. C. Cookson, W. Klyne, and C. W. Shoppee, *Chem. Ind. (London)*, **1954,** 21.

the behavior of substituents at positions 3, 4, 5, and 6 we may call on derivatives of tetralin (1,2,3,4-tetrahydronaphthalene) as well as cyclohexene, as the geometry of the two molecules is similar.[185]

The *cis* (*e',a* or *a',e*) and *trans* (*e,e'*) isomers of 3-cyclohexene-1,2-dicarboxylic acid are known:[186] the *cis* is converted to the *trans* by base (with concomitant isomerization to the $\Delta^{2,3}$ acid). A similar isomerization has been observed in the 4-cyclohexene-1,2-dicarboxylic acid series[187] where the dimethyl ester of the *cis* isomer is isomerized to the *trans* acid (*e,e*) by base. Analogous isomerizations have been observed in the tetralin-1,2-dicarboxylic acids (*cis e'a* or *a'e* to *trans e',e*)[188] in the tetralin-2,3-dicarboxylic acids (*cis e,a* to *trans e,e*)[189] and in the 2-phenyl-3-cyclohexenecarboxylic acids.[190] In none of these acids has the exact position of equilibrium been pinpointed; hence comparison with the saturated analogs (e.g., cyclohexane-1,2-dicarboxylic acid[191]) is not possible. Even were such a comparison possible, it would be complicated by the dipolar repulsions of the carboxyl groups when next to each other; data on 1,3- and 1,4-dicarboxycyclohexenes are lacking, although the two isomers of 2-cyclohexene-1,4-dicarboxylic acid have been characterized.[192] These compounds are of potential interest because the *cis* isomer is *e',a'* and the *trans* is *e',e'*. The 1-phenyl-2,3-dicarboxytetralins[189] might also be mentioned here; of the four isomers the *cis-trans* (*e',a,a* or *a',e,e*) is quite unstable, being transformed even on ordinary saponification into the *trans-trans* (*e',e,e*).

Another probe into cyclohexene geometry comes from intramolecular hydrogen bonding and glycol cleavage of suitable diols. The ratios of the rates of lead tetraacetate cleavage for the *cis* and *trans* isomers (k_{cis}/k_{trans}) are as follows: 1,2-cyclohexanediol,[172] 22.4; 1,2-tetralindiol,[172] 21.6; 2,3-tetralindiol,[193] 9.75. The differences between free and intramolecularly bonded O—H stretching in the infrared are 1,2-cyclohexanediol,[194] *cis*, 39 cm.$^{-1}$; *trans*, 32 cm.$^{-1}$; 1,2-tetralindiol,[194] *cis*, 43 cm.$^{-1}$,

[185] Cf. W. H. Mills and I. G. Nixon, *J. Chem. Soc.*, **1930**, 2510; M. A. Lasheen, *Acta Cryst.*, **5**, 593 (1952).

[186] K. Alder and M. Schumacher, *Ann.*, **564**, 96 (1949).

[187] I. N. Nazarov and V. F. Kucherov, *Izv. Akad. Nauk SSSR Otd. Khim. Nauk*, 329 (1954) [*Chem. Abstr.*, **49**, 5328i (1955)]. See also V. F. Kucherov, V. M. Andreev, and N. Y. Grigorieva, *Bull. Soc. Chim. France*, **1960**, 1406.

[188] K. Alder and K. Triebeneck, *Chem. Ber.*, **87**, 237 (1954).

[189] R. D. Haworth and F. H. Slinger, *J. Chem. Soc.*, **1940**, 1321.

[190] K. Alder, K.-H. Decker, and R. Lienau, *Ann.*, **570**, 214 (1950).

[191] W. Hückel and E. Goth, *Ber.*, **58**, 447 (1925).

[192] W. H. Mills and G. H. Keats, *J. Chem. Soc.*, **1935**, 1373. The authors assume a boat form for the *cis* isomer.

[193] M. E. Ali and L. N. Owen, *J. Chem. Soc.*, **1958**, 1066.

[194] L. P. Kuhn, *J. Am. Chem. Soc.*, **74**, 2492 (1952).

trans, 33 cm.$^{-1}$; 2,3-tetralindiol,[193] *cis*, 32 cm.$^{-1}$, *trans* 32 cm.$^{-1}$; 4-cyclo-hexene-1,2-diol,[193] *cis*, 35 cm.$^{-1}$, *trans*, 32 cm.$^{-1}$. These data lend mild support to a picture for cyclohexene in which the e',a (or a',e) bonds are somewhat closer together (or can approach each other more readily) than ordinary e,a bonds; they also suggest that the e,a bonds in the 4,5-positions of cyclohexene or the 2,3-positions of tetralin can*not* approach each other as readily as ordinary e,a bonds, implying enhanced rigidity of the buckled part of the half-chair.

In general, the conformational aspects of cyclohexene chemistry have not been explored so extensively as one might hope. It has been pointed out[195] that in a 1-methylenecyclohexane the axial bond at C_2 is better situated for orbital overlap with the pi electrons of the double bond than the corresponding equatorial bond. A number of illustrations from the steroid field have been given; they will be pointed out in Chap. 5. To a somewhat lesser extent, it is also true that a *pseudo*axial hydrogen in cyclohexene is better disposed for pi-orbital overlap than a *pseudo*-equatorial one, since the plane of the double bond does not bisect the a'-e' dihedral angle but is more nearly (though not quite) perpendicular to the a' bond.* As a result, one might expect that formation of carbanions, radicals, and carbonium ions at the allylic position may involve departure of an a' group more readily than departure of an e' group.

Among other consequences of the peculiar geometry of cyclohexene are the greater stability of $\Delta^{2,3}$ over $\Delta^{3,4}$ unsaturated steroids with A/B ring fusion *trans* (Chap. 5) and the fact that strain relief is incurred when a double bond is introduced in the 2,3-position relative to an axial group at C_1, since the axial group will be converted to a (less crowded) pseudoaxial one. The latter point has been called on[180] to explain the already mentioned fact (Fig. 42) that menthyl chloride undergoes E_2 elimination with base more rapidly than the expected population of the all-axial conformation might suggest. The argument previously given assumed that k_a, the specific rate of elimination of axial Cl, is the same in menthyl and neomenthyl chloride, but in fact it may be considerably greater in menthyl chloride because of strain relief of the axial methyl and isopropyl groups in the transition state.

b. Cyclohexene Oxide. Cyclohexene oxide has a half-chair conformation quite analogous to that of cyclohexene, as established by electron

---

[195] E. J. Corey and R. A. Sneen, *J. Am. Chem. Soc.*, **78**, 6269 (1956).

* Corey and Sneen (ref. 180) calculate ca. 67° for the angle to the a' bond and thus 53° for the angle to the e' bond. On a Dreiding model (which, in view of the slight distortion of the molecule, may not be completely representative) the angles are closer to 75° and 45°.

diffraction measurements.[196] It might be noted that only the *cis* epoxide is known, for, whereas adjacent *e,a* bonds in a cyclohexane can be fairly readily brought into a plane to make the required half-chair (cf. Fig. 33), this is virtually impossible for the *e,e* bonds. The ring opening of epoxides has already been discussed, and further examples will be given in Chap. 6. Apart from ring opening, the geometry of cyclohexene oxide does not seem to have been extensively explored. It might be noted that bicyclo-[0.1.4]heptane and bicyclo[0.2.4]octane* should be in half-chair conformation similar to that of cyclohexene oxide, but the same is no longer true of bicyclo[0.3.3]nonane (hydrindane) which can exist in *cis* and *trans* forms; in the *cis* form the six-membered ring is a slightly distorted chair rather than a true half-chair (see Chap. 4).

c. Cyclohexanone

General Remarks. Cyclohexanone is the best-studied of the cyclohexane systems containing trigonal atoms. The molecule is essentially free of angle strain,[180] but there is appreciable bond opposition strain due to eclipsing of the carbonyl oxygen with the equatorial hydrogens at C_2 and C_6. With CH_2—CH_2 bond lengths of 1.545 Å, CH_2—CO bond lengths of 1.50 Å, and a C—CO—C angle of 116°, the dihedral angle between C=O and C_2—H_e is only 4°.[197] This eclipsing probably accounts for the high reactivity of cyclohexanone in addition reactions where the hybridization of the carbonyl carbon is changed from sp^2 to sp^3 and the eclipsing is thus alleviated.[198,†] Thus the equilibrium constant for cyclohexanone cyanohydrin (ketone + HCN → cyanohydrin) is 70 times as great as that for di-*n*-octyl ketone cyanohydrin,[199] and the rate of reduction of cyclohexanone with sodium borohydride is 355 times as great as that of di-*n*-hexyl ketone.[200]

It was believed at one time that replacement of the equatorial hydrogen by methyl at C_2 in cyclohexanone would set up additional steric repulsion,

[196] B. Ottar, *Acta Chem. Scand.*, **1**, 283 (1947).

[197] W. Moffitt, R. B. Woodward, A. Moscowitz, W. Klyne, and C. Djerassi, *J. Am. Chem. Soc.*, **83**, 4013 (1961). See also ref. 205a.

[198] H. C. Brown, R. S. Fletcher, and R. B. Johannesen, *J. Am. Chem. Soc.*, **73**, 212 (1951); see also H. C. Brown, *J. Chem. Soc.*, **1956**, 1248.

[199] V. Prelog and M. Kobelt, *Helv. Chim. Acta*, **32**, 1187 (1949).

[200] H. C. Brown and K. Ichikawa, *Tetrahedron.* **1**, 221 (1957).

* The bicyclo[0.2.4]octan-3-one system, surprisingly, can be synthesized photochemically in the highly strained *trans* form: E. J. Corey, R. B. Mitra, and H. Uda, *J. Am. Chem. Soc.*, **85**, 362 (1963). See also Sec. 4-5a.

† It has been pointed out, however (W. D. Cotterill and M. J. T. Robinson, *Tetrahedron Letters*, **1963**, 1833), that the eclipsing strain in cyclohexanone may be ephemeral, inasmuch as aliphatic carbonyl compounds prefer the carbonyl-eclipsed conformation (Sec. 1-3), and that therefore other reasons for the high reactivity of cyclohexanone must be postulated.

as a result of which the difference in energy between 2-methylcyclohexanones with equatorial and axial methyl should be less than the normal difference of 1.7 kcal./mole between equatorial and axial methyl groups in cyclohexane. This supposed reduction in ΔG was called the 2-alkylketone effect and was estimated to amount to about 1 kcal./mole, reducing the normal difference between Me_{eq} and Me_{ax} by this amount.[201,202] It has been pointed out, however,[203] that the distance between the carbonyl oxygen and the adjacent equatorial methyl is so large that van der Waals repulsion is negligible, and bond opposition strain is essentially the same for hydrogen and methyl (cf. Chap. 1). Equilibrium data for the 2-methyl-4-t-butylcyclohexanones[204] and 2,6-dimethylcyclohexanones[205] do, in fact, indicate a ΔG°_{Me} of -1.56 and -1.82 kcal./mole, respectively, for 2-methylcyclohexanone with ΔH°_{Me} -1.57 and -1.95 kcal./mole. These data bracket the accepted range for the corresponding values for methylcyclohexane and thus indicate that there is, in fact, no 2-methylketone effect.* The effect does, however, come into play in 2-alkylketones with alkyl groups larger than methyl, as shown in Table 12, being about 0.7 kcal./mole for ethyl and 1.8 kcal./mole for isopropyl [2-alkylketone effect $= -\Delta G^{\circ}_{R}$ (in cyclohexane) minus $-\Delta G^{\circ}_{R}$ in 2-alkylcyclohexanone]. No 2-alkylketone effect can be assessed for the t-butyl group, since both $trans$-2,4-di-t-butylcyclohexanone[204] and $trans$-2,6-di-t-butylcyclohexanone[205] appear to exist predominantly in the flexible form.

Another effect postulated to exist in cyclohexanones is the "3-alkylketone effect."[202,206] It was pointed out that in the axial conformation

[201] P. A. Robins and J. Walker, *J. Chem. Soc.*, **1955,** 1789.

[202] W. Klyne, *Experientia*, **12,** 119 (1956).

[203] L. F. Fieser and M. Fieser, *Steroids*, Reinhold Publishing Co., New York, 1959, p. 213.

[204] N. L. Allinger and H. M. Blatter, *J. Am. Chem. Soc.*, **83,** 994 (1961).

[205] B. Rickborn, *J. Am. Chem. Soc.*, **84,** 2414 (1962). A correction of $R \ln 2$ is, of course, used to take account of the fact that the cis isomer is $meso$ whereas the $trans$ is a dl pair.

[205a] W. D. Cotterill and M. J. T. Robinson, *Tetrahedron*, **20,** 765, 777 (1964).

[206] P. A. Robins and J. Walker, *Chem. Ind. (London)*, 772 (1955).

* Klyne's argument (ref. 202) was based on an old value (80–20) for the carvomenthone-isocarvomenthone equilibrium. The corrected value (91:9) (ref. 205) is consistent with a ΔG°_{Me} of about 1.8 kcal./mole (i.e., the absence of a 2-methylketone effect) if one considers that isocarvomenthone will exist about equally in the two possible chair conformations; even that is only a very approximative assumption. More recent data (ref. 205a) suggest that ΔH for 2-methylcyclohexanone (2.16–2.18 kcal./mole) is actually $larger$ than for 2-methylcyclohexane. The $inverse$ 2-alkylketone effect is ascribed to the extra preference of the methyl-carbonyl eclipsed over the methyl-carbonyl skew conformation (cf. the conformation of propionaldehyde, p. 20). The effect is, to some extent, offset by the tendency of the axial bonds at C_2 and C_6 in a cyclohexanone to diverge (cf. p. 115).

of 3-methylcyclohexanone one of the normal 1:3 diaxial $CH_3:H$ inter-actions was missing and that, as a result, the energy difference between this conformation and the corresponding equatorial form should be reduced from the normal value of 1.7 kcal./mole to one-half this value or 0.85 kcal./mole, the difference of 0.85 kcal./mole being called the "3-alkyl-ketone effect." However, in this case also a more detailed consideration[207]

Table 12. 2-Alkylketone Effects[a,b]

Alkyl	ΔG_R° (Ketone)	ΔS_R° (Ketone)	ΔH_R° (Ketone)	2-Alkylketone Effect
CH_3	−1.56; −1.82	−0.1;[b] −0.4[b]	−1.57;[b] −1.95[b]	−0.5[c] − 0
C_2H_5	−1.09; −1.21	0;[d] −0.2[b]	−1.09; −1.28[b]	0.7 ± 0.1
i-C_3H_7	−0.59; −0.56	+0.5; +0.8[c]	−0.44; −0.32[c]	1.7
t-C_4H_9	−1.62; −1.52	−2.5; −1.7	−2.39; −2.02	—

[a] In kcal./mole. First figure in each set is derived from 2-alkyl-4-t-butyl-cyclohexanone equilibrium (ref. 204), second figure from 2,6-dialkylcyclo-hexanone equilibrium (ref. 205) corrected by entropy of mixing of dl pair.

[b] See also ref. 205a.   [c] See ref. 205a.   [d] Estimated.

of van der Waals repulsions leads to a reduction of this difference to a calculated 0.6 kcal./mole. Experimental study of the 3,5-dimethyl-cyclohexanone equilibrium (attained over palladium at 220°C.) indicates $\Delta H° = -1.36$ kcal./mole for the *trans* to *cis* epimerization; comparing this with the normal liquid phase value of ΔH_{Me}° of −1.7 kcal./mole suggests a value of about 0.4 kcal./mole for the 3-methylketone effect. Unfortunately, the equilibrium of the 3,5-dimethylcyclohexanones as established over hot palladium is not clean, and there is also some question about the intervention of boat forms at the temperature of the experiment (see below). Consideration of the menthone-isomenthone equilibrium[205] also indicates that the 3-methylketone effect is small, however, and equilibration data for 2,5-dimethylcyclohexanone[205a] support a value of 0.5 kcal./mole.

Although there is no special interaction between the carbonyl oxygen in cyclohexanone and an equatorial methyl substituent in the alpha position, there will be a corresponding interaction in 2-halocyclohexanones, for, although steric interaction may be assumed to be negligible here also, there remains a considerable repulsion of the nearly parallel dipoles. As a result, 2-bromocyclohexanones will generally be more stable in the axial form. Equilibration studies of the 2-bromo-4-t-butylcyclohexanones[208]

[207] N. L. Allinger and L. A. Freiberg, *J. Am. Chem. Soc.*, **84**, 2201 (1962).
[208] N. L. Allinger and J. Allinger, *J. Am. Chem. Soc.*, **80**, 5476 (1958).

support this prediction: the *trans* (axial) isomer predominates at equilibrium to the extent of 78:22 in carbon tetrachloride and to the extent of 63:37 in dioxane. These findings and the observed solvent effect will be discussed in more detail in Sec. 7-3. Suffice it to say at this point that the preference for the axial position is less for chlorine than for bromine—the equilibrium constant for 2-chloro-4-*t*-butylcyclohexanone[209] being 2.64 *in favor of the cis* (equatorial) isomer in dioxane, 1.75 in benzene, and ca. 1.0 in carbon tetrachloride—and still less for fluorine which, in 2-fluorocyclohexanone, appears to occupy mainly the equatorial position.[210]

The intervention of flexible forms in certain substituted cyclohexanones has been mentioned above. It has been estimated[211] that the energy difference between the chair and the flexible form in cyclohexanone is only 2.7 kcal./mole, i.e., about half as much as in cyclohexane. The barrier height to chair-chair and chair-flexible form interconversion is not known (see, however, Sec. 3-10), but it might be expected to be less than in cyclohexane. Cyclohexanones with bulky substituents, such as the 2,4- and 2,6-di-*t*-butyl derivative (see above) and *cis*-2-*t*-butyl-5-methylcyclohexanone[212] appear to exist to a considerable extent in the flexible form. In the latter case, a combination of the 1:3 diaxial Me:H interaction and the *t*-butyl-carbonyl eclipsing interaction apparently forces the molecule out of the chair conformation. Other examples will be considered in Sec. 7-4.

A 4-alkylketone effect has also been defined.[99] This effect depends on the relief of compression of an axial 4-alkyl group by axial hydrogens at C_2 and C_6 when C_1 is converted from the tetrahedral configuration to the trigonal. (This conversion leads to an outward motion of the axial hydrogens at positions 2 and 6.) The effect has been called on to explain the greater rate of chromic acid oxidation of 4,4-dimethylcyclohexanol as compared to that of cyclohexanol[99] (Table 8) and the larger dissociation constant of 4,4-dimethylcyclohexanone cyanohydrin as compared to that of cyclohexanone cyanohydrin.[213]

Reactions at the Carbonyl Function. In a rigid or biased system a reaction at the carbonyl carbon may involve equatorial attack to give an axial alcohol, or axial attack to give an equatorial alcohol. The former mode of reaction involves attack on the less hindered side, and the result

[209] N. L. Allinger, J. Allinger, L. A. Freiberg, R. F. Czaja, and N. A. LeBel, *J. Am. Chem. Soc.* **82,** 5876 (1960).

[210] A. S. Kende, *Tetrahedron Letters,* **14,** 13 (1959); N. L. Allinger and H. M. Blatter, *J. Org. Chem.,* **27,** 1523 (1962).

[211] N. L. Allinger, *J. Am. Chem. Soc.,* **81,** 5727 (1959).

[212] C. Djerassi, E. J. Warawa, J. M. Berdahl, and E. J. Eisenbraun, *J. Am. Chem. Soc.,* **83,** 3334 (1961).

[213] R. A. Benkeser and E. W. Bennett, *J. Am. Chem. Soc.,* **80,** 5415 (1958).

may be said to be due to "steric approach control."[214] The latter mode often produces the more stable product, and it is said to be governed by "product development control."[214]

Before looking at stereochemistry or at rate data, it may be well to assess the effect of steric compression on equilibria. Pertinent data[215] are shown in Table 13 which speaks largely for itself. The effect of an

Table 13. Relative Dissociation Constants of Cyanohydrins[215]

No.	Compound	R	R'	R''	R'''	K_{rel}
1	Cyclohexanone	H	H	H	H	1.0
2	3-Methylcyclohexanone	H	Me	H	H	0.93
3	cis-3,5-Dimethylcyclohexanone	H	Me	H	Me	4.4
4	3,3-Dimethylcyclohexanone	Me	Me	H	H	30
5	3,3,5-Trimethylcyclohexanone	Me	Me	H	Me	38
6	3,3,5,5-Tetramethylcyclohexanone	Me	Me	Me	Me	800

equatorial group at C_3 is small; the same is true for equatorial or axial groups at C_2 and C_4.[213,215] An axial group at C_3, however, has a large effect on the dissociation equilibrium, owing to 1:3 diaxial interaction in the cyanohydrin.

Table 14 gives the over-all rates of reduction of various steroidal ketones.[216] The retarding effect of a *syn*-axial methyl group (in cholestan-6-one) and of an adjacent equatorial substituent (in cholestan-7-one) is not very large. A larger effect is found with both adjacent equatorial and axial substituents (hecogenin); the biggest reduction in specific rate (by a factor of 800) is found in an 11-ketosteroid where both angular methyl groups interfere through *syn*-axial interactions.

[214] W. G. Dauben, G. J. Fonken, and D. S. Noyce, *J. Am. Chem. Soc.*, **78**, 2579 (1956). The terminology was originally applied only to hydride reduction of cyclohexanones, but it may be generalized. See also the extensive review of the stereochemistry of addition to cyclohexanones by A. V. Kamernitzky and A. A. Akhrem, *Tetrahedron*, **18**, 705 (1962); *Usp. Khim.*, **30**, 142 (1961). [*Russ. Chem. Rev. (Engl. Transl.)*, **1961**, 43].

[215] O. H. Wheeler and J. Z. Zabicky, *Can. J. Chem.*, **36**, 656 (1958).

[216] O. H. Wheeler and J. L. Mateos, *Can. J. Chem.*, **36**, 1049 (1958); J. L. Mateos, *J. Org. Chem.*, **24**, 2034 (1959).

The question may well be asked whether the reduced ease of reduction of 2-substituted and axially 3-substituted cyclohexanones is a consequence of steric approach control (i.e., interference with the access of the reducing agent), or of product development control (i.e., steric crowding in the developing reduction product). The following results regarding the stereochemistry of ketone reduction are pertinent: (1) When the ketone, such as 4-t-butylcyclohexanone[43,75] or cholestan-3-one,[217] is unhindered,

Table 14. Relative Rates of Reduction[216] with Sodium Borohydride in Isopropyl Alcohol at 25°

Ketone	Relative Rate
3-Cholestanone	1.00
6-Cholestanone	0.60
7-Cholestanone	0.28
Hecogenin (12-ketone)	0.106
11-Ketotigogenin	0.00126

the product of lithium aluminum hydride reduction is largely the equatorial alcohol (trans-4-t-butylcyclohexanol, 89–91%; cholestan-3β-ol, 88–91%). (2) When there is one 3-axial methyl group, the equatorial and axial alcohols are produced in roughly equal amounts; thus 3,3,5-trimethylcyclohexanone, with lithium aluminum hydride, gives 55% of trans (axial) alcohol[218,219] and cholestan-2-one gives cholestan-2β- (axial) and 2α-ol (equatorial) in a ratio of 58:42.[220] Finally, when there are two axial methyl groups, as in an 11-ketosteroid, only the crowded axial alcohol (11β-stanol) is obtained,[221] although in this product there are two 1:3 diaxial CH_3-OH interactions. It has been said, therefore, that in unhindered ketones "product development control" (giving the less hindered, i.e., equatorial, alcohol) is operative, whereas in cyclohexanones containing axial groups at C_3 and (a fortiori) at C_3 and C_5 "steric approach control" is operative so that the hydride approaches from the less encumbered equatorial side to give the axial alcohol. From the phenomenological point of view this classification is unexceptional, but its

[217] H. R. Nace and G. L. O'Connor, J. Am. Chem. Soc., 73, 5824 (1951); E. L. Eliel and J. C. Richer, J. Org. Chem., 26, 972 (1961).

[218] H. Haubenstock and E. L. Eliel, J. Am. Chem. Soc., 84, 2363 (1962).

[219] See also K. D. Hardy and R. J. Wicker, J. Am. Chem. Soc., 80, 640 (1958).

[220] W. G. Dauben, E. J. Blanz, J. Jiu, and R. A. Micheli, J. Am. Chem. Soc., 78, 3752 (1956).

[221] L. H. Sarett, M. Feurer, and K. Folkers, J. Am. Chem. Soc., 73, 1777 (1951); S. Bernstein, R. H. Lenhard, and J. W. Williams, J. Org. Chem., 18, 1168 (1953).

mechanistic implications (viz., that product stability reflects itself in the transition state in the unhindered ketones whereas interference with approach of hydride comes to the fore in hindered ones) remain to be demonstrated.*

Information on the reaction of cyclohexanones with nucleophiles other than cyanide and hydride is limited. The reaction of 4-*t*-butylcyclohexanone with sodium acetylide[91] gives mainly 1-ethynyl-*trans*-4-*t*-butylcyclohexanol in which the ethynyl group is axial and the hydroxyl group equatorial. The conformational energy of ethynyl is not known, but it is probably small, similar to that of cyanide (0.15–0.25 kcal./mole), so that the isomer formed is one of product development control. As the alkyl group becomes bigger, it might thus be expected that it would prefer the equatorial position to give the *trans*-1-alkyl-*cis*-4-*t*-butylcyclohexanol. In fact, with the methyl Grignard the ratio of *trans*-1-methyl (CH_3 equatorial) to *cis*-1-methyl (CH_3 axial) varies from 1:1 to about 2:1 depending on conditions,[222] and even with the ethyl Grignard the ratio of *trans* to *cis* is[91] only 2.7:1. The major isomer (equatorial alkyl) may be the one of steric approach control in these cases, whereas the minor product (equatorial hydroxyl) may be product development controlled. It must be remembered that although OH has a rather small $-\Delta G°$ value (Table 1), the corresponding $-\Delta G°$ value for an —OM group in ether (where M is a metal) is probably quite large, even larger than that of ethyl (cf. Fig. 20).

Metal-proton donor combinations (sodium in ethanol, lithium in ammonia) are often used to reduce ketones and, under the normal basic conditions, appear to give the more stable equatorial alcohol in predominance.[219] Thus sodium-in-ethanol reduction of *trans*-1-decalone gives mainly *trans*-1β-decalol,[223] and reduction of cholestanone with sodium and amyl alcohol gives largely cholestan-3β-ol.[224] Oximes are similarly reduced to equatorial amines; thus *trans*-1-decalone oxime gives[225] *trans*-1β-decalylamine with sodium in ethanol, and 4-*t*-butylcyclohexanone oxime similarly gives *trans*-4-*t*-butylcyclohexylamine.[125,226]

[222] B. Cross and G. H. Whitham, *J. Chem. Soc.*, **1960**, 3892; C. H. DePuy and R. W. King, *J. Am. Chem. Soc.*, **83**, 2743 (1961); W. J. Houlihan, *J. Org. Chem.*, **27**, 3860 (1962).

[223] W. Hückel, *Ann.*, **441**, 1 (1925).

[224] O. Diels and E. Abderhalden, *Ber.*, **39**, 884 (1906).

[225] W. Hückel, R. Danneel, A. Gross, and H. Naab, *Ann.*, **502**, 99 (1933).

[226] D. V. Nightingale, J. D. Kerr, J. A. Gallagher, and M. Maienthal, *J. Org. Chem.*, **17**, 1017 (1952).

* The effect of solvent and reagent (e.g., sodium borohydride vs. lithium aluminum hydride) cannot be entered upon here. For leading references see H. Haubenstock and E. L. Eliel, *J. Am. Chem. Soc.*, **84**, 2368 (1962); W. M. Jones and H. E. Wise, *ibid.*, **84**, 997 (1962); E. L. Eliel, *Record Chem. Progr.*, **22**, 129 (1961). See also C. D. Ritchie, *Tetrahedron Letters*, **1963**, 2145.

In contrast, catalytic reduction of ketones in strongly acidic medium involves steric approach control (absorption on the catalyst and approach of hydrogen from the less hindered equatorial side) and thus yields axial alcohols. (This generalization is sometimes called the von Auwers-Skita hydrogenation rule, although it has been somewhat modified since its original enunciation by von Auwers and Skita.) Examples are the hydro-·genation of trans-1-decalone over platinum in acetic acid-hydrochloric acid to trans-1α-decalol,[227] reduction of cholestan-3-one over active platinum to give cholestan-3α- and -3β-ol in a ratio of 3:1,[228] and reduction of 4-t-butylcyclohexanone over platinum in acetic acid-hydrochloric acid to give cis- and trans-4-t-butylcyclohexanol in a ratio of 4:1.[43] Similar principles apply to oximes; thus trans-1-decalone oxime hydrogenated over platinum in acetic acid gives mainly trans-1α-decalylamine,[225] and 4-t-butylcyclohexanone oxime similarly gives cis-4-t-butylcyclohexyl-amine, though only in very low yield.[102,125]

Few if any of the reductions mentioned above are completely stereo-selective; in most instances the desired isomer is obtained in predominance and needs to be purified further if it is desired pure. Catalytic hydrogenation appears to give the greatest amount of axial alcohol when carried out over active catalysts under conditions where hydrogenation is fast; thus hydrogenation of 4-t-butylcyclohexanone is more stereoselective in the presence of hydrochloric acid (4:1) than in its absence (2:1); use of an aged and not very active platinum catalyst reduces the selectivity even more (1.2:1).[43] An exception to the hydrogenation rule is cholestan-1-one which is reduced to the (equatorial)1-β-ol over platinum in acetic acid in over 90% yield.[229]

It was believed at one time that catalytic hydrogenation of cyclohexanones over nickel catalysts (basic or neutral) gave predominantly the equatorial alcohols except in the case of strongly hindered ketone groups.[230] It appears,[231] however, that under mild conditions, with active Raney nickel, axial alcohols are obtained just as with platinum (e.g., trans-3,3,5-trimethylcyclohexanol from isophorone via 3,3,5-trimethyl-cyclohexanone) and that equatorial alcohols are produced only under more stringent conditions as a result of epimerization of the initial product to the more stable epimer. Reduction of phenols to cyclohexanols over nickel,[231] ruthenium,[43] rhodium,[76] or platinum[43] is usually of low

[227] W. Hückel, O. Neunhoeffer, A. Gercke, and E. Frank, Ann., **477**, 150 (1930).
[228] G. Vavon and B. Jakubowicz, Bull. Soc. Chim. France, **53**, 581 (1933).
[229] P. Striebel and C. Tamm, Helv. Chim. Acta, **37**, 1094 (1954).
[230] D. H. R. Barton, J. Chem. Soc., **1953**, 1027.
[231] E. G. Peppiat and R. J. Wicker, J. Chem. Soc., **1955**, 3122; R. J. Wicker, ibid., **1956**, 2165.

stereoselectivity, as is the reduction of ketones with aluminum isopropoxide.[214,219]

Reactions at the Alpha-Methylene.   It has already been stated that an axial atom in the ring position next to an exocyclic double bond (e.g., the 2-position in a cyclohexanone) is better disposed to orbital overlap of its sigma bond with the pi-electrons of the double bond than is a corresponding equatorial atom. It might therefore be expected that in the formation of an enol or enolate ion from a cyclohexanone the axial alpha-proton should depart, and in a reaction of the enol or enolate ion (such as protonation, deuteration, or bromination) the incoming group should come predominantly from the axial side. The data, on the whole, appear to support this view. Thus bromination of 3β-acetoxycholestan-7-one gives mainly 3β-acetoxy-6β-bromocholestan-7-one (axial bromine, 3–4 parts) with but a minor amount of the 6α-bromo compound (equatorial bromine, 1 part) despite the fact that the axial 6β-compound suffers from a 1:3 diaxial CH₃-Br interaction and is readily epimerized, on prolonged heating with acid, to the more stable 3β-acetoxy-6α-bromocholestan-7-one.[195] Debromination of either bromoketone with zinc and deuteroacetic acid gives predominantly 3β-acetoxy-6β-deuterocholestan-7-one (9 parts) with but a little of the 6α-deutero epimer. Thus introduction of either bromine or deuterium into the enol leads predominantly to the axially substituted ketone.   Enolization of 3β-acetoxy-6β- and -6α-deuterocholestan-7-one also indicates that, after correction for the isotope effect, abstraction of the axial proton or deuteron is preferred.[195] However, the picture is not so clean-cut in all cases, and there are instances where one must assume either equatorial attack on the chair form of the enolate or axial attack on a flexible form with over-all results contrary to those expected on the basis of the principles postulated above.[232,233] A number of such cases will be considered in Chaps. 5 and 7.

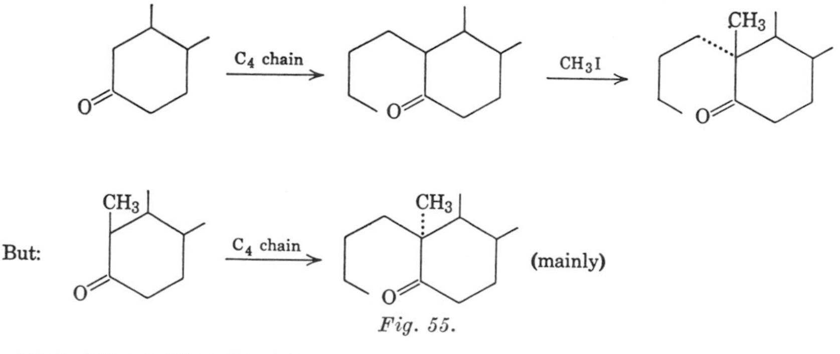

Fig. 55.

[232] R. Villotti, H. J. Ringold, and C. Djerassi, *J. Am. Chem. Soc.*, **82**, 5693 (1960).
[233] J. Valls and E. Toromanoff, *Bull. Soc. Chim. France*, **1961**, 758.

Axial substitution in enolates appears to occur also in alkylation (e.g., with methyl iodide) and in Michael addition.[234] Thus in the total synthesis of steroids, if one introduces the four carbon atoms of ring A into a precursor having, in addition to rings B, C, and D, the methyl group at C_{10}, the carbons of ring A will attach themselves predominantly axially and the methyl group at C_{10} will be forced into the unnatural equatorial position. The difficulty can be avoided by first introducing the carbons of ring A and *then* the angular methyl group; under these circumstances the angular methyl group will be introduced largely axially with respect to the future A/B *trans* ring system (Fig. 55).[233,235]

d. Cyclohexyl Carbonium Ions, Radicals, and Anions.

The geometry of cyclohexyl carbonium ions corresponds to sp^2 hybridization and thus resembles that of cyclohexanone. What little is known about the behavior of such ions has been discussed in connection with the solvolysis of cyclohexyl tosylates and the reaction of cyclohexylamines with nitrous acid.

Cyclohexyl radicals may be either sp^2 with the odd electron occupying a p-orbital or sp^3 with one half-filled orbital; in the latter case there would be rapid movement of the odd electron from the equatorial to the axial position and vice versa, the hydrogen attached to the radical carbon moving in the opposite direction. Reactions involving such radicals would be non-stereospecific, but it could not be said *a priori* whether they would or would not be stereoselective and, if so, in what direction. The Hunsdiecker reaction of *cis*- and *trans*-4-*t*-butylcyclohexanecarboxylic acids (reaction of the silver salts with bromine) to give 4-*t*-butylcyclohexyl bromide appears to involve the 4-*t*-butylcyclohexyl radical as an intermediate. The two acids give the same or nearly the same bromide mixture in which the *trans* (equatorial) isomer predominates by a ratio of $2:1$.[236] This is what one would expect if the two starting isomers reacted by way of a common intermediate radical whose further fate was subject to product development control. Similarly, the reaction of the 4-methylcyclohexylmercuric bromides with bromine in non-polar solvents appears to proceed via a radical path; either isomer gives [237] a mixture of *trans*- and *cis*-4-methylcyclohexyl bromide in a ratio of $1.1:1$. Again the decomposition of either 4-*t*-butylcyclohexanecarboxylic acid peroxide in 1,1,2,2–tetrabromoethane gave, among other products, the same

[234] Cf. W. S. Johnson, *Chem. Ind. (London)*, **1956**, 167.

[235] L. B. Barkley, W. S. Knowles, H. Raffelson, and Q. E. Thompson, *J. Am. Chem. Soc.*, **78**, 4111 (1956); see also L. Velluz, G. Nomine, and J. Mathieu, *Angew. Chem.*, **72**, 725 (1960).

[236] E. L. Eliel and R. V. Acharya, *J. Org. Chem.*, **24**, 151 (1959).

[237] F. R. Jensen and L. H. Gale, *J. Am. Chem. Soc.*, **82**, 148 (1960).

mixture of *trans*- and *cis*-4-*t*-butylcyclohexyl bromide (ratio 1.1–1.2:1).[238] On the other hand, the decomposition of either *cis*- or *trans*-4-*t*-butylcyclohexyldimethylcarbinyl hypochlorite, $(CH_3)_3CC_6H_{10}C(CH_3)_2OCl$, while giving the same mixture of 4-*t*-butylcyclohexyl chlorides (in addition to acetone) gave the *cis* isomer in predominance (2:1).[239] Finally, the Hunsdiecker reaction of either *cis*- or *trans*-cyclohexane-1,2-dicarboxylic acid, while again giving the same product, yielded exclusively *trans*-1,2-dibromo-cyclohexane.[240] Since, as implied earlier, the thermodynamic stability of the two dibromides is similar so that product development control is not responsible for the result, it appears that some special factor, such as neighboring group participation at some intermediate stage of the reaction, accounts for the observed stereochemistry.[240a]

The reaction of cholestanylmagnesium bromide (prepared from either 3α- or 3β-bromocholestane) with oxygen to give a nearly 50:50 mixture of cholestan-3α- and 3β-ol may also involve radical intermediates.[241]

Carbanions probably have sp^3 hybridization, the free pair occupying one of the orbitals. It has been suggested[150] that the free pair is more space-consuming than hydrogen, though less so than carbon, and will therefore occupy the equatorial position in C—$\overline{C}$H—C but the axial position in C—$\overline{C}$—C, as a result of which reactions of the former species
|
C
will lead to introduction of an electrophile in the equatorial position and reactions of the latter will place the electrophile in the axial position. Among the examples cited to support this view are the reaction of 3-cholestanylmagnesium bromide with carbon dioxide to give exclusively cholestane-3β-carboxylic acid[242] (introduction of the CO_2 group in the equatorial position), reduction of 5α-chlorocholestane with lithium and ammonia to 5α-cholestane[243] (axial approach of proton to the intermediate 5-cholestanyl carbanion), reduction of 4-cholesten-3-one to 5α-cholestan-3-one,[224] and reduction of 3β-chloro-3α-methylcholestane to 3β-methylcholestane[151] (axial approach of the proton to the 3-methyl-3-cholestanyl carbanion). It has also been pointed out that in all these examples the more stable product results.[243] Unfortunately, this finding is not general

[238] H. H. Lau and H. Hart, *J. Am. Chem. Soc.*, **81**, 4897 (1959).

[239] F. D. Greene, C.-C. Chu, and J. Walia, *J. Org. Chem.*, **29**, 1285 (1964).

[240] P. I. Abell, *J. Org. Chem.*, **22**, 769 (1957).

[240a] For an example, see P. S. Skell, D. L. Tuleen, and P. D. Readio, *J. Am. Chem. Soc.*, **85**, 2849 (1963).

[241] G. Roberts and C. W. Shoppee, *J. Chem. Soc.*, **1964**, 3418.

[242] G. Roberts, C. W. Shoppee, and H. J. Stephenson, *J. Chem. Soc.*, **1954**, 2705.

[243] D. H. R. Barton and C. H. Robinson, *J. Chem. Soc.*, **1954**, 3045.

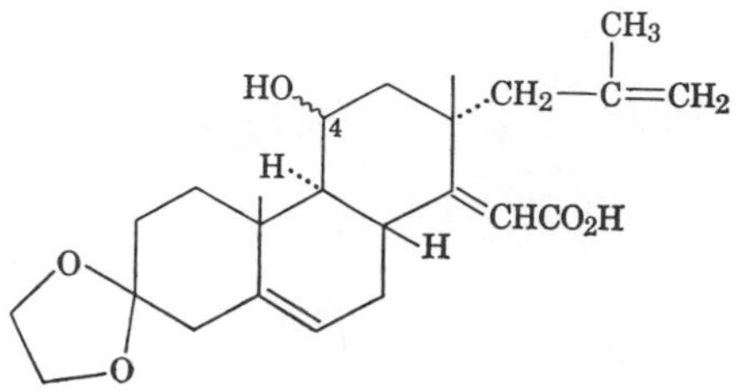

Fig. 56.

and the interpretation given above is an oversimplified one. To begin with, the stereochemical result of the metal-proton donor reduction of α,β-unsaturated carbonyl compounds is strongly dependent on conditions and system. Thus reduction of the cyclohexylideneacetic acid shown in Fig. 56 to the corresponding cyclohexylacetic acid with potassium, ammonia, and 2-propanol gave the acetic acid side chain exclusively in the β (equatorial) position when the hydroxyl at C_4 was α (axial), but when the hydroxyl was β (equatorial) the side chain would end up in the α (axial) position.[244] Again, in the same case where the C_4-OH group was α, the proportion of the acetic acid side chain in the product could be changed from all β with potassium, ammonia, and 2-propanol to 4:3 in favor of α with lithium and ammonia (either with or without 2-propanol).[244] It has furthermore been shown that in some systems the product of reduction is the less stable isomer; thus lithium-ammonia-alcohol reduction of 7-methoxy-5-methyl-$\Delta^{1,10}$-2-octalone (Fig. 57) gives only derivatives of the corresponding *trans*-decalin system.[245] Here, because of the diaxial methyl-methoxyl interaction in the *trans* decalinoid product, the *cis*-decalin derivative would be more stable, but it is not formed. It has been suggested[245] that the stereochemistry of reduction in this case is governed by the need of the electron pair of the developing carbanion to overlap with the pi-orbitals of the carbonyl group. This requires that the orbital of the carbanion be axial and not equatorial; it will lead to a *trans* ring fusion in the final product. It should be noted

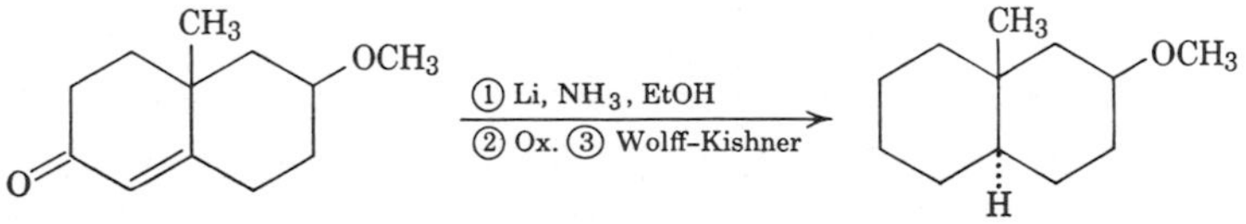

Fig. 57.

[244] G. E. Arth, G. I. Poos, R. M. Lukes, F. M. Robinson, W. F. Johns, M. Feurer, and L. H. Sarett, *J. Am. Chem. Soc.*, **76**, 1715 (1954).

[245] G. Stork and S. D. Darling, *J. Am. Chem. Soc.*, **82**, 1512 (1960).

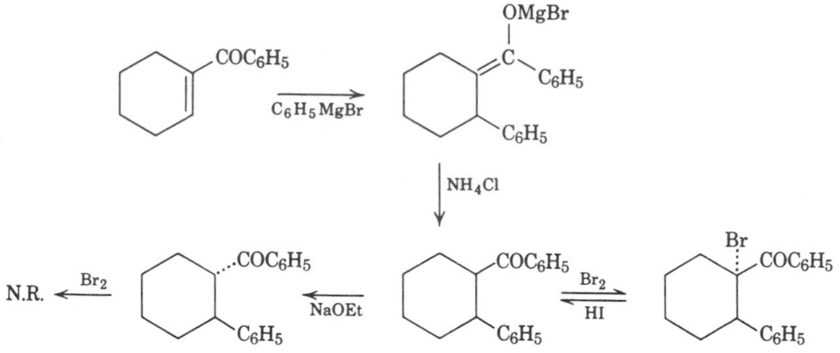

Fig. 58.

that the previous assertion that the free pair of electrons occupies the axial position is not disputed but that the reason given is one of orbital overlap rather than one of spatial requirements of the pair. Clearly, the new argument is not directly pertinent to the stereochemistry of carbanions which do not overlap with carbonyl double bonds.

A fairly extensive study has been made of carbanions in which the negatively charged carbon is part of the ring but the stabilizing group (C=O, phenyl, etc.) is exocyclic. In this case the system can readily adjust itself to the stereoelectronic demands of carbanion formation through rotation of the exocyclic substituent. Usually it has been found that the incoming proton approaches the enolate from the less hindered equatorial side, forcing the substituent into the axial position, so that the less stable epimer is obtained. A typical example is provided by 1-benzoyl-2-phenylcyclohexane (Fig. 58).[246] When phenylmagnesium bromide is added to benzoylcyclohexene in the presence of cuprous chloride (to facilitate 1,4-addition) and the product is acidified with ammonium chloride, the 2-benzoyl-1-phenylcyclohexane is largely *cis*, indicating that protonation of the enolate has occurred from the equatorial side. Removal and replacement of the proton from the product by bromine in acetic acid to give 1-phenyl-2-bromo-2-benzoylcyclohexane occurs readily. The product probably has its bromine axial. Removal of the bromine by hydriodic acid is facile and returns pure *cis*-2-benzoyl-1-phenylcyclohexane via the enolate. Treatment of the *cis* isomer with base converts it to the more stable *trans*-2-benzoyl-1-phenylcyclohexane, proving that the result of protonation of the enol is not thermodynamically controlled. The *trans* isomer, being more stable, is less prone to revert to the enol and, accordingly, is not attacked by bromine in acetic acid. Preferential

[246] H. E. Zimmerman, *J. Org. Chem.,* **20,** 549 (1955). See, however, S. K. Malhotra and F. Johnson, *J. Am. Chem. Soc.,* **87,** 5493 (1965).

equatorial attack has also been found in the enolate of 2-acetyl-1-phenyl-cyclohexane,[247] 2-phenylcyclohexanecarboxylic acid,[248] 2-nitro-1-phenyl-cyclohexane,[249] and, though to a lesser extent, in the enolate of 4-phenyl-cyclohexanecarboxylic acid.[71] Experiments concerned with the ketoni-zation of 1-decalone[250] indicated that here, as in one of the cases mentioned earlier, the nature of the proton donor has an important effect on the stereochemistry of the product. Significantly, protonation of the anion of 1-benzenesulfonyl-2-phenylcyclohexane under all conditions tried led to predominance of axial approach yielding mainly *trans*-2-phenyl-1-ben-zenesulfonylcyclohexane.[251] The reason is that here the anion is stabilized by an inductive effect and *p-d* orbital overlap rather than the *p*-pi overlap found in the other conjugated anions studied by Zimmerman and co-workers.

In conclusion, attention must be drawn to certain electrophilic reactions which appear to involve an S_E2 mechanism and are stereospecific, pro-ceeding with complete retention of configuration. Thus *cis*-4-methyl-cyclohexylmercuric bromide when treated with bromine *in pyridine* gives exclusively *cis*-4-methylcyclohexyl bromide, whereas the *trans* isomer gives the corresponding *trans* bromide in pure form.[237]

e. Cyclohexadienes and Cyclohexanediones.   Little appears to be known about the geometry of cyclohexadienes. The 1,4-isomer would appear to be boat-shaped and is generally so considered. The 1,3-isomer, in a model, appears to be a half-chair, but it is clear that to this end the pi-electron system must be skewed, so that there will be a slight loss of overlap energy. Optical rotatory dispersion measurements[252] suggest that the system is in fact twisted out of the plane.

The cyclohexanediones may all assume chair shapes, although, in view of the effect of the carbonyl carbon of reducing the energy difference between chair and boat, boat or flexible forms may also contribute appreciably. Only the 1,4-dione seems to have been considered; calcu-lations[10b] indicate that the chair and skew-boat form may have approxi-mately equal energies. The small but finite dipole moment (1.3 D)[253] of

[247] H. E. Zimmerman, *J. Am. Chem. Soc.*, **79,** 6554 (1957).

[248] H. E. Zimmerman and H. J. Giallombardo, *J. Am. Chem. Soc.*, **78,** 6259 (1956); H. E. Zimmerman and T. W. Cutshall, *ibid.*, **81,** 4305 (1959).

[249] H. E. Zimmerman and T. E. Nevins, *J. Am. Chem. Soc.*, **79,** 6559 (1957).

[250] H. E. Zimmerman and A. Mais, *J. Am. Chem. Soc.*, **81,** 3644 (1959).

[251] H. E. Zimmerman and B. S. Thyagarajan, *J. Am. Chem. Soc.*, **80,** 3060 (1958).

[252] A. W. Burgstahler, H. Ziffer, and U. Weiss, *J. Am. Chem. Soc.*, **83,** 4661 (1961); A. Moscowitz, E. Charney, U. Weiss, and H. Ziffer, *ibid.*, **83,** 4660 (1961).

[253] C. G. Le Fèvre and R. J. W. Le Fèvre, *J. Chem. Soc.*, **1935,** 1696; N. L. Allinger and L. A. Freiberg, *J. Am. Chem. Soc.*, **83,** 5028 (1961).

the compound is in agreement with the assumption that the skew-boat form contributes substantially; in fact it may be the only conformation in both solid and liquid.[253a] Equilibrium data for the *cis*- and *trans*-2,5-di-*t*-butyl-1,4-cyclohexanediones[254] suggest that at least the *cis* and possibly also the *trans* isomer of this compound exist mainly in the flexible form, although in this case the true boat may be preferred over the skew boat.

Little seems to be known about the conformation of α,β-unsaturated ketones, although optical rotatory dispersion may well throw light on this problem eventually.[255]

2-7. Deformations in Cyclohexane Rings[76]

The assumption is usually made that cyclohexane rings are perfect chairs. It has already been pointed out that bulky axial substituents (as in *trans*-1,3-di-*t*-butylcyclohexane) may force the molecule into the flexible

Fig. 59.

form. Much lesser deformations may suffice to distort the molecule out of the perfect chair form, for, as has been pointed out in the classical work of Westheimer,[256] any kind of van der Waals compression in a molecule will tend to relieve itself by deformation, in as much as angle and torsional deformations are generally much less expensive, in terms of energy, than are van der Waals interactions. Examples have already been pointed out for *trans*-1,1,3,5-tetramethylcyclohexane in which there are two *syn*-axial methyl groups and in which the experimental interaction energy (3.7 kcal./mole) of these groups is vastly less than the 200 kcal./mole calculated on the basis of van der Waals repulsion between two truly axial methyl groups.[50] In a 3,3,5,5-tetramethylcyclohexanone the repulsion between the axial methyl groups leads to a distortion of the molecule as the result of which the axial positions at C_2 and C_6 (the alpha positions) are compressed

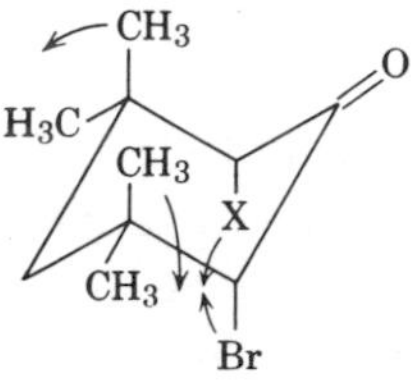

[253a] See also A. Mossel, C. Romers, and E. Havinga, *Tetrahedron Letters*, **1963,** 1247; P. Groth and O. Hassel, *Proc. Chem. Soc.*, 218 (1963).

[254] R. D. Stolow and C. B. Boyce, *J. Am. Chem. Soc.*, **83,** 3722 (1961). See also R. D. Stolow and M. M. Bonaventura, *Tetrahedron Letters*, **1964,** 95, for the corresponding dimethyl analogs.

[255] Cf. C. Djerassi, R. Records, E. Bunnenberg, K. Mislow, and A. Moscowitz, *J. Am. Chem. Soc.*, **84,** 870 (1962).

[256] Cf. F. H. Westheimer, "Calculation of the Magnitude of Steric Effects," in M. S. Newman, ed., *Steric Effects in Organic Chemistry*, John Wiley and Sons, New York, 1956, pp. 523–555.

(Fig. 59). This effect has been called the reflex effect.[257] A manifestation is seen in the comparison of the equilibria of the epimeric 2,6-dibromocyclohexanones on one hand and the 2,6-dibromo-3,3,5,5-tetramethylcyclohexanones on the other. In the former, the *trans* isomer (*e,a*) is favored by a ratio of 85:15, presumably because of the previously mentioned unfavorable dipole repulsion in the *cis* isomer (*e,e*).*[,258] In contrast, in the tetramethyl-substituted compound,[257] equilibrium favors the *cis* dibromide by over 90:10; in the *trans* isomer the axial bromine would be compressed against the *syn*-axial hydrogens by virtue of the reflex effect. It is possible to force one (but not two) bromines into this axial position; thus 3,3,5,5-tetramethylcyclohexanone can be tribrominated (to the 2,2,6 derivative) but not tetrabrominated, in contrast to cyclohexanone which can be tetrabrominated to the 2,2,6,6 derivative.

Among other deformations, that involving a comparison of *cis*- and *trans*-1,2-disubstituted cyclohexanes has already been mentioned (p. 77). The *trans* substituents will tend to bend away from each other and in the process deform the ring; the *cis* substituents cannot do this readily (Fig. 33). It must be pointed out again that even a single axial substituent will tend to bend outward by deforming the ring (p. 47). There is some evidence that this deformation is easier when there is an equatorial substituent in the 4-position than when there is one in the 3-position[76,99], and as a result the *cis* 1,4-isomer will be less destabilized than the conformationally analogous corresponding *trans* 1,3-isomer. The methylcyclohexanols (p. 65) provide an example where equilibrium lies more on the side of the *e,e* isomer in the 3-isomer than in the 4-isomer.

Deformation effects will be discussed further in Secs. 5–9 and 7–3.

General References

Angyal, S. J., and J. A. Mills, "The Shape and Reactivity of the Cyclohexane Ring," *Rev. Pure Appl. Chem.*, **2**, 185 (1952).

Barton, D. H. R., "The Stereochemistry of Cyclohexane Derivatives," *J. Chem. Soc.*, **1953**, 1027.

Barton, D. H. R., "Some Recent Progress in Conformational Analysis," *Experientia*, *Suppl.* **11**, 121 (1955).

Barton, D. H. R., "Some Recent Progress in Conformational Analysis," in *Theoretical Organic Chemistry (Kekulé Symposium)*, Butterworth & Co., Ltd., London, 1959, pp. 127–143.

[257] C. Sandris and G. Ourisson, *Bull. Soc. Chim. France*, **1958**, 1524; J.-F. Biellmann, R. Hanna, G. Ourisson, C. Sandris, and B. Waegell, *ibid.*, **1960**, 1429.

[258] E. J. Corey, *J. Am. Chem. Soc.*, **75**, 3297 (1953).

* The *a,a* form of the *cis* isomer is, of course, disfavored because of both steric and polar repulsion of the two axial bromines.

Barton, D. H. R., "Some Recent Progress in Conformational Analysis," *Svensk Kem. Tidskr.*, **71**, 256 (1959).

Barton, D. H. R., and R. C. Cookson, "The Principles of Conformational Analysis," *Quart. Rev. (London)*, **10**, 44 (1956).

Dauben, W. G., and K. S. Pitzer, "Conformational Analysis," in M. S. Newman, ed., *Steric Effects in Organic Chemistry*, John Wiley and Sons, New York, 1956, Chap. 1.

Eliel, E. L., "Conformational Analysis in Mobile Systems," *J. Chem. Educ.*, **37**, 126 (1960).

Hassel, O., "Stereochemistry of Cyclohexane," *Quart. Rev. (London)*, **7**, 221 (1953).

Hückel, W., "Der gegenwärtige Stand der Spannungstheorie," *Fortschr. Chem. Physik u. physik. Chem.*, **19**, 243 (1927).

Klyne, W., "The Conformations of Six-membered Ring Systems," in *Progress in Stereochemistry*, Vol. 1, Butterworth and Co., Ltd., London, 1954, Chap. 2.

Lau, H. H., "Prinzipien der Konformationsanalyse," *Angew. Chem.* **73**, 423 (1961).

Mizushima, S., *The Structure of Molecules and Internal Rotation*, Academic Press, Inc., New York, 1954.

Orloff, H. D., "The Stereoisomerism of Cyclohexane Derivatives," *Chem. Rev.*, **54**, 347 (1954).

Wasserman, A., "Spannungstheorie and physikalische Eigenschaften ringförmiger Verbindungen," in K. Freudenberg, ed., *Stereochemie*, Franz Deuticke, Leipzig, 1932, pp. 781–802.

Chapter 3

The Use of Physical Methods in
Conformational Analysis

3–1. Introduction

Conformational analysis may be described as the interpretation of the reactions and properties of compounds in terms of their conformations. Since these conformational arrangements are almost invariably determined by physical methods, it seems appropriate to discuss at this point the use of the available methods, the general area of applicability, and the limitations of each.

There are a number of very important physical methods which have been used in conformational analysis that will be covered only briefly here. These are X-ray diffraction, electron diffraction, and microwave spectra, which are methods that yield the total structure of a molecule rather than a simple piece of conformational information. Although these methods are of extreme value in structural determination, they are laborious to apply, and they have been used only by specialists in those areas and not by organic chemists in general. Excellent reviews on each of these methods are available.

3–2. Methods of Determining Total Structure

a. X-ray Diffraction.   The first of these total methods is X-ray crystallography,[1] which in principle will give the total structure of any compound which can be obtained in crystalline form. The method has been used to deduce total structures, including all conformational details, for

---

[1] J. M. Robertson, in E. A. Braude and F. C. Nachod, eds., *Determination of Organic Structures by Physical Methods*, Vol. I, Academic Press, New York, 1955, p. 463; *Proc. Chem. Soc.*, **1963**, 229.

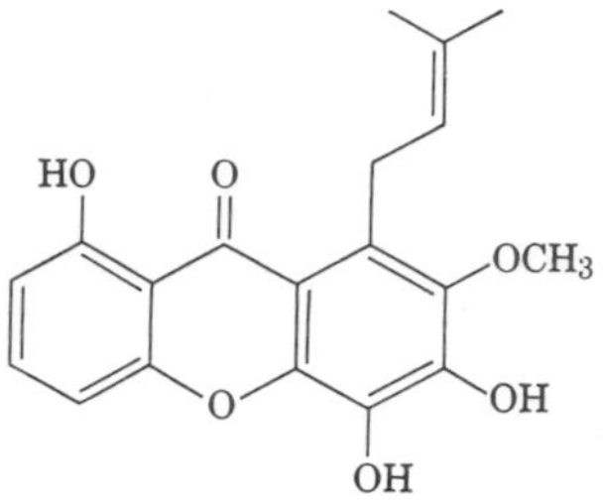

Fig. 1. Structures of celebixanthone (1) and vitamin B$_{12}$ (2) from X-ray studies.

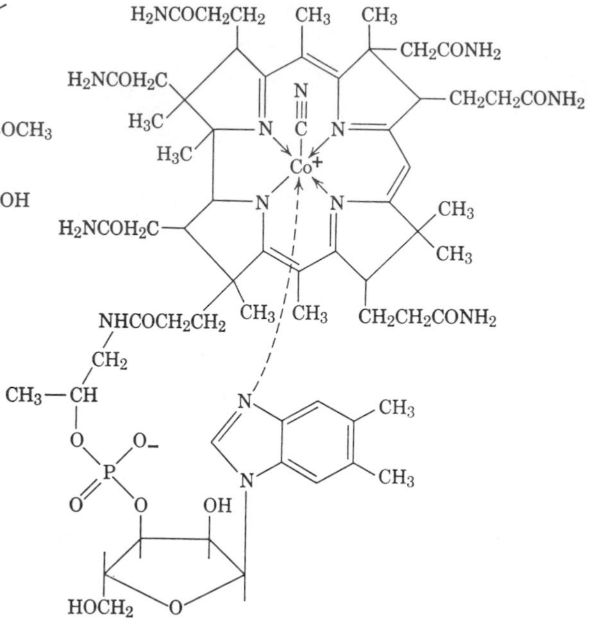

very complicated molecules such as celebixanthone[2] and vitamin B$_{12}$.[3] (See Fig. 1.) Although tremendous progress in this area has been made since about 1950, it is still not possible to determine the structure of a typical organic compound without benefit of chemical information unless the molecule happens to belong to one of several kinds of special cases. The basic difficulty that prevents solution of the problem in the general case is what is referred to as the "phase problem" by crystallographers. Radiation being scattered from different centers will be more or less out of phase, leading to the familiar interference pattern. The amplitude of the resulting wave will tell how much out of phase the component waves are, but it will not tell which wave is ahead of the others, the phase constants which determine this being unknown. These phase constants are essential for solution of the problem, but they cannot be determined from the experimental data.

It is possible that a general solution to this problem may be reached in the future, but at present it is necessary to solve the problem by coercing

[2] G. H. Stout, V. F. Stout, M. J. Welsh, and L. H. Jensen, *Tetrahedron Letters*, **1962**, 541.

[3] D. C. Hodgkin, *Fortschr. Chem. Org. Naturstoffe*, **15**, 167 (1958).

the molecule to fit into one or another of the special cases which can be treated, or by using chemical information. When chemical information is used, the approach is as follows. While it is not in general possible to calculate the structure which gave rise to a diffraction pattern, it is comparatively easy to do the reverse. Hence, if sufficient chemical information is available, an approximate or partial[4] structure can be assumed, and the corresponding diffraction pattern can be calculated. The atomic coordinates can then be adjusted to improve the agreement between the calculated and experimental diffraction patterns, and the process can be repeated until good agreement is obtained. Celebixanthone is one of the largest molecules which has had its structure determined in this way without recourse to the heavy atom method.

With very complicated structures such as vitamin B_{12}, or with structures for which chemical data are unavailable, the heavy atom method (or the closely similar isomorphous replacement method) is required. The heavy atom method, even when not required, usually greatly facilitates solution of the problem. The idea is to insert a heavy atom, which has a large scattering power, directly into the molecule. The heavy atom essentially overwhelms the phase contributions of the remaining atoms. Then it becomes possible actually to work the structural problem in the desired direction and calculate the structure from the diffraction pattern. The location of the heavy atom in the molecule, and exactly what atom is used, are not important. Bromine is perhaps most often used, and it can be inserted almost anywhere into most organic molecules (with some synthetic effort). If the molecule is an acid or a base, the lead and hydrobromide salts have been found useful, respectively, and they are easy to prepare.

Alternatively, if it is known that two compounds which are isomorphous (have identical crystal structures) differ only by the change of one atom (for instance, a steroidal chloride and the corresponding bromide), the differences in the diffraction patterns of the two compounds can be used to solve the phase problem. This variation is a very powerful one, but limited by the necessity of having available the isomorphous compounds.

When carried sufficiently far, the X-ray method can yield very accurate bond lengths and angles.

Perhaps the ultimate limitation of the method lies in the fact that the structures are determined for the compound in the crystal. The conformation is not necessarily the same in the crystal as in solution, and it is the latter in which chemists tend to be more interested.

[4] C. E. Nordman and K. Nakatsu, *J. Am. Chem. Soc.*, **85**, 353 (1963).

b. Electron Diffraction. The second method of total structural determination is electron diffraction.[5] Although electron diffraction studies have been carried out on solids, the method is best suited to the study of gases. Any substance which is thermally stable and which will give a vapor pressure of at least a few millimeters of mercury when heated can be examined in the gas phase.

The diffraction pattern obtained when electrons strike the gas being investigated is radially symmetric, since all possible spatial orientations of the molecule exist. There is scattering from both the individual atoms, called atomic scattering, and from all possible pairs of atoms, called molecular scattering. The latter pattern is what one wishes to observe. It appears as a small variation on the strong atomic scattering curve, and this makes it somewhat difficult to observe the part of the curve that is of interest. The radial distribution curve obtained from the data is essentially the relative probability of finding a pair of atoms at the distance in question. A typical curve, that of 1,2-dichloroethane, is shown in Fig. 2. Since there are only two carbons and two chlorines, there is only one carbon-carbon distance, but there is one chlorine-chlorine distance for the *anti* conformation (about 4.2 Å) and one for the *gauche* conformation (about 3.4 Å). For hydrogen-chlorine distances there is one (about 2.4 Å) between the two hydrogens on the same carbon as the chlorine, and one between any hydrogen and a chlorine on the other carbon in the *anti* conformation (about 2.9 Å). There is a long distance (3.8 Å) between the *anti* pair in the *gauche* conformation, and one might expect the same

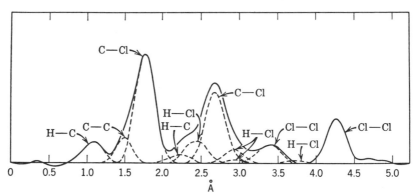

Fig. 2. Radial distribution curve for 1,2-dichloroethane. From Karle and Karle.[5a]

[5] (a) J. Karle and I. L. Karle, in E. A. Braude and F. C. Nachod, eds., *Determination of Organic Structures by Physical Methods*, Vol. I, Academic Press, New York, 1955, p. 427; (b) O. Hassel, *Quart. Rev. (London)*, **7**, 221 (1953); (c) O. Bastiansen and P. N. Skancke, *Advan. Chem. Phys.*, **3**, 323 (1961).

hydrogen-chlorine distance between any *gauche* arrangement in the *gauche* conformation, but two such distances are found (near 2.9 Å and 3.0 Å), and this indicates that the dihedral angle between the carbons is not exactly 120° in the *gauche* arrangement, but rather that the groups are rotated slightly.

The problem here is generally solved by working backwards. That is, a structure is assumed, and its diffraction pattern is calculated. Small (or large) adjustments in structure are then made, and the pattern is recalculated. The process is repeated until the calculated and experimental curves coincide, and the structure is then considered to be known. The reason why the probability of finding a pair of atoms varies around a certain distance, rather than being exactly at a certain distance, is that the atoms are in vibrational and rotational motion. Information on vibrational amplitudes can also be obtained from electron diffraction data, at least in principle.

The method is obviously not going to be of much use in a really complicated molecule like a steroid, because in such a molecule there are so many carbon-carbon distances that they form almost a continuum, and a number of completely erroneous structures might very well fit the experimental data.

Cycloöctatetraene is an example of a compound which was thoroughly studied by a number of investigators who obtained divergent results. The first application of electron diffraction to the problem led to the conclusion that the molecule was crown shaped,[6] but subsequent studies have shown that it is in fact a tub.[7] The radial distribution curves for the two structures differ but slightly. Similarly, early electron diffraction data were perhaps a little misleading in indicating the equatorial position for the halogen in chlorocyclohexane,[8] when in fact the compound contains some 30% of the axial conformer. Hassel[8] was, however, careful to state that the presence of some of the axial form was not excluded by his data.

The main usefulness of the electron diffraction method to date has been in deciding between alternative possible structures, and in measuring distortions in molecules. The ring in cyclohexane has been found to be slightly flattened, the C—C—C angles do not have regular tetrahedral values[9] but are widened slightly to 111.5°. Such bond angle widening has

[6] O. Bastiansen, O. Hassel, and A. Langseth, *Nature*, **160**, 128 (1947).

[7] (a) I. L. Karle, *J. Chem. Phys.*, **20**, 65 (1952); (b) O. Bastiansen, L. Hedberg, and K. Hedberg, *ibid.*, **27**, 1311 (1957).

[8] O. Hassel and H. Viervoll, *Tidsskr. Kjemi Bergvesen Met.*, **3**, 35 (1943).

[9] V. A. Atkinson and O. Hassel, *Acta Chem. Scand.*, **13**, 1737 (1959); M. Davis and O. Hassel, *ibid.*, **17**, 1181 (1963).

also been observed in aliphatic systems[9a] and confirmed by microwave spectra,[10] and it is a general phenomenon. In an undistorted axial halocyclohexane, a sizable van der Waals repulsion would exist between the halogen and the *syn*-axial hydrogens, and the C—X bond bends to reduce this interaction. Such bending has been found[9] to amount to 6.3° for chlorine and 7.7° for bromine.

The electron diffraction patterns of *n*-hexane and homologs show that these molecules are not constrained to a rigid all-*anti* conformation, nor are they highly crumpled, but they must in fact show a distribution between *gauche* and *anti* conformations similar to that calculated statistically[9a] (see Sec. 1-3a).

c. Microwave Spectra. These are electromagnetic spectra determined in the wavelength range of approximately 1 mm. to 10 cm., and from a study of the microwave spectrum of a compound it may be possible to determine accurately its total structure.[11] The principal type of absorption observed in this region results from a transition of the molecule from one rotational state to another. Although all molecules have various rotational states, the excitation from one to another corresponds to a forbidden transition unless the molecule has a dipole moment. Hence a molecule which lacks a dipole moment cannot ordinarily be studied by the method, but even a small moment such as exists in propane or isobutane is sufficient. In order that the individual lines of the spectrum be observed, it is necessary to operate in the gas phase at a pressure of not more than about 0.1 mm., and the compound must be stable under these conditions. For various reasons it would seem that it will not be possible in the near future to study molecules containing more than perhaps ten to fifteen atoms.

Briefly, the procedure employed is to calculate by analysis of the spectrum what the three moments of inertia of the molecule are. This calculation is easy if the molecule is highly symmetrical, it is difficult if there is no symmetry, but it can ordinarily be done for molecules which meet the other criteria that determine the feasibility of this type of study. The moments of inertia arise in a simple way from the masses of the atoms and their arrangement, that is, their bond lengths and bond angles. Since the masses of the atoms are known, the three moments of inertia are sufficient to determine three parameters (bond lengths or angles). Obviously,

[9a] R. A. Bonham, L. S. Bartell, and D. A. Kohl, *J. Am. Chem. Soc.*, **81**, 4765 (1959).

[10] D. R. Lide, Jr., *J. Chem. Phys.*, **33**, 1514 (1960).

[11] E. B. Wilson, Jr., and D. R. Lide, Jr., in E. A. Braude and F. C. Nachod, eds., *Determination of Organic Structures by Physical Methods*, Vol. I, Academic Press, New York, 1955, p. 503.

most molecules have more than three such parameters to be determined; hence one is faced with only three equations containing many unknowns. A trick is used which depends on the fact that substitution of one isotope for another in the molecule changes the moments of inertia and gives in general three additional equations. It does not result in any change in the molecular geometry and hence introduces no more unknowns. Substitution of a deuterium for a hydrogen is a change that is frequently made. For many elements, including carbon and nitrogen, sufficient quantities of a second isotope occur naturally that the spectra due to molecules containing them arc often seen directly without isotopic enrichment.

Since the procedure here involves solution of simultaneous equations, usually it is necessary to solve them all, or else nothing can be learned. Sometimes it is possible to assume certain parameters and solve for others, but in general the total molecular structure is determined, or nothing is determined. For molecules which can be properly treated, the results are very accurate, bond lengths are determined to within a few thousandths of an angstrom and bond angles to a fraction of a degree.

Various kinds of additional information can also be obtained from microwave spectra, one of the most useful being the dipole moment. Dipole moments obtained in this way are in general quite accurate, and, if the moment is very small, say 0.1 D, this is the only method presently available by which it may be accurately measured (Sec. 3-5). By this method the dipole moments of propane[10] and isobutane[12] were found to be 0.083 and 0.132 D, respectively. Rotational barriers can also be determined from microwave spectra (see Sec. 3-3a).

d. Vibrational Spectra. Vibrational spectra have been quite useful in establishing the total and exact structures of many simple molecules. A molecule in a ground vibrational state may be excited to a higher state by the absorption of electromagnetic radiation in the wavelength region of roughly 1–100 μ (10^{-3} to 10^{-1} mm.). Such a transition may be observable in the infrared spectrum[13] or in the Raman spectrum[13,14].or in

[12] D. R. Lide, Jr., and D. E. Mann, *J. Chem. Phys.*, **29,** 914 (1958).

[13] (a) M. K. Wilson, in F. C. Nachod and W. D. Phillips, eds., *Determination of Organic Structures by Physical Methods*, Vol. II, Academic Press, 1962, p. 181; (b) R. P. Bauman, *Absorption Spectroscopy*, John Wiley and Sons, New York, 1962; (c) G. Herzberg, *Infrared and Raman Spectra of Polyatomic Molecules*, Van Nostrand, Princeton, N.J., 1945; (d) E. B. Wilson, J. C. Decius, and P. C. Cross, *Molecular Vibrations, The Theory of Infrared and Raman Vibrational Spectra*, McGraw-Hill Book Co., New York, 1955.

[14] (a) F. F. Cleveland, in E. A. Braude and F. C. Nachod, eds., *Determination of Organic Structures by Physical Methods*, Vol. I, Academic Press, New York, 1955, p. 231; (b) J. H. Hibben, *The Raman Effect and Its Chemical Applications*, Reinhold Publishing Corp., New York, 1939. (c) B. P. Stoicheff, *Advan. Spectr.*, **1,** 91 (1959).

both or in neither. For a transition to be allowed in the infrared, the requirement is that the dipole moment of the vibrating system be different in the ground and excited states. Thus vibration across a center of symmetry is always a forbidden transition, and H_2 and Cl_2 do not show stretching frequencies in the infrared whereas H—Cl does. For Raman absorption the selection rules are different. Here it is not the dipole moment but the polarization which must change in going from one state to the other, and the H_2 and Cl_2 stretching transitions are allowed.

Vibrational spectra can be used to establish a total structure. Here again the probelm is solved backwards; that is, possible structures are assumed and spectral predictions are made for each. Often the correct structure of a simple molecule can be chosen unambiguously.

The starting point for a structural determination by this method is the determination of the symmetry classes to which the different possible structures belong. From the symmetry class, by straightforward operations it is possible to predict which kinds of vibration correspond to allowed and which to forbidden transitions in the infrared and in the Raman. In addition, it can be predicted which vibrations correspond in the Raman spectrum to polarized lines and which correspond to depolarized lines (referring to the intensities of the scattered light in incident and perpendicular planes when the exciting beam is polarized). Also, infrared bands have characteristic shapes (at low pressures in the gas phase) depending on whether the absorption corresponds to a vibration parallel or perpendicular to the axis of the principal moment of inertia.

A vibrating non-linear molecule containing n atoms will have $3n - 6$ degrees of vibrational freedom, and hence a maximum of $3n - 6$ fundamental frequencies. A molecule like benzene would therefore have 30 fundamental frequencies, and because of the selection rules dictated by the symmetry class it is to be expected that some of these frequencies will appear in the Raman, some in the infrared and some in neither. Using the other data mentioned, together with additional facts known about vibrations, for example that absorption from a C—H stretching vibration will always be found near 2950 cm.$^{-1}$, it was possible to show[13,14] that benzene is a regular hexagon, that cyclohexane is a chair,[15] that methane is a regular tetrahedron, and so on. The long series of papers by Mizushima and co-workers on the spectroscopic investigations of ethane and its halogenated derivatives is a classic example of the determination of conformation by this method. An excellent summary of the work is available.[16]

[15] (a) D. A. Ramsay and G. B. B. M. Sutherland, *Proc. Roy. Soc.*, **A190,** 245 (1947); (b) H. E. Bellis and E. J. Slowinski, Jr., *Spectrochim. Acta*, **15,** 1103 (1959).

[16] S. Mizushima, *Structure of Molecules and Internal Rotation*, Academic Press, New York, 1954, p. 18.

On the other hand, studies of this kind led to the assignment of a crown structure to cycloöctatetraene[17] which is known to be wrong,[7] and to the assignment of a peculiar butterfly-like structure to cyclooctane[15b] which is probably wrong. With a molecule the size of cyclooctane (66 fundamental frequencies), many of the bands overlap, and many bands may accidentally appear to be coincident in the Raman and infrared when they really correspond to different vibrations and the assignment of the various bands is not always unambiguous.

The data needed to assign a complete structure from vibrational spectra are high resolution infrared and Raman spectra from 100 to 4000 cm.$^{-1}$, together with depolarization data for the latter. From such data it is usually possible to assign exactly and unambiguously the structure of any simple molecule (about six atoms) or a sufficiently symmetrical larger molecule (cyclohexane), but for most molecules which are currently of structural interest to organic chemists studies of this type are not definitive. The organic chemist, of course, uses vibrational spectra in quite a different way, the conformational aspects of which are discussed later (Sec. 3-4).

A very useful additional quantity, the entropy of the substance, can also be obtained once the fundamental frequencies have been correctly assigned, if the principal moments of inertia are also known.[18] The entropy obtained in this way, referred to as the spectroscopic entropy, is compared with the value from measurements of heat capacity (which does not depend on structure or assignment) whenever possible, and very often the two values for simple compounds agree to a fraction of one entropy unit. The usual cause for disagreement between the spectroscopic and thermodynamic entropies is the presence of hindered rotation (see Sec. 3-3).

3-3. Thermodynamic Methods

a. Calorimetry. Perhaps the most direct (at least in principle) way of determining the conformational structure of a compound is by the determination of the enthalpy and entropy by calorimetric measurements.[19] Many fundamental conformational facts have been learned or confirmed

[17] E. R. Lippincott, R. C. Lord, and R. S. McDonald, *J. Am. Chem. Soc.*, **73**, 3370 (1951).

[18] J. G. Aston and J. J. Fritz, *Thermodynamics and Statistical Thermodynamics*, John Wiley and Sons, New York, 1959, p. 248.

[19] J. G. Aston, in E. A. Braude and F. C. Nachod, eds., *Determination of Organic Structures by Physical Methods*, Vol. I, Academic Press, New York, 1955, p. 525.

in this way, for example that cyclopentane is non-planar,[20] that cyclo-hexane is a chair and not a boat[21] and that the *cis* and *trans* structures of 1,3-dimethylcyclohexane had been assigned incorrectly.[21,22]

Relatively little modern accurate calorimetric work has been done on compounds other than hydrocarbons, mainly because of the tedious nature of the measurements, and this is an area in which more effort could profitably be expended. An authoritative treatment on the experi-mental aspects of the subject is now available.[23]

The heats of combustion of most compounds are determined by placing a sample of the compound in a bomb under a pressure of oxygen and igniting it. The bomb is enclosed in a calorimeter so that its change in temperature can be determined. From the known heat capacity of the system the observed temperature increase corresponds to a known amount of heat released by combustion of the sample.

Heats of combustion are of general applicability, but they suffer from a disadvantage when one wishes to measure conformational energy differences of the order of 1 kcal./mole. To do this by measuring a total heat of combustion of the order of 1000 kcal./mole means that only the most precise measurements on highly purified samples will give the desired quantity with a reasonable accuracy, and this requires exceedingly tedious work.

Entropy is determined calorimetrically by measuring the heat capacity of a substance. Most substances form crystals at absolute zero which are perfectly ordered, and hence the entropy of such a crystal is zero. The compound is cooled to as near absolute zero as convenient (an extrapo-lation to absolute zero is made), then heat is added in measured incre-ments and the temperature rise for each increment is noted. The heat capacity at constant pressure, C_p, is the molar increment of heat added (q), divided by the temperature change:

$$C_p = q/\Delta T \tag{1}$$

The entropy of a substance increases according to the amount of heat added and the temperature:

$$d\,(\Delta S) = C_p(dT/T) \tag{2}$$

[20] (a) J. G. Aston, S. C. Schumann, H. L. Fink, and P. M. Doty, *J. Am. Chem. Soc.*, **63**, 2029 (1941); (b) J. G. Aston, H. L. Fink, and S. C. Schumann, *ibid.*, **65**, 341 (1943).

[21] C. W. Beckett, K. S. Pitzer, and R. Spitzer, *J. Am. Chem. Soc.*, **69**, 2488 (1947).

[22] K. S. Pitzer and C. W. Beckett, *J. Am. Chem. Soc.*, **69**, 977 (1947).

[23] F. D. Rossini, ed., *Experimental Thermochemistry*, Interscience Division, John Wiley and Sons, New York, 1956.

Ordinarily the entropy of a compound is desired in the gas phase at 25°C, and this means that the measurements must be made from near absolute zero up to at least 25°, and the heat of all phase changes (melting, change in crystal form, vaporization) which occur over this range must also be measured. The entropy of a substance can be determined in this way to perhaps ±0.2 e.u. (at 25°), which is adequate for all ordinary purposes. The measurements are quite tedious, however.

When the calorimetric entropy, determined as above, and the spectroscopic entropy (Sec. 3-2) are compared, they are often found to be identical. Frequently, however, it is found that the calorimetric value is somewhat lower. These occurrences are usually the result of assuming free rotation about a single bond in the statistical calculation, whereas in the real molecule a barrier to rotation in fact exists. Such a barrier restricts the molecule to small segments of the otherwise available 360° of rotation, and such a restriction decreases the entropy. The amount of restriction depends on the height of the rotational barrier. The thermodynamic method for determining rotational barriers consists in calculating the barrier height which will bring the calorimetric and spectroscopic entropies into agreement. Thus Teller and Topley suggested[24] as early as 1935 that in ethane a three-fold potential barrier to rotation would reconcile most of the then available experimental data (also see Sec. 1-1), and the heights of many such barriers have subsequently been determined in this way[25] (Table 1). The heights of such barriers can also be found from microwave spectra (Table 1). Ordinarily the values found by the two methods agree pretty well. These methods are best suited for determining rather small barriers, up to about 5 kcal./mole. Those in the vicinity of 10 kcal./mole and higher are more easily determined from nuclear magnetic resonance spectra (Sec. 3-4d), while ordinary kinetic methods have often been used above about 20 kcal./mole. There are other methods that also could in principle be used.[25]

Heats of hydrogenation of unsaturated compounds can be determined with high accuracy. They have been very useful for the quantitative determination of strain in cyclic olefins,[26] for determining the stability of *endo*-cyclic vs. *exo*-cyclic double bonds,[27] and for measurement

[24] E. Teller and B. Topley, *J. Chem. Soc.*, **1935**, 876.

[25] For a review see E. B. Wilson, Jr., *Advan. Chem. Phys.*, **2**, 367 (1959). Also see D. J. Millen, in P.B.D. de la Mare and W. Klyne, eds., *Progress in Stereochemistry*, Vol. III, Butterworth, Inc., Washington D.C., 1962, p. 138.

[26] (a) R. B. Turner and W. R. Meador, *J. Am. Chem. Soc.*, **79**, 4133 (1957); (b) R. B. Turner, W. R. Meador, and R. E. Winkler, *ibid.*, **79**, 4116 (1957); (c) R. B. Turner, W. R. Meador, W. von E. Doering, L. H. Knox, J. R. Mayer, and D. W. Wiley, *ibid.*, **79**, 4127 (1957).

[27] R. B. Turner and R. H. Garner, *J. Am. Chem. Soc.*, **80**, 1424 (1958).

Table I. Experimental Rotational Barriers

| Compound | Barrier, kcal./mole | | References |
	Microwave	Thermodynamic	
CH_3CH_3	3.0	2.9	25, 28
CH_3CH_2F	3.30	—	25
CH_3CHF_2	3.18	—	25
CH_3CF_3	—	3.45	19
CH_3CH_2Cl	3.56	2.7, 4.7	25
CH_3CH_2Br	3.57	—	25
CH_3OH	1.07	1.6	25
CH_3SH	1.26	1.5	25
CH_3NH_2	1.94	1.9	25
CH_3CHO	1.15	1.0	25
CH_3CFO	1.08	—	25
CH_3CClO	1.35	—	25
CH_3COCN	1.27	—	25
CH_3COOCH_3	1.17	—	25
CH_3OCH_3	—	2.7	19
$CH_3CH_2CH_3$	—	3.3	19
$(CH_3)_2CHCH_3$	—	3.87	19
$(CH_3)_4C$	—	4.80	19
$CH_3CH{=}CH_2$	1.98	1.95	25
$CH_3CH{=}CHF$-*trans*	2.20	—	25
$CH_3CF{=}CH_2$	2.62	—	25
$CH_3CH{=}C{=}CH_2$	1.59	1.65	25
CH_3COOH	0.48	2.5 ± 0.7	25
CH_3SiH_3	1.70	—	25
CH_3SiH_2F	1.56	—	25
CH_3SiHF_2	1.28	—	25
CH_3CHOCH_2	2.56	—	25
CH_3GeH_3	1.2	—	25
CH_3NO_2	0.006 (6-fold)	<0.25	25
CH_3BF_2	0.014 (6-fold)	—	25
$(CH_3)_2CO$	0.8	$0.56 - 1.24$	29

[28] D. R. Lide, Jr., *J. Chem. Phys.*, **29**, 1426 (1958).

[29] Available (and conflicting) conclusions are discussed by N. L. Allinger, *J. Am. Chem. Soc.*, **81**, 5727 (1959); also see J. D. Swalen and C. C. Costain, *J. Chem. Phys.*, **31**, 1562 (1959).

of the changes in stability which result from the substitution of various alkyl groups on the double bonds.[30]

b. Equilibrations. Because of the experimental difficulty in determining conformational energies by direct calorimetry, most of the available data have in fact been obtained by equilibration methods of one sort or another. Here the idea is that one measures the energy difference between isomers, rather than the total energy of a molecule, and in this way fairly accurate results can usually be obtained from rather crude measurements.

Since chemists are ordinarily interested in the differences in the thermodynamic properties of two isomers, rather than in the absolute values of these properties, equilibration measurements give directly the quantities desired. In addition, one is usually interested primarily in differences in free energy. To determine free energy differences by calorimetry, one must independently determine the entropy and enthalpy differences and then calculate the free energy difference. If the determination is carried out by the equilibration method, the difference in free energy is directly available from experiment and the differences in enthalpy and entropy are found if desired by determining the free energy difference as a function of temperature.

After equilibrium has been established, the equilibrium mixture is analyzed and the equilibrium constant K is calculated at each temperature. The free energy change at each temperature can be found from

$$\Delta G^\circ = -RT \ln K \tag{3}$$

A plot is then made graphically (or the equivalent analytic treatment, usually a least squares fitting of the data) of $\ln K$ vs. $1/T$ is made, and the intercept at $1/T = 0$ gives $\Delta S^\circ / R$ while the slope of the line gives $-\Delta H^\circ / R$. The method is such that K must be determined quite accurately to get moderate accuracy in the values of ΔH° and ΔS°, but ΔG° is ordinarily found quite exactly. There are any number of methods available by which one may determine K within an accuracy of 1–5%. Infrared spectra, for example, have been used occasionally for the analysis of equilibrium mixtures,[13b,31] but the most generally useful method has proved to be vapor phase chromatography, which often gives an accuracy of 0.1–1%. Under favorable circumstances ΔH° and ΔS° obtained in this way are comparable in accuracy to the values obtained by the more laborious calorimetric method, and ΔG° is usually much more accurate over the

---

[30] (a) R. B. Turner, W. R. Meador, and R. E. Winkler, *J. Am. Chem. Soc.*, **79**, 4122 (1957); (b) R. B. Turner, D. E. Nettleton, Jr., and M. Perelman, *ibid.*, **80**, 1430 (1958).

[31] W. E. Bachmann, A. Ross, A. S. Dreiding, and P. A. S. Smith, *J. Org. Chem.*, **19**, 222 (1954).

experimental temperature range. The equilibration method has failed to give useful results in a few special instances where the satisfactory separation of the epimers by gas phase chromatography was not possible.

Compounds so far studied by the equilibration method include two general classes, those which contain epimerizable hydrogens at the center which is to be isomerized, and those which do not. Compounds which do contain epimerizable hydrogens can be equilibrated in the presence of base, and these include such compounds as the 4-t-butylcyclohexanecarboxylic acid esters,[32] the corresponding nitriles,[33] and the 2-alkyl-4-t-butylcyclohexanones.[34] When epimerizable hydrogens are not present, one may be able to apply a special technique, as for the hydroxyl group, which can be equilibrated in the presence of an oxidation-reduction system such as an Oppenauer-Meerwein-Ponndorf mixture.[35] If such methods are inapplicable, as with hydrocarbons, equilibrium may be established in the presence of a hydrogenation catalyst at elevated temperatures.[36,37] In a modification of this method aluminum chloride is employed as catalyst, and hence lower temperatures can be used. Aluminum chloride is capable of causing skeletal isomerizations,[38] however, and its use has led to erroneous conclusions in the past. When the aluminum chloride is deactivated by complexing with an olefin, it appears that it may be possible to isomerize hydrocarbons selectively in the desired way. The equilibration of the hydrindanes by this method appears to have given good results.[39]

Applications of the equilibration method are discussed in detail in Sec. 2-4.

3–4. Spectroscopic Methods

a. Infrared Spectra.   Although the total analysis of the infrared spectrum of a simple compound may be sufficient to establish its complete structure

[32] E. L. Eliel, H. Haubenstock, and R. V. Acharya, *J. Am. Chem. Soc.*, **83**, 2351 (1961).

[33] (a) N. L. Allinger and W. Szkrybalo, *J. Org. Chem.*, **27**, 4601 (1962); (b) B. Rickborn and F. R. Jensen, *ibid.*, **27**, 4606 (1962).

[34] N. L. Allinger and H. M. Blatter, *J. Am. Chem. Soc.*, **83**, 994 (1961).

[35] (a) E. L. Eliel and R. S. Ro, *J. Am. Chem. Soc.*, **79**, 5992 (1957); (b) G. Chiurdoglu and W. Masschelein, *Bull. Soc. Chim. Belges*, **70**, 767 (1961).

[36] N. L. Allinger and J. L. Coke, *J. Am. Chem. Soc.*, **81**, 4080 (1959).

[37] N. L. Allinger and J. L. Coke, *J. Am. Chem. Soc.*, **82**, 2553 (1960).

[38] W. G. Dauben, J. B. Rogan, and E. J. Blanz, Jr., *J. Am. Chem. Soc.*, **76**, 6384 (1954).

[39] K. R. Blanchard and P. von R. Schleyer, *J. Org. Chem.*, **28**, 247 (1963).

unambiguously (Sec. 3-2d), the method is not commonly used in this way by organic chemists. The organic chemist thinks of infrared spectra in terms of group vibrations, and, if the molecule absorbs at a certain frequency, this indicates the presence of a group which is known empirically to be associated with that frequency.[40]

For conformational analysis such information is useful if the group in question absorbs at different frequencies when in different conformations, and it has been found that such is indeed the case for many atoms or groups. It is known that in general an equatorial group (X) attached directly to a cyclohexane ring will show a (C—X) stretching vibration at a higher frequency than will the corresponding axial group. (Note that when X = O—H it is the C—O not the O—H band to which this statement applies.) This has been proved for a number of groups by the shapes of the infrared bands,[41] and by studies on compounds of known and unambiguous conformation. Examples are given in Table 2. (The band assignments as stretching frequencies are only tentative for some groups.)

Within a limited group of compounds the frequency ranges into which the axial and equatorial substituents fall are quite narrow. These ranges vary when, for example, a monocyclic compound is compared with a steroid. Within a given group, however, the equatorial vibration is almost always found at a higher frequency than is the axial. It has been suggested that the reason for this consistent difference is that when the C—Y bond is stretched there is a small restoring force acting on the carbon when Y is axial, and the vibration is essentially perpendicular to the plane of the ring.[42] When Y is equatorial, the motion of the carbon forces a ring expansion, the restoring force is greater and the frequency of the motion is therefore higher.

Cases are known which are exceptions to this general rule, for instance, in the triterpenoids in which there is a *gem*-dimethyl group next to the C-3 hydroxyl the axial epimer consistently shows the C—O stretching absorption at higher frequency than does the equatorial.[43] In these molecules ring A is rather badly deformed, however, and under such

[40] (a) L. J. Bellamy, *The Infra-Red Spectra of Complex Molecules*, John Wiley and Sons, New York, second edition, 1958; (b) F. A. Miller, in H. Gilman, *Organic Chemistry, An Advanced Treatise*, Vol. III, John Wiley and Sons, New York, 1953, p. 122; (c) a useful index to the literature has been published: H. M. Hershenson, *Infrared Absorption Spectra*, Academic Press, New York, 1959, and supplements. (d) A very large collection of the infrared spectra of organic compounds is the *Stadtler Standard Spectra*, Stadtler Research Laboratories, Philadelphia, 1959, and supplements.

[41] M. Larnaudie, *J. Phys. Radium*, **15**, 650 (1954).

[42] A. R. H. Cole, R. N. Jones, and K. Dobriner, *J. Am. Chem. Soc.*, **74**, 5571 (1952).

[43] I. L. Allsop, A. R. H. Cole, D. E. White, and R. L. S. Willix, *J. Chem. Soc.*, **1956**, 4868.

Table 2. Stretching Frequencies of Secondary C—Y Bonds
Attached to Cyclohexane Rings

Bond	Frequency, cm.$^{-1}$		References
	Axial	Equatorial	
C—D	2146	2174	41
	2114–2138	2155–2162	44
	2139–2164	2171–2177	
C—F	1020	1053	a
C—Cl	646–730	736–856	46
	688	742	41
C—Br	658	685	41
	542–692	682–833	46
	658	687	47
C—I		672	46
	638	654	41
C—O (of C—OH)	996–1036	1037–1044	42, 48
C—O (of C—O—CH$_3$)	1086–1090	1100–1104	45
C—O (of C—O—Ac)^{b,c}	1013–1022	1025–1031	45

[a] Larnaudie assigned equatorial and axial frequencies at 1062 and 1129 cm.$^{-1}$. Analogy with the other halogens suggests that the assignment shown is correct. The band at 1053 cm.$^{-1}$ is about 50% stronger than the 1020 cm.$^{-1}$ band.

[b] The axial and equatorial values were interchanged in ref. 49 by typographical error. The values quoted here are as they appear in the original paper (ref. 45).

[c] Also see ref. 50.

[44] E. J. Corey, R. A. Sneen, M. G. Danaher, R. L. Young, and R. L. Rutledge, *Chem. Ind.* (*London*), **1954**, 1294.

[45] J. E. Page, *J. Chem. Soc.*, **1955**, 2017.

[46] D. H. R. Barton, J. E. Page, and C. W. Shoppee, *J. Chem. Soc.*, **1956**, 331.

[47] E. L. Eliel and R. G. Haber, *J. Org. Chem.*, **24**, 143, 2074 (1959).

[48] (a) A. Fürst, H. H. Kuhn, R. Scotoni, Jr., and H. H. Günthard, *Helv. Chim. Acta*, **35**, 951 (1952); (b) H. Rosenkrantz and L. Zablow, *J. Am. Chem. Soc.*, **75**, 903 (1953); (c) W. G. Dauben, E. Hoerger, and N. K. Freeman, *ibid.*, **74**, 5206 (1952).

[49] D. H. R. Barton, *Experientia Suppl.*, **2**, 121 (1955).

[50] R. N. Jones and F. Herling, *J. Am. Chem. Soc.*, **78**, 1152 (1956).

circumstances the restoring forces must be rather different from those in the simple cases.[51]

An alternative band which can often be used as a check for conformational assignment of an alcohol is the O—H stretching frequency.[43] For a group of triterpenes the axial alcohol consistently shows a higher frequency (3637–3639 cm.$^{-1}$) than does the equatorial (3629–3630), and, while the differences are small, they are easily detectable with the resolving power of a fluoride prism. The band shape is also informative.[51a]

Infrared spectra are useful as a general analytical tool and can be used to determine the relative amounts of any two substances in a mixture if each of these substances shows absorption at a frequency not shown by the other substance.[52] Then the relative areas of the absorption bands involved, corrected for any differences in the (integrated) extinctions of the pure substances, give directly the relative ratios of the substances. This is an empirically useful method which is often employed and does not depend on any understanding at all regarding the nature of the absorption being examined.

Infrared spectra can usually be determined in the solid phase (in a mull or as a pellet) as well as in the liquid phase. In the solid phase, substances are in general crystalline, and a crystal ordinarily contains only one conformation. It is therefore usually possible to determine whether or not the spectrum of the liquid contains bands not shown by the solid. If the liquid and solid spectra are identical, one has good evidence that the compound exists in a single conformation in the liquid as well as in the solid state. Such measurements led, for example, to the conclusion that cyclooctane exists in a single conformation, both in the solid and the liquid.[15b] On the other hand, bands often appear in the liquid state which were not present in the solid, and this indicates that a second conformation not present in the solid is in equilibrium with the first conformation in solution. Chlorocyclohexane is an example of such a compound.[53,54] It freezes to a solid at one temperature, but the infrared spectrum of this solid is essentially that of the liquid. At a still lower temperature there is a transition of the first solid to another solid, and the latter is conformationally pure. The solid obtained at higher temperature is such that the molecules retain their conformational freedom in the crystal lattice, and examination of the solid at too high a temperature would have led to an

[51] N. L. Allinger and M. A. DaRooge, *J. Am. Chem. Soc.*, **84**, 4561 (1962); *Tetrahedron Letters*, **1961**, 676.

[51a] H. S. Aaron and C. P. Rader, *J. Am. Chem. Soc.*, **85**, 3046 (1963).

[52] R. P. Bauman, *Absorption Spectroscopy*, John Wiley and Sons, New York, 1962, p. 364.

[53] P. Klaeboe, J. J. Lothe, and K. Lunde, *Acta Chem. Scand.*, **10**, 1465 (1956).

[54] K. Kozima and K. Sakashita, *Bull. Chem. Soc. Japan*, **31**, 796 (1958).

incorrect conclusion. Fluorocyclohexane is equally interesting. The infrared spectra of the liquid and of the solid down to very low temperatures are essentially identical. This fact has led to the conclusion that the compound exists as a single conformation, even in the liquid.[53] This conclusion, although consistent with the spectrum, is implausible. The actual situation must surely be that the two conformations of fluorocyclohexane are isomorphous; that is, they fit equally well into the crystal lattice. Such a situation is not unexpected, since fluorine and hydrogen are so nearly the same in size. It could be checked, for example, by determining the residual entropy at $0°K$. The liquid would be expected to be a mixture of conformations also, and nuclear magnetic resonance data (see Sec. 3-4d) indicate that it is.[55]

If it is found that there are two conformations (1 and 2) in equilibrium in solution, it may be desirable to determine the energy difference between them. If there is in the infrared region one unique band characteristic of each conformation, and if the (integrated) molar extinctions α_1 and α_2 are known for each of these bands, then the equilibrium constant can be calculated from the measured extinctions (A_1 and A_2) by

$$K = N_1/N_2 = A_1\alpha_2/A_2\alpha_1 \qquad (4)$$

where N is the mole fraction. The standard free energy change from 1 to 2 is obtained from K with the aid of equation (3). If desired, $\Delta H°$ and $\Delta S°$ can be found from measurements at two or more temperatures (Sec. 3-3b).

Unfortunately it is not usually possible to apply equation (4) directly, because the molar coefficients (α's) are not known. As an approximation the following trick has been used.[56] Suppose we wish to study the axial $\rightleftharpoons$ equatorial equilibrium in cyclohexanol by this method. What is needed are the frequencies and α's for a band characteristic of each conformation. Now it is known that in the *cis*- and *trans*-4-*t*-butylcyclohexanols the C—O stretching vibrations occur at 955 and 1062 cm.$^{-1}$, respectively, and that cyclohexanol itself shows a strong band at 1069 cm.$^{-1}$. Assuming that the latter corresponds to the 1062 cm.$^{-1}$ band of the *t*-butyl derivative, and assuming that the two bands have the same molar extinction in the two compounds, the ratio of the A's of these two bands in spectra of the two compounds determined at the same concentration gives the mole fraction of the equatorial form directly. The

[55] A. J. Berlin and F. R. Jensen, *Chem. Ind.* (*London*), **1960**, 998.

[56] (a) R. A. Pickering and C. C. Price, *J. Am. Chem. Soc.*, **80**, 4931 (1958); (b) G. Chiurdoglu and W. Masschelein, *Bull. Soc. Chim. Belges*, **70**, 29 (1961); (c) W. Hückel and Y. Riad, *Ann.*, **637**, 33 (1960); (d) F. R. Jensen and L. H. Gale, *J. Org. Chem.*, **25**, 2075 (1960).

value found (0.65) is in reasonable agreement with the values obtained in other ways, and it shows that the method may be satisfactory as long as no change in the coupling of the vibration examined is brought about by the addition of the substituent.

A more general method of getting around the fact that α_1 and α_2 are not known is to repeat the determination at two or more temperatures.[57,58] The difference in enthalpy between the conformations is given by

$$\Delta H = [RT_1 T_2/(T_2 - T_1)][\ln (A_1/A_2)_{T_2} - \ln (A_1/A_2)_{T_1}] \qquad (5)$$

and this method has had occasional use in the determination of conformational enthalpy.[59] Often the change in entropy can be simply estimated or neglected, and the conformational free energy then can be found. An attempt to apply this method to cyclohexyl mercaptan led, however, to the unlikely conclusion that the S—H group preferred to be axial,[60] which is now known to be incorrect.[61]

The extent of hydrogen bonding in hydroxylic substances has also been useful in establishing their conformation.[40,62] Usually such hydrogen bonding is conveniently measured by the infrared method, since a hydrogen bonded group, for example hydroxyl, absorbs at a lower frequency than does an unbonded one. Intramolecular hydrogen bonding has proved to be the most informative. The spectrum must be obtained in dilute solution so that the intermolecular hydrogen bonding is suppressed. Under these conditions the O—H···O region gives one or the other of two general patterns. Either there is just one peak (near 3630 cm.$^{-1}$ if there is no hydrogen bonding, or near 3590 if there is), or there are two separate peaks with an intensity ratio which depends on the ratio of conformers present. Kuhn showed that the separation between the bonded and unbonded frequencies was larger the stronger the hydrogen bond.[63,64] Thus

[57] K. Kozima and Y. Yamanouchi, J. Am. Chem. Soc., **81,** 4159 (1959).

[58] N. L. Allinger, J. Allinger, L. A. Freiberg, R. F. Czaja, and N. A. LeBel, J. Am. Chem. Soc., **82,** 5876 (1960).

[59] The method appears to have first been applied by Y. Morino, I. Watanabe, and S. Mizushima, Sci. Papers Inst. Phys. Chem. Res. (Tokyo), **39,** 396 (1942), in connection with Raman spectra. Also see Y. Morino, S. Mizushima, K. Kuratani, and M. Katayama, J. Chem. Phys., **18,** 754 (1950).

[60] G. Chiurdoglu, J. Reisse, and M. Vander Stichelen Rogier, Chem. Ind. (London), **1961,** 1874.

[61] E. L. Eliel and B. P. Thill, Chem. Ind. (London), **1963,** 88.

[62] W. Masschelein, Ind. Chim. Belge, No. 10, 1193 (1960).

[63] (a) L. P. Kuhn, J. Am. Chem. Soc., **76,** 4323 (1954); (b) L. P. Kuhn, ibid., **74,** 2492 (1952).

[64] Comparisons refer to the same solvent [A. Allerhand and P. von R. Schleyer, J. Am. Chem. Soc., **85,** 371 (1963)].

cis-1,2-cyclopentanediol ($\Delta \tilde{\nu}$ 61 cm.$^{-1}$) has the hydroxyls closer and the hydrogen bond stronger than does *cis*-1,2-cyclohexanediol ($\Delta \tilde{\nu}$ 39 cm.$^{-1}$).

The *trans*-cyclopentane-1,2-diol showed only an unbonded frequency; the hydroxyls simply could not get together. The *trans*-cyclohexane-1,2-diol, on the other hand, showed that the bond was not so strong as in the *cis* isomer ($\Delta \tilde{\nu}$ 32 cm.$^{-1}$). The reason for the difference between the *cis* and *trans* isomers is that the cyclohexane ring can undergo a deformation to bring the hydroxyls closer together, and this deformation requires less energy with an axial and an equatorial substituent (*cis*) than with diequatorial substituents (Sec. 2-5b).

Aliphatic glycols have also been studied. For molecules of the type RCHOH—RCHOH, the *dl* forms are hydrogen bonded.[65] The *meso* forms are also if R is normal, but only partially so for *i*-Pr and not at all for *t*-Bu. The compound (t-Bu)$_2$COHCOH(t-Bu)$_2$ shows a very large shift ($\Delta \tilde{\nu}$ 170 cm.$^{-1}$), suggesting rather serious bond angle deformation. Presumably the C—C—C angles are all widened appreciably, and thus the C—C—O angles, are decreased and the hydrogens in the *syn* conformation are pushed very close together.

The ratio of the intensities of the bonded and free O—H bands has been used to obtain conformational information about a number of 1-alkylcyclohexane-1,2-diols.[66] When the hydroxyls are *trans*, hydrogen bonding is possible in the form with the hydroxyls diequatorial, but not when they are diaxial (Fig. 3.). When R was methyl the bonded and free absorption bands were of approximately equal intensity, while if R was ethyl the ratio was reduced somewhat, indicating that ethyl is slightly less favorable in the axial position than is methyl. When R was isopropyl only one band was observed, the non-bonded one. The interpretation given was that the isopropyl group was effectively very much larger than the methyl or ethyl group.[66] It is now known from other work[67,68] that

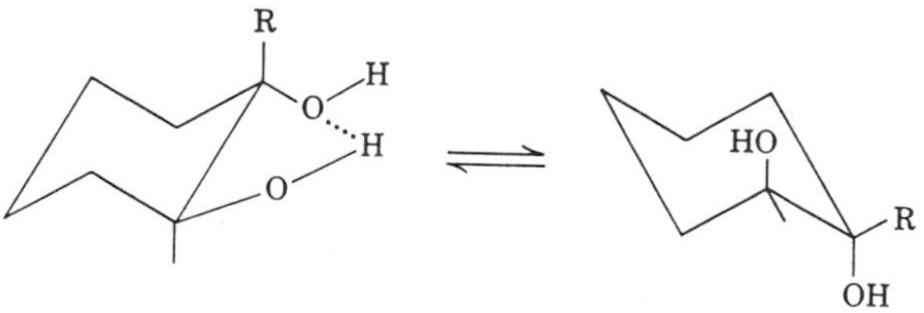

Fig. 3. Conformations of 1-alkyl-*trans*-1,2-cyclohexanediol.

[65] L. P. Kuhn, *J. Am. Chem. Soc.*, **80**, 5950 (1958).

[66] A. R. H. Cole and P. R. Jefferies, *J. Chem. Soc.*, **1956**, 4391.

[67] (a) N. L. Allinger and S. Hu, *J. Org. Chem.*, **27**, 3417 (1962); (b) N. L. Allinger, L. A. Freiberg, and S. Hu, *J. Am. Chem. Soc.*, **84**, 2836 (1962).

[68] A. H. Lewin and S. Winstein. *J. Am. Chem. Soc.*, **84**, 2464 (1962).

the isopropyl group is essentially of the same "size" as the ethyl group, and an alternative interpretation of the data seems required. An examination of models shows that in the hydrogen-bonded conformation one of the hydrogens on one of the methyls of the isopropyl group necessarily comes closer than the van der Waals distance from the oxygen of the hydroxyl at C-2, and it would appear that the energy of this interaction, added to that of the axial isopropyl, is responsible for the absence of this conformation.

Hydrogen bonding has been a useful tool for the detection of boat forms in appropriate systems (Sec. 7-4).

Often the resolution in the H—Y fundamental region is insufficient to enable all the analytically useful information present to be extracted. It has been found that it is frequently more useful to study the overtones, rather than the fundamental vibrations, of hydrogen atoms bonded to various other atoms. Many of these overtones occur in the near infrared, that is, from 0.7 to 3.5 μ. Some combination bands also occur in this region. So far, this region of the spectrum has been used mainly for quantitative functional group analysis and is capable of yielding more qualitative information than is commonly realized. Reviews covering the theory and analytical applications of near infrared spectra are available.[69]

While a rotation about the C—O bond in methanol or t-butanol yields three conformations of the same minimum energy, in a less symmetrical molecule like ethanol there are *anti* and *gauche* conformations. It is reasonable to suppose that the O—H stretching frequencies will be different in the two conformations, and hence two bands are expected and found (but resolvable with difficulty) in the 3600 cm.$^{-1}$ region and in the first overtone region.[70] More complicated alcohols in general show a similar splitting of the O—H band, and the conformational basis of the effect seems well established,[71] though not yet thoroughly exploited.[51a]

b. Raman Spectra. Raman spectra[13,14] measure the energies of vibrational transitions as do infrared spectra. The selection rules, however, are different for the Raman than for the infrared, and therefore, depending on the symmetry characteristics of the vibration, the transition may be allowed in the Raman but not the infrared, or vice versa, or it may be allowed or forbidden in both. The Raman and infrared spectra taken together have been very useful in establishing the symmetry classes of

[69] (a) W. Kaye, *Spectrochim. Acta*, **6,** 257 (1954); (b) R. F. Goddu, *Advan. Anal. Chem. Instr.*, **1960,** 347.

[70] R. M. Badger and S. H. Bauer, *J. Chem. Phys.*, **4,** 711 (1936).

[71] (a) R. Piccolini and S. Winstein, *Tetrahedron Letters*, No. 13, 4 (1959); (b) F. Dalton, G. D. Meakins, J. H. Robinson, and W. Zaharia, *J. Chem. Soc.*, **1962,** 1566.

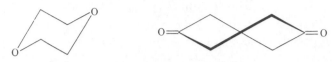

Fig. 4. Stable conformations of 1,4-dioxane and 1,4-cyclohexanedione.

many simple molecules. The selection rules are such that for a molecule with a center of symmetry any absorption which is allowed in the Raman is forbidden in the infrared, and any absorption allowed in the infrared is forbidden in the Raman.

A comparison of the spectra of p-dioxane showed that, of twenty-one observed frequencies in the Raman, only two had coincidences in the infrared.[72] This small number of accidental coincidences suggests that the molecule has a chair form (which has a center of symmetry) rather than a boat form (Fig. 4). The observed Raman spectrum of 1,4-cyclo-hexanedione, on the other hand, showed seven bands, all of which showed coincidences in the infrared.[73] That so many coincidences are accidental is highly unlikely, and it suggests that the compound has the boat rather than the chair form (Sec. 7-4). Similarly, it was shown that the observed Raman spectra of the solid paraffins were only consistent with an extended (all *anti*) conformation. Many additional bands appear in the spectra of the liquid paraffins, however, which were attributed to the various con-formations containing one or more *gauche* arrangements.[73a] The change in the Raman spectrum of a compound in going from a crystal to a liquid or solution can be used to detect or measure conformational equilibria in the same way as described for the corresponding infrared measurements (equation 4). Likewise the temperature dependence of band intensities in Raman spectra have also been used to obtain energy differences between conformations (equation 5).[59]

c. Ultraviolet Spectra.[13b,74]

While infrared spectra have proved to be of general use in the solution of conformational problems, ultraviolet spectra have been of rather limited use. The reason for the difference is simple. All classes of compounds show infrared absorption, which in general con-sists of a large number of bands and hence contains a large amount of in-formation, whereas only a few selected classes of compounds absorb in the

[72] F. E. Malherbe and H. J. Bernstein, *J. Am. Chem. Soc.*, **74**, 4408 (1952).

[73] N. L. Allinger and L. A. Freiberg, *J. Am. Chem. Soc.*, **83**, 5028 (1961).

[73a] S. Mizushima and T. Simanouti, *J. Am. Chem. Soc.*, **71**, 1320 (1949).

[74] (a) H. H. Jaffé and M. Orchin, *Theory and Applications of Ultraviolet Spectro-scopy*, John Wiley and Sons, New York, 1962. Useful indices to the literature have been published: (b) H. M. Hershenson, *Ultraviolet and Visible Absorption Spectra*, Academic Press, New York, 1956, and supplements; (c) M. J. Kamlet, ed., *Organic Spectral Data*, Interscience Division, John Wiley and Sons, New York, 1960.

accessible part of the ultraviolet region, and the spectra of these compounds are usually the result of only one or two electronic transitions. One important application of ultraviolet spectra to conformational problems has been made with ketones which contain on the α-carbon a substituent capable of accepting electrons, that is, a substituent which contains an empty orbital of relatively low energy such as halogen,[75] phenyl,[76] or hydroxyl.[77] A ketone shows weak absorption at about 280 mμ, which is· usually referred to as an $n \rightarrow \pi^*$ transition and which results from the excitation of a non-bonding electron on oxygen to an anti-bonding π^* orbital.[78] An electron-accepting group in the axial position on the α-carbon has an empty orbital (for example, the $3s$ orbital of fluorine) which overlaps slightly with the anti-bonding π^* orbital on carbon. The excited state of the ketone is consequently stabilized by the axial neighboring halogen.[79] No such stabilization is available for the non-bonding electrons on oxygen, since they are too far from the axial group for significant overlap to occur. Because the axial substituent stabilizes the excited state and not the ground state, the transition involves less energy, and hence it is found that such a substituent causes a shift of the $n \rightarrow \pi^*$ transition to longer wavelength.

An equatorial electron-accepting substituent, on the other hand, is in a favorable position to overlap with an orbital containing non-bonding electrons on oxygen, and here a stabilization of the ground state is found. The n-orbital on oxygen already has such a low energy, however, that the resulting shift in the ultraviolet spectrum is small. Qualitatively these shifts have been understood for some time, and detailed calculations (Pariser-Parr approximation[80]) have been carried out for the 2-halocyclohexanones.[81] These calculations predict that the presence of an electron acceptor in the axial position will lead to a red shift, while the effect of the group in the equatorial position will be small, as is observed.[75] Additional data on the α-haloketones are given in Sec. 7-3. These ideas can be extended to electronically comparable conjugated systems.[82]

[75] R. C. Cookson, *J. Chem. Soc.*, **1954**, 282.

[76] R. C. Cookson and J. Hudec, *J. Chem. Soc.*, **1962**, 429.

[77] (a) R. C. Cookson and S. H. Dandegaonker, *J. Chem. Soc.*, **1955**, 352; (b) G. Baumgartner and C. Tamm, *Helv. Chim. Acta*, **38**, 441 (1955).

[78] J. W. Sidman, *Chem. Rev.*, **58**, 689 (1958).

[79] This interpretation of the known facts was first suggested to N. L. A. by Dr. R. B. Hermann in 1960. Subsequently, on the basis of data furnished by N. L. A., Dr. E. M. Kosower reached the same conclusion [E. M. Kosower, G. Wu, and T. S. Sorensen, *J. Am. Chem. Soc.*, **83**, 3147 (1961)].

[80] R. Pariser and R. G. Parr, *J. Chem. Phys.*, **21**, 466, 767 (1953).

[81] M. A. Miller and J. Tai, unpublished work.

[82] A. Bowers and H. J. Ringold, *Experientia*, **17**, 65 (1961).

Ultraviolet spectra have been widely used to indicate conformational disruption of conjugated systems (steric inhibition of resonance).[83] Perhaps the best known examples are the biphenyls. 3,3'-Dimethylbiphenyl shows absorption at 255 mμ, ϵ 14,000. The 2,2'-dimethylbiphenyl, on the other hand, shows absorption about 260, with ϵ 800. The interpretation is that the planar conjugated system which is present in the first compound is not possible in the second because of the interference between the *ortho* substituents on the two rings. Extensive studies[84] on α-substituted benzaldehydes and benzophenones have shown that the extinction coefficient in such a compound is proportional to the square of the cosine of the angle of twist of the carbonyl from the benzene plane, and information on the effective sizes of different groups has been so obtained. Other applications of the method have been used to obtain conformational information about large rings (Sec. 4-1).

d. Nuclear Magnetic Resonance Spectra.[85,86] To date virtually all conformational applications of NMR which have been made have utilized the hydrogen nucleus, and only those cases will be discussed here. There are basically two methods which have been used; one makes use of chemical shifts, the second uses coupling constants.

The utilization of chemical shifts for the solution of conformational problems can be conveniently illustrated with bromocyclohexane.[87] The proton on the carbon carrying the bromine is deshielded by the bromine and is found at a very low field compared to the other protons in the system. It can therefore be easily identified and studied. At room temperature in chloroform solutions the *cis-* and *trans-*4-*t*-butylbromocyclohexanes showed chemical shifts of 160.5 and 198 c.p.s., respectively (60 Mc.). It has been found to be most often true that an equatorial proton is less shielded than the corresponding axial one,[88] and an

[83] E. A. Braude and F. C. Nachod, in *Determination of Organic Structures by Physical Methods*, Vol. I, Academic Press, New York, 1955, p. 131.

[84] E. A. Braude, F. Sondheimer, and W. F. Forbes, *Nature*, **173**, 117 (1954).

[85] (a) J. D. Roberts, *Nuclear Magnetic Resonance*, McGraw-Hill Book Co., New York, 1959; (b) L. M. Jackman, *Applications of Nuclear Magnetic Resonance Spectroscopy in Organic Chemistry*, Pergamon Press, New York, 1959; (c) J. D. Roberts, *An Introduction to the Analysis of Spin-Spin Splitting in High Resolution Nuclear Magnetic Resonance Spectra*, W. A. Benjamin, New York, 1961; (d) H. Conroy, *Advan. Org. Chem.* **2**, 265 (1960); (e) W. D. Phillips, in F. C. Nachod and W. D. Phillips, eds., *Determination of Organic Structures by Physical Methods*, Vol. II, Academic Press, New York, 1962, p. 401.

[86] J. A. Pople, W. G. Schneider, and H. J. Bernstein, *High-Resolution Nuclear Magnetic Resonance*, McGraw-Hill Book Co., New York, 1959.

[87] E. L. Eliel, *Chem. Ind. (London)*, **1959**, 568.

[88] R. U. Lemieux, R. K. Kullnig, H. J. Bernstein, and W. G. Schneider, *J. Am. Chem. Soc.*, **80**, 6098 (1958).

interpretation of this fact in terms of ring currents has been given.[89] The 4-t-butylbromocyclohexanes are seen to follow the generalization. It was also found that bromocyclohexane showed a shift of 191.5 c.p.s. At room temperature the equilibration of axial to equatorial bromocyclohexane is very fast compared to the rate of the nuclear transition being examined, and the familiar Franck-Condon principle of optical spectra does not apply here. Thus one does not observe two peaks due to the two conformers, but rather a single peak, the location of which is related to the proportion of the conformers in the equilibrium. In the present case (if it is assumed that the alkyl group has no effect on the chemical shift) the percent conformer with axial hydrogen (and equatorial bromine) in the equilibrium is $31/37.5 = 83\%$. This method as so far outlined can be used only if the appropriate model compounds are available. Under these circumstances the method is convenient, and the equilibrium constants obtained are accurate as long as the conformers do not differ in energy by more than about 1 kcal./mole.[90] In a modification of this method the spectrum is obtained at low temperature. If the temperature is sufficiently low (ca. $-80°$ to $-95°$ for cyclohexane systems[55]), the rate of conformational inversion becomes much lower than the rate of the nuclear transition observed. Under these conditions (in carbon disulfide with tetramethylsilane reference) a compound such as bromocyclohexane shows two peaks, one due to the equatorial proton (at -278.3 c.p.s.) and one due to the axial proton (at -235.4 c.p.s.). The ratio of the peak areas gives the ratio of conformers present. The same compound shows an average resonance at -245.7 c.p.s. (extrapolated from higher temperatures). The values obtained for the conformational free energy of the bromine were 0.48 kcal./mole from the areas and 0.44 kcal./mole from the chemical shifts. The low temperature method has the advantage that model compounds are not required. It has been applied to the determination of the conformational energies of all the halogens.[55]

An additional piece of information can also be obtained from low temperature NMR spectra. If there are two conformations in equilibrium, the compound will show a low temperature spectrum which is really the separate spectra of the two conformations superimposed, and a high temperature spectrum which is a single spectrum of an averaged conformation. The transition from the low temperature spectrum to the high temperature spectrum occurs gradually but over a relatively small temperature interval, from which a coalescence temperature can be

[89] Ref. 86, p. 183; for an alternative interpretation see W. C. Neikam and B. P. Dailey, *J. Chem. Phys.*, **38**, 445 (1963).

[90] E. L. Eliel and M. H. Gianni, *Tetrahedron Letters*, **1962**, 97.

estimated. This temperature is related to the rate constant for interconversion as in[91]

$$k = 2^{-\frac{1}{2}}\pi \, |\nu_a - \nu_e| \tag{6}$$

where the ν's are the frequencies of the axial and equatorial lines at the coalescence temperature. The barrier in cyclohexane was found[92] by noting that at $-105°$ the compound in carbon disulfide showed two nearly equal peaks at -71.8 and -99.1 c.p.s. due mainly to the axial and equatorial protons, respectively (but complicated somewhat by coupling). The coalescence temperature was $-66.7°$, which gave a rate constant for the chair-chair interconversion of 52.5 sec.$^{-1}$. From the Eyring equation

$$k = [(\kappa k_B T)/h]e^{-\Delta G^{\ddagger}/RT} \tag{7}$$

(where k is the rate constant, κ the transmission coefficient, k_B Boltzmann's constant and h Planck's constant), the energy barrier ($\Delta G^{\ddagger}$) that must be surmounted to convert a chair to a boat was calculated to be 10.1 kcal./mole.

The other method of utilizing NMR spectra in conformational analysis makes use of the magnitude of the coupling constant, which varies with the dihedral angle between the coupled protons. A study of a number of acylated sugars and related compounds[88] gave values of the coupling constants between an axial and an equatorial proton ($J_{a,e}$) or between two equatorial protons ($J_{e,e}$) on adjacent carbons as 2–3.5 c.p.s., whereas the values are much larger for diaxial protons ($J_{a,a}$ is 5–9 c.p.s.; Sec. 6-5c). By means of a valence bond calculation Karplus[93] showed that as the dihedral angle (θ) between two protons on adjacent carbons increases from $0°$ to 90 to $180°$, the coupling constant J_{HH} should go from a medium value to near zero to a large value, according to the relation

$$J_{\mathrm{HH}} = a \cos^2 \theta \tag{8}$$

The value of a appears to be dependent on the system,[93a] but it is somewhat larger from $90°$ to $180°$ than it is from $0°$ to $90°$. Williamson and Johnson[94] used this method to study conformations of ring A in a series of acetoxylated 3-keto steroids. The 2α-acetoxyl compound (A, Fig. 5) showed the axial proton at C-2 as a quartet. The coupling constants measured were

[91] H. S. Gutowsky and C. H. Holm, *J. Chem. Phys.*, **25**, 1228 (1956).

[92] (a) F. R. Jensen, D. S. Noyce, C. H. Sederholm, and A. J. Berlin, *J. Am. Chem. Soc.*, **82**, 1256 (1960); (b) F. R. Jensen, D. S. Noyce, C. H. Sederholm, and A. J. Berlin, *ibid.*, **84**, 386 (1962); (c) R. K. Harris and N. Sheppard, *Proc. Chem. Soc.*, **1961**, 418.

[93] M. Karplus, *J. Chem. Phys.*, **30**, 11 (1959).

[93a] K. L. Williamson, *J. Am. Chem. Soc.*, **85**, 516 (1963).

[94] K. L. Williamson and W. S. Johnson, *J. Am. Chem. Soc.*, **83**, 4623 (1961).

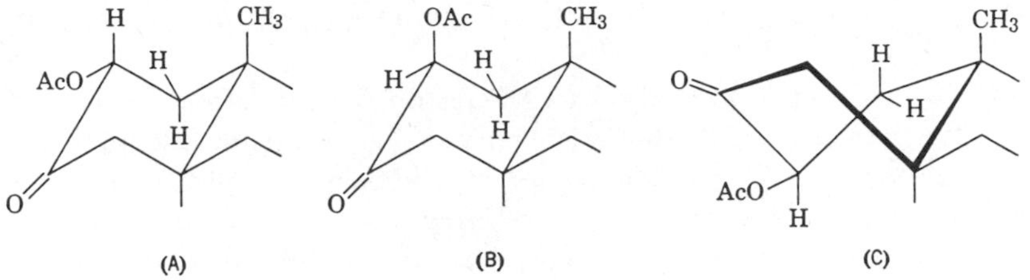

Fig. 5. Ring A in 2-acetoxy-3-ketosteroids.

13.1 c.p.s. for $J_{a,a}$ and 6.6 for $J_{a,e}$, which values can be shown to correspond to a slightly deformed chair. The 2β isomer might have been expected to have a chair conformation as in B, Fig. 5. In this conformation the protons at C-1 are nearly equivalent, and a triplet would be predicted for the α proton at C-2, with a J of 3–4 c.p.s. Actually a quartet was found, with $J = 7.4$ and 9.5 c.p.s. The interpretation given was that these values correspond to dihedral angles of 13° and 133°. Such an arrangement is possible only if ring A is in the boat form (C). The repulsion between the oxygen axial at C-2 and the C-19 methyl is presumably sufficient to cause ring A to exist in this form.

This conclusion was then checked[95] by replacing the 1α hydrogen by deuterium, thus eliminating the *cis* coupling, so the hydrogen at C-2 then appeared as a doublet with $J = 7.0$ c.p.s. as expected. The simplification of an NMR spectrum by the elimination of long range or unwanted couplings through the use of deuterium substitution is a generally useful technique.[96]

An alternative way of eliminating an unwanted coupling between the proton being observed and another nucleus, which is often useful, is referred to as the "double resonance" technique.[96a] The principle involves irradiating the atom causing the unwanted coupling at its nuclear magnetic resonance frequency. The troublesome nucleus can thus be raised to its excited magnetic state and held there by the irradiation. Since all the nuclei of the given magnetic type are then in a single magnetic state, no splitting of the observed signal occurs—the signal is simply shifted by the small change in field strength at the observed nucleus. The technique can be applied to decoupling a number of nuclei simultaneously. The principal limitation of the method is that the nucleus one wishes to observe and the one being decoupled must have sufficiently

[95] F. J. Schmitz and W. S. Johnson, *Tetrahedron Letters*, **1962**, 647.

[96] F. A. L. Anet, *J. Am. Chem. Soc.*, **84**, 1053 (1962); E. A. Allan, E. Premuzic, and L. W. Reeves, *Can. J. Chem.*, **41**, 204 (1963).

[96a] J. D. Baldeschwieler and E. W. Randall, *Chem. Revs.*, **63**, 81 (1963).

different chemical shifts to make it possible to saturate one and not the other.

Analogous studies were also carried out on the 2-fluoro compounds.[97] Here the proton at C-2 in the 2β isomer was seen as a pair of triplets (the fluorine splits the C-2 proton triplet into a pair of triplets) with $J = 3.9$ c.p.s., which corresponds to an ordinary chair form. Since fluorine is only slightly smaller than oxygen (Sec. 2-2b), the difference between the 2β-fluoride and the corresponding acetate was unexpected.

This method is somewhat limited by the fact that it is necessary that the proton being examined resonates at a field strength which separates it from other protons in the molecule; otherwise with such complicated structures the fine structure of the peaks cannot be determined. Furthermore, if the chemical shifts of the two protons coupled to the one being examined are not quite different, the coupling constants are not simply given by the separation of the individual peaks but must be found by more complicated methods.[86] In the steroidal systems mentioned the presence of the electronegative atom (oxygen or fluorine) on C-2 shifts the C-2 proton away from the rest of the absorption of the molecule sufficiently to make its study easy. The chemical shifts of the protons coupled to the one being studied cannot be observed, and it was necessary to assume that they were quite different. The assumption has been shown to be justified in the case of the acetoxy compounds.[95]

Bond angles and hybridizations have also been measured by an NMR method (Sec. 4-1).

3–5. Dipole Moments[98]

When two non-identical atoms are bonded together, the centers of negative and positive charge will not ordinarily coincide and an electric dipole moment will result. The dipole moment (μ) is related to the amount of charge separated (e) and the distance between positive and negative centers (d) by

$$\mu = ed \tag{9}$$

The charge of an electron is 4.8×10^{-10} esu., and, if a unit negative charge is separated by 1 Å (10^{-8} cm.) from a unit positive charge, the

[97] N. L. Allinger, M. A. DaRooge, M. A. Miller, and B. Waegell, *J. Org. Chem.*, **28**, 780 (1963).

[98] (a) L. E. Sutton, in E. A. Braude and F. C. Nachod, eds., *Determination of Organic Structures by Physical Methods*, Vol. I, Academic Press, Inc., New York, 1955, p. 373; (b) J. W. Smith, *Electric Dipole Moments*, Butterworth & Co., Ltd., London, 1955.

resultant moment is 4.8×10^{-18} esu. or 4.8 D. For the hydrogen chloride molecule in the gas phase, $\mu = 1.03$ D. Since the distance between atomic centers is 1.28 Å, separation of a full charge would correspond to $1.28 \times 4.8 = 6.15$ D. The actual moment is therefore said to correspond to 17% ionic character.[99]

The theory behind the determination of a dipole moment is briefly as follows.[98] If a non-polar molecule is introduced between the plates of a condenser to which a unit field has been applied, the molecule is deformed and the electrons tend to shift their average position so as to be a little nearer the positive plate, leaving the nuclei a little nearer the negative plate. The magnitude of this shift depends on α, the polarizability of the molecule. The total induced molar polarization is given by $P_\alpha = (4\pi/3)N\alpha$, where N is Avogadro's number.

If a molecule containing a dipole moment is now placed in the same field, it similarly experiences an induced polarization. In addition, the dipole will also try to line up so as to oppose the field. The latter tendency leads to further polarization, called the permanent molar polarization, which is given by $P_\mu = (4\pi/3)N(\mu^2/3kT)$, where k is the Boltzmann constant and T is the absolute temperature. The observed polarization P is then given by $P = P_\alpha + P_\mu$. The dipole moment can obviously be determined from the variation of P with temperature. Such measurements are sometimes made on gases, and they yield true, or gas phase, moments. Most of the measurements on organic compounds, however, have been made in solution, and this can be regarded as the ordinary method. The moments so obtained are somewhat affected by solvation, but for most purposes this is of no consequence.

In practice, P is found from the measured dielectric constant (ϵ) using the Clausius-Mosotti relationship:

$$P = [(\epsilon - 1)/(\epsilon + 2)] \cdot (M/\rho) \tag{10}$$

where M is the molecular weight and ρ is the density. Because intermolecular interactions would render these relationships useless in a polar liquid, measurements are consequently made in dilute solution in non-polar solvents. Benzene is commonly used as the solvent, sometimes heptane or dioxane. It is assumed that the solutions are ideal, and the solvent and solute contribute to the polarization according to their mole

[99] Most of the commonly encountered single functional groups give a compound a dipole moment in the range of roughly 0.5 to 5 D. A moment smaller than 0.5 D cannot be measured very accurately by the method discussed here but can sometimes be determined by a similar method[100] or, if the molecule is sufficiently simple, from its microwave spectrum (see Sec. 3–2c).

[100] A. J. Petro and C. P. Smyth, *J. Am. Chem. Soc.*, **80**, 73 (1958).

fractions (N): $P = N_1P_1 + N_2P_2$, where $N_1 + N_2 = 1$. The measurements are extrapolated to obtain the molar polarization of the solute at infinite dilution $(P_{2\infty})$.

To obtain the dipole moment of a compound the quantity needed from experiment is P_μ. This quantity cannot be determined directly, but P and P_α can, and their difference gives P_μ. P is determined with an electric field which alternates at radiofrequencies. As the field alternates, the molecules turn over and the electrons readjust their positions. P_α is determined by having the field alternate so rapidly that the molecule cannot turn fast enough to keep up with it. The electrons move much more rapidly, and in this case P_α alone is observed. In practice the electric vector of a light beam is used as the high frequency field. The Maxwell relation, $\epsilon = n^2$ (where n is the refractive index) leads to

$$P_\alpha = [(n^2 - 1)/(n^2 + 2)] \cdot (M/\rho) \tag{11}$$

For most purposes P_α need not be measured at all, since it is equal to the molar refraction, which can be evaluated from tables.[101]

The actual numerical calculations as usually done[102] are tedious to carry out and require several hours with a desk calculator, whereas an electronic computer can do the entire calculation in less than one minute.[103] Simplified (approximate but rather accurate) calculations are also possible.[98] The dipole moments reported in the literature before 1962 have been compiled in a comprehensive book.[104]

Dipole moments are vectors, and as such they are conveniently useful for the solution of geometric problems. As a result, dipole measurements have been widely used in the past for the determination of chemical structure, mainly by physical chemists. Recently precision apparatus for the accurate determination of dipole moments has become commercially available, and the method is now being applied rather extensively in organic chemistry, and to the solution of conformational problems in particular. For example, one isomer of 2-fluorocholestan-3-one was prepared some years ago, but it was not completely clear from the available data whether the configuration of the fluorine was α or β in the compound. The moments of the individual C=O and C—F groups were known from model studies, and therefore, by adding these values vectorially, one could calculate the moments that should have been obtained for the

[101] A. I. Vogel, W. T. Cresswell, G. J. Jeffrey, and J. Leicester, *Chem. Ind. (London)*, **1950**, 358.

[102] I. F. Halverstadt and W. D. Kumler, *J. Am. Chem. Soc.*, **64**, 3988 (1942).

[103] N. L. Allinger and J. Allinger, *J. Org. Chem.*, **24**, 1613 (1959).

[104] A. L. McClellan, *Tables of Experimental Dipole Moments*, W. H. Freeman and Company, San Francisco, 1963.

different configurations of the fluorine. Since, if the fluorine is α it is equatorial and almost in the plane of the carbonyl, a large resultant moment is expected. For the β-axial fluorine, on the other hand, the moment points at about right angles to the plane of the carbonyl, and hence a much smaller moment is expected. The values calculated for the two alternative configurations were 4.28 D and 2.95 D, respectively. The experimental value was determined and found to be 4.39 D, which unequivocally established the configuration as α.[105]

Conformational equilibria have also been widely studied by this method. The *trans* isomer of 1,2-dibromocyclohexane would be expected to exist as a mixture of two conformational isomers, the diaxial and the diequatorial.[106] Since in the former the dipoles are oriented at approximately 180° to one another, the resultant moment of that conformer is near 0. The resultant moment estimated for the diequatorial conformation is the same as that of the *cis* isomer, 3.13 D. The experimental dipole moment determined was 1.76 D in carbon tetrachloride solution at 30°, and from the equation

$$\mu^2 = N_e \mu_e^2 + N_a \mu_a^2 \tag{12}$$

it was calculated that the molecules existed 32% in the diequatorial form and 68% in the diaxial form under these conditions. When the solvent was changed to benzene (which is an effectively more polar solvent), the observed moment increased to 2.16 D, corresponding now to 48% of the diequatorial form. The moment of the *cis* isomer was invariant with solvent as anticipated. These numerical values were confirmed by infrared measurements (Sec. 3-4a) which gave 21% and 42% diequatorial in carbon tetrachloride and benzene, respectively.[107] The interpretation of the shift in equilibrium with solvent is as follows. If the compound is placed in a solvent of dielectric constant D, the classical electrostatic energies of both forms are reduced according to

$$E_{\text{solvent}} = E_{\text{vacuum}}/D \tag{13}$$

The more polar solvent (larger D) brings about a greater lowering of the electrostatic energy. Since the larger dipole has the larger electrostatic energy, the equatorial form is stabilized more by increasing solvent polarity, and the equilibrium is therefore shifted toward the diequatorial conformer with increasing solvent polarity. The conformational equilibria

[105] N. L. Allinger, M. A. DaRooge, H. M. Blatter, and L. A. Freiberg, *J. Org. Chem.*, **26**, 2550 (1961).

[106] (a) W. Kwestroo, F. A. Meijer, and E. Havinga, *Rec. Trav. Chim.*, **73**, 717 (1954); also see (b) K. Kozima, K. Sakashita, and S. Maeda, *J. Am. Chem. Soc.*, **76**, 1965 (1954); (c) P. Bender, D. L. Flowers, and H. L. Goering, *ibid.*, **77**, 3463 (1955); and (d) E. C. Wessels, Ph. D. Dissertation, University of Leiden, Holland, 1960.

[107] P. Klaeboe, J. J. Lothe, and K. Lunde, *Acta. Chem. Scand.*, **11**, 1677 (1957).

which exist in α-halocyclohexanones have also been studied in detail by this method (see Sec. 7-3), as have the halogenated derivatives of ethane.[16]

The geometries of more complicated systems have also been studied by means of dipole moments for the determination of conformation and configuration, for the measurement of conformational equilibria and for the detection of interactions between distant groups.[108] With complicated structures such as 5β-cholestan-3, 17-dione, the determination of the coordinates of the atoms needed for the dipole moment calculation could be carried out by vector analysis, but it would be very laborious. What was therefore done[109,110] was to construct a scale model of the system and to measure the angle in question. The value calculated is limited in accuracy by the faithfulness of the representation of the actual structure by the model. In this case the observed moment (3.5 D) differed from that calculated (3.04 D) from the model of the regular chair system (A in Fig. 6) by 0.5 D, an amount considered to be significantly in excess of experimental error. It was therefore proposed[109] that there was some

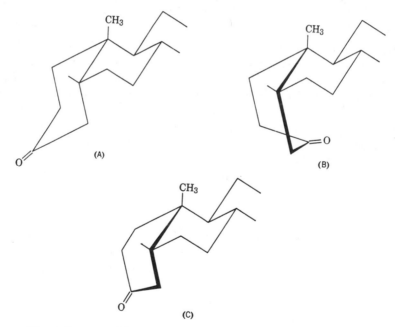

Fig. 6. Some possible conformations of ring A in 3-keto-5β-steroids.

[108] (a) H. R. Nace and R. B. Turner, *J. Org. Chem.*, **25,** 1403 (1960); (b) N. J. Leonard, D. F. Morrow, and M. T. Rogers, *J. Am. Chem. Soc.*, **79,** 5476 (1957).

[109] H. R. Nace and R. B. Turner, *J. Am. Chem. Soc.*, **75,** 4063 (1953).

[110] M. A. DaRooge and C. L. Neumann, unpublished work.

16% of B present in equilibrium with A. In the light of present-day knowledge, however, the structure B puts C-3 much too close to the 9α hydrogen. It seems likely that ring A is in fact in the chair form, which is flattened somewhat by the repulsion between the α hydrogens at C-7 and C-9 with those at C-2 and C-4. A small amount of conformation C may exist in equilibrium with A also (see Sec. 7-4).

With the aid of Dreiding models, which can be projected conveniently onto graph paper, the coordinates of the dipoles can be found rather easily and with fair accuracy. The resultant moment can then be calculated. By using generalized methods this procedure can be applied to a molecule containing any number of dipoles. In this way the configurations and conformations were determined for the 2-fluorocholestan-3,17-diones epimeric at C-2.[97] The calculations become more tedious as the number of dipoles increases, and the possiblility of arithmetic error also increases. The electronic computer has proved to be helpful here also.

Dipole moments have proved to be useful in establishing the side chain conformation of the 20-ketopregnanes,[111] in demonstrating the hindered rotation about the C—O bond in hydroxylated steroids[112] and especially in studies of boat forms and unusual molecular distortions as discussed in Sec. 7-4.

3–6. Optical Rotation

Until about 1956 "optical rotation" without further qualification implied rotation at the sodium D line.[113] Major steps forward in the relation of rotation at the D line to structure have been made since 1959 by J. H. Brewster.[114–116] His fundamental assumption is that the polarizabilities of the groups attached to the asymmetric center can be arranged in clockwise or counterclockwise order, and the contributions to the rotatory power of the center can be made up algebraically from the polarizabilities of the groups. All possible conformations should properly be included in the calculation in proportion to their mole fractions. Since

[111] N. L. Allinger and M. A. DaRooge, *J. Am. Chem. Soc.*, **83,** 4256 (1961).

[112] W. D. Kumler and G. M. Fohlen, *J. Am. Chem. Soc.*, **67,** 437 (1945).

[113] W. Klyne, in E. A. Braude and F. C. Nachod, eds., *Determination of Organic Structures by Physical Methods*, Vol. I, Academic Press, New York, 1955, p. 73.

[114] J. H. Brewster, *J. Am. Chem. Soc.*, **81,** 5475 (1959).

[115] (a) J. H. Brewster, *J. Am. Chem. Soc.*, **81,** 5483 (1959); (b) *ibid.*, 5493; (c) *Tetrahedron*, **13,** 106 (1961); also see D. H. Whiffen, *Chem. Ind. (London)*, **1956,** 964.

[116] For a thorough summary of the general method see E. L. Eliel, *Stereochemistry of Carbon Compounds*, McGraw-Hill Book Co., New York, 1962, p. 398.

such calculations are exceedingly laborious, a number of simplifying assumptions have been made. It was found that for a number of simple molecules the sign and approximate magnitude of the rotation could be predicted. The next step will be to carry out the calculations more accurately with less simplifying assumptions; such calculations are eminently feasible with the aid of an electronic computer.[81] The method would appear to be potentially more useful for the assignment of absolute configuration from known conformational data, rather than the reverse. Applications have been particularly simple and successful in carbohydrate chemistry (Sec. 6-5a).

Another method for obtaining conformational and configurational information which is essentially complementary to Brewster's method has been developed since the late 1950's and is based mainly on the extensive experimental work of Carl Djerassi and his co-workers. Djerassi has made use of optical rotation, not at the D line, but as a function of wavelength, and this is usually referred to as the "optical rotatory dispersion method." [117,118]

The experimental method is simple in principle. Light from an intense source is passed through a prism, and the monochromatic beam is passed through a polarimeter containing the sample to be studied. The two beams emerging from the polarimeter are balanced as for visual polarimetry, except that a photocell is used as the detector. For instrumental reasons the wavelength range commonly examined is approximately from 230 to 700 mμ, although measurements to or below 200 mμ are sometimes reported.

Most compounds show a simple curve (called a "plain curve") when [α] is plotted against λ, for example deoxytigogenin (curve II, Fig. 7). If the compound absorbs light as a result of an electronic transition in the region being examined, an anomalous curve (called a Cotton effect curve) is usually seen (I, Fig. 7). Much more information is usually available from a curve of this type than from a plain curve. Brewster's method is best applied to compounds which have plain curves, whereas the rotatory dispersion method has been most informative for compounds which show a Cotton effect in the wavelength region examined. The sign of the Cotton effect curve gives sufficient information for many purposes. If the peak is at longer wavelength than the trough, the Cotton effect is said to be positive; if the trough is at the longer wavelength, it is negative. For

[117] C. Djerassi, *Optical Rotatory Dispersion*, McGraw-Hill Book Co., New York, 1960.

[118] G. G. Lyle and R. E. Lyle, in F. C. Nachod and W. D. Phillips, eds., *Determination of Organic Structures by Physical Methods*, Vol. II, Academic Press, New York, 1962, p. 1.

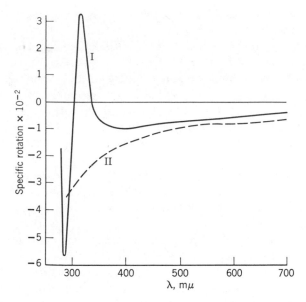

Fig. 7. Positive Cotton effect curve of tigogenone (curve I) superimposed on plain negative "background" curve of deoxytigogenin (II). From Djerassi.[117]

quantitative work the amplitude (measured vertically between the trough and peak with due regard to sign) is desired. Because the amplitudes are so large if expressed as specific rotations, they are ordinarily recorded as molecular values [A], where [A] is equal to the specific rotation at the first extremum minus that at the second extremum, times the molecular weight, times 10^{-4}.

A Cotton effect curve results when there is light absorption by a chromophore which has some sort of asymmetric environment. The wavelength of the absorption maximum corresponds approximately to the wavelength between the two extrema at which the optical rotation is zero. As a practical matter, the compound should not have an extinction coefficient of much more than 100 in the region under examination. If it does, it absorbs so much light that that which finally reaches the photocell may not be sufficient to allow accurate balancing of the two halves of the field. Few of the common organic functional groups meet these requirements, the ketone function being one that does. Other groups such as xanthate, nitrite, nitro and disulfide have been examined.[119] An alcohol, for example, may show an uninformative plain curve, but it may be conveniently converted to a xanthate ester, and the latter will show a Cotton effect

[119] For a summary see ref. 117, p. 191.

curve. The configurations of the C-20 hydroxylated steroids were determined in this way.[120]

The rotatory dispersion method of establishing configuration or conformation has in the past been applied mainly to ketones. This has turned out not to be a serious limitation in practice, since a ketone group can ordinarily be inserted at any desired place in a system without too much trouble. Although the method has proved to have many applications other than the determination of conformations, it is the latter use which is of interest here.

The underlying principle for determining configuration and conformation from rotatory dispersion measurements is the octant rule.[121] This rule enables one to predict qualitatively what kind of a shift of the Cotton effect curve will occur when a substituent is introduced into a carbonyl-containing system at a specified point. It seems likely that it will be possible before long to put this rule on at least a semiquantitative basis, and some success has already been attained along these lines.[51,122-126]

The octant rule was first uncovered for an axial halogen on the α-carbon of a cyclohexanone, and the empirical "axial haloketone rule" was formulated.[127] Corey and Ursprung had noted earlier that the presence of an axial halogen adjacent to a keto group in the triterpene series allowed the absolute configuration at the halogenated carbon to be assigned from the rotation at the D line.[128] Djerassi and Klyne showed that the presence of an axial halogen resulted in an enormous Cotton effect which overwhelmed the effects of the other substituents, and the tailing-off of this Cotton effect curve was what Corey and Ursprung had studied at the D line.

The "axial haloketone rule" states:[117]

1. Introduction of equatorial halogen in either adjacent position of a

[120] C. Djerassi, I. T. Harrison, O. Zagneetko, and A. L. Nussbaum, J. Org. Chem., 27, 1173 (1962).

[121] W. Moffitt, R. B. Woodward, A. Moscowitz, W. Klyne, and C. Djerassi, J. Am. Chem. Soc., 83, 4013 (1961).

[122] C. Beard, C. Djerassi, T. Elliott, and R. C. Tao, J. Am. Chem. Soc., 84, 874 (1962).

[123] C. Djerassi, E. Lund, and A. A. Akhrem, J. Am. Chem. Soc., 84, 1249 (1962).

[124] C. Beard, C. Djerassi, J. Sicher, F. Šipoš, and M. Tichý, Tetrahedron, 19, 919 (1963); see also C. Djerassi, P. A. Hart, and E. J. Warawa, J. Am. Chem. Soc., 86, 78 (1964), and ref. 135a.

[125] N. L. Allinger, J. Allinger, L. E. Geller, and C. Djerassi, J. Org. Chem., 25, 6 (1960).

[126] J. Allinger, N. L. Allinger, L. E. Geller, and C. Djerassi, J. Org. Chem., 26, 3521 (1961).

[127] C. Djerassi and W. Klyne, J. Am. Chem. Soc., 79, 1506 (1957).

[128] E. J. Corey and J. J. Ursprung, J. Am. Chem. Soc., 78, 5041 (1956).

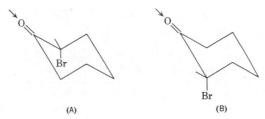

Fig. 8. The two configurations of axial-2-bromocyclohexanone.

keto group in a cyclohexanone does not alter the sign of the Cotton effect of the halogen-free ketone.

2. The effect of introducing an axial chlorine or bromine (and probably also iodine, but not fluorine) atom next to the keto group of a cyclohexanone may change the sign of the Cotton effect from that of the parent ketone. This can be predicted by placing a model of the molecule in such a manner that the carbonyl group occupies the "head" of the chair (or boat) as in A or B (Fig. 8). By looking down the O=C axis as indicated by the arrow, a cyclohexanone derivative with chlorine or bromine to the left (A) of the observer will exhibit a negative Cotton effect whereas a positive curve will be found when the halogen atom is to the right (B).

Theoretical studies[129] indicated that the axial haloketone rule is really a special case of the "octant rule." The latter indicates[121] that the change in a Cotton effect curve which results from a substituent in a given position increases with the increasing polarizability or rotating power of the substituent. Some numerical values are given in Table 3. The high polarizabilities of the halogens (except that of fluorine) means that if the halogen is in the proper position it will dominate completely the Cotton effect curve and for a qualitative prediction the other substituents may be ignored. The conformation of the substituent is critical; substitution at the axial position at C-2 in cyclohexanone leads to a large Cotton effect, while substitution at the equatorial position leads to a negligible effect.

The theory underlying the octant rule is qualitatively easy to understand in terms of the orbitals involved in the $n \rightarrow \pi^*$ transition (Fig. 9). The n orbital is a $2p$ orbital localized on oxygen and lying in the YZ plane. It has a nodal plane (the XZ plane) at which the wave function changes sign and the electron density is zero. During the $n \rightarrow \pi^*$ transition, an electron is excited from this orbital to the antibonding π orbital, which is made up of two $2p$ orbitals, one on carbon and one on oxygen, both of which are parallel to the X-axis. The resulting π^* orbital has

[129] W. Moffitt and A. Moscowitz, *J. Chem. Phys.*, **30**, 648 (1959); ref. 117, p. 150.

Table 3. Rotatory Powers of Some Common Groups[114] (degrees)

Atom or Group	Empirical Value	Calculated Value
I	250	268
Br	180	192
SH	—	174
Cl	170	139
CN	160	87
C_6H_5	140	82
CO_2H	90	82
CH_3	60	60
NH_2	55	53
OH	50	23
H	0	0
F	—	−10

two nodal surfaces: the YZ plane, and a surface (not quite a plane) between the carbon and oxygen and perpendicular to the C—O bond where it cuts the latter. There are thus four lobes in this orbital, and the sign of the wave function in each is shown in Fig. 9. Then n and π^* orbitals are said to be orthogonal; for present purposes this simply means that they do not interact. If a substituent is interjected into a system at any point other than on a nodal surface, it destroys the symmetry of the system and allows both the n and the π^* orbitals to interact with it, and hence with each other. The sign of the Cotton effect curve caused by the substituent will simply be the sign of the product of the

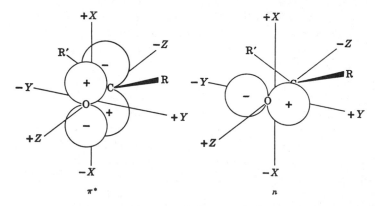

Fig. 9. The n and π^* orbitals of the carbonyl group.

two orbitals (n and π^*) at the position occupied by the substituent. The three nodal surfaces of the carbonyl group divide it into eight octants, and the cyclohexanone may conveniently be projected as in Fig. 10. To predict a qualitative Cotton effect from the location of a symmetrical substituent, it is only necessary to multiply together the coordinates of the substituent. If the substituent lies on one of the planes mentioned, one coordinate is zero, and hence the coordinate product is zero and the substituent does not contribute to the Cotton effect curve. If the product is not zero, the sign of the product is the sign of the Cotton

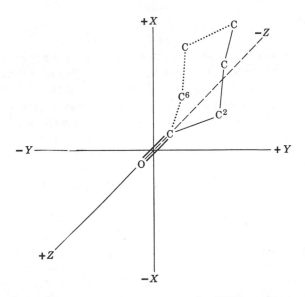

Fig. 10. The orientation of a cyclohexanone on the XYZ coordinates for "octant rule" treatment. From Lyle and Lyle.[118]

effect curve. Thus Fig. 10 shows that a bromine axial at C-2 has a positive Cotton effect $(-X, +Y, -Z)$, while at C-6 the axial bromine has a negative contribution $(-X, -Y, -Z)$. At the equatorial position at either C-2 or C-6, X is approximately zero, so the substituent exerts essentially no effect. It is convenient to project the cyclohexanone against the XY plane looking from the $+Z$ position (Fig. 11). Since Z is negative back of the C—O nodal plane, a substituent in octant II or IV makes a positive contribution, while that of a substituent in octant I or III is negative. A substituent at C-4 does not contribute to the Cotton effect.

To calculate a quantitative value for the amplitude of the curve, such a simple multiplication of the coordinates does not suffice because the contribution to the amplitude made by a substituent varies with its location

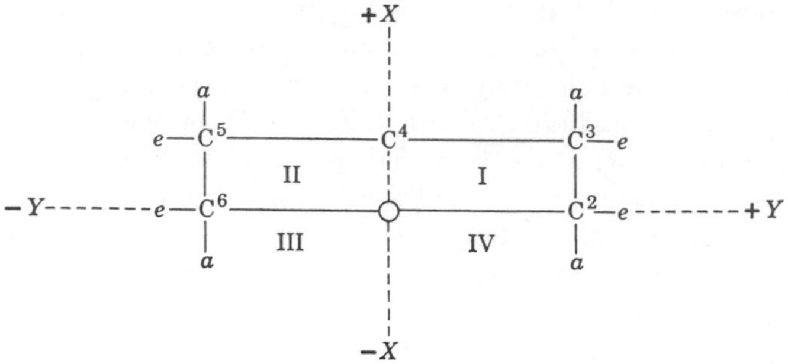

Fig. 11. Octant projection for cyclohexanone.

in a complex way.[129] So far, therefore, an empirical approach has been used. Thus in an ordinary cyclohexanone a methyl axial at C-2 contributes about ± 56 to the molar amplitude,[121] while an equatorial methyl at C-3 contributes ± 26.[123] These effects are approximately additive, barring distortion, and from an accumulation of such values the amplitude corresponding to a given conformational arrangement in an arbitrary molecule can be calculated. Any deviation between the calculated and observed amplitudes leads one to question the assumed conformation, and the method has been useful in quite complicated cases.[51,124]

From these simple considerations an enormous number of conformational facts have been determined. For example, *trans*-2-chloro-5-methylcyclo-hexanone (of the absolute configuration indicated in Fig. 12) has a Cotton effect curve which is highly solvent dependent (Fig. 13).[130] This compound would be expected to be a mixture of diaxial and diequatorial conformations. The dipole moment of the equatorial form is much greater than that of the axial form since the angle between the individual dipoles is much smaller. Hence the equilibrium is shifted to the left as solvent polarity is increased[131] (Sec. 3-5). In octane the strong negative curve is as predicted for axial chlorine, and it corresponds to the diaxial

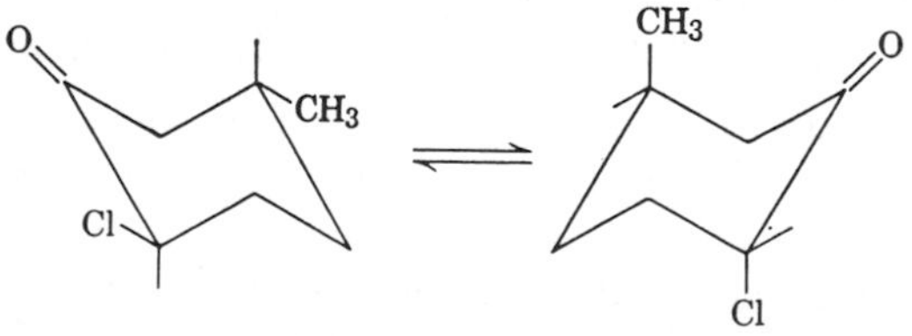

Fig. 12. Conformational equilibrium of *trans*-2-chloro-5-methylcyclohexanone.

[130] Ref. 117, p. 125.
[131] J. Allinger and N. L. Allinger, *Tetrahedron*, **2**, 64 (1958).

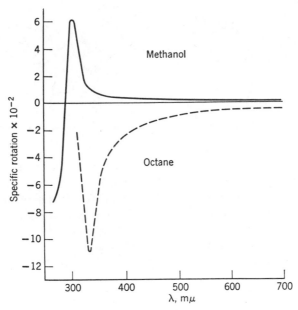

Fig. 13. The rotatory dispersion curves of (−)-*trans*-2-chloro-5-methylcyclohexa-none in methanol and octane solutions. Adapted from Djerassi.[117]

arrangement. When the solvent is changed to methanol, a positive curve is obtained; this corresponds to the diequatorial arrangement.[132]

More interesting perhaps are cases in which unexpected conformational arrangements have been uncovered by rotatory dispersion measurements. The kinetically controlled monobromination product of 2-methylcholestan-3-one was shown by chemical means to have the bromine at C-2. Infrared and ultraviolet spectra indicated that the bromine was in the axial position, suggesting that the compound was 2β-bromo-2α-methylcholestan-3-one. The axial haloketone rule predicts a strong positive Cotton effect for this compound; however, a negative one was observed experimentally.[133] The only way the data can be reconciled is if the compound is the 2α-bromo derivative with ring A in the flexible form, approximately as in (2) in Fig. 14. Confirmation of this interpretation was obtained by the preparation of authentic 2β-bromo-2α-methylcholestanone, which possessed the geometry (1) and showed the predicted negative Cotton effect.

[132] This type of conclusion cannot be drawn indiscriminately when the changes are small however, since cases are known of rigid systems (e.g., cholestanone) which show marked change in the amplitudes of their rotatory dispersion curves with changing solvent (J. G. D. Carpenter, unpublished).

[133] (a) C. Djerassi, N. Finch, R. C. Cookson, and C. W. Bird, *J. Am. Chem. Soc.*, **82**, 5488 (1960); (b) C. Djerassi, N. Finch, and R. Mauli, *ibid.*, **81**, 4997 (1959).

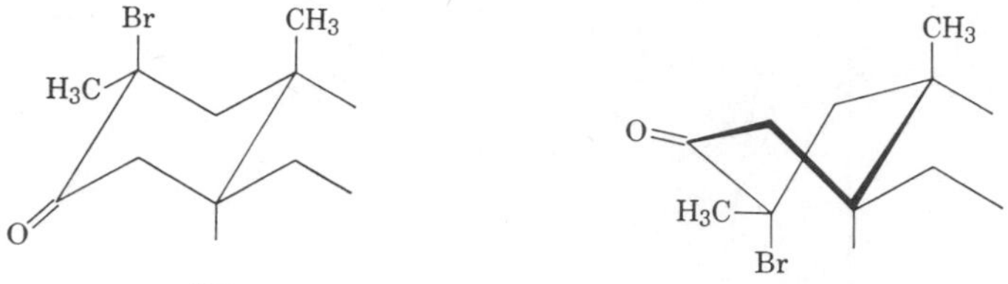

Fig. 14. The 2-axial bromo-2-equatorial methylcholestan-3-ones.

Another example which occasioned some surprise when it was uncovered is the conformation of isomenthone.[117] This molecule may be assumed to consist of a mixture of largely A and B (Fig. 15). It is now known that the conformational energy of an isopropyl group is quite small when it is adjacent to a ketone[34,134] (Sec. 2-6c), but in 1960 it was thought that this energy was very large.[135] By application of the octant rule to the rotatory dispersion curve of the compound, Djerassi suggested[117] that the predominant conformation was A, rather than B, a conclusion which is now known to be correct.[135a]

Another interesting case is the conformation of cis-10-methyl-2-decalone. Here there is in principle an equilibrium between the "steroid" (A) and "non-steroid" (B) conformations (Fig. 16). Conformational analysis indicates that in B (but not in A) there is an axial alkyl group at the carbon β to the carbonyl, and hence the 3-alkylketone effect (Sec. 2-6c) should favor this conformation. The latter effect is now considered to be fairly small (0.5 kcal./mole or less),[134,136] but the prediction required appears to be that A and B should occur together in a nearly 1:1 ratio, with B

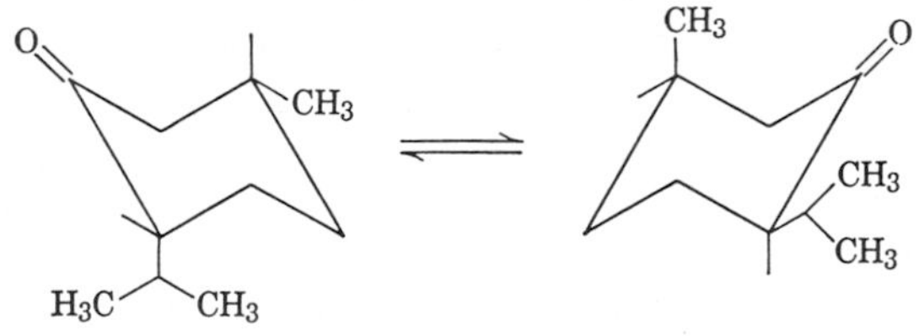

Fig. 15. Conformational equilibrium of isomenthone.

[134] B. Rickborn, J. Am. Chem. Soc., **84**, 2414 (1962).

[135] S. Winstein and N. J. Holness, J. Am. Chem. Soc., **77**, 5562 (1955).

[135a] C. Djerassi, P. A. Hart, and C. Beard, J. Am. Chem. Soc., **86**, 85 (1964).

[136] N. L. Allinger and L. A. Freiberg, J. Am. Chem. Soc., **84**, 2201 (1962).

predominating slightly. The rotatory dispersion curve, however, was originally interpreted[137] as indicating that A in fact predominates, and the matter is not settled at this writing.

Beginning about 1960, optical rotatory dispersion (O.R.D.) curves have been supplemented by circular dichroism (C.D.) curves which ordinarily yield the same general kinds of information.[137a] Optical rotatory dispersion is a result of unequal refractive indices of the medium for right and left circularly polarized light, while circular dichroism results from unequal absorption of right and left circularly polarized light. For the most part, the same qualitative information is available from either type of measurement. The major difference between the two is that Cotton

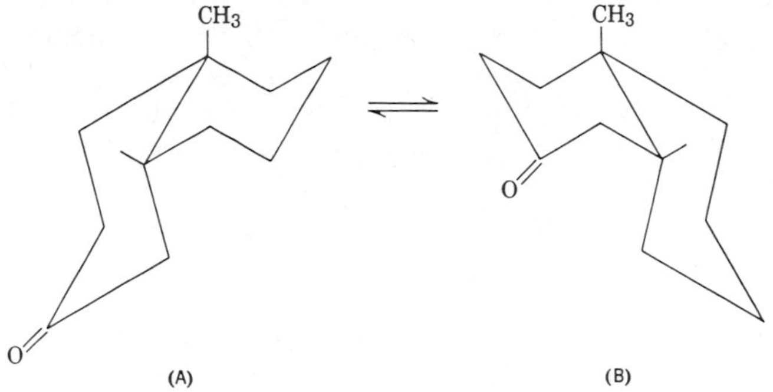

Fig. 16. Conformations of cis-10-methyl-2-decalone.

effect curves die off slowly with distance, so that, in the ultraviolet region ordinarily studied, the Cotton effect curve being examined is superimposed on a plain curve, which is the tailing off of transitions much further down in the ultraviolet. Circular dichroism curves do not have these long tails but are much sharper (like ultraviolet absorption spectra). These properties mean that O.R.D. curves are generally more useful for qualitative studies, since the sign of the background curve is usually apparent and it is often useful, while C.D. curves are quantitatively more useful because the extraneous tailings do not interfere with quantitative measurements pertaining to the transition being studied.

Conformational equilibria have also been studied by means of the variation of C.D. with temperature,[137b] the same principles that were discussed in Sec. 3-4a in application to infrared spectra being used.

[137] C. Djerassi and D. Marshall, J. Am. Chem. Soc., **80**, 3986 (1958).

[137a] A concise summary of the literature and uses of C.D. is given by C. Djerassi, H. Wolf, and E. Bunnenberg, J. Am. Chem. Soc., **84**, 4552 (1962).

[137b] A. Moscowitz, K. Wellman, and C. Djerassi, J. Am. Chem. Soc., **85**, 3515 (1963); K. M. Wellman, E. Bunnenberg, and C. Djerassi, ibid., **85**, 1870 (1963).

3–7. von Auwers' Rule (Physical Properties; Energy Relationships)

Organic chemists have long felt that there should exist simple relationships between the stereochemistry of compounds and their gross physical properties such as boiling point, density and refractive index.* Properties of this sort are bulk properties; that is, they depend on an assemblage of molecules, and not only on the molecule itself. Such properties are consequently determined by intermolecular interactions, while the stereochemistry has a direct effect on the intramolecular interactions. It is not clear *a priori* whether or not there will be any connection between the two types of interactions. Because our knowledge of intermolecular forces is in a very primitive state, any interpretation of bulk properties seems doomed to failure at the outset. As usual though, the organic chemist is able to discern and exploit empirical relationships that are found to exist. Some empirical relationships, such as the laws of thermodynamics and the Schroedinger equation, have been found to be highly useful, most others less so. The most important thing to know in applying empirical relationships to any problem is the boundaries within which the relationship will hold.

von Auwers' rule suggested that with *cis-trans* isomeric hydroaromatic compounds the *cis* had the higher density and refractive index and lower molecular refractivity.[138] This rule sometimes failed, and its use in earlier times led to the misassignment of stereochemistry to a number of molecules. The original rule and subsequent modifications[139] do not in general attempt to calculate a bulk molecular property from first principles, but rather compare the compound in question with a stereoisomer. It was found that the *cis* isomer of 1,2-dimethylcyclohexane had a higher boiling point, greater density and greater refractive index than did the *trans* isomer, and the same relationship was seen to hold for the 1,4-dimethylcyclohexanes. It seemed logical at the time to assume that the isomer of 1,3-dimethylcyclohexane which had the higher values for these quantities would similarly be *cis*, and the geometric assignment was made on that basis. Many years later it was shown by preparation of the supposedly *cis* isomer in optically active form[140] that the original assignment was incorrect, and the proper interpretation of the data was that the

---

[138] K. von Auwers, *Ann.*, 420, **84** (1920).

[139] A review is given by H. van Bekkum, A. van Veen, P. E. Verkade, and B. M. Wepster, *Rec. Trav. Chim.*, **80**, 1310 (1961).

[140] M. Mousseron and R. Granger, *Bull. Soc. Chim. France*, **5**, 1618 (1938).

* Melting points may also relate to stereochemistry, but the relationship is quite different and will be dealt with separately (Sec. 4-4c).

important relationship in determining the physical properties was whether the methyls were axial or equatorial, and not whether they were *cis* or *trans*.[21,141,142]

Rather than to consider separately the various properties discussed above, it is convenient to examine only the molecular volume. In most cases (but not all) the isomer which has the smaller molecular volume also has the greater refractive index and the greater boiling point, because the smaller molecular volume means that the molecules are packed more tightly together and the dispersion forces tending to keep them together are greater. A more modern statement of von Auwers' rule,[143] which has been called the conformational rule, is: "The isomer with the smaller molar volume has the greater heat content." Rather than to attempt to delineate empirically the exact region of applicability of the conformational rule, it is preferable to try to understand where it succeeds and why it fails.

When one considers the epimeric pairs of the dimethylcyclohexanes, the conformational rule is seen to hold for each of them. It only relates the molecular volume to the heat content, however, and does not tell anything about geometry. Yet, since it is easy to relate heat content and geometry by conformational analysis, the conformational rule provides, in effect, a way to deduce the geometry from an easily determined physical quantity, the molar volume of the liquid.

Why does the observed relationship hold? If a methyl is axial, it suffers from an overlapping of its van der Waals radius with those of the *syn*-axial hydrogens. Certainly, if two or more atoms are pressed closer together than the sum of their van der Waals radii, the enthalpy of such an arrangement will be increased. Simultaneously the volume of the arrangement will be decreased, if by volume we mean that amount of space displaced by the atoms. To measure this displaced volume, a continuous and indifferent medium is desirable, and the nearest thing to such a medium is a hydrocarbon solvent. Hence the molar volumes of the pure dimethylcyclohexanes should be as good a measure of their enthalpies as one might hope to obtain by this method. More generally, it would be expected that any *gauche* arrangement of a paraffin chain would be of higher enthalpy and smaller volume that the corresponding *anti* arrangement, and this seems indeed to be the case.[144]

For a number of simple molecules, enthalpy data are available for both the liquid and gas phases, and one should ask which values are to be used for conformational analysis and why. Because the compounds which appear

[141] N. L. Allinger, *Experientia*, **10**, 328 (1954).
[142] R. B. Kelly, *Can. J. Chem.*, **35**, 149 (1957).
[143] N. L. Allinger, *J. Am. Chem. Soc.*, **79**, 3443 (1957).
[144] N. L. Allinger, M. Nakazaki, and V. Zalkow, *J. Am. Chem. Soc.*, **81**, 4074 (1959).

to be most carefully studied in this respect are the dimethylcyclohexanes, the 1,3-isomers are considered specifically here.

The most precise data on the difference between the *cis* and *trans* isomers come from equilibration[145] and calorimetric[21] studies. From the former, the enthalpy of the *trans* isomer was found to be greater than that of the *cis* by 1.80 ± 0.06 kcal./mole in the liquid phase at $274°$. When the flexible and diaxial conformations were taken into account, this gave the conformational enthalpy of the methyl group as 1.74 ± 0.19 kcal./mole. Calorimetric measurements gave a value of 1.72 ± 0.22 kcal./mole in the liquid phase ($25°$), and 1.96 ± 0.22 kcal./mole in the gas phase, and at this temperature these values correspond to the conformational enthalpy of the methyl group. The gas phase value compares the isolated molecules when there are no intermolecular forces acting. In the liquid phase the *trans* isomer has the smaller molecular volume, the molecules are therefore more closely packed, and the intermolecular forces are greater. This means that more energy will be required to overcome these forces during vaporization, and the boiling point and heat of vaporization will be higher for the *trans* isomer (Trouton's rule). Hence it is found that the enthalpy of the *trans* isomer is greater than that of the *cis* by a larger amount in the gas phase, since the enthalpy of vaporization adds to the total enthalpy. In solution, the intermolecular forces acting are more like those in the liquid hydrocarbon than in the vapor, and for properties or reactions in solution, which is what the organic chemist is ordinarily interested in, the conformational enthalpy of the methyl group is properly taken to be 1.7 ± 0.1 kcal./mole. The value for a *gauche* interaction in an open chain would be expected to be a little smaller than one-half this, since rotation about the central bond will cause introduction of a small amount of torsional strain but substantial relief of the van der Waals repulsion. Such a rotation is more difficult if the atoms are part of a cyclohexane ring, because then concomitant bond angle distortion occurs. The best value for the *gauche* interaction in open chains appears to be 0.8 kcal./mole.[146]

A comparison may now be made of the volume-enthalpy relationships between the isomers of the dimethylcyclohexanes, other than epimers.[147] When the 1,1-dimethyl compound was omitted, it was found that the molar volume was inversely proportional to the enthalpy.[144] The 1,1-dimethyl isomer was found to be about 0.6 kcal./mole more stable than was

[145] W. Szkrybalo and F. A. Van Catledge, unpublished results.

[146] K. S. Pitzer, *Chem. Rev.*, **27**, 39 (1940).

[147] The data on hydrocarbons which follow are taken from *Physical and Thermodynamic Properties of Hydrocarbons and Related Compounds* (American Petroleum Institute), National Bureau of Standards.

indicated by its molar volume. Similarly, it was noted that ethylcyclo-hexane was about 0.6 kcal./mole less stable. It therefore seems likely that the "branched chain effect"[148] which serves to lower the enthalpy of branched chain relative to normal compounds does not affect molar volume. If the enthalpies of these two isomers are corrected for the difference in chain branching by this amount, they also fit in the linear relationship.

Application of the conformational rule to open chain systems appeared also to be possible. When the relative heat contents were plotted against the molar volumes for all the octanes (omitting one which is a solid), little correlation was apparent (Fig. 17). However, when the chain branching effect was taken into account, sixteen of the seventeen available experimental points fell (within the probable experimental error in the combustion measurements) on a straight line[144] (Fig. 18), and this suggests that an experimental reinvestigation of the seventeenth compound is now in order.

Strain energy in a molecule may result from van der Waals interactions (such as the *gauche* interactions) which appear to be related to the physical properties, or from unfavorable torsional angles or from deformed bond angles. Bond angle strain and torsional strain appear to raise the energy of a molecule without the corresponding decrease in the molar volume. Thus, if *n*-propylcyclopentane were to be plotted on a graph with the dimethyl- and ethylcyclohexanes, it would be seen that the energy of the

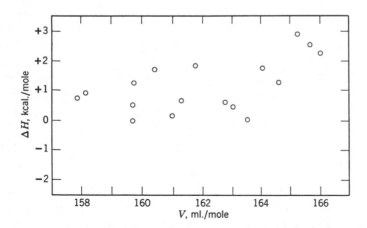

Fig. 17. The relationship between the heat content and molar volume for the octanes in the liquid phase at 25°. From Allinger, Nakazaki, and Zalkow.[144]

[148] G. E. K. Branch and M. Calvin, *The Theory of Organic Chemistry*, Prentice-Hall, Inc., New York, 1941, p. 274.

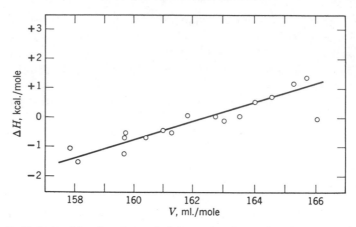

Fig. 18. Relationship of molar énthalpies and volumes for the octanes corrected for chain branching in the liquid phase at 25°. From Allinger, Nakazaki, and Zalkow.[144]

compound is much too high in proportion to its volume. The interpretation is that the strain energy of the cyclopentane ring is largely angular and torsional rather than van der Waals, and hence it is not reflected in the molar volume. The six-membered ring is unique in that the substituents are in general positioned in nicely staggered arrangements, as with open chains, while rings of other sizes do not have such a simple geometry. It has nonetheless proved possible to apply the conformational rule to three-,[149] five-,[147,37] seven-,[150,151] and eight-[152] membered rings in specific cases, but the generality of the rule when applied to rings other than six-membered remains to be established. The only case which appears to be known of a hydrocarbon which positively does not fit the conformational rule is that of the [3.3.0]bicyclooctanes.[147] The molar volume of the *trans* isomer is definitely greater, but the heat content of that isomer is also definitely greater. The interpretation of this apparent anomaly is that the situation at the ring juncture of the *cis* isomer is more analogous to a *gauche* interaction than in the *trans* isomer, and hence the *trans* has the larger molar volume. The *trans* isomer, however, suffers from much more severe angle strain than is present in the *cis*, and this raises the enthalpy of the *trans* isomer very considerably. Hence, if one wishes to compare isomers in which angular strain is present, one can do so only if the angular strain remains reasonably constant from one isomer

[149] W. von E. Doering and W. Kirmse, *Tetrahedron*, **11**, 272 (1960).

[150] N. L. Allinger and V. B. Zalkow, *J. Am. Chem. Soc.*, **83**, 1144 (1961).

[151] N. L. Allinger, *J. Am. Chem. Soc.*, **81**, 232 (1959).

[152] N. L. Allinger and S. Hu, *J. Am. Chem. Soc.*, **83**, 1664 (1961).

to the other. Thus one would expect, and would find, that the dimethyl-cyclopentanes fall reasonably well on a line when a plot is made of their molar volumes vs. their heat contents.[147]

The *cis* and *trans* isomers of decalin, which do not differ in angular strain, also fit the rule very well.[152a] The *cis* and *trans* isomers of hydrindane do differ slightly in angular strain but not sufficiently to make the rule inapplicable to them.[152b] This fact could not have been firmly predicted in advance, however.

The conformational rule is on safest ground when applied to hydrocarbons, and with such compounds (in the absence of angle and torsional strain) it seems to be as good a method for stereochemical assignments as one could want. It is certainly related to the rather small dispersion or London forces that hold a liquid together. If the molecules of the liquid contain dipoles, then the dipole-dipole interaction, a very strong force relative to the intermolecular dispersion forces in hydrocarbons, comes into play. A larger dipole moment tends to hold a liquid more tightly together, and, if two isomers show a significant difference in dipole moment, the one with the higher moment will ordinarily show the higher boiling point.[143] In the few cases studied where the two epimers have the same moment, the one with the equatorial dipole has the higher boiling point, although the molecular volume follows the conformational rule.[153,154]

The equatorial group is less shielded by the surrounding hydrocarbon part of the molecule than is the axial, and it is therefore better able to associate intermolecularly with other dipoles. Such an association must be overcome to vaporize the substance, and consequently the boiling point is increased. [While such association is expected to lead to a slight volume contraction (electrostriction), in the cases studied the observed molar volumes show that this contraction is small compared to the effect of overlapping van der Waals radii on the volume.] This means that an equatorial polar group, like a hydroxyl[153] or carboethoxyl[154] group, is effectively more polar than its axial counterpart. This difference is quite consistent, and of enormous practical value in the separation of epimeric mixtures. Thus a molecule with a large dipole moment has a longer retention time on an alumina column than does its geometric isomer,[154a]

[152a] N. L. Allinger and J. L. Coke, *J. Am. Chem. Soc.*, **81**, 4080 (1959); D. M. Speros and F. D. Rossini, *J. Phys. Chem.*, **64**, 1723 (1960).

[152b] N. L. Allinger and J. L. Coke, *J. Am. Chem. Soc.*, **82**, 2553 (1960); F. D. Rossini and C. C. Browne, *J. Phys. Chem.*, **64**, 927 (1960).

[153] E. L. Eliel and R. G. Haber, *J. Org. Chem.*, **23**, 2041 (1958).

[154] N. L. Allinger and R. J. Curby, Jr., *J. Org. Chem.*, **26**, 933 (1961).

[154a] E. R. Ward, W. H. Poesche, D. Higgens, and D. D. Heard, *J. Chem. Soc.*, **1962**, 2374.

and, similarly, an equatorial alcohol is eluted more slowly than is its axial epimer.[155] Furthermore, the more polar isomer is more soluble in a polar solvent and will have a longer retention time when subjected to partition chromatography with a polar stationary phase;[156] this is also true for simple cases in gas phase chromatography.[157] The correlation between conformation and retention times in the more complicated cases such as the inositol series is obscure, however.[158] The equatorial epimer will also tend to concentrate in the polar phase during countercurrent distribution.[159] These trends seem to be fairly general for polar molecules, but any conformational assignment based on them must be regarded as tentative.[139]

3–8. Kerr Constants[160]

When a voltage is impressed on a dielectric, the latter becomes doubly refracting; that is, the refractive index at a given wavelength λ is different in directions parallel and perpendicular to the field. This phenomenon is known as the Kerr effect. The relationship

$$n_p - n_s = B\lambda E^2 \tag{14}$$

is found to hold, where n_p and n_s are, respectively, the refractive indices parallel and perpendicular to the field, E is the electric field strength, and B is a constant characteristic of the substance, the so-called Kerr constant. For most purposes it is the molecular Kerr constant, $_mK$ which is desired, and this quantity is in principle determinable for a gas, a liquid, or a substance in solution. The physical significance of this quantity may be taken as the difference between the molecular refractions taken parallel and perpendicular to an applied unit field.

To make use of the observed $_mK$ of a substance, it is necessary to calculate the values of this quantity for various assumed conformations, and, if all but one of the possible conformations show values which deviate appreciably from the experimental value, the conformation can be considered reasonably established. It may happen that several conformers predict the same $_mK$, in which case no decision is possible. It may also happen that the correct conformer was not considered as a possibility.

[155] G. H. Alt and D. H. R. Barton, *J. Chem. Soc.*, **1954**, 4284.

[156] K. Savard, *J. Biol. Chem.*, **202**, 457 (1953).

[157] (a) E. L. Eliel and R. S. Ro, *J. Am. Chem. Soc.*, **79**, 5992 (1957); (b) R. Komers, K. Kochloefl, and V. Bazant, *Chem. Ind. (London)*, **1958**, 1405; R. Komers and K. Kochloefl, *Collection Czech. Chem. Commun.*, **28**, 46 (1963).

[158] Z. S. Krzeminski and S. J. Angyal, *J. Chem. Soc.*, **1962**, 3251.

[159] G. Kortüm and A. Bittel, *Chem. -Ing.-Tech.*, **30**, 95 (1958).

[160] C. G. Le Fèvre and R. J. W. Le Fèvre, *Rev. Pure Appl. Chem.*, **5**, 261 (1955).

The method is consequently most suited to showing which conformer is most likely when there are only a few very definitely understood possibilities from which to choose.

In practice the approximate values of the polarizabilities of various bonds are known along the longitudinal and transverse axes. In an electric field a polar molecule will align its resultant moment with respect to the field, whereas a non-polar molecule will align its axis of greatest polarizability with the field. If a given conformation and its alignment with the field are assumed, the orientation of every bond with respect to the field is known, and addition of the bond polarizability tensors gives the molecular polarizabilities along the three axes, from which $_mK$ can be computed.

As an example, the molecular Kerr constants (times 10^{12}) calculated for cyclohexyl bromide were 179 and 82 for the conformations with equatorial and axial bromine, respectively.[160] The experimental value of 181 shows that the equatorial bromine predominates, but the values were not sufficiently accurate for determination of the equilibrium constant. Later, more refined calculations gave results more in line with those obtained by other methods.[161] A more difficult problem, the conformation of the ring in cyclopentane, was also attacked with the aid of Kerr constants.[162] Agreement between calculation and experiment was found only for the "half-chair" form which has one carbon atom above and one below the plane of the other three.[162] Other workers consider that the weight of the evidence favors a pseudorotating conformation which is not completely in agreement with the "half-chair" form (Sec. 4-3).[163]

One of the rather surprising conclusions drawn from Kerr constant studies concerned the "size" of the lone pair of electrons on nitrogen. It was concluded[164] that in N-methylpiperidine (Fig. 19) a conformational equilibrium existed for which the equilibrium constant was about unity, and that with piperidine itself the hydrogen on nitrogen was very largely

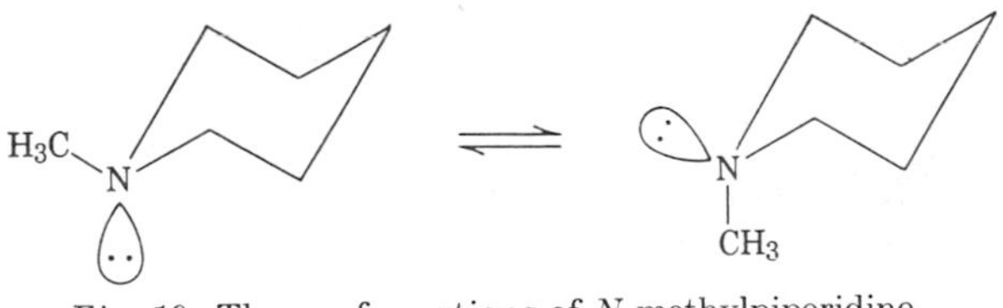

Fig. 19. The conformations of N-methylpiperidine.

[161] C. G. Le Fèvre, R. J. W. Le Fèvre, R. Roper, and R. K. Pierens, *Proc. Chem. Soc.*, **1960**, 117.

[162] R. J. W. Le Fèvre and C. G. Le Fèvre, *Chem. Ind. (London)*, **1956**, 54.

[163] K. S. Pitzer and W. E. Donath, *J. Am. Chem. Soc.*, **81**, 3213 (1959).

[164] M. Aroney and R. J. W. Le Fèvre, *J. Chem. Soc.*, **1958**, 3002.

in the axial position, the lone pair preferring to be equatorial. The available theory is inconsistent with these conclusions, however. First of all, simple van der Waals calculations (Sec. 7-2) indicate[165] that the methyl in *N*-methylpiperidine will be equatorial to the extent of more than 90%, and they also indicate that in piperidine itself the hydrogen will be axial to the extent of about 50%. The "size" of the axial methyl is really concerned with the repulsions between it and the *syn*-axial hydrogens, which are similar to the carbocyclic case. The piperidine case is different, however. The equatorial hydrogen on nitrogen here is just about at the van der Waals distance from the four hydrogens on the α carbons and is hence stabilized by the attractions. The axial hydrogen on nitrogen is just about at the van der Waals distance from the equatorial hydrogens on the α carbons and from the axial hydrogens on the β carbons. The hydrogen on nitrogen should therefore show little conformational preference.

In addition, a detailed quantum mechanical treatment[165] has indicated that when a closed shell particle (a helium atom) approaches an ammonia molecule, for a given nitrogen-helium distance, the repulsion energy is much less when the helium comes in along the axis of the lone pair than when it comes in along an N-H axis.

Recently the axial-equatorial equilibrium constants for the *N*-hydrogen and *N*-methyl of piperidine and *N*-methylpiperidine have been determined by dipole moment measurements on a variety of systems, and for each the data indicated that the hydrogen had a small preference for the equatorial position, while the methyl was found to be very largely in the equatorial position, the free energies varying somewhat, but being about 0.5 kcal./mole for hydrogen and 1.5 kcal./mole for methyl.[166] The same (qualitative) conclusion has also been reached from a variety of other studies, particularly from the proton magnetic resonance spectra of a series of methylated quinolizidines.[167] Hence it would seem that the conclusions drawn from the Kerr constants are not correct. The error appears to stem from the fact that the molecular quantities (bond polarizabilities, angles, etc.) which go into the calculation of the Kerr constants are not known with the accuracy that one would desire, and the resulting inaccuracies in the calculated Kerr constants may be sufficient to make it often impossible to draw reliable conclusions.[167a]

[165] N. L. Allinger and J. C. Tai, *J. Am. Chem. Soc.,* **87,** 1227 (1965).

[166] N. L. Allinger, J. G. D. Carpenter, F. Karkowski, and S. P. Jindal, presented at the National Meeting of the American Chemical Society, April, 1963.

[167] T. M. Moynehan, K. Schofield, R. A. Y. Jones, and A. R. Katritzky, *J. Chem. Soc.,* **1962,** 2637.

[167a] For example, R. F. Zürcher, *J. Chem. Phys.,* **37,** 2421 (1962).

Although the Kerr effect was discovered as early as 1875, it was not used to any extent for conformational studies until the late 1940's, and since that time it has been used almost exclusively by R. J. W. Le Fèvre, C. G. Le Fèvre, and their co-workers.[160] The method has not been widely used, because the equipment is not commercially available but must be constructed and the calculations involved are very lengthy. Beyond this, however, there are a number of assumptions and approximations that must be made, the validity of which is not always certain. The method at best will furnish only one piece of information about the molecule the interpretation of which may be ambiguous. It seems unlikely that the use of Kerr constants as a method of conformational analysis will ever be as important as most of the previously discussed methods. Recently, however, applications of the method to the determination of configurations in complex molecules (cholesteryl halides) have been reported,[168] and some of the computation has been taken over by an electronic computer. With the accumulation of more data on frameworks (like the cholesteryl skeleton) and the development of computer programs for the calculations, the method may become of greater importance in the future.

3–9. Polarographic Reduction[169]

This is a method which in principle might be applied to any reducible group, but which in fact appears only to have been applied (conformationally speaking) to the reductive removal of halogen from carbon systems. For the reduction of an alkyl halide at the dropping mercury electrode, the accepted reaction mechanism is

(1) $\qquad R—X + e^\ominus \rightarrow (R\cdot—X)^\ominus \rightarrow R\cdot + X^\ominus$

(2) $\qquad R\cdot + e^\ominus \rightarrow R\colon^\ominus$

(3) $\qquad R\colon^\ominus + H—\text{solvent} \rightarrow R\colon H + \text{solvent}^\ominus$

with one of the first two processes rate and potential determining.[170] For primary and secondary halides there is a qualitative correlation with S_N2 substitution processes;[171] thus ethyl reduces at a lower potential than propyl, which is lower than isobutyl, which is lower than neopentyl (Table 4). A change in mechanism is indicated for tertiary halides, as

[168] J. M. Eckert and R. J. W. Le Fèvre, *J. Chem. Soc.*, **1962**, 1081.

[169] (a) I. M. Kolthoff and J. J. Lingane, *Polarography: Polarographic Analysis and Voltammetry, Amperometric Titrations*, Interscience Division, John Wiley and Sons, New York, 1941; (b) L. Meites, *Polarographic Techniques*, Interscience Division, John Wiley and Sons, New York, 1955.

[170] N. S. Hush, *Z. Elektrochem.*, **61**, 734 (1957).

[171] F. L. Lambert and K. Kobayashi, *J. Am. Chem. Soc.*, **82**, 5324 (1960).

t-butyl is between ethyl and isopropyl. According to Elving,[172] the highly nucleophilic electrode adds the initial electron from the back side; hence the analogy to S_N2 processes. In the tertiary case the stability of the radical formed seems to speed up the reaction, and the reduction of cyclic

Table 4. Polarographic Reduction Potentials of Selected Compounds[a]

Compound	$-E_{1/2}$, volts	Reference
Ethyl bromide	2.13	171
Isopropyl bromide	2.26	171
Neopentyl bromide	2.37	171
t-Butyl bromide	2.19	171
Cyclobutyl bromide	2.36	171
Cyclopentyl bromide	2.19	171
Cyclohexyl bromide	2.29	171
Cycloheptyl bromide	2.27	171
cis-4-*t*-Butylcyclohexyl bromide	2.28	174
trans-4-*t*-Butylcyclohexyl bromide	2.43	174
2-Fluorocyclohexanone	2.00	175
cis-2-Fluoro-4-*t*-butylcyclohexanone	2.08	175
trans-2-Fluoro-4-*t*-butylcyclohexanone	1.85	175
2-Chlorocyclohexanone	1.40	175
cis-2-Chloro-4-*t*-butylcyclohexanone	1.57	175
trans-2-Chloro-4-*t*-butylcyclohexanone	1.43	175
cis-1,2-Dibromocyclohexane	1.64	178
trans-1,2-Dibromocyclohexane	1.04	178
Ethylene dibromide	1.23	178
erythro-5,6-Dibromodecane	1.14	178
threo-5,6-Dibromodecane	1.24	178
4-*t*-Butyl-*trans*-3-*cis*-4-dibromocyclohexane	0.86	178
4-*t*-Butyl-*cis*-3-*trans*-4-dibromocyclohexane	1.67	178

[a] Because of experimental differences, the values from any one reference are not directly comparable with those from another reference.

halides seems more in accord with the I-strain concept for an S_N1, rather than an S_N2 process,[173] except for the fact that cyclodecyl bromide requires the same potential as does cyclohexyl. The *cis*-4-*t*-butylcyclohexyl bromide reduces at -2.28 volts, while for the *trans* isomer the corresponding value is -2.43 volts.[174] The concept[172] of a more hindered

[172] (a) P. J. Elving, *Record Chem. Progr.*, **14**, 99 (1953); (b) P. J. Elving and J. T. Leone, *J. Am. Chem. Soc.*, **79**, 1546 (1957).

[173] H. C. Brown, R. S. Fletcher, and R. B. Johannesen, *J. Am. Chem. Soc.*, **73**, 212 (1951).

[174] F. L. Lambert, private communication.

back side approach for the equatorial isomer is in general consistent with the observed results; so is the idea that, since the same radical is obtained from both epimers, the less stable epimer should react more rapidly.

When the 2-halo-4-t-butylcyclohexanones (fluorides or chlorides) were reduced,[175] the potentials were found in the same order, that is, the axial epimer reduced the more easily, and this was also found to be true in a number of steroidal haloketones.[176] With the haloketones the reduction potentials are much lower than in the simple halides (Table 4), and here the addition of the first electron is into the π^* orbital of the carbonyl rather than the σ^* orbital of the halogen-carbon link.[177]

A compound like cyclohexyl bromide contains both the axial and the equatorial conformations in sizable amount. Since the equilibrium between the conformers is very fast compared to the rate of reduction, the observed potential for such a compound is that of the most easily reduced form.[174] A compound like 2-fluorocyclohexanone, on the other hand, is substantially all in the equatorial conformation in the highly polar solutions used for polarographic reduction, and it reduces at the same potential as does cis-2-fluoro-4-t-butylcyclohexanone in which the halogen is fixed in the equatorial position.[175]

It thus seems to be generally true that axial halides are more easily reduced than are their epimers, and stereochemistry can probably be assigned accordingly. Since the number of cases so far studied has not been large, the conclusions should be drawn with some reservations.

Another series of compounds which has been examined is the 1,2-dibromides.[178] Here again the required potentials are much lower than for ordinary halides, suggesting that these reactions have some E_2 character.

Among those studied were the isomeric $trans$-dibromides in Fig. 20.

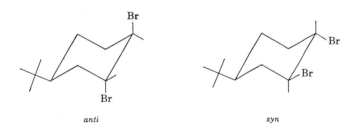

Fig. 20. The isomeric 1,2-$trans$-dibromo-4-t-butylcyclohexanes.

[175] A. M. Wilson and N. L. Allinger, *J. Am. Chem. Soc.*, **83**, 1999 (1961).

[176] P. Kabasakalian and J. McGlotten, *Anal. Chem.*, **34**, 1440 (1962).

[177] P. J. Elving and R. E. VanAtta, *J. Electrochem. Soc.*, **103**, 676 (1956).

[178] J. Závada, J. Krupička, and J. Sicher, *Collection Czech. Chem. Commun.*, **28**, 1664 (1963).

The diaxial isomer was reduced at -0.86 volts, while the diequatorial required -1.67 volts. cis-1,2-Dibromocyclohexane was also reduced with difficulty (-1.64 volts). From a study of a number of additional compounds it was concluded that halogens in an *anti* arrangement were most easily reduced (-0.8 volt), in a *syn* arrangement less easily ($-.1.2$ volts), and most difficultly with a dihedral angle of 90° (-1.9 volts). The general pattern follows rather closely that of the typical E_2 reaction (see Sec. 7-4).

3–10. Acoustical Measurements

The absorption of sound in most organic liquids is proportional to the square of the frequency as predicted by classical formulas.[179] A few organic liquids are known, however, in which the absorption has instead been found to obey the laws of relaxation processes.[180] It is usually assumed that such an acoustical relaxation process is the result of the perturbation by the sound wave of a chemical equilibrium in the liquid. If the equilibrium is between two states 1 and 2, the frequency of maximum absorption f_c is related to the rate constants in the forward and reverse directions (k_f and k_b) by [181]

$$f_c = [(k_f + k_b)/2\pi] \tag{15}$$

Since it is necessary that the two states have different energies for a relaxation to occur, usually one of the energies (state 2) is considerably higher than the other (state 1), and $k_b \gg k_f$. Then, combining equation (15) with the Eyring equation (equation 7), one obtains

$$f_c = (k_{2 \to 1}/2\pi) = (\kappa k_b T/h) \exp(-\Delta G_2/RT) \tag{16}$$

As an example methylcyclohexane may be considered. For this compound there are two principal conformations; the chair with the equatorial methyl, and the chair with the axial methyl. By this method the value of ΔG_2 (Fig. 21) may be found from the measured value of f_c (0.22 Mc./sec.).[182] For the equatorial methyl to change over to the axial position, the path of lowest energy must go through a skew boat conformation, so that in this case the energy diagram in fact is as shown in Fig. 21. The barriers separating the boat from the two chairs must be

[179] J. M. M. Pinkerton, *Proc. Phys. Soc.* (*London*), **B62,** 129 (1949).
[180] J. Lamb and J. H. Andreae, *Nature*, **167,** 898 (1951).
[181] J. Karpovich, *J. Chem. Phys.*, **22,** 1767 (1954).
[182] J. Karpovich, *J. Acoust. Soc. Am.*, **26,** 819 (1954).

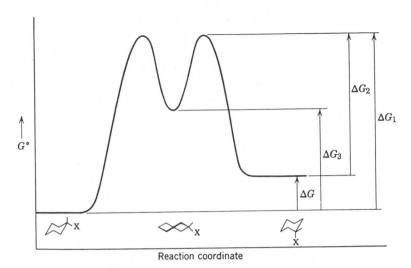

Fig. 21. Energy diagram for methylcyclohexane.

very similar in height, since the interaction of the methyl with the other axial hydrogens is relatively small, and most of the repulsion which is present in the chair form is relieved in the transition state. The presence of the skew boat and the apparent lack of a mechanism for using momentum to carry past it mean that there is an equal probability of the skew boat going to either chair. Hence the transmission coefficient κ is $\frac{1}{2}$. Ordinarily ΔG_2 is not the quantity that is desired, however. Usually one wants to know ΔG, the difference in energy between the axial and equatorial groups, or ΔG_1, the energy barrier to the change from the stable form. If one knows ΔG for the methyl independently and measures ΔG_2, one can calculate ΔG_1. For methylcyclohexane the original experimental value of ΔG_2 was 8.7 kcal./mole.[183] If the energy of the axial methyl (ΔG) is taken as 1.7 kcal./mole, the value of ΔG_1 is 10.4 kcal./mole. This value constitutes an experimental determination of the energy barrier between the boat (flexible) and chair forms of cyclohexane (although it was not originally so interpreted) and is in good agreement with the values obtained by NMR measurements.[92]

If it is assumed that for any simple cyclohexane the value for the ΔG_1 is approximately the same (and this seems to be the case[183]), measurements of ΔG_2 permit the calculation of ΔG. The method is not very accurate because the desired value is the difference between two large

[183] Similar values have recently been reported for this and related compounds. See (a) J. E. Piercy, *J. Acoust. Soc. Am.*, **33**, 198 (1961); (b) J. Lamb and J. Sherwood, *Trans. Faraday. Soc.*, **51**, 1674 (1955).

numbers. The available data give the reasonable values of 1.2 kcal./mole for the conformational energy of the hydroxyl group in liquid cyclo-hexanol and 1.6 kcal./mole for the amino group in liquid cyclohexyl amine (see Sec. 7-1). Cyclohexane does not show absorption in the range investigated (0.02–0.60 Mc./sec.), from which fact one can calculate that the boat form must have a free energy of more than 2.2 kcal./mole above that of the chair; this is known to be true (Sec. 2-1). Because cyclo-hexanone likewise fails to show absorption in this region, for it ΔG_2 is less than 8.2 kcal./mole. It has been calculated[184] that the difference in enthalpy between the boat and the chair in cyclohexanone should be 2.7 kcal./mole, but no experimental value is yet available. The value of ΔG_1 for cyclohexanone is also unknown but is probably no more than 6 kcal./mole.

From measurement of both the velocity and the absorption of sound as a function of frequency at various temperatures it is possible to obtain the difference in enthalpy, entropy, and free energy between the two conformations directly.[185] The accuracy of the determinations so far reported[183] has been fair with regard to free energy (±0.3 kcal./mole), but less satisfactory for the other quantities.

The ultrasonic method is potentially useful for ascertaining whether or not equilibria exist between conformers in complicated structures where most other methods tend to be of limited use. It has not been so used to date, however, the lack of generally available equipment being the principal obstacle.

3–11. pK Measurements[186]

In general, the ionization constant of an acidic or basic functional group varies with the axial or equatorial nature of the group.[187–190] Thus, by the determination of the pK values for the cis- and trans-4-t-butylcyclo-hexanecarboxylic acids, these quantities were determined for axial and equatorial carboxyls, and proved to be considerably different (8.23 and

[184] N. L. Allinger, J. Am. Chem. Soc., **81,** 5727 (1959).

[185] (a) R. O. Davies and J. Lamb, Quart. Rev., **11,** 134 (1957). (b) R. O. Davies and J. Lamb, Proc. Phys. Soc. (London), **73,** 767 (1959).

[186] H. C. Brown, D. H. McDaniel, and O. Häfliger, in E. A. Braude and F. C. Nachod, eds., Determination of Organic Structures by Physical Methods, Vol. I, Academic Press, New York, 1955, p. 567.

[187] M. Tichý, J. Jonáš, and J. Sicher, Collection Czech. Chem. Commun., **24,** 3434 (1959).

[188] J. F. J. Dippy, S. R. C. Hughes, and J. W. Laxton, J. Chem. Soc., **1954,** 4102.

[189] R. D. Stolow, J. Am. Chem. Soc., **81,** 5806 (1959).

[190] P. F. Sommer, V. P. Arya, and W. Simon, Tetrahedron Letters, No. 20, 18 (1960); P. F. Sommer, C. Pascual, V. P. Arya, and W. Simon, Helv. Chim. Acta, **46,** 1734 (1963).

7.79, respectively[189]). Since the axial anion is more hindered and difficult to solvate (Sec. 3-7), the axial carboxyl would be predicted to be less acidic, as found. (It has in fact been shown[190] that there is a quantitative relationship between the number of *syn*-axial hydrogens and the pK_a.) From the pK of cyclohexanecarboxylic acid (7.82), a simple calculation showed the value of the axial to equatorial ratio of the carboxyl group and allowed the conformational energy of the carboxyl to be calculated.[187,189] Since the probable errors in the pK measurements involve about 0.01 unit, it is clear that when the equilibrium is so far to one side the numerical results are not very accurate. Obviously, the determination of a conformational equilibrium by pK measurements can be of high accuracy only when the difference in free energies between the conformations is not more than about 1 kcal./mole. What can be done is to balance the group to be studied (carboxyl) against one for which numerical data are available (methyl) so that the equilibrium constant is not too far from unity. In this way the conformational free energies of the carboxyl and carboxylate anion were found[187] to be 1.6 $\pm$ 0.3 and 2.2 $\pm$ 0.3 kcal./mole, respectively (in 80% methyl Cellosolve). Once these quantities were known, the pK's of *cis*-4-alkylcyclohexanecarboxylic acids were used to measure the conformational energies of the alkyl groups.[191]

The difference between the first and second ionization constants of a dicarboxylic acid is a function of (among other things) the distance apart of the carboxyl groups.[192] From the dissociation constants it was shown as long ago as 1928 that the carboxyls are farther apart in *cis*-1,3-cyclohexane dicarboxylic acid than in the *trans* isomer.[193] Concerning the *cis*- and *trans*-1,2-cycloalkanedicarboxylic acids it was concluded that the *trans* isomer in the six-membered ring is diequatorial, and the seven- and eight-membered rings must be similarly staggered.[194]

3–12. Mass Spectra

A small number of examples suggests that the axial or equatorial nature of the hydroxyl in an assortment of compounds,[195] and the *cis/trans* nature of the A/B ring juncture in 1-, 3- and 4-keto steroids[196] exert a

[191] H. van Bekkum, P. E. Verkade, and B. M. Wepster, *Konikl. Ned. Akad. Wetenschap Proc., Ser. B*, **64**, 161 (1961).

[192] E. S. Gould, *Mechanism and Structure in Organic Chemistry*, Henry Holt and Co., New York, 1959, p. 200; see, however, L. L. McCoy and G. W. Nachtigall, *J. Am. Chem. Soc.*, **85**, 1321 (1963).

[193] R. Kuhn and A. Wasserman, *Helv. Chem. Acta*, **11**, 50 (1928).

[194] J. Sicher, F. Šipoš, and J. Jonáš, *Collection Czech. Chem. Comm.*, **26**, 262 (1961).

[195] K. Biemann and J. Seibl, *J. Am. Chem. Soc.*, **81**, 3149 (1959).

[196] H. Budzikiewicz and C. Djerassi, *J. Am. Chem. Soc.*, **84**, 1430 (1962).

systematic influence on the mass spectra of these compounds.[197] These differences appear to have a conformational basis, and they suggest that mass spectrometry may be of very considerable future use in conformational analysis. Currently it is only possible to decide which of two epimers is which when both are available. The method has the advantage of requiring only a trace of compound.

3–13. Kinetic Methods[198]

Just as an equilibrium method for a chemical reaction (such as pK measurements) can be used for conformational studies, kinetic methods can be used similarly.[135,199,200] Thus the oxidation rates of the *cis*- and *trans*-4-*t*-butylcyclohexanols gave rates for axial and equatorial groups, and the corresponding rate for the parent compound, cyclohexanol, allowed the conformational energy of the hydroxyl to be calculated.[135] The kinetic method appears unfortunately to be quite sensitive to small ring deformations brought about by the presence of substituents, especially if the reaction takes place directly on a ring atom as in tosylate solvolysis, and it would seem therefore to be of limited use for quantitative work.[201-204] The method has already been discussed in detail in Sec. 2-2c.

General References

Braude, E. A., and F. C. Nachod, eds., *Determination of Organic Structures by Physical Methods*, Academic Press, New York, 1955.

Lau, H. H., *Angew. Chem.*, **73**, 423 (1961).

Nachod, F. C., and W. D. Phillips, eds., *Determination of Organic Structures by Physical Methods*, Vol. II, Academic Press, New York, 1962.

Pinder, A. R., *Chem. Ind. (London)*, **1961**, 1180.

Weissberger, A., ed., *Physical Methods of Organic Chemistry*, Interscience Division, John Wiley and Sons, New York.

Williams, D., ed., *Methods of Experimental Physics*, Vol. 3, "Molecular Physics," Academic Press, New York, 1962.

---

[197] Also see J. Momigny and P. Natalis, *Bull. Soc. Chim. Belges*, **66**, 26 (1957).

[198] See (a) E. L. Eliel, *Stereochemistry of Carbon Compounds*, McGraw-Hill Book Co., New York, 1962, p. 234; (b) W. G. Dauben and K. S. Pitzer, in M. S. Newman, ed., *Steric Effects in Organic Chemistry*, John Wiley and Sons, New York, 1956, p. 44, for general reviews.

[199] E. L. Eliel and R. S. Ro, *J. Am. Chem. Soc.*, **79**, 5995 (1957).

[200] E. L. Eliel and R. G. Haber, *J. Am. Chem. Soc.*, **81**, 1249 (1959).

[201] W. Hückel, *Bull. Soc. Chim. France*, **1960**, 1369.

[202] E. L. Eliel, *J. Chem. Educ.*, **37**, 126 (1960).

[203] J. C. Richer, L. A. Pilato, and E. L. Eliel, *Chem. Ind. (London)*, **1961**, 2007.

[204] F. Biros, Ph.D. Dissertation, University of Notre Dame, Notre Dame, Ind., 1964.

Chapter 4

Conformational Analysis in Ring Systems
Other Than Cyclohexane

4–1. Introduction

The cyclohexane ring system differs, conformationally speaking, from the other monocyclic carbon ring systems in two important ways. First, in cyclohexane all the ring atoms have the same torsional angles.* A result of this equivalence is that there are only two kinds of positions on the cyclohexane ring which a substituent may occupy: axial or equatorial. The second important conformational feature of the cyclohexane ring is that a substituent in either kind of position corresponds to a conformation at an energy minimum. These features make the treatment of such systems relatively simple. In ring systems other than cyclohexane, the ring atoms are generally not equivalent, and, instead of only two positions which substituents may take up, there is a larger number. The situation in these other rings is further complicated by the fact that when the ring conformation is at an energy minimum the substituent itself will not generally be located in a perfectly staggered position with respect to the ring, and the balancing of the substituent's interaction energy against the conformational energy of the ring may be expected to lead to distortion of some type.

On the basis of the acceptance of the tetrahedral carbon atom, von Baeyer proposed his theory of ring strain in 1885[1] (see p. 2). That the three- and four-membered rings were strained was evidenced by the difficulty in preparing them synthetically, and by the rarity of their occurrence in nature.[2] von Baeyer also thought that the medium and large rings (then unknown) would be unstable, because if they were planar regular polygons they would have their interior bond angles expanded

---

[1] A. von Baeyer, *Ber.*, **18,** 2277 (1885).
[2] C. K. Ingold, *Ann. Rept.*, **23,** 112 (1926).
* This is also true in cyclopropane and cyclobutane.

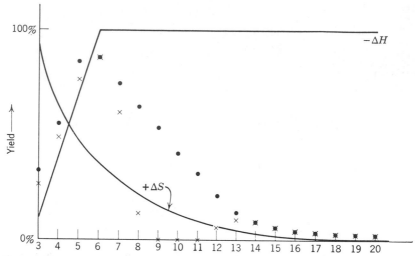

Fig. 1. Thermodynamic parameters for ring closure as a function of ring size (schematic). Dots represent predicted yields; crosses, experimental yields.

above the tetrahedral value. After Hückel showed (1925) that the six-membered ring was not planar,[3] the commonly held view was that the medium and large rings would in fact show no special instability. It was also recognized,[4,5] however, that these compounds were not easy to synthesize, and an analysis of the synthetic difficulty was made in this way: For a typical simple ring closure reaction in solution, both the enthalpy and the entropy of ring closure must be considered. In Fig. 1 the effects of these two quantities on the yield of the ring closure reaction as a function of ring size are indicated. The small rings are strained, but, since it was thought that rings of six members or more would be unstrained, the enthalpy effect would be predicted to be unfavorable for the small rings, and then constant for rings of six or more atoms. The entropy change would have a maximum value for the closure of a three-membered ring. For larger rings, as the ends required to react during the closure got farther and farther apart, the entropy of ring closure should become less and less favorable, leading to a slowly falling curve. A combination of these two effects then should lead to clear-cut predictions of yield as a function of ring size. Low yields in the small rings were then predicted because of the very unfavorable enthalpy change. The five- and six-membered rings had less favorable entropies of ring closure than

[3] W. Hückel, *Ann.*, **441**, 1 (1925).

[4] L. Ruzicka, W. Brugger, M. Pfeiffer, N. Schinz, and M. Stoll, *Helv. Chem. Acta*, **9**, 499 (1926).

[5] L. Ruzicka, *Chem. & Ind.*, (*London*), **54**, 2 (1935).

did the small rings, but very much more favorable enthalpies, and are both obtained in good yield. (The six-membered ring is usually obtained in a somewhat better yield than is the five-membered, however, the reverse of the early prediction.) The rings containing seven or more atoms were expected to be obtained in steadily decreasing yields. The ring closure reactions investigated at the time[4,6] appeared to fulfill the simple predictions quite well, with the glaring exception of rings containing from eight to twelve carbons. With these compounds the yields were exceedingly poor, indicating a previously unexpected effect, a "medium ring effect," the correct explanation of which was not immediately forthcoming.

It may be noted that, while azelaonitrile was cyclized to cyclooctanone in 30% yield by a Thorpe-Ziegler reaction, the 5-t-butyl derivative gave an 89% yield of the corresponding ketone,[7] and α, α'-dimethylazelaonitrile[8] (either diastereomer) could be cyclized in a nearly quantitative yield. The effect of substituents in promoting cyclizations was noted as early as 1915;[9] it has been referred to as the "*gem*-dimethyl effect." A suggestion regarding the observed facts which is commonly referred to as the "Thorpe-Ingold effect" is that, in propane, for example, the C—C—C angle is larger than tetrahedral, while the H—C—H angle at carbon 2 is smaller. When the hydrogens at C-2 are replaced by methyls, the C—C—C angles all become tetrahedral, and the two methyl groups in the original propane are thus brought closer together in neopentane. Recent microwave data have shown (Sec. 3-2c) that the early deductions regarding these angle changes are qualitatively correct, changes of 1–2° being found in the direction indicated. This change in angle has definite consequences; thus, whereas in 1,3-dihydroxypropane the intramolecular hydrogen bond shows[10] $\Delta\tilde{v} = 78$ cm.$^{-1}$ (difference in O—H stretching frequency between free and bonded hydroxyl), in the 2,2-dimethyl derivative $\Delta\tilde{v} = 88$ cm.$^{-1}$. The "Thorpe-Ingold effect" is thus found to be small in this case, and, although the angular deformations described above are undoubtedly real, they probably contribute little to the experimentally observed *gem*-dimethyl effect, even in the small rings where they are most important. It has been shown more recently that, at least in the six-membered ring, the "*gem*-dimethyl effect" can be quantitatively explained

[6] K. Ziegler, H. Eberle, and H. Ohlinger, *Ann.*, **504**, 94 (1933).

[7] N. L. Allinger and S. Greenberg, *J. Am. Chem. Soc.*, **84**, 2394 (1962).

[8] N. L. Allinger and S.-E. Hu, *J. Am. Chem. Soc.*, **83**, 1664 (1961).

[9] (a) R. M. Beesley, C. K. Ingold, and J. F. Thorpe, *J. Chem. Soc.*, **107**, 1080 (1915); (b) C. K. Ingold, *ibid.*, **119**, 305 (1921); (c) G. S. Hammond, in M. S. Newman, ed., *Steric Effects in Organic Chemistry*, John Wiley and Sons, New York, 1956, p. 463.

[10] P. von R. Schleyer, *J. Am. Chem. Soc.*, **83**, 1368 (1961).

in terms of the thermodynamics of the conformations involved.[11] An ordinary unsubstituted hydrocarbon chain exists predominantly in an extended (zig-zag) form, and energy is required to introduce the *gauche* interactions necessary to put the molecule into the proper conformation for cyclization. The energy difference between the ground state (zig-zag) and the transition state thus includes this conformational energy. If one or more substituents are present on the hydrocarbon chain, the transition state may contain no additional *gauche* interactions (or it may contain some small number) beyond those present in the unsubstituted case. The ground state necessarily will contain some (or many) additional *gauche* interactions, and hence upon substitution the energy of the ground state is raised in relation to that of the transition state. Substitution consequently encourages cyclization,[11,12] with regard to both rate and equilibrium.

In the 1920's the medium rings presented two interesting chemical problems; the practical synthetic problem of preparing such compounds; and the theoretical problem of the reason for the existence of the strain in such compounds. The first problem was not solved until 1947, when it was found that the acyloin ring closure gave the medium rings in good yields through a fortunate mechanistic quirk (high dilution at an interface).[13,14] Nor was the theoretical problem solved during the 1920's, since it was thought at the time that there was free rotation about single bonds. Instead it was thought that the instability of the medium rings arose from repulsions between transannular hydrogens.[15] More recently it was recognized that, while there are transannular repulsions in certain cases, the basic difficulty in the medium ring series is that in order to have normal bond angles the compounds would necessarily have more or less unfavorable dihedral angles which would lead to considerable ring strain.[16,17] The strained rings would, of course, distort their bond angles if they could lower their total energy by doing so, and some compromise between various kinds of distortions would be expected in practice.

From a consideration of the geometries of the various rings, some general predictions can be made. For carbocyclic rings containing five or fewer atoms, if the bond angles were kept as nearly tetrahedral as possible

[11] (a) N. L. Allinger and V. Zalkow, *J. Org. Chem.*, **25.** 701 (1960); (b) H. Yumoto, *J. Chem. Phys.*, **29,** 1234 (1958).

[12] J. Dale, *J. Chem. Soc.*, **1963,** 93.

[13] M. Stoll and J. Hulstkamp, *Helv. Chim. Acta*, **30,** 1815 (1947).

[14] V. Prelog, L. Frenkiel, M. Kobelt, and P. Barman, *Helv. Chim. Acta*, **30,** 1741 (1947).

[15] M. Stoll and G. Stoll-Compte, *Helv. Chim. Acta*, **13,** 1185 (1930).

[16] R. Spitzer and H. M. Huffman, *J. Am. Chem. Soc.*, **69,** 211 (1947).

[17] V. Prelog, *J. Chem. Soc.*, **1950,** 420.

to minimize angle strain, the rings would be planar with a maximum of torsional strain. In fact, it is found for both cyclobutane[18-20] and cyclopentane[18,21,22] that the minimum of total energy is obtained by having these rings non-planar. The strain energies of these rings are clearly evident from the heat of combustion data listed in Table 1. The next to

Table 1. Properties of Carbocyclic Rings

Ring Size	Dihedral Angle[a]	$(H_c/n)^c$	$(H_c/n - 157.4)$	$n(H_c/n - 157.4)$
2[b]	—	168.7	11.3	22.6
3	0	166.6	9.2	27.6
4	0	164.0	6.5	26.4
5	0	158.7	1.3	6.5
6	59°	157.4	0.0	0.0
7	—	158.3	0.9	6.3
8	96°	158.6	1.2	9.6
9	—	158.8	1.4	12.6
10	115°	158.6	1.2	12.0
11	—	158.4	1.0	11.0
12	126°	157.7	0.3	3.6
13	—	157.8	0.4	5.2
14	134°	157.4	0.0	0.0
15	—	157.5	0.1	1.5
16	—	157.5	0.1	1.6
17	—	157.2	−0.2	−3.4

[a] Assumes a regular crown conformation and tetrahedral bond angles insofar as possible (ref. 23).

[b] Ethylene.

[c] The heat of combustion of the gaseous cycloalkane in kilocalories per mole, divided by the number of methylene groups. Data from refs. 24–26 (p. 194).

the last column in the table gives the strain per methylene group, and this quantity is a measure of the distortion present. The last column gives the total strain in the molecule, which is a measure (in part) of the

[18] A. Almenningen, O. Bastiansen, and P. N. Skancke, *Acta Chem. Scand.*, **15**, 711 (1961).

[19] J. D. Dunitz and V. Schomaker, *J. Chem. Phys.*, **20**, 1703 (1952).

[20] G. W. Rathjens, Jr., N. K. Freeman, W. D. Gwinn, and K. S. Pitzer, *J. Am. Chem. Soc.*, **75**, 5634 (1953).

[21] J. E. Kilpatrick, K. S. Pitzer, and R. Spitzer, *J. Am. Chem. Soc.*, **69**, 2483 (1947).

[22] K. S. Pitzer and W. E. Donath, *J. Am. Chem. Soc.*, **81**, 3213 (1959).

difficulty to be anticipated in ring closure. The angular strain is seen to be quite severe in cyclopropane, less severe in cyclobutane, and still less in cyclopentane. Cyclohexane lies at the energy minimum, and for the larger rings the strain increases to a maximum at cyclononane and then falls off. For the rings containing an even number of atoms (six or greater), if tetrahedral bond angles are assumed, the dihedral angle corresponding to a regular symmetrical crown structure can be calculated geometrically.[23] These calculated dihedral angles are also given in Table 1, and they show that, if these molecules were regular crowns, the torsional energy per methylene group would increase to a maximum at C-10 (an angle of 120° corresponding to an eclipsed conformation) and then fall off. The heat of combustion of cyclooctane[16,24] was found to be inconsistent with the regular crown conformation, and subsequently it was shown[27] that by skewing or stretching the crown the dihedral angles could be changed (some becoming larger and some smaller) but in such a way that the total energy of the ring would be lowered appreciably. When a small bond angle expansion was also allowed to occur, the calculated heat of combustion could be brought into agreement with experiment (see Sec. 7-2). These changes were a preview of what was to be expected from the larger rings. Clearly, if the rings from C-9 to C-12 were to have more or less regular crown conformations, they would be very highly strained. It was also clear that the energies of these rings could be lowered by changing the dihedral angles and bond angles, but, with so many possible variables, the calculation of the conformation of minimum energy for rings larger than cyclooctane posed a very difficult problem. It turned out that X-ray crystallography in fact revealed the conformations of these compounds[28] before the theoretical approach was able to make much progress.

After the significance of the unfavorable dihedral angles of the regular crown structures for the medium rings was recognized, it was realized that rings of different sizes could undergo a variety of different contortions to minimize this strain, but of course even this minimum amount of strain is sufficient to have a profound effect on the chemical and physical properties. Some chemical aspects of this basic idea were recognized and

[23] L. Pauling, *Proc. Natl. Acad. Sci. U.S.*, **35**, 495 (1949).

[24] S. Kaarsemaker and J. Coops, *Rec. Trav. Chim.*, **71**, 261 (1952).

[25] J. W. Knowlton and F. D. Rossini, *J. Res. Natl. Bur. Std.*, **43**, 113 (1949).

[26] J. Coops, H. van Kamp, W. A. Lambregts, B. J. Visser, and H. Dekker, *Rec. Trav. Chim.*, **79**, 1226 (1960); cf. H. van Kamp, Ph.D. Dissertation, Vrije Universiteit te Amsterdam, Amsterdam, Netherlands, 1957.

[27] N. L. Allinger, *J. Am. Chem. Soc.*, **81**, 5727 (1959).

[28] J. D. Dunitz and V. Prelog, *Angew. Chem.*, **72**, 896 (1960).

effectively utilized by H. C. Brown in 1951,[29] although no real conformational approach to the problem was possible at that time. The total ring strain was referred to by Brown as I-(for internal) strain. It was pointed out that, in any reaction at a ring carbon, changes in the coordination number and hybridization would accompany the motion along the reaction coordinate from the ground to the transition state. If these changes led to more strain in the transition state than was present in the ground state, compared to an acyclic analog, then there would be a retardation of the reaction (I-strain). Conversely, if there was more strain in the ground state than in the transition state, there would be a steric acceleration (relief of I-strain). The same concepts were also applied to equilibria which involved changes in coordination number, for example, cyanohydrin formation from cycloalkanones or the hydrogenation of olefins.

The I-strain concept has been and is a very useful one, and it can be applied to different classes of compounds, depending on whether there is a single ring atom involved, or two as with an endocyclic olefin (single-constraint and double-constraint processes).[30,31] The interpretation of the rate of reduction of the cycloalkanones by sodium borohydride as a function of ring size may be considered as a typical example of the usefulness of the I-strain theory (Fig. 2).[32] The relative rate constant (k_2) is 15.1 for acetone and much less, 0.45, for di-n-hexyl ketone. The difference can be accounted for, at least qualitatively,* by the fact that in the latter case there are two *gauche* interactions between hydroxyl and methylene in the product which are missing in the former. Cyclopentanone has a relative rate of 7.01, appreciably slower than acetone. While there would be a considerable decrease in angular deformation at the carbonyl upon reduction, the increased torsional strain in the product more than outweighs it, and the rate is decreased. This behavior is typical;

[29] (a) H. C. Brown, R. S. Fletcher, and R. B. Johannesen, *J. Am. Chem. Soc.*, **73,** 212 (1951); (b) H. C. Brown, *Record Chem. Progr.* (*Kresge-Hooker Sci. Lib.*), **14,** 83 (1953); *J. Chem. Soc.*, **1956,** 1248; *Bull. Soc. Chim. France*, **1956,** 980; (c) E. L. Eliel, *Stereochemistry of Carbon Compounds*, McGraw-Hill Book Co., New York, 1962, pp. 265–269.

[30] J. Sicher and M. Svoboda, *Collection Czech. Chem. Commun.*, **23,** 2094 (1958).

[31] J. Sicher, *Progr. Stereochem.*, **3,** 202 (1962).

[32] H. C. Brown and K. Ichikawa, *Tetrahedron*, **1,** 221 (1957).

* A quantitative explanation would have to separate differences in enthalpy (I-strain) and entropy and would depend in part on the fact that the open chain carbonyl compound is a mixture of conformations, some of which have the alkyl group coiled around the oxygen and exerting a hindrance effect as well as changing the nature of the medium in which the reaction is occurring.

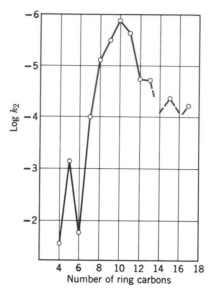

Fig. 2. Effect of ring size on the rate constant (k_2) for the reduction of cycloalkanones. From Brown and Ichikawa.[32]

cyclopentanone is usually (but not always)* less reactive than acetone in carbonyl additions.

The six-membered ring ketone has the carbonyl eclipsed by the equatorial hydrogens, a situation believed to be unfavorable relative to an open chain (Sec. 7-4), and the addition reaction is much faster ($k_2 = 161$). The carbonyls in rings containing 14 or more members have rates similar to dihexyl ketone, while those from C-7 to C-13 are more reluctant to undergo addition than are those in an open chain. In each of the medium rings at least one of the dihedral angles is close to being eclipsed, a condition which is very unfavorable for two methylenes. Replacement of the two hydrogens on one methylene (without any motion of the carbons) by a carbonyl puts the latter in a favorable, nearly staggered arrangement with respect to the methylene. Such changes pretty well account for the observed differences in the rate constants for all the cycloalkanones. Many other processes, such as cyanohydrin formation equilibria[33,34] or solvolysis of the tosylates of the corresponding alcohols,[35] are accounted for in a similar manner.

The conformations of the medium rings are such that many of the

[33] A. Lapworth and R. H. F. Manske, *J. Chem. Soc.*, **1928**, 2533; **1930**, 1976.

[34] V. Prelog and M. Kobelt, *Helv. Chim. Acta*, **32**, 1187 (1949).

[35] R. Heck and V. Prelog, *Helv. Chim. Acta*, **38**, 1541 (1955).

* The equilibrium constants for cyanohydrin formation are reported as 67 and 32.8, respectively.[33]

relationships between diastereomers are diminished or reversed with respect to those found in the common rings. For example, in the six-membered ring it is easier to bring together two substituents if one is axial and one is equatorial than if both are equatorial (Sec. 2-5). With cyclodecane-1,2-diol, on the other hand, the diequatorial hydroxyls are spatially closer together than the equatorial-axial ones, and it is found that the stronger hydrogen bond which characterizes a *cis*-1,2-cyclo-hexanediol in the infrared (Sec. 3-4a) is characteristic of a *trans* isomer in the ten-membered ring.[36] Similarly, although the *cis*-1,2-diol reacts more rapidly with lead tetraacetate than does the *trans* in the six-membered ring, the reverse is true in the ten-membered ring.[37]

For many years, use has been made of the fact that the carbonyl stretching frequencies of cyclic ketones vary with ring size;[38] thus cyclohexanones in non-polar solvents absorb in the range 1709–1714 cm.$^{-1}$, whereas for cyclopentanones and cyclobutanones the absorptions are near 1745 and 1775 cm.$^{-1}$, respectively. Halford derived an inverse relationship between the stretching frequency and the C—C—C bond angle at the carbonyl carbon.[39] It was noted many years ago that the carbonyl stretching frequencies of the medium rings were quite low, the 7–10 membered cycloalkanones absorbing uniformly at 1701–1702 cm.$^{-1}$. An early suggestion was made that these low frequencies were due to hydrogen bonding between a transannular hydrogen and the carbonyl oxygen.[17] In view of Halford's work it seems more likely that the observed frequencies are due to bond angle expansion in the medium rings.[27,40] Such expansion would allow for better dihedral angles and would reduce the torsional strain of the ring. Other systems in which there are bulky groups about the carbonyl also give evidence for a similar bond angle expansion. In 2,2,4,4-tetramethylcholestan-3-one and related compounds for example,[41] the absorption is found near 1698 cm.$^{-1}$. That transannular hydrogen bonding in the medium ring ketones is important in determining the carbonyl stretching frequencies has not been disproved, but there appears to be no evidence for it.*

[36] L. P. Kuhn, *J. Am. Chem. Soc.*, **76,** 4323 (1954).

[37] V. Prelog, K. Schenker, and W. Kung, *Helv. Chim. Acta*, **36,** 471 (1953).

[38] L. J. Bellamy, *The Infra-Red Spectra of Complex Molecules*, John Wiley and Sons, New York, 2nd ed., 1958.

[39] J. O. Halford, *J. Chem. Phys.*, **24,** 830 (1956).

[40] C. Castinel, G. Chiurdoglu, M. L. Josien, J. Lascombe, and E. Vanlanduyt, *Bull. Soc. Chim. France*, **1958,** 807.

[41] (a) Y. Mazur and F. Sondheimer, *J. Am. Chem. Soc.*, **80,** 5220 (1958); (b) D. T. Cropp, B. B. Dewhurst, and J. S. E. Holker, *Chem. & Ind. (London)*, **1961,** 209.

* A sizable displacement of a C—H stretching overtone was offered as such evidence many years ago, but the data do not appear to have been published (see ref. 17).

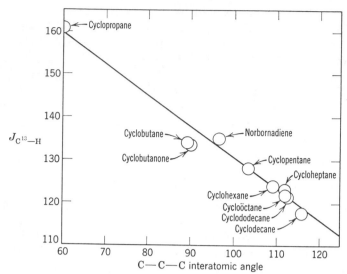

Fig. 3. The relationship between $J_{C^{13}-H}$ and the C—C—C interatomic angle. From Foote.[42a]

A recently developed method[42a] which can be used to measure bond angles makes use of the fact that, whereas the carbon-12 nucleus has no magnetic moment, the carbon-13 nucleus does, and hence its nuclear spin couples with that of the proton to which it is attached. It is convenient that the natural abundance of carbon-13 is sufficient that its effect may be detected on ordinary compounds without isotopic enrichment. The coupling constant between the carbon-13 and its attached proton appears to be a linear function of the amount of s character in the carbon orbital from which the C—H bond is formed, and thus the simple measurement of $J_{C^{13}-H}$ gives the hybridization of the carbon atom.[42] It appears that the bond angle is linearly related to the coupling constant, although it is not apparent that this should be so (Fig. 3).[42a] Although not yet tested extensively, the method shows great promise because of its simplicity.

A number of medium and large ring compounds are known whose conformations as a whole are uncertain, but whose properties are such that certain conformational facts may be inferred.[43] Compounds such as A (X = H) in Fig. 4, for example, show systematic variations in the

[42] (a) C. S. Foote, *Tetrahedron Letters*, **1963,** 579; (b) N. Muller and D. E. Pritchard, *J. Chem. Phys.*, **31,** 768, 1471 (1959); (c) J. N. Shoolery, *ibid.*, **31,** 1427 (1959); (d) C. Juan and H. S. Gutowsky, *ibid.*, **37,** 2198 (1962).

[43] (a) R. Huisgen, I. Ugi, E. Rauenbusch, V. Vossius, and H. Oertel, *Ber.*, **90,** 1946 (1957); (b) E. S. Jones, Thesis, Wayne State University, submitted 1960; (c) E. A. Braude and E. S. Waight, *Progr. Stereochem.*, **1,** 126 (1945); (d) R. Huisgen and H. Ott, *Tetrahedron*, **6,** 253 (1959); (e) R. Huisgen and H. Walz, *Ber.*, **89,** 2616 (1956).

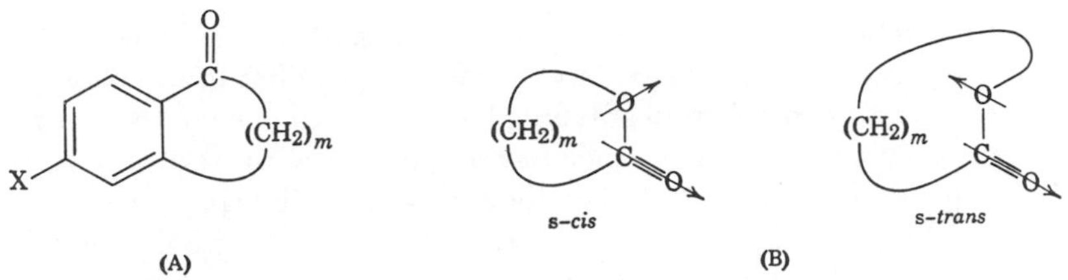

Fig. 4.

carbonyl stretching frequencies and in the ultraviolet extinction coefficients as m is increased from 2 to 5.[43a-c] The chemical shift of the fluorine in the NMR spectrum of compound A ($X = F$) similarly varies,[43b] as do the pK's for A ($X = NH_2$ or COOH).[43b] From these properties the angle between the C=O and phenyl planes may be calculated.

The dipole moments of esters[43d] (also amides[43e]) of the type B have been measured, and they were found to have values of 3.7–4.4 D for $m = 4$–8, and 1.9–2.2 D for $m = 9$–16. For a simple ester the s-*trans* is the stable conformation (Sec. 1–3b) and this form is also present in large rings. If the ring is too small for the s-*trans* form to exist comfortably, the s-*cis* form is found instead. In the latter there is an unfavorable repulsion between the dipoles, but the only alternative would be a non-planar system in which the ester resonance would be diminished or lost, and this is even less favorable.

4–2. Small Rings

Cyclopropane is necessarily a planar molecule, and, although bulky substituents attached to it may suffer some bond angle bending, this ring system appears to present no conformational problems.

Cyclobutane has been shown by electron diffraction[18,19] and by spectroscopic and thermodynamic measurements[20] to be puckered (Fig. 5). Apparently the relief in torsional strain which results from a modest amount of puckering is sufficient to induce the necessary additional bond angle distortion into the molecule. The chemically interesting consequences of this non-planarity have been only slightly explored.[44-46]

Fig. 5. Cyclobutane.

[44] (a) J. M. Conia, J.-L. Ripoll, L. A. Tushaus, C. L. Neumann, and N. L. Allinger, *J. Am. Chem. Soc.*, **84**, 4982 (1962); (b) J.-M. Conia and J. Gore, *Tetrahedron Letters*, **1963**, 1379; (c) J.-M. Conia and J.-L. Ripoll, *Bull. Chim. Soc. France*, **1963**, 768.

[45] (a) M. Takahashi, D. R. Davis, and J. D. Roberts, *J. Am. Chem. Soc.*, **84**, 2935 (1962); (b) J. B. Lambert and J. D. Roberts, *ibid.*, **85**, 3710 (1963).

[46] N. L. Allinger and L. A. Tushaus, *J. Org. Chem.*, **30**, 1945 (1965).

Scale models show that there are two kinds of positions in cyclobutane, somewhat analogous to the axial and equatorial positions of cyclohexane. If there are two non-polar substituents located 1,3 to one another on the cyclobutane ring, the *trans* isomer has an equatorial-axial arrangement while the *cis* isomer may be either diequatorial or diaxial. Simple van der Waals' calculations (Sec. 7-2) indicate that a methyl group would be more comfortable in the equatorial than in the axial position here, analogous to the situation in cyclohexane. One would therefore predict that a compound such as methyl 3-methylcyclobutanecarboxylate would be more stable in the *cis* than in the *trans* arrangement, and this has been found to be true.[46] Dimethyl 1,3-cyclobutanedicarboxylate turns out to be slightly more stable in the *trans* form; this anomaly does not appear to result from a steric effect, but rather from an electrostatic repulsion between the dipoles.[46]

4–3. Cyclopentane

The most important ring system smaller than cyclohexane, from the conformational point of view, is undoubtedly cyclopentane. For this molecule a planar model can be constructed which is nearly free of angle strain, since the angle between the sides of a regular pentagon (108°) is very close to the tetrahedral angle (109° 28'). Planar cyclopentane would, however, have no less than five eclipsed ethane units, leading to a Pitzer strain of 14.0 kcal./mole (Sec. 7-2). In order to minimize this eclipsing strain, cyclopentane actually assumes a puckered form, the increase in angle strain thus arising (4.3 kcal./mole) being more than compensated for by the considerable drop in eclipsing strain (7.8 kcal./mole). The over-all reduction in strain, estimated[21,22] at about 3.6 kcal./mole, brings the calculated residual strain in cyclopentane down to 10.4 kcal./mole, a value which is in fair agreement with the heat of combustion[22,47] (Table 1).* The puckering of cyclopentane has been detected experimentally by several methods, of which the earliest[48] was based on entropy measurements, and the most recent[18] on electron diffraction. It is important to note that the puckering is not fixed but rotates around the ring by an up

[47] F. D. Rossini, K. S. Pitzer, R. L. Arnett, R. M. Braun, and G. C. Pimentel, *Selected Values of Physical and Thermodynamic Properties of Hydrocarbons and Related Compounds*, Carnegie Press, Pittsburgh, Pa., 1953.

[48] J. G. Aston, S. C. Schumann, H. L. Fink, and P. M. Doty, *J. Am. Chem. Soc.*, **63**, 2029 (1941); J. G. Aston, H. L. Fink, and S. C. Schumann, *ibid.*, **65**, 341 (1943); J. P. McCullough, *J. Chem. Phys.*, **29**, 966 (1958); J. P. McCullough, R. E. Pennington, J. C. Smith, I. A. Hossenlopp. and G. Waddington, *J. Am. Chem. Soc.*, **81**, 5880 (1959).

* Still better agreement is obtained when the van der Waals' attraction between non-bonded atoms (Chap. 7) is included in the calculation.

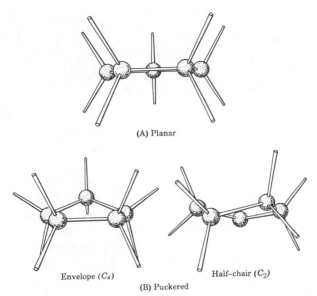

(A) Planar

Envelope *(Cs)*   Half-chair *(C₂)*

(B) Puckered

Fig. 6. Cyclopentane. From E. Eliel, *Stereochemistry of Carbon Compounds*, copyright 1962, McGraw-Hill Book Company, Inc., by permission of the publishers.

and down motion of the five methylene groups in what has been termed[21] a "pseudorotation." In the course of this pseudorotation the internal energy of the molecule changes little (by less than RT, i.e., 600 cal./mole at room temperature), so that, unlike in the chair form of cyclohexane (Chap. 2), no definite energy minima (corresponding to stable conformations) and maxima come into evidence.

The situation may change in substituted cyclopentanes. It has been pointed out[21] that of the various puckered forms of cyclopentane the so-called C_s or "envelope" form* and the C_2 or "half-chair" form* (Fig. 6) have greater symmetry than all the others and there is evidence that certain substituted cyclopentanes exist preferentially with a geometry close to one or the other of these forms, which then correspond to real energy wells or stable conformations, similar to the equatorially and axially substituted chair forms in cyclohexane (see below).

The difference in the heats of combustion[47] of cyclopentane and methylcyclopentane (155.3 kcal./mole) is only a little greater than the corresponding difference betwen those of cyclohexane and methylcyclohexane

[49] F. V. Brutcher, T. Roberts, S. J. Barr, and N. Pearson, *J. Am. Chem. Soc.*, **81**, 4915 (1959).

* The symbols C_s and C_2 (ref. 22) refer to the existence of a plane of symmetry and a two-fold symmetry axis in these respective forms, whereas the names (ref. 49) are based on analogies obvious from Fig. 6.

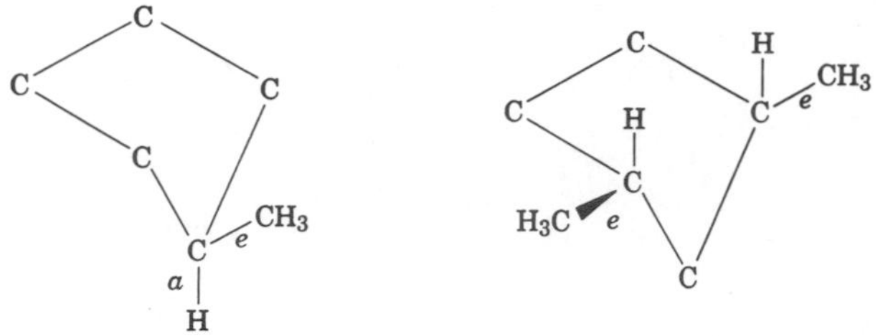

Fig. 7. Methylcyclopentane and *cis*-1,3-dimethylcyclopentane. From E. Eliel, *Stereochemistry of Carbon Compounds*, copyright 1962, McGraw-Hill Book Company, Inc., by permission of the publishers.

(154.8 kcal./mole). It appears[22] that the stable conformation of methyl-cyclopentane is an envelope form in which the methyl group occupies an "equatorial" position at the tip of the envelope (Fig. 7); in this form there is a minimum of eclipsing. A more convincing example of envelope conformation is presented by the 1,3-dimethylcyclopentanes; the *cis* isomer is thermochemically more stable by 0.53 kcal./mole than is the *trans* isomer.[50] This finding is not reasonable on the basis of a planar model, but it is readily explained if *cis*-1,3-dimethylcyclopentane is a diequatorially substituted envelope[51] as shown in Fig. 7. (The *trans* isomer would more closely correspond to equatorial-axial substitution.) The *difference* in enthalpy between *cis*- and *trans*-1,3-dimethylcyclopentane is, however, considerably less than the difference between the corresponding cyclohexanes (1.96 kcal./mole), suggesting that in cyclopentane the staggering is not so pronounced as in cyclohexane. (If it were, angle strain would become excessive.) The difference in enthalpy between *cis*- and *trans*-1,2-dimethylcyclopentane (1.71 kcal./mole) is also less than the corresponding difference for the cyclohexane analogs (1.87 kcal./mole); again it might be noted that the contrary should occur if *cis*-1,2-dimethylcyclopentane presented eclipsed methyl groups, as it would if the ring were planar. It is of interest that the physical properties of the geometrically isomeric 1,2- as well as 1,3-dimethylcyclopentanes are in accordance with the conformational rule (Sec. 3-7).

Calculations on cyclopentane derivatives containing sp^2 hybridized atoms (such as methylenecyclopentane, cyclopentanone) or hetero-atoms (e.g., tetrahydrofuran, tetrahydrothiophene, pyrrolidine) suggest[22] that such molecules exist in the half-chair form with the maximum puckering occurring at carbon atoms 3 and 4, i.e., away from the sp^2 hybridized or hetero-atom. It might be noted that an sp^2 hybridized atom in planar surroundings is not seriously troubled by bond eclipsing (unlike an sp^3

[50] Cf. S. F. Birch and R. A. Dean, *J. Chem. Soc.*, **1953**, 2477, for assignment of configuration, and ref. 48 for the enthalpy measurements.

[51] J. N. Haresnape, *Chem. & Ind.* (*London*), **1953**, 1091.

hybridized atom), and it is therefore reasonable that whatever puckering occurs should involve mainly other parts of the molecule. That atoms with free pairs, such as O, S, and N, should behave similarly to an sp^2 hybridized atom implies that the pairs give rise to little or no eclipsing of their own, an assumption which is in accordance with the low barriers to internal rotation in molecules such as methanol, methyl mercaptan, and methylamine (as compared to ethane) referred to in Secs. 1-2 and methylamine (as compared to ethane) referred to in Secs. 1-2 and 3-3; the point will be returned to in Sec. 4-6.

Relatively little consideration has been given to the correlation of conformation with physical and chemical properties in cyclopentane derivatives, and treatment of this subject must thus necessarily be brief; there is little doubt in the authors' minds, however, that in view of the importance of cyclopentane rings in natural products (e.g., Chap. 5) and in view of the impetus given to the subject by recent considerations of the detailed shape of cyclopentane rings,[22,49,52] considerable progress in this area is likely in the near future.

Cyclopentane-cis-1,2-diol in dilute solution in carbon tetrachloride shows intramolecular hydrogen bonding,[53] the separation of the unbonded and bonded hydroxyl stretching frequencies in the infrared being 61 cm.$^{-1}$. The corresponding separation in cyclohexane trans-1,2-diol (torsional angle 60°) is 32 cm.$^{-1}$, whereas in a variety of bicyclo[2.2.1]-heptane-cis-2,3-diols[54–56] in which the hydroxyl groups are necessarily eclipsed (torsional angle 0°), the separation is of the order of 100 cm.$^{-1}$. It may thus be estimated that the torsional angle in cyclopentane-cis-1,2-diol is of the order of 20–30° (Sec. 3-4a). The predicted torsional angle for cis-1,2 substituents may be as low as 0° (in the envelope form, at the carbon atoms distant from the flap) or as high as 48° (in the half-chair form, at the carbon atoms involved in maximum staggering).[22] The minimum torsional angle for trans-1,2 substituents is thus 72°, which is too large for intramolecular hydrogen bonding in diols; in fact, cyclopentane-trans-1,2-diol shows no intramolecular hydrogen bond in the infrared.

Measurements of infrared spectra and dipole moments of 2 halocyclopentanones indicate an angle of 77° between the C—X and C=O bond, in good agreement with the calculated 78° for the half-chair, but not with that calculated for the envelope form (94°) or the planar form (60°).[49,52]

The change in infrared spectrum of 1,1,2-trichlorocyclopentane with

[52] F. V. Brutcher, Jr., and W. Bauer, Jr., J. Am. Chem. Soc., **84**, 2233 (1962).
[53] L. P. Kuhn, J. Am. Chem. Soc., **74**, 2492 (1952).
[54] H. Kwart and W. G. Vosburgh, J. Am. Chem. Soc., **76**, 5400 (1954).
[55] H. Kwart and G. C. Gatos, J. Am. Chem. Soc., **80**, 881 (1958).
[56] S. J. Angyal and R. J. Young, J. Am. Chem. Soc., **81**, 5467 (1959).

temperature and state of aggregation has been interpreted[57] in terms of two distinct contributing conformations in equilibrium with each other.

The distinctive chemical properties of cyclopentyl systems, compared to, say, cyclohexyl systems or acyclic systems may usually be explained in terms of the fact that even in the puckered form there is substantial bond eclipsing giving rise to torsional strain, in contrast to the situation in cyclohexane or open chain derivatives which tend to be nearly perfectly staggered. Reactions of cyclopentane derivatives in which there is relief of torsional strain and accompanying angle strain (I-strain) will tend to be abnormally fast, whereas reactions in which strain is built up in a cyclopentanoid system will be abnormally slow. Among the former are the solvolysis of 1-methylcycloalkyl chlorides in 80% ethanol,[29a] the acetolysis of cycloalkyl tosylates in acetic acid,[58] and the hydrolysis of cycloalkyl bromides in 41% dioxane.[59] The ratios of rate constants of the cyclopentyl:acyclic:cyclohexyl compounds are 43.7:1:0.35 in the first case, 10.5:1:0.75 in the second, and 13.3:1:0.45 in the third. One might have expected a similar trend in the S_N2 reaction, since in it, as in the S_N1 reaction, the remaining hydrogen at the reaction site has its eclipsing with

the adjacent methylene groups ($-CH_2-\overset{+}{CH}-CH_2$ or $-CH_2-CH-CH_2$

configuration) relieved. In fact, however, the rate of reaction of cyclopentyl bromide with iodide ion is nearly the same as for an acyclic bromide (ratio 1.2:1), although cyclohexyl bromide (0.015 on the same scale) shows the expected slow reaction.[59] Evidently, a factor is at work which counteracts the expected I-strain acceleration in the five-membered ring (but reinforces the retardation to be expected in the six-membered ring). It has been suggested[59] that the factor which slows down S_N2 reactions in ring systems as compared to acyclic ones is interference of the ring structure with the optimum bipyramidal transition state which requires collinearity of the incoming and outgoing groups with the carbon atom at which reaction occurs.

Solvolysis rates of 2-alkylcyclopentyl p-toluenesulfonates have been studied by Hückel and co-workers.[60] The ratio k_{cis}/k_{trans} is of the order of 7 (depending somewhat on the substituent), i.e., barely larger than the corresponding ratio in 4-alkylcyclohexyl p-toluenesulfonates* (Sec. 2-5c).

[57] K. Kozima and W. Suetaka, *J. Chem. Phys.*, **35,** 1516 (1961).

[58] H. C. Brown and G. Ham, *J. Am. Chem. Soc.*, **78,** 2735 (1956).

[59] L. Schotsmans, P. J. C. Fierens, and T. Verlie, *Bull. Soc. Chim. Belges*, **68,** 580 (1959); P. J. C. Fierens and P. Verschelden, *ibid.*, **61,** 609 (1952).

[60] W. Hückel et al., *Ann.*, **624,** 142 (1959); W. Hückel and E. Mögle, *ibid.*, **649,** 13 (1961).

* Or 2-t-butylcyclohexyl tosylates. For other 2-alkylcyclohexyl tosylates, possibly because of participation of neighboring hydrogen, the ratio is much larger.

This suggests the absence of strong eclipsing effects in the *cis* isomer as well as the absence of neighboring hydrogen participation; what difference there is may well be explained on conformational grounds by assuming that the puckered chair gives rise to axial-like and equatorial-like substituents.

The reactivities of cyclopentanone and cyclohexanone have already been compared in Sec. 4-1. Similar strain considerations are involved in the relative stability of cyclopentene and cyclohexene and their respective epoxides. Models suggest that cyclopentene and cyclopentene oxide are more strained than cyclohexene and cyclohexene oxide. Nevertheless, hydrogenation of cyclopentene[61] is 1.66 kcal./mole less exothermic than hydrogenation of cyclohexene, and the heat of hydrogenation of 1-methylcyclopentene, 23.01 kcal./mole, is considerably less than that of 1-methylcyclohexene, 25.70 kcal./mole, or of 2,4-dimethyl-2-pentene, 25.15 kcal./mole.[62,63] Clearly the greater angle strain in the cyclopentenes is more than offset by the greater residual eclipsing strain in the cyclopentane hydrogenation products. That cyclohexene oxide is less readily formed from the *trans*-chlorohydrin than is cyclopentene oxide[64] but, on the other hand, is more readily opened by hydrogen chloride[65] or potassium thiocyanate[66] may be a manifestation of the same phenomenon of relatively greater bond eclipsing strain in the saturated as compared to the unsaturated five-membered cycle (and vice versa in the six-membered ring).

The relative rates of ring opening of cyclohexene oxide and cyclopentene oxide are contrary to what one might have expected on the basis of data cited earlier regarding S_N2 reactions in cyclohexyl and cyclopentyl systems. This is an illustration of the fact that considerations of bond eclipsing tend to assume an overriding importance in cyclopentane systems. One of the most remarkable illustrations of this fact comes from consideration of E_2 eliminations. It has been stated in Chaps. 1 and 2 that E_2 eliminations proceed most easily if the groups eliminated are *anti*, i.e., *trans* and axial. Since the *anti* relationship can be more easily achieved in *trans*-1,2-disubstituted cyclohexanes, provided that the substituents are axial, than in correspondingly substituted cyclopentanes, it might have been expected that the cyclohexane systems would undergo

[61] Cf. R. B. Turner and W. R. Meador, *J. Am. Chem. Soc.*, **79**, 4133 (1957).

[62] R. B. Turner and R. H. Garner, *J. Am. Chem. Soc.*, **80**, 1424 (1958).

[63] R. B. Turner, D. E. Nettleton, and M. Perelman, *J. Am. Chem. Soc.*, **80**, 1430 (1958).

[64] F. V. Brutcher and T. Roberts, Abstracts, Meeting of the American Chemical Society, Cincinnati, Ohio, 1955, p. 39N.

[65] F. V. Brutcher and T. Roberts, Abstracts, Meeting of the American Chemical Society, New York, 1954, p. 64-0.

[66] E. E. van Tamelen, *J. Am. Chem. Soc.*, **73**, 3444 (1951). Cyclopentene oxide fails to react.

bimolecular ionic elimination faster. The contrary is the case, however. The relative rates of elimination of trans-1,2-dibromocyclopentane and -cyclohexane with iodide ion at 110° are 3.9:1,[67] and, even if it is considered that only half the cyclohexyl compound may be in the favorable (for elimination) diaxial conformation (Sec. 3-5) and if no corresponding allowance is made for the cyclopentane derivative, the latter reacts faster. Clearly, the overriding consideration is the tendency of the cyclopentane derivative to escape the torsionally unfavorable saturated state. Similar results are found in the cis-2-p-toluenesulfonylcycloalkyl p-toluenesulfonates for which the relative rates of elimination (trans) for the five- and six-membered rings are 2.9:1 with hydroxide, 5.45:1 with trimethylamine and 7.25:1 with triethylamine.[68] Another point of considerable interest is that the ratio of trans elimination (from the cis isomers) to cis elimination (from the trans isomers) range from only 1.2 to 14.3 in the cyclopentyl compounds, whereas the corresponding ratio in the cyclohexyl compound (depending on the base) ranges from 25.4 to 426.[68] The anti geometry preferred for E_2 eliminations (Sec. 2-5d) cannot be achieved in either the cyclopentyl or the cyclohexyl series when the substituents are cis. Theoretical considerations (Sec. 7-4) indicate that when anti elimination is impossible the next most favorable transition state for E_2 elimination involves a syn-cis (syn-periplanar) arrangement of the leaving groups. Experimentally, it has been found that the ratio of elimination rates with alkoxide of cis- and trans-2-phenylcyclopentyl tosylate is only 14, whereas the corresponding ratio for the 2-phenylcyclohexyl tosylates was extremely high[69] (the trans isomer, which requires cis elimination, was inert).

4–4. Medium and Large Rings

The chemistry of the medium rings shows a number of novel features which must for the most part be attributed to conformational factors peculiar to these systems. Since a number of excellent reviews on the chemistry of the medium rings are available (refs. 17, 28, 31, 70–72), this section will be limited to a discussion of those examples in which the conformational aspects of the situation are clearly of paramount importance.

[67] J. Weinstock, S. N. Lewis, and F. G. Bordwell, J. Am. Chem. Soc., **78**, 6072 (1956).

[68] J. Weinstock, R. G. Pearson, and F. G. Bordwell, J. Am. Chem. Soc., **78**, 3468, 3473 (1956).

[69] C. H. DePuy, R. D. Thurn, and G. F. Morris, J. Am. Chem. Soc., **84**, 1314 (1962).

[70] V. Prelog and A. R. Todd, Perspectives in Organic Chemistry, Interscience Division, John Wiley and Sons, New York, 1956.

[71] V. Prelog, Record Chem. Progr. (Kresge-Hooker Sci. Lib.), **18**, 247 (1957).

[72] R. A. Raphael, Proc. Chem. Soc., **1962,** 97.

a. **Cycloheptane.** The seven-membered ring shows a flexibility analogous to that found in the boat form of cyclohexane. Because the symmetry is considerably lower in the seven-membered ring, a conformational study is more involved than in the cases previously considered. Clearly, the solution of such a complicated problem requires either the use of an electronic computer or making some drastic simplifying assumptions. The latter approach was undertaken first, independently by two research groups.[27,73] In one case[27] it was assumed that the conformation could be approximately predicted solely on the basis of the torsional energy of the cycloheptane ring, the van der Waals' repulsions and bond angle deformations being neglected. Such an approach was found to give a conformational energy which was in fair agreement with the experimental heat of combustion. The second approach considered only the interaction potential between hydrogen atoms, neglecting torsion, bond angle deformations, and van der Waals' interactions involving carbon.[73] This approximation was quantitatively unsatisfactory, but qualitatively it gave the same prediction as the alternative approach. It was recognized by both groups that the proper approach to the problem was in fact to consider all the interactions (Sec. 7-2) and make use of an electronic computer, since computers were just then making their debut in the field of organic chemistry.[74] A detailed calculation on cycloheptane, with the help of an electronic computer, was reported in 1961.[75] The conclusions reached were qualitatively the same as those reached by the earlier workers, but now a somewhat greater quantitative significance could be attached to them. The details of this type of calculation will be considered in Sec. 7-2.

Cycloheptane has two types of conformations, which we might call boat and chair by analogy with the six-membered ring (Fig. 8). Both of these conformations are flexible, similar to the boat form of cyclohexane. Thus the energies of the conformations are not fixed numbers but rather

Fig. 8. Cycloheptane.

[73] R. Pauncz and D. Ginsburg, *Tetrahedron*, **9**, 40 (1960).

[74] (a) C. G. Swain, R. B. Mosely, and D. E. Bown, *J. Am. Chem. Soc.*, **77**, 3731 (1955); (b) A. Streitwieser, Jr., and P. M. Nair, *Tetrahedron*, **5**, 149 (1959); (c) N. L. Allinger and J. Allinger, *J. Org. Chem.*, **24**, 1613 (1959); (d) B. Waegell and G. Ourisson, *Bull. Soc. Chim. France*, **1961**, 2443.

[75] J. B. Hendrickson, *J. Am. Chem. Soc.*, **83**, 4537 (1961).

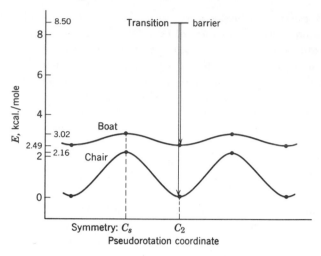

Fig. 9. The energy function for cycloheptane. From Hendrickson.[75]

are functions of the pseudorotational coordinate. The most detailed calculation yet reported[75] gave the energy function as illustrated in Fig. 9. The chair form was predicted to be more stable than the boat, and a particular arrangement of the chair (referred to as a skew chair or twist chair) was considerably preferred over other alternatives. The barrier to the interconversion of the chair and boat forms was calculated to be 8.5 kcal./mole, somewhat lower than that calculated for cyclohexane by the same method (12.7 kcal./mole).

In the most stable arrangement of cycloheptane there are, conformationally speaking, four different types of carbon atoms (Fig. 10). The two positions at C-1 are equivalent, but C-2, C-3, and C-4 show non-equivalent positions which might be referred to as axial and equatorial by analogy with the six-membered ring. Thus there are seven kinds of positions, in

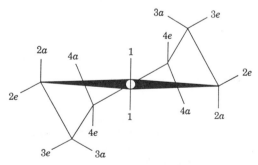

Fig. 10. The geometry of cycloheptane. From Hendrickson.[75]

contrast with the two kinds found for cyclohexane. To complicate matters further, the barrier to pseudorotation of the chair is only 2.16 kcal./mole. This means that, if the substituents on the seven-membered ring are unfavorably placed, the ring can easily pseudorotate to relieve this unfavorable interaction, and the exact rotational arrangement of a substituted ring is therefore quite difficult to specify without a detailed calculation. Certain qualitative predictions can be easily made, however. First, it is clear that the positions labeled equatorial afford a more comfortable location for a substituent than do those labeled axial. The exact order of increasingly unfavorable energy for a substituent when the ring itself is at the conformation of minimum energy is given as $2e \approx 3e \approx 4e < 1 < 4a < 2a$, $3a$. Thus one can predict that for ordinary non-polar substitutents the 1,2-*trans* isomer (e.g., $2e$, $3e$) will be more stable than the *cis* (e.g., $2e$, 1), while for both the 1,3- and 1,4-isomers the *cis* should be the more stable ($2e$, $4e$ rather than 1, $3e$ or $2e$, $5e$ rather than 1, $4e$). Be these substituents *cis* or *trans*, however, there is at least one conformation in which they can both simultaneously take up a position other than axial. The second and perhaps most important prediction to be made is, therefore, that the *cis* and *trans* isomers should differ in energy rather little here, compared to what is found with the six-membered ring. Very few data are available to test this prediction, but for the *cis* and *trans* isomers of 3,5-dimethylcycloheptanone the *cis* isomer was found[76] to have a more negative enthalpy than the *trans* by only 0.8 kcal./mole, as predicted.

Another pair of compounds for which experimental data are available are the perhydroazulenes.[77] It was predicted that the energy difference between the *cis* and *trans* isomers should be quite small, and experimentally it was found that the *trans* had a lower enthalpy by 0.3 ± 0.2 kcal./mole. A very detailed series of applications of the conformational analysis of cycloheptane to the perhydroazulenoid sesquiterpenes has been used to derive the stereochemistry of many of these compounds.[77a]

The conformation of cycloheptene has also been studied.[73,78] The problem is simpler here than for cycloheptane, because the introduction of the double bond into the ring system affords a rigidity which prohibits the pseudorotation. Cycloheptene, therefore, can exist in boat and chair conformations, but each of these is rigid and their geometries can be represented unambiguously with models. Pauncz and Ginsburg, applying their method of calculation which neglected all interactions other than

[76] N. L. Allinger, *J. Am. Chem. Soc.*, **81**, 232 (1959).

[77] N. L. Allinger and V. B. Zalkow, *J. Am. Chem. Soc.*, **83**, 1144 (1961).

[77a] J. B. Hendrickson, *Tetrahedron*, **19**, 1387 (1963).

[78] N. L. Allinger and W. Szkrybalo, *J. Org. Chem.*, **27**, 722 (1962).

hydrogen-hydrogen, concluded that the boat form of cycloheptene would be more stable than the chair by 7 kcal./mole.[73] More detailed calculations[78] showed, however, that in this form a repulsion existed between the carbon atom at the prow of the boat and those of the double bond, and it was concluded that the chair form should in fact be the more stable. A determination of the dipole moment of a 4-cycloheptene-1-one showed that the molecule was largely in the chair form.

The conformations of cycloheptanone have scarcely been mentioned in the literature.[27,79] Starting from the known structure of minimum energy for cycloheptane, we can speculate about where the most comfortable location for the ketone group would be. From models it would seem that it would be at either C-1, C-2, or C-7. The strain appears rather small for any of these possibilities (after small readjustments), and it does not seem possible to predict which conformation is the most favorable without a detailed calculation. In fact, it seems likely that the compound will exist as a conformational mixture.

b. Cyclooctane. The conformations of cyclooctane and its derivatives have been discussed rather extensively. Spitzer and Huffman[16] considered it likely that cyclooctane would be a regular crown, but they showed from heat of combustion measurements that the compound had a lower heat content than anticipated. A satisfactory interpretation of the data was made on the basis of a distorted crown (Fig. 11, D).[27] It is worthwhile to note here that, for rings containing eight or more ring atoms, the conformation of the ring itself appears to be essentially unchanged when a —CH$_2$— group is replaced by —NH—, —O—, or —S—, etc. Similarly cyclooctanone, cyclooctanol, and other simply substituted derivatives all appear to have the same ring conformation.[12,80,81] A detailed analysis of the Raman and infrared spectra of cyclooctane showed that the molecule exists in a single conformation rather than as a mixture, and the same conformation persists in the solid and in the liquid over a wide range of temperatures. This study led to the assignment of conformation as a butterfly structure (Fig. 11, A), but the assignment was not conclusive.[81] Additional data from dipole moments,[82,83] infrared spectra[36,80-82,84] and

[79] C. G. Le Fèvre, R. J. W. Le Fèvre, and B. P. Rao, *J. Chem. Soc.*, **1959**, 2340.

[80] G. Chiurdoglu, T. Doehaerd, and B. Tursch, *Bull. Soc. Chim. France*, **1960**, 1322.

[81] H. E. Bellis and E. J. Slowinski, Jr., *Spectrochim. Acta*, **15**, 1103 (1959).

[82] N. L. Allinger, S. P. Jindal, and M. A. DaRooge, *J. Org. Chem.*, **27**, 4290 (1962).

[83] (a) N. J. Leonard, T. W. Milligan, and T. L. Brown, *J. Am. Chem. Soc.*, **82**, 4075 (1960); (b) A. E. Yethon, Ph.D. Thesis, University of Illinois, 1963.

[84] J. Sicher, M. Horak, and M. Svoboda, *Collection Czech. Chem. Commun.*, **24**, 950 (1959).

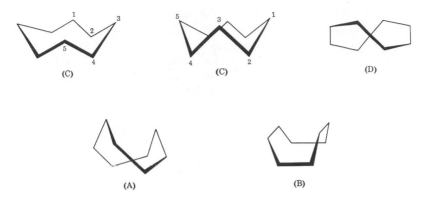

Fig. 11. Possible conformations for cyclooctane.

equilibrations[8] all were consistent with a distorted crown formulation, and the dipole moment data appear inconsistent with A. Models show that there are serious van der Waals repulsions in A, and it does not appear that they can be alleviated without prohibitive bond angle expansion. The infrared spectra of *cis-* and *trans-*1,2-cyclooctanediol show hydrogen bonds of similar strength[36] for the two isomers, which is a little hard to rationalize on the basis of C. The latter conformation appears to require a stronger hydrogen bond in the *trans* isomer. This is in fact what is observed in the corresponding amino alcohol.[84] Conformation B for cyclooctane was excluded by chemical evidence. X-ray crystallographic study of aza-cyclooctane hydrobromide[28] showed that at least with that particular derivative the ring had a conformation which might best be described as a stretched crown, i.e., a crown which is distorted by pulling apart two carbons on opposite sides of the ring. With cyclooctane itself this would lead to three conformationally different kinds of carbons[7]; 1 and 5 are an equivalent pair, 3 and 7 another equivalent pair, and 2, 4, 6, and 8 are all equivalent. Each of these carbons has an axial and an equatorial position, so there are six kinds of positions a substituent may occupy, as compared with two in the chair form of cyclohexane. A monosubstituted cyclooctane may be expected to exist in solution as a mixture of six conformers. Any axial conformation is separated by a substantial energy barrier from any equatorial conformation, and a ring inversion is required for their interconversion in analogy with cyclohexane. (Low temperature proton magnetic resonance spectra [Sec. 3-4d] gave a barrier height of 7.7 kcal./mole.[85]) The barriers separating one kind of axial position from other kinds of axial positions appear to be much smaller.[85] It might be

[85] F. A. L. Anet, *J. Am. Chem. Soc.*, **85,** 1204 (1963).

thought that the axial positions, although differing from one another, would be similar in various properties relative to the differences found between an axial position and an equatorial position. In the brief chemical studies so far carried out, no evidence regarding this situation has been found.[7] This means that the simple Winstein-Holness type of treatment[86] (Sec. 2-2c) is inapplicable, because introduction of any substituent into the ring, depending on its size and location, will shift the equilibrium between the three kinds of axial forms in different ways. This approach to the conformations of the eight-membered ring is clearly complicated, and at this writing it has not been very fruitful. One ray of hope remaining is that the nuclear magnetic resonance spectrum may suffice to separate the axial and equatorial groups of protons, and the theory appears to encourage such a hope (Sec. 3-4d). The limited available evidence adds further encouragement.[7]

The crown structure for cycloöctane (C or D), which is favored by the bulk of the evidence, allows a prediction to be made regarding the sign of the Cotton effect for the 3- and 4-methylcycloöctanones.[87] For compounds of the same configuration as the 3-methylcyclohexanone used for the synthesis (+ Cotton effect), the Cotton effects of the 3- and 4-methyl derivatives should be (+) and (−), respectively. The synthetic intermediate reported originally as 3-methylcycloheptanone was subsequently shown to be mainly the 4-methyl derivative.[87] On conversion to the cycloöctanone by ring expansion, it would have given mainly a mixture of 4- and 5-methylcycloöctanone, and the latter contains no asymmetric center. The observed (−) Cotton effect reported for 3-methylcyclo-öctanone[87] therefore appears, in fact, to be that of the 4-methyl compound, and, if this interpretation is correct, the data are consistent with conformation C or D (as well as A) for the eight-membered ring (Sec. 7-2).

The *cis* and *trans* isomers of cycloöctene are both known[88] and are of conformational interest. The *cis* isomer has a very low heat of hydrogenation[61] as predicted (I-strain).[29] The *trans* isomer, on the contrary, has an exceptionally high heat of hydrogenation.[61] The double bond and the two attached methylenes in the latter compound form a *trans*-butene system, in which all four carbons would like to lie in the same plane. With only four more carbons to bridge the ends of the butene system, some kind of distortion in the molecule is required. It is probable that, as

[86] S. Winstein and N. J. Holness, *J. Am. Chem. Soc.*, **77**, 5562 (1955).

[87] (a) C. Djerassi, B. F. Burrows, C. G. Overberger, T. Takekoshi, C. D. Gutsche, and C. T. Chang, *J. Am. Chem. Soc.*, **85**, 949 (1963); (b) C. Djerassi and G. W. Krakower, *ibid.*, **81**, 237 (1959).

[88] A. C. Cope, R. A. Pike, and C. F. Spencer, *J. Am. Chem. Soc.*, **75**, 3212 (1953).

usual, all the various bond angles show some distortion from their normal values. It has been shown[89] that the molecule possesses a dipole moment which is extraordinarily large for an olefin (0.8 D), and this large moment has been interpreted in terms of a unique hybridization at the olefinic carbons which puts considerable s character into the π bond.[89] The tight belt of saturated carbons joining the ends of the olefin leads to a molecular asymmetry which cannot be destroyed by rotation of the belt around the olefin, as is possible in larger rings, and trans-cyclooctene has been prepared in optically active form.[90] Similar molecular asymmetries brought about by restriction of an internal rotation have also been found in the ansa-[91] and para-cyclophane[92] series of compounds.

The strain on the double bond in trans-cyclooctene leads to some remarkable reactions. The addition of chlorine,[46] for example, under conditions which convert the cis isomer cleanly to the trans-1,2-dichloride, gives a mixture of no less than six compounds, the expected cis-1,2-dichloride being a minor product.

Transannular reactions and interactions are very important phenomena in the eight-membered ring, and they will be discussed in more detail in Sec. 4-4e.

The conformation of cyclooctatetraene is of some theoretical interest,[93] and early spectroscopic and electron diffraction work (Sec. 3-2b) indicated that it was non-planar but did not immediately lead to the correct conformation. It is now established that the molecule possesses a tub shape[94] (Fig. 12, A). There is a significant barrier to inversion of this tub, and, although the inversion appears to occur rapidly at room temperature in cyclooctatetraene itself,[95] compounds of this geometric type (Fig. 12, B[96] and C[97]) have been obtained optically active.

c. Larger Rings.

The X-ray crystallographic study on cyclononylamine

[89] N. L. Allinger, J. Am. Chem. Soc., **80**, 1953 (1958).

[90] (a) A. C. Cope, C. R. Ganellin, H. W. Johnson, Jr., T. V. Van Auken, and H.J. S. Winkler, J. Am. Chem. Soc., **85**, 3276 (1963); (b) A. C. Cope and A. S. Mehta, ibid., **86**, 1268 (1964).

[91] A. Lüttringhaus and H. Gralheer, Ann., **550**, 67 (1942); **557**, 108, 112 (1947).

[92] D. J. Cram, W. J. Wechter, and R. W. Kierstead, J. Am. Chem. Soc., **80**, 3126 (1958).

[93] (a) R. A. Raphael, in D. Ginsburg, ed., Non-Benzenoid Aromatic Compounds, Interscience Division, John Wiley and Sons, New York, 1959, p. 465; (b) N. L. Allinger, J. Org. Chem., **27**, 443 (1962).

[94] O. Bastiansen, L. Hedberg, and K. Hedberg, J. Chem. Phys., **27**, 1311 (1957).

[95] F. A. L. Anet, J. Am. Chem. Soc., **84**, 671 (1962).

[96] K. Mislow and H. D. Perlmutter, J. Am. Chem. Soc., **84**, 3591 (1962).

[97] N. L. Allinger, W. Szkrybalo, and M. A. DaRooge, J. Org. Chem., **28**, 3007 (1963).

(A) 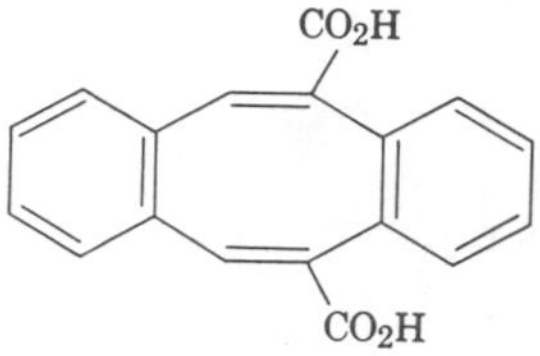   (B) 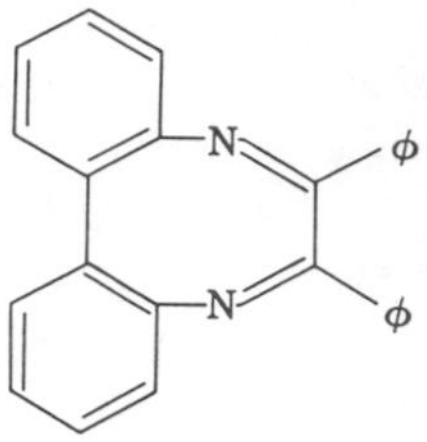   (C)

Fig. 12.

hydrobromide[98] has yielded the conformation of the ring (actually two different but closely similar conformations) and it shows that the ring can minimize its strain by expanding the C—C—C angles to an average value of 117° and adopting the arrangement shown in Fig. 13, A. In this arrangement the torsional angles are for the most part about 10° away from optimum, and there exist substantial van der Waals repulsions between three of the methylenes on the top of the ring and three on the bottom, the transannualar H—H distances being 1.8 Å (Sec. 7-2). The strain in the system can be seen to be divided into many individual interactions and distortions of several kinds. A calculation of the approximate strain energy of this system from the known geometry was not difficult, and it was carried out as described for cyclodecane (Sec. 7-2). The calculation of the conformation of minimum energy *a priori* would, on the other hand, have been exceedingly laborious.

The conformation shown in Fig. 13 is quite suggestive as to just which hydrogens might be expected to migrate during transannular reactions, but studies along these lines have not yet been reported.

While the symmetrical crown (D_{5d}) conformation for cyclodecane was calculated to have a total strain energy of 27.5 kcal./mole, a deformation of the type observed for cycloöctane gave a predicted strain energy of 18.9 kcal./mole.[27] The subsequently published strain energy (12.0 kcal./mole)[26] shows that some other kind of deformation must in fact have occurred. More recently the molecule was studied crystallographically as the 1,6-diamine dihydrochloride.[99] It was found that the molecule had very much reduced the torsional strain of the ordinary crown (to about 4 kcal.) at the expense of a considerable amount of bond angle bending, the observed C—C—C angles varying from 113° to 120° (corresponding to a total bending energy of about 5 kcal.). There is also some van der Waals compression present, three hydrogens above the plane and three

[98] R. F. Bryan and J. D. Dunitz, *Helv. Chim. Acta*, **43**, 1 (1960).
[99] J. D. Dunitz and K. Venkatesan, *Helv. Chim. Acta*, **44**, 2033 (1961).

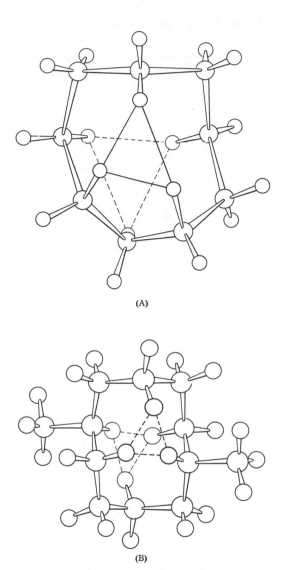

(A)

(B)

Fig. 13. The conformations of the rings in cyclononylamine hydrobromide (A) and 1,6-diaminocyclodecane dihydrochloride (B). (A) From Bryan and Dunitz.[98] (B) From Dunitz and Venkatesen.[99]

below being about 1.8 Å apart* (Fig. 13, B), which accounts for about 3 kcal. of strain. The strain energy is thus seen to be divided between a large number of small interactions of various kinds. The details of this type of calculation are given in Sec. 7-2. Cyclodecane can be looked upon

as having an idealized conformation which is part of a diamond lattice (Fig. 14) which is then distorted somewhat to allow more separation between the excessively close pairs of transannular hydrogens.

Cyclododecane, the smallest cycloalkane which is crystalline at room temperature, has also been studied by X-ray crystallography.[100] Again the observed conformation deviates significantly from a crown, being perhaps best described as a large

Fig. 14.

square, each side being composed of a butane segment and the corner atoms being common to two segments (Fig. 15). Again the strain is to be found partly in angular deformation, partly in torsional deformation. The strain calculated from the known geometry agrees well with that found experimentally.

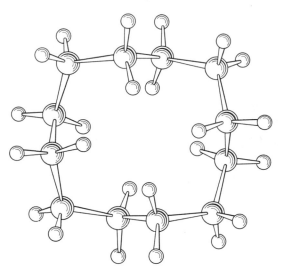

Fig. 15. The conformation of cyclododecane. From Dunitz and Shearer.[100]

---

[100] J. D. Dunitz and H. M. M. Shearer, *Helv. Chim. Acta*, **43,** 18 (1960).

* This distance was calculated from the known geometry of the carbon skeleton, assuming normal C—H bond lengths but modifying the H—C—H angle in accord with the increased C—C—C angle. The interfering hydrogens will doubtless tend to bend away from one another; hence this value may be too small.

For rings larger than cyclododecane, conformational information is rather sparse. The conformation of cyclododecane suggests that the higher even-membered rings will follow the same pattern but will be rectangular instead of square, that is, composed of two long parallel zig-zag poly-methylene chains, joined (*gauche*) by butane-like end segments. As long ago as 1930, Stoll and Stoll-Comte predicted[15] that the very large rings would consist of parallel chains. Torsional strain was not understood at that time, and to account for the known strain in the medium rings they envisioned unrealistically large C—H bond lengths and correspondingly large transannular hydrogen-hydrogen repulsions, which led them to believe that pentane, rather than butane, end segments would be found. An X-ray investigation carried out[101] on 24- and 28-membered diketones showed that the unit cells were long and thin, quite similar to those found for aliphatic hydrocarbons. The interpretation given was that the data were consistent with pentane end segments, but the agreement was in fact much better for butane segments.

That the very large rings are large squares rather than rectangles seems most unlikely, as the square would contain a large hole and would not allow for the intramolecular close packing (nor, for that matter, the intermolecular close packing in the condensed phases) which would yield a maximum van der Waals attraction. There appears to be little doubt that the rings larger than about C-20 are rectangular in shape. This suggests that the large even-membered rings would have enthalpies corresponding approximately to eight *gauche* interactions. Since there are six such interactions in cyclohexane, the total enthalpy per methylene in sufficiently large rings should be less than in cyclohexane, and this is found with the seventeen-membered ring, the largest for which data are available (Table 1, Sec. 4-1). The enthalpies of the very large rings should also be appreciably less than for open chains at room temperature, since the latter compounds can, by mixing in conformations which contain *gauche* arrangements, increase their entropy at the expense of some increase in enthalpy, the result being a minimizing of the free energy.

The even-membered rings from cyclododecane up should fall into two classes, the first class being the rings with 12, 16, 20, . . . ring atoms, and the second being those with 14, 18, 22, . . . , etc. Considering the diones in which the carbonyl groups are as far apart from each other as possible in these rings [the 1, $(n/2) + 1$ diones, Fig. 16], those in the C_2 class should have their carbonyl dipoles more or less parallel while those in the C_{2h} class have them more nearly opposed, and the dipole moments should therefore be greater in the former family than in the latter. The

[101] A. Muller, *Helv. Chim. Acta*, **16**, 155 (1933).

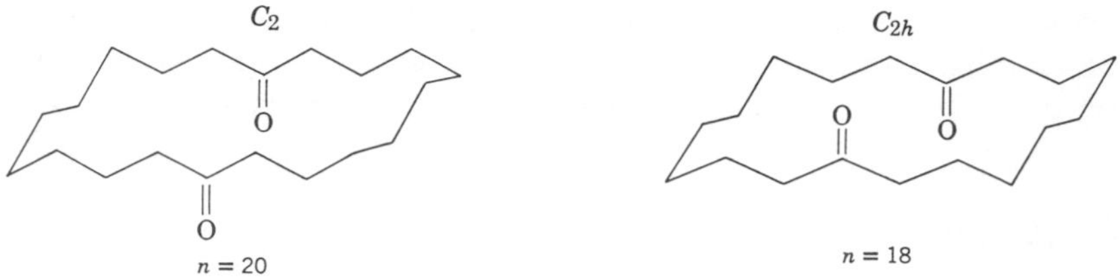

Fig. 16. Probable conformations of the eighteen- and twenty-membered ring diketones.

predicted relationship was found to hold for the cases examined;[102] 1,9-cyclohexadecanedione (C_2 class) has a moment of 4.28 D, while 1,10-cyclooctadecanedione (C_{2h} class) has a moment of 2.78 D. Other predictions of this type can be made, but none has been tested as yet. Thus the *trans* isomer of 1,4-dimethylcyclohexane has the lower enthalpy, but with 1,5-dimethylcyclooctane the *cis* isomer should have the lower value. Similarly, 1,7-dimethylcyclododecane and 1,8-dimethylcyclotetradecane, respectively, should have their *cis* and *trans* isomers of lower enthalpy. This effect would appear to extend to large rings, but it is expected to diminish with increasing ring size and flexibility.

The even-membered rings through C-12 with the possible exception of cyclooctane appear to be fairly well understood conformationally, as are the very large rings (upwards of about C-20). The fourteen-membered ring appears to have one conformation which is much more favorable than other possibilities; it is rectangular.[12] The sixteen-membered ring can have a rectangular conformation, but at the expense of a little torsional or angular strain, or it can have a strainless square conformation, which lacks the large van der Waals attraction between the parallel chains of the rectangular conformation.[12] The rings containing in the neighborhood of 16 to 20 ring members have conformations which are still open to discussion.

The melting points of the series of cycloalkanes are potentially informative.[12] At the melting point of a substance, $\Delta G_m = 0$. Thus the melting point is related to the thermodynamic quantities of fusion as follows:

$$T_m = \Delta H_m / \Delta S_m$$

One might expect that ΔH_m would slowly increase in the homologous series, and it often does, leading in a rough way to an increase in melting point with molecular size. The ΔS_m term can have a dramatic influence on the melting point; in general, a high ΔS_m means a low melting point,

---

[102] B. Gordon, unpublished work.

and vice versa.[103] The ΔS_m for cycloalkanes is composed of three parts: (1) the change in translational entropy on melting (which does not vary much throughout the homologous series); (2) the orientational ΔS, which is a measure of the orientational freedom in the liquid compared to the lack of such freedom in the crystal; (3) the conformational entropy change. This term will be small if the molecule has a single rigid conformation in both the liquid and the solid, and it may be very large if there is only one conformation in the solid and more than one in the liquid.

The even-membered rings can, at least in principle, have a single relatively stable conformation, whereas in the odd-membered rings there is an essential irregularity which would be expected to rotate around the ring somewhat like the pseudorotation found in cyclopentane. The series

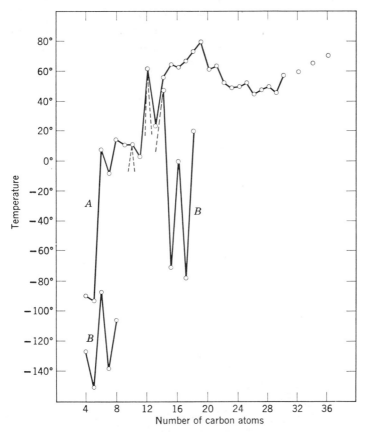

Fig. 17. Melting points (*A*) and transition points (*B*) of the cyclanes. From Dale.[12]

[103] A. R. Ubbelohde, *Quart. Rev.* (*London*), **4,** 356 (1950).

of odd-membered rings would therefore be expected to show greater increases in conformational entropy on melting, and hence lower melting points than the even-membered series. Figure 17 is a plot of the melting points versus ring size for the cycloalkanes; the expected alternation is not observed. Some of these compounds have anomalously small heats of fusion, however; and on further investigation it was found that they undergo phase transitions at temperatures below their melting points.[12,24,103,104] These transitions usually have rather high entropies, and the solid which exists at the melting point can be looked upon as already partially melted. It is therefore these transition points which should show the even-odd alternation and they do (Fig. 17). Very low transition temperatures are observed for the rings containing six, seven, and eight members. These are molecules which are close to being spherical. The interpretation of the phenomena is that below the transition temperature the molecules exist in a regular ordinary crystal, while between the transition temperature and the melting point, although the molecules lack translational freedom, they are free to rotate. The ten- and twelve-membered rings do not show the lower transition temperatures, and presumably they deviate sufficiently from being spherical to form ordinary crystals. The fifteen- to eighteen-membered rings again show transition temperatures well below their melting points. Consistent with this interpretation, the infrared spectra of the compounds between the transition temperatures and the melting points are generally almost identical with those of the liquid, while there is some simplification of the spectra below the transition point.[104] With the larger rings, the higher temperature solid phase is anisotropic and thus there is at least some orientational restriction in this phase; whereas, with the C-4 to C-9 rings, the upper solid phase is isotropic, consistent with rotational freedom. It has been suggested[12] that in the larger rings the conformation can be looked upon as approximately that of a disk, and at the transition temperature rotation of the disk about its principal axis sets in, but rotation about the other axes does not occur below the melting point, and hence the solid remains anisotropic. Alternatively it was suggested[12] that there existed a "lattice imperfection" which travels around the ring (analogous to the pseudo-rotation in cyclopentane). It seems to the present authors that, if the rings containing sixteen to twenty ring atoms were large squares rather than parallel chains, the molecules should occupy a relatively large volume, and hence the liquid should show a low density. The densities of the liquid hydrocarbons and certain of their derivatives show a rather smooth decrease in density from C-14 up to about C-30, and then a slow increase,

[104] E. Billeter and H. H. Günthard, *Helv. Chim. Acta*, **41**, 338 (1958).

the density at this point still being quite a bit below that for the normal alkane series, but slowly converging toward it.[105] The two-dimensional rotation of a large sixteen-membered square is especially hard to understand because the smaller twelve-membered square does not rotate. Rather than an imperfection traveling around the ring (in the even-membered rings), the facts would seem in better accord with a lattice

Table 2. Thermodynamic Data for the trans ⇌ cis-*Cycloalkene Equilibrium in Acetic Acid at 373.6°K*[a]

Ring Size	$\Delta G°$	$\Delta H°$	$\Delta S°$
8	(−10.4)	−9.3	(+3.0)
9	−4.04	−2.9	+3.0
10	−1.86	−3.6	−4.7
11	+0.67	+0.12	−1.5
12	+0.49	−0.41	−2.4

[a] From ref. 106, except for cyclooctene, for which the entropy value is assumed and the enthalpy is from ref. 61. The units are kcal./mole and cal./deg. mole.

imperfection of the crystal, where the imperfection travels around the crystal. The ordinary crystal element and the imperfection could be the rectangular and square conformations, not necessarily respectively.

Above about C-20, the interior hole of the disk conformation would become too large and van der Waals forces would be expected to collapse the molecule into two parallel straight chains.

The cyclic olefins from cyclooctene to cyclododecene are known as both *cis* and *trans* isomers. *trans*-Cycloheptene must be very strained, and it has not yet been isolated. *trans*-Cyclooctene is also very strained, but isolable. The heat of hydrogenation of the *trans* isomer is 9.3 kcal./mole greater than that of the *cis* isomer.[61] The corresponding differences in the cyclononenes and cyclodecenes have similarly been measured, and they have also been determined from the variation of the *cis* ⇌ *trans* equilibrium constant with temperature.[106] These results are summarized in Table 2. With respect to the free energies, it can be seen that *trans*-cyclooctene is highly unstable with respect to the *cis* isomer. For the cyclononenes the difference is smaller, and for the cyclodecenes smaller still. The eleven- and twelve-membered rings, and presumably the still

[105] (a) L. Ruzicka, M. Stoll, H. W. Huyser, and H. A. Boekenogen, *Helv. Chim. Acta*, **13**, 1152 (1930); (b) M. Kobelt, P. Barman, V. Prelog, and L. Ruzicka, *ibid.*, **32**, 256 (1949).

[106] A. C. Cope, P. T. Moore, and W. R. Moore, *J. Am. Chem. Soc.*, **81**, 3153 (1959).

larger homologs, have a *trans* isomer more stable by roughly 1 kcal./mole, similar to simple open chain analogs.[107] It may be noted, however, that the enthalpies and entropies for the isomerization vary with ring size in complicated ways. There is, of course, no reason to expect that these quantities should vary in a smooth systematic manner. The entropy of the isomerization favors *cis*-cyclononene, but it favors the *trans* isomer for the larger homologs. It is known that the *cis* isomers of cyclooctene[89] and cyclodecene[108] have dipole moments of the order of 0.4 D, typical for *cis*-olefins. While *trans*-cyclodecene has a dipole moment of zero[108] like an ordinary *trans*-olefin, that observed for *trans*-cyclooctene[89] was 0.8 D. It seems reasonable that the solvation of the larger dipole moment is attended by the greater entropy loss, and the suggestion was made[106] that *trans*-cyclononene might have a larger moment than its isomer, for the same reason as does *trans*-cyclooctene. The subsequent measurement of the moment of *trans*-cyclononene[109] showed that it did in fact have a moment of 0.6 D. An additional effect to be considered is the rigidity of the system. The *trans* isomers of cyclooctene and cyclononene are a good deal more stretched and hence probably more rigid than their *cis* counterparts, but this difference is expected to become small in the larger rings. In any case the entropy of the isomerization of *trans* → *cis*-cyclooctene is probably more positive than for cyclononene, but the amount of *trans*-cyclooctene at equilibrium is too small to measure by ordinary methods. The equilibrium might be determined by isotopic dilution, or the entropies might be measured directly by calorimetry (Sec. 3-3a), but no such experiments have yet been reported. It can be seen from Table 2 that the *cis* isomer of cyclododecene actually has the lower heat content, but this is outweighed by the entropy effect and the *trans* isomer is the stable one (under the conditions cited). Entropy effects are often overlooked in discussions of "stability," and such neglect is usually permissible *as an approximation* because entropy effects are ordinarily small. Occasionally, however, the entropy of a reaction is very significant; hence the possible error introduced by its neglect should always be considered.[110]

A number of macrocyclic dienes are known, and in several cases the positional isomers have been equilibrated by treatment with acid or base. Cyclooctadiene exists almost exclusively as the 1,3-isomer at equilibrium,[111] as would be expected for dienes in which the double bonds can

---

[107] S. W. Benson and A. N. Bose, *J. Am. Chem. Soc.*, **85,** 1385 (1963).

[108] N. L. Allinger, *J. Am. Chem. Soc.*, **79,** 3443 (1957).

[109] C. L. Neumann, unpublished work.

[110] R. M. Gascoigne, *J. Chem. Soc.*, **1958,** 876.

[111] D. Devaprabhakara, C. G. Cardenas, and P. D. Gardner, *J. Am. Chem. Soc.*, **85,** 1553 (1963).

conjugate, i.e., become more or less coplanar.[112] With cyclononadiene, on the other hand, the 1,5 isomer predominates over the 1,3 by a factor of over 15 to 1.[111] The 1,3 isomer does not have much conjugation energy, the double bonds lie in planes which are nearly perpendicular, but the strong predominance of the 1,5 isomer over the others must be attributed to a particularly favorable conformation for that isomer. At equilibrium, the predominant diene in the twelve- and thirteen-membered rings also has the double bonds separated by two methylene groups, while in the larger rings studied the conjugated isomer is the favored one.[113] In these large rings there is present at equilibrium at least 5 percent of each possible position isomer.

d. Catenanes. Chemists have speculated for many years about whether or not there might exist compounds in which cyclic moieties are joined together as links in a chain (Fig. 18). If these moieties are aliphatic in nature, the

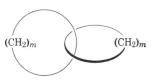

$(CH_2)_m$ $(CH_2)_m$

Fig. 18. The catenanes.

van der Waals radii of the atoms involved require that the rings be rather large, a minimum value for m being of the order of 20. Evidence for the preparation of such a compound (called a catenane) has been reported,[114] and it has been noted that it is an example of a topological isomer, i.e., an isomer differing in topology but not in bonding from the other possible topological isomer (separated rings). Other kinds of topological isomers can be imagined, e.g., an ordinary cyclic chain and one containing a knot, but such compounds have not as yet been prepared (1963).

e. Transannular Reactions and Interactions. A number of reactions not previously known were uncovered in the medium rings independently by two groups, one led by Prelog which worked in the cyclodecane series,[115] and the other led by Cope which worked in the cyclooctane series.[116] The reactions are transannular in nature, and they result from the spatial proximity of groups on opposite sides of the ring. Striking examples are the 1,5-transannular hydride transfer which often occurs when a carbonium ion is generated in the cyclooctyl ring[7,116,117] (more or less 1,3-hydride

[112] E. A. Braude, *Chem. & Ind. (London)*, **1954,** 1557

[113] A. J. Hubert and J. Dale, *J. Chem. Soc.*, **1963,** 4091.

[114] E. Wasserman, *J. Am. Chem. Soc.*, **82,** 4433 (1960); G. Schill and A. Lüttringhaus, *Angew. Chem. (Intern. Ed. Engl.)*, **3,** 546 (1964).

[115] V. Prelog and K. Schenker, *Helv. Chim. Acta*, **35,** 2044 (1952).

[116] A. C. Cope, S. W. Fenton, and C. F. Spencer, *J. Am. Chem. Soc.*, **74,** 5884 (1952).

[117] A. C. Cope, G. A. Berchtold, P. E. Peterson, and S. H. Sharman, *J. Am. Chem. Soc.*, **82,** 6366, 6370 (1960).

transfer may also occur), and both 1,5- and 1,6-hydride transfers which occur similarly in the cyclodecyl ring.[115] Thus the formolysis of trans-1,2-epoxycyclooctane gave trans-cyclooctane-1,4-diol as the principle product, together with the expected trans-1,2-diol. The cis-1,4-diol is similarly formed from the cis-epoxide. Two different hydride migrations can be imagined which will lead to the observed product, either 1,3 or 1,5. With the use of deuterium, both of these pathways have been shown to be important.[117] These hydride transfers have been investigated in some

Table 3. The Relative Solvolysis Rates of Some Cyclooctyl Tosylates[120]

Tosylate	Relative Rate
2-Pentyl	1
Cyclooctyl	173
trans-5-t-Butylcyclooctyl	69
cis-5-t-Butylcyclooctyl	3700

detail.[118] Models show that the appropriate hydrogens are situated quite close to the carbons to which they migrate, and these are spectacular examples of reactions in which conformational factors determine the outcome.

The thermal decompositions of cyclic diazo compounds to give medium ring carbenes also lead to transannular hydride transfers, followed by various elimination or cyclization reactions.[117,119]

An early idea was that anchimeric acceleration resulted from these transannular migrations, which would contribute in part to the extremely fast rates in, for example, the solvolysis of the medium ring tosylates. The I-strain hypothesis qualitatively accounts for most of the known facts, however, and it seems now that anchimeric assistance may or may not be important, depending on the particular case. The relative rates of solvolysis of the tosylates of some cyclooctyl derivatives are given in Table 3. There can be little participation of the 5-hydrogen in trans-5-t-butyl-cyclooctyl tosylate for steric reasons (Fig. 19), and yet the solvolysis rate of the compound is 69 times that of an aliphatic analog. Cyclooctyl tosylate itself can exhibit such participation to the extent that the tosyl group is in the equatorial position. Without participation the rate would be expected to be somewhat slower for the equatorial tosylate than for the axial (Sec. 2-5c), but the observed rate of cyclooctyl is in fact nearly

[118] A. C. Cope, J. M. Grisar, and P. E. Peterson, J. Am. Chem. Soc., 81, 1640 (1959).
[119] L. Friedman and H. Shechter, J. Am. Chem. Soc., 83, 3159 (1961).
[120] W. Szkrybalo, Ph.D. Thesis, Wayne State University, April, 1964.

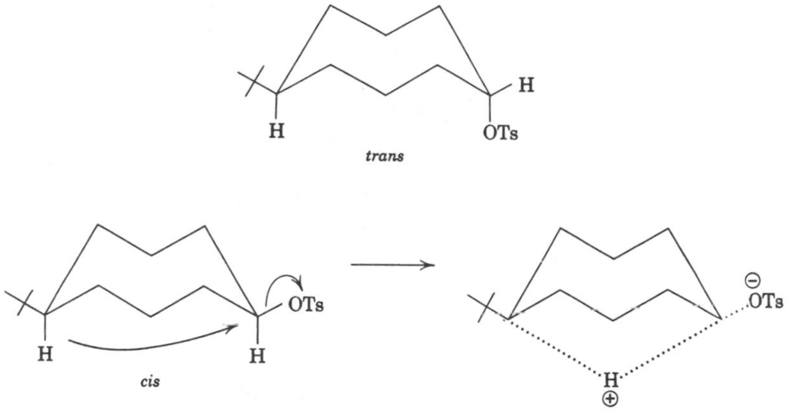

Fig. 19.

three times that of *trans*-5-*t*-butylcyclooctyl. It therefore appears likely that there is some acceleration due to the participation of the 5-hydrogen in cyclooctyl tosylate solvolysis, but it would seem that this acceleration is rather small compared to the steric acceleration present in the *trans*-5-*t*-butylcyclooctyl case, which seems best attributed to I-strain. The *cis*-5-*t*-butylcyclooctyl tosylate is dramatically faster than cyclooctyl itself, and it appears that this acceleration results from the increased participation of the 5-hydrogen due to the stabilization of the positive charge in the transition state by the *t*-butyl group (Fig. 19). Support for this interpretation comes from an analysis of the products of these solvolysis reactions. The *trans*-5-*t*-butyl tosylate gives 88% of 5-*t*-butylcyclooctene (no hydride migration) and 12% of the 1-olefin (migration, probably via a free carbonium ion), while the *cis* isomer gives a 99% yield of the 1-olefin. The driving force for the migration here, in contrast to cyclooctyl tosylate itself, appears to result from the conversion of the incipient secondary carbonium ion to a tertiary one.

Other evidence of interest is that a transannular deuterium, even though it migrates during solvolysis of 5,6-tetradeuterocyclodecane tosylate, does not show an appreciable isotope rate effect.[121] This fact suggests that the involvement of the deuterium in the rate-determining step is small. In addition, transannular phenyl migration occurs only to a very slight extent,[122] and this must be a consequence of the steric disadvantage of the participating phenyl outweighing any potential electronic advantage.

[121] S. Borčić, unpublished work, quoted by V. Prelog, *Record Chem. Progr.* (*Kresge-Hooker Sci. Lib.*), **18**, 247 (1957).
[122] A. C. Cope, P. E. Burton, and M. L. Caspar, *J. Am. Chem. Soc.*, **84**, 4855 (1962).

Fig. 20.

Careful investigation of rings of other sizes has shown that transannular hydride migration occurs to a detectable extent under proper conditions in rings having as few as six[123,124] or seven members[125] and is often important in rings containing eight to eleven members. Such migration has not yet been detected in rings containing more than eleven members.[126] Hydrogen transfers in medium ring polyenes have also been observed[127] and most probably occur by transannular mechanisms.

Cycloöctatriene very easily undergoes a transannular valence bond isomerization to give the bicyclic isomer (bicyclo[4.2.0]octadiene, Fig. 20). Both isomers have been isolated, and they are easily interconvertible.[128]

Transannular interactions have also been extensively studied. Interactions (differing from reactions) can exist in the resting molecule and may be detected by various physical means. Such interactions have been studied by a number of groups, the work by Leonard[129] and by Cram[130] being particularly extensive and noteworthy. A number of reviews are available.[17,31,70–72]

4–5. Fused Ring Systems

a. Configuration and Conformation. Small fused ring systems, such as bicyclo[1.1.0]butane,[131] bicyclo[2.1.0]pentane,[132] bicyclo[2.2.0]hexane,[133]

[123] A. C. Cope, H. E. Johnson, and J. E. Stephenson, J. Am. Chem. Soc., **78**, 5599 (1956).

[124] N. L. Wendler, R. P. Graber, C. S. Snoddy, Jr., and F. W. Bollinger, J. Am. Chem. Soc., **79**, 4476 (1957).

[125] A. C. Cope, T. A. Liss, and G. W. Wood, J. Am. Chem. Soc., **79**, 6287 (1957).

[126] V. Prelog and M. Speck, Helv. Chim. Acta, **38**, 1786 (1955).

[127] D. S. Glass, J. Zirner, and S. Winstein, Proc. Chem. Soc., **1963**, 276, and references therein.

[128] A. C. Cope, A. C. Haven, Jr., F. L. Ramp, and E. R. Trumbull, J. Am. Chem. Soc., **74**, 4867 (1952).

[129] N. J. Leonard and C. R. Johnson, J. Am. Chem. Soc., **84**, 3701 (1962), and earlier papers.

[130] D. J. Cram and L. A. Singer, J. Am. Chem. Soc., **85**, 1075 (1963), and earlier papers.

[131] D. M. Lemal, F. Menger, and G. W. Clark, J. Am. Chem. Soc., **85**, 2529 (1963); K. B. Wiberg and G. M. Lampman, Tetrahedron Letters, **1963**, 2173.

[132] R. Criegee and A. Rimmelin, Chem. Ber., **90**, 414 (1957).

[133] S. Cremer and R. Srinivasan, Tetrahedron Letters, No. 20, 24 (1960).

bicyclo[3.1.0]hexane,[134] bicyclo[3.2.0]heptane,[135] and bicyclo[4.1.0]heptane,[134] and their heterocyclic analogs such as the epoxides derived from cycloalkenes of up to six members, can, for reasons of strain, exist only in the *cis* configuration. However, two of the three possible fused bicyclooctane systems, namely, the [4.2.0] and [3.3.0] system, have been obtained in both *cis* and *trans* forms. Derivatives of *trans*-bicyclo[4.2.0]octane have been obtained by various routes;[136] their existence demonstrates that the adjacent *trans* equatorial bonds in cyclohexane can be deflected sufficiently from their normal torsional angle of 60° to accommodate the nearly (but not quite; cf. Sec. 4-2) flat cyclobutane ring. However, as shown earlier (p. 77, and Fig. 33 in Chap. 2), such deflection is much more difficult than the corresponding deflection of the *cis*-1,2 (equatorial-axial) bonds into a plane, and it is therefore not surprising that *cis*-bicyclo[4.2.0]octane is more readily accessible[135b] and presumably more stable than its *trans* epimer. Successful preparation of bicyclo[3.3.0]octane[137] in the *trans* as well as the *cis* form provides added evidence that the five-membered ring is not flat (Sec. 4-3); nevertheless, the *trans* isomer, as might be expected, is considerably strained; its enthalpy exceeds that of the *cis* by 6.0 kcal./mole.[137] The 3-oxa and 3-thia derivatives of bicyclo[3.3.0]-octane[138] (Fig. 21) seem to be more easily accommodated in the *trans* form than the carbocyclic analog. The reason for this is not obvious from model considerations; it may reflect the greater tendency of the tetrahydrofuran and tetrahydrothiophene rings to exist in the half-chair form (Sec. 4-3) in which carbon atoms 3 and 4 of the heterocycles are sufficiently staggered to make *trans* fusion at this point facile. Biotin, a heterocyclic analog

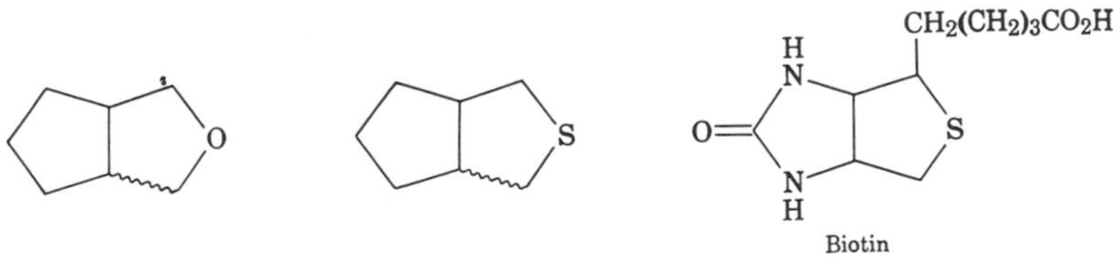

Biotin

Fig. 21.

[134] H. E. Simmons and R. D. Smith, *J. Am. Chem. Soc.*, **81**, 4256 (1959).

[135] (a) A. T. Blomquist and J. Kwiatek, *J. Am. Chem. Soc.*, **73**, 2098 (1951); (b) O. L. Chapman and D. J. Pasto, *Chem. & Ind.* (*London*), **1961**, 53.

[136] M. P. Cava and E. Moroz, *J. Am. Chem. Soc.*, **84**, 115 (1962); J. Meinwald, G. G. Curtis, and P. G. Gassman, *ibid.*, **84**, 116 (1962); E. J. Corey, R. B. Mitra, and H. Uda, *ibid.*, **85**, 362 (1963); P. de Mayo, R. W. Yip, and (in part) S. T. Reid, *Proc. Chem. Soc.*, **1963**, 54.

[137] J. W. Barrett and R. P. Linstead, *J. Chem. Soc.*, **1936**, 611.

[138] L. N. Owen and A. G. Peto, *J. Chem. Soc.*, **1955**, 2383.

containing the [3.3.0] system (Fig. 21) exists in nature with *cis* ring fusion but can be synthesized with the rings fused *trans* as well as *cis*.

[4.3.0]Bicyclononane (hydrindane) has long been known to exist with both *cis* and *trans* fused rings.[139] As might be expected from the analogy of the 1,2-dimethylcyclohexanes (Sec. 2-3b), the *trans* isomer (*e,e*) is thermochemically more stable than the *cis* (*e,a*). However, the difference in enthalpy—determined as 1.04 ± 0.53 kcal./mole from differences in heats of combustion (vapor state at 25°C.)[140]—is less than the corresponding difference (1.87 kcal./mole) for the 1,2-dimethylcyclohexanes. The liquid phase enthalpy difference, 0.74 ± 0.52 kcal./mole as determined by heat of combustion[140] or 1.07 ± 0.09 kcal./mole from temperature dependence of the equilibrium of the two isomers determined at elevated temperatures (193–365°) over a palladium catalyst[141] or 0.58 ± 0.05 kcal./mole as determined by temperature dependence of equilibrium established over aluminum bromide,[142] is also less than the corresponding difference in the 1,2-dimethylcyclohexanes (1.54 kcal./mole). This is undoubtedly due to the fact, evident from models, that *trans*-hydrindane is appreciably strained, the torsional angle of 72° of the *trans*-1,2 bonds[22] in the five-membered ring (in the half-chair form which is most favorable) being larger than the normal torsional angle of 60° in a six-membered ring. In *cis*-hydrindane the difficulty is much less,[22] since the torsional angle of the *cis* bonds in the half-chair of cyclopentane at C_3-C_4 (48°) can readily be accommodated within the six-membered ring by a slight flattening of the latter* (bending of adjacent *e,a* bonds toward each other; cf. Fig. 33 in Chap. 2).[143,144]

It is interesting that Granger et al.[144] were able to obtain both A and B,

---

[139] (a) W. Hückel and H. Friedrich, *Ann.*, **451**, 132 (1926); (b) W. Hückel, M. Sachs, J. Yantschulewitsch, and F. Nerdel, *ibid.*, **518**, 155 (1935).

[140] C. C. Browne and F. D. Rossini, *J. Phys. Chem.*, **64**, 927 (1960).

[141] N. L. Allinger and J. L. Coke, *J. Am. Chem. Soc.*, **82**, 2553 (1960).

[142] K. R. Blanchard and P. von R. Schleyer, *J. Org. Chem.*, **28**, 247 (1963).

[143] E. L. Eliel and C. Pillar, *J. Am. Chem. Soc.*, **77**, 3600 (1955).

[144] R. Granger, P. F. G. Nau, J. Nau, and C. François, *Bull. Soc. Chim. France*, **1962**, 496.

* At one time, it was believed that the six-membered ring in *cis*-hydrindane was boat-shaped: H. G. Derx, *Rec. trav. chim.*, **41**, 318 (1922), and ref. 139a. Then the geometric relation of the hydroxyl groups in the *cis*-hydrindane *cis*-5,6-diols (Fig. 22, A and B) should have resembled that in *cis*-1,2-cyclopentanediol. Experimentally, however, it was shown that both A and B (ref. 144) as well as one of their 2-oxa analogs (Fig. 22, C) (ref. 143) resemble *cis*-1,2-cyclohexanediol with respect to both intramolecular hydrogen bonding (as shown by infrared spectrum; Sec. 3-4a) and lead acetate cleavage rate. This is compatible with a six-membered ring in a slightly deformed chair form in Fig. 22, A–C, but not with a boat or skew-boat shaped six-membered ring. See also R. A. Wohl, *Chimia*, **18**, 219 (1964).

Fig. 22.

Fig. 22, by hydrogenation of the corresponding indanediol (along with some *cis*-hydrindane-*trans*-5,6-diol). They were not definite about the configurational assignment, but it appears to us that such assignment can be made on fairly firm grounds on the basis of evidence given in ref. 144. The diol of m.p. 79° (63% crude yield) is undoubtedly the all-*cis* diol A, which would be expected to be the major product of catalytic hydrogenation. In contrast, the diol of m.p. 134°, now assigned structure B, is obtained in only 7.5% yield. In agreement with this assignment, the acetonide of A which is strained by 1,3-diaxial crowding is more readily hydrolyzed than the acetonide of B. Finally, both intramolecular hydrogen bonding (as indicated by infrared spectrum) and lead tetraacetate cleavage rates suggest that the hydroxyl groups are closer together in A than in B. This, also, may be readily explained by the *syn*-axial interaction of the axial linkage of the five-membered ring and the axial hydroxyl group which will promote bending of the axial hydroxyl toward the equatorial hydroxyl with corresponding diminution of the dihedral angle between the two.

The spectral behavior and the lead tetraacetate cleavage behavior of C (ref. 143) suggest that its configuration corresponds to that of B. This is reasonable because C was obtained by permanganate oxidation of the corresponding olefin which should have led to approach of the hydroxyl groups from the less hindered side, i.e., the side away from the five-membered ring.

We return now to the thermodynamic properties of the hydrindanes. It is known that the entropy of the *cis* isomer exceeds that of the *trans*, the difference having been determined as 1.68 ± 0.10 cal./deg. mole by heat capacity measurements,[145] and as 2.30 ± 0.10 cal./deg. mole[141] and 1.00 ± 0.06 cal./deg. mole[142] from the temperature dependence of equilibrium. A maximum of 1.4 cal./deg. mole in favor of the *cis* isomer might have been expected from symmetry considerations similar to those discussed earlier for the 1,2-dimethylcyclohexanes (Sec. 2-3b), on the assumption that *trans*-hydrindane actually has a symmetry number of 2 and *cis*-hydrindane a symmetry number of 1.

[145] J. P. McCullough, data cited in ref. 140. The data refer to the liquid phase.

The energy barrier to ring inversion in cis-hydrindane has been determined[146] by low temperature NMR measurements (Sec. 3-4d) to be of the order of 6.5 kcal./mole and is certainly appreciably lower than the barrier in cyclohexane (p. 41). The reason for this is evident from an inspection of Dreiding models: the six-membered ring in cis-hydrindane is flattened and strained, leading to an elevated ground state energy level; however, no corresponding elevation of energy occurs in the transition state for ring inversion, which as already indicated (p. 42) is half-chair shaped and lends itself readily to the fusion of the cyclopentane ring (envelope form). Clearly, if the ground state energy is elevated but the transition state energy for ring inversion is not, the activation energy for ring inversion in cis-hydrindane is diminished below the corresponding energy in cyclohexane.

The difference in free energy of cis- and trans-hydrindane near room temperature is very small; from the thermodynamic parameters cited above, $\Delta G°$ (favoring the trans isomer) may be calculated as 0.24–0.38 kcal./mole at 25°; however, at temperatures much above 200°C. the cis isomer becomes more stable because of its greater entropy. Because of the small intrinsic difference in free energy between the two isomers, small changes in structure—such as the introduction of an angular methyl group, the introduction of a ketone function in various parts of the ring system, or the rigidification of the otherwise mobile cis-hydrindane system (e.g. when it forms part of a steroid)—may change the picture sufficiently to override any relative stability considerations based on the parent isomers, and predictions in such systems are not easy to make.[147–149] For example, 1-hydrindanone is more stable (by a factor of 3) in the cis configuration than in the trans.[149] The reason in this particular instance may be that, as mentioned earlier (Sec. 4-3), the preferred conformation for cyclopentanone is C_2 and the normal dihedral angle at C_2-C_3 in this form[22] is 39.5°, appreciably smaller than the maximum angle in the C_s form[22] (46.1°), which makes trans fusion of the six-membered ring (normal torsional angle 60°) correspondingly more difficult. Introduction of a 5,6 double bond in 1-hydrindanone makes the trans isomer relatively more stable (cf. the case of the cyclohexene-4,5-diols, p. 111), as does the introduction of a methyl group at C_4.[149a]

[146] W. B. Moniz and J. A. Dixon, J. Am. Chem. Soc., **83**, 1671 (1961).

[147] Cf. G. Quinkert, Experientia, **13**, 381 (1957); L. F. Fieser and M. Fieser, Steroids, Reinhold Publishing Co., New York, 1959, pp. 211–216; D. H. R. Barton and G. A. Morrison, Fortschr. Chem. Org. Naturstoffe, **19**, 165 (1961).

[148] N. L. Allinger, R. B. Hermann, and C. Djerassi, J. Org. Chem., **25**, 922 (1960); N. L. Allinger and S. Greenberg, ibid., **25**, 1399 (1960).

[149] (a) H. O. House and G. H. Rasmusson, J. Org. Chem., **28**, 31 (1963); (b) W. Hückel, W. Egerer, and (in part) F. Mossner, Ann., **645**, 162 (1961).

Probably the most important fused carbocyclic system is decalin ([4.4.0]bicyclodecane). Counting of butane-*gauche* interactions in *cis*- and *trans*-decalin reveals three more such interactions in the former than in the latter, the situation being similar to that in the 1,2-dimethylcyclo-hexanes (Sec. 2-3b). Computing the butane-*gauche* interaction as 0.85 kcal./mole in the liquid phase or 0.95 kcal./mole in the vapor phase leads to a calculated difference in enthalpy between *cis*- and *trans*-decalin (favoring the latter) of 2.55 kcal./mole in the liquid state or 2.85 kcal./mole in the vapor. Thermochemical studies[150] in fact indicate a liquid phase difference in enthalpy of 2.69 ± 0.31 kcal./mole in favor of the *trans* isomer, in excellent agreement with the calculated value and with another experimental value of 2.72 ± 0.20 kcal./mole obtained by direct equil-ibration of the decalins over a palladium catalyst in the temperature range 268–378°.[151] The thermochemically determined enthalpy differ-ence in the vapor state of 3.09 ± 0.77 kcal./mole[150] is also in good agree-ment with the calculated value. The entropy difference between the decalins might be computed as $R \ln 2$ or 1.38 cal./deg. mole in favor of the *cis* isomer, because it is a (rapidly interconvertible) *dl* pair, similar to *cis*-1,2-dimethylcyclohexane (Sec. 2-3a) whereas the *trans* isomer is a *meso* form. (Both isomers have symmetry numbers of 2; thus symmetry does not give rise to any entropy difference.) In fact, however, the entropy difference seems to be less than calculated, having been reported as 0.55 ± 0.3 cal./deg. mole in one determination[151] and essentially zero in another.[152] The position of equilibrium of the 1-decalones at 250° corre-sponding to 5–10% of *cis* isomer[153] is in agreement with a calculated $\Delta G°$ at that temperature of 2.4 kcal./mole (calculated equilibrium com-position 91% *trans*).

The calculated energy difference between *cis*- and *trans*-9-methyldecalin is less than that between the parent compounds because the 9-methyl substituent introduces four extra butane-*gauche* interactions in the *trans* isomer but only two in the *cis*, thus diminishing the difference between the two from three butane-*gauche* interactions to one; this gives the *trans* isomer a calculated lower enthalpy of only 0.85 kcal./mole in the liquid state. The experimental difference is 1.39 ± 0.64 kcal./mole from

[150] D. M. Speros and F. D. Rossini, *J. Phys. Chem.*, **64**, 1723 (1960).

[151] N. L. Allinger and J. L. Coke, *J. Am. Chem. Soc.*, **81**, 4080 (1959).

[152] J. P. McCullough, H. L. Finke, J. F. Messerly, S. S. Todd, T. C. Kincheloe, and G. Waddington, *J. Phys. Chem.*, **61**, 1105 (1957). Since the two determinations were not at the same temperature, complete agreement is not be to expected, even though both sets of data refer to the liquid phase.

[153] H. E. Zimmerman and A. Mais, *J. Am. Chem. Soc.*, **81**, 3644 (1959).

heat of combustion data[154] or 0.55 ± 0.28 kcal./mole from temperature dependence of equilibrium.[155] $\Delta G°$ at 250° is calculated to be about 0.3 kcal./mole, corresponding to a predominance of the *trans* isomer of 57:43. In 9-methyl-1-decalone the experimentally established equilibrium[156] at $250\overline{°}$C. is 41% *trans*, 59% *cis*.

Because of rapid interconversion of the chair forms the hydrogens in *cis*-decalin are equivalent in the NMR spectrum not only at room temperature[157] but also at temperatures as low as −121°,[146] showing that the barrier to chair inversion is much lower than in cyclohexane and lower even than in *cis*-hydrindane, possibly owing to the large number of *gauche* interactions in the ground state. The rigid *trans*-decalin shows the expected broad, partly resolved band indicating distinct equatorial and axial protons.[146,157]

Perhydroazulene, [5.3.0]bicyclodecane, isomeric with decalin, has been studied through thermal equilibration of the *cis* and *trans* isomers over a palladium catalyst,[77] through *a priori* calculation,[158] and, as a heterocyclic analog, through equilibration of the lactones of the 2-hydroxycyclo-heptanecarboxylic acids.[159] The results all indicate that the difference in free energy, enthalpy, and entropy between the *cis* and the *trans* isomers in the 7-5 fused ring system is very small. This is in contrast to the 5-5 system where, because of strain, the *cis* isomer is appreciably more stable, and to the 6-5 system, where, because of conformational considerations, the *trans* isomer has somewhat greater stability. The reasons for the difference have been discussed in Sec. 4-4, p. 207.

Among the tricyclic systems, the perhydrophenanthrene system (Fig. 23) which forms part of the skeleton of numerous steroids and terpenes (Chap. 5) is the best studied. The configurations of the six diastereo-isomers have been assigned through the elegant work of Linstead, Doering, and co-workers[160,161] which has been summarized in detail elsewhere.[162] The relative stability of the diastereoisomers as indicated in Fig. 23 can

[154] W. G. Dauben, O. Rohr, A. Labbauff, and F. D. Rossini, *J. Phys. Chem.*, **64**, 283 (1960).

[155] N. L. Allinger and J. L. Coke, *J. Org. Chem.*, **26**, 2096 (1961).

[156] A. Ross, P. A. S. Smith, and A. S. Dreiding, *J. Org. Chem.*, **20**, 905 (1955).

[157] J. Musher and R. E. Richards, *Proc. Chem. Soc.*, **1958**, 230.

[158] J. B. Hendrickson, *J. Am. Chem. Soc.*, **83**, 4537 (1961).

[159] W. Herz and L. A. Glick, *J. Org. Chem.*, **28**, 2970 (1963).

[160] R. P. Linstead, W. E. Doering, S. B. Davis, P. Levine, and R. R. Whetstone, *J. Am. Chem. Soc.*, **64**, 1985, 1991, 2003, 2006, 2009, 2014 (1942).

[161] S. B. Davis, W. E. Doering, P. Levine, and R. P. Linstead, *J. Chem. Soc.*, **1950**, 1423; S. B. Davis and R. P. Linstead, *ibid.*, **1950**, 1425.

[162] E. L. Eliel, *Stereochemistry of Carbon Compounds*, McGraw-Hill Book Co., New York, 1962, pp. 282–284.

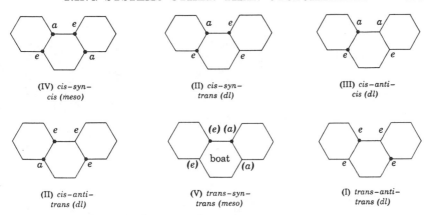

(IV) *cis—syn—cis (meso)*

(II) *cis—syn—trans (dl)*

(III) *cis—anti—cis (dl)*

(II) *cis—anti—trans (dl)*

(V) *trans—syn—trans (meso)*

(I) *trans—anti—trans (dl)*

Fig. 23. The perhydrophenanthrenes and their relative stability (indicated by Roman numerals). From E. Eliel, *Stereochemistry of Carbon Compounds,* copyright 1962, McGraw-Hill Book Company, Inc., by permission of the publishers.

be predicted on the basis of the three following rules:[163]

1. Considering the equatorial or axial fusion of the outer rings to the central ring (Fig. 23), the system with the larger number of equatorial bonds is the more stable.

2. For systems with two axial bonds, the one having the axial bonds disposed 1,4 or 1,2 (at the backbone) will be more stable than the one having axial bonds 1,3 (i.e., *syn*-axial).

3. Systems having two axial bonds (1,2) at the same ring fusion evidently cannot exist in that particular conformation; if the alternate (inverted) conformation also leads to an axial-axial ring fusion, the system necessarily exists in the boat rather than the chair form. Such a system (V in Fig. 23) has built into it the instability of the boat form (Sec. 2-1).

Experimental evidence regarding the relative stability of the perhydrophenanthrenes is, unfortunately, limited.[161] Perhaps the most elegant support of at least some of the predictions comes from a consideration of some of the intermediates in W. S. Johnson's steroid synthesis.[164] Hydrogenation of intermediate A (Fig. 24) gives two alcohols (B, C) which, upon oxidation, in turn give two ketones. If it is assumed that catalytic reduction leads, mainly, to all-*cis* addition of hydrogen, the ketones must be represented by stereoformulas D and E. The fact that E is epimerized by base to F but D cannot be epimerized is consistent with the stereochemical assignment shown, namely, E = *trans-anti-cis* epimerized to F

[163] W. S. Johnson, *Experientia,* **7,** 315 (1951); *J. Am. Chem. Soc.,* **75,** 1498 (1953).

[164] W. S. Johnson, E. R. Rogier, and J. Ackerman, *J. Am. Chem. Soc.,* **78,** 6278, 6322 (1956).

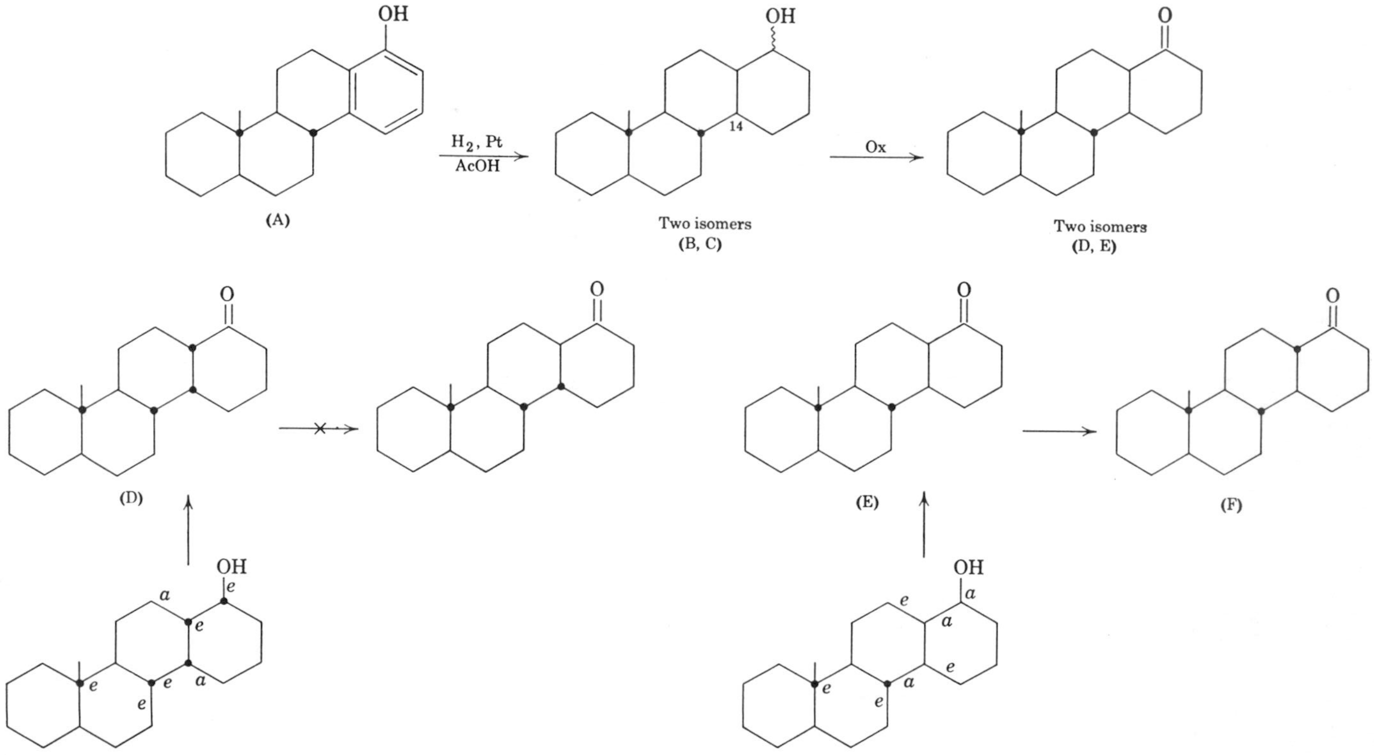

Fig. 24. Relative stability of perhydrophenanthrene systems (according to Johnson[164]). From E. Eliel, *Stereochemistry of Carbon Compounds*, copyright 1962, McGraw-Hill Book Company, Inc., by permission of the publishers.

(*trans-anti-trans*) and D = *trans-syn-cis*, not epimerizable to *trans-syn-trans* which would have the central ring in the boat form. The additional experimental fact that alcohol C is oxidized by chromic acid more rapidly than B suggests that C is axial and B equatorial (cf. Sec. 2-5c) and permits the complete stereochemical assignments of the alcohols B and C shown in Fig. 24, which, incidentally, is consistent with the original assumption of all-*cis* addition in the hydrogenation of A.

The perhydroanthracenes are shown in Fig. 25. The assignment of relative stability of I, II, and III is easily understood on the basis of Johnson's rules. IV is relatively unstable because of the existence of the central ring in the skew-boat form.* Isomer V is also unstable because of the *syn*-axial methylene interaction. In the diaxial form of 1,3-dimethylcyclohexane, this interaction costs about 5.4 kcal./mole (see Tables 1 and 2 in Chap. 2) and thus would be similar in magnitude to the instability of the twist boat (IV), but it has been suggested[165] that, because of the greater rigidity of V as compared with an axial 1,3-dimethylcyclohexane, V will, in fact, be less stable than IV.

All the perhydroanthracenes are known,[166,167] and the relative stability of I, II, and IV has been unequivocally established by equilibration over aluminum bromide;[167] the equilibrium mixture contains 96% I, 4% II, and no detectable IV. Somewhat less conclusive evidence for the relative stability of the remaining isomers comes from the observation that

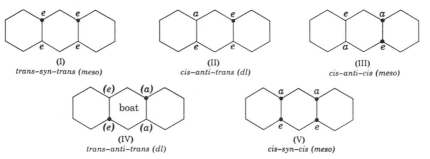

(I)
trans–syn–trans (meso)

(II)
cis–anti–trans (dl)

(III)
cis–anti–cis (meso)

(IV)
trans–anti–trans (dl)

(V)
cis–syn–cis (meso)

Fig. 25. Relative stability of the perhydroanthracenes. From E. Eliel, *Stereochemistry of Carbon Compounds*, copyright 1962, McGraw-Hill Book Company, Inc., by permission of the publishers.

[165] W. G. Dauben and K. S. Pitzer, in M. S. Newman, ed., *Steric Effects in Organic Chemistry*, John Wiley and Sons, New York, 1956, p. 34.

[166] R. L. Clarke, *J. Am. Chem. Soc.*, **83**, 965 (1961), and earlier references cited therein.

[167] R. K. Hill, J. G. Martin, and W. H. Stouch, *J. Am. Chem. Soc.*, **83**, 4006 (1961).

* The perhydrophenanthrene V (Fig. 23) must have its central ring in a classical (rigid) boat form, but the central ring in perhydroanthracene IV (Fig. 25) is a skew boat.

aluminum chloride epimerization of V gives II (which, on further treatment, is converted to I)[168] and that III is similarly epimerized to I.[169] Unfortunately, in these instances a total material balance was not achieved; hence it is not certain that the product isolated was, in fact, the only perhydroanthracene present at equilibrium. Finally, the heats of combustion of I and IV have been determined,[170] and it has been found that I is more stable, thermochemically, than IV by 5.39 ± 0.86 kcal./mole (in the vapor phase).

Many fused ring systems occur in natural products, such as steroids and terpenoids, and a number of these will be considered in Chap. 5.

b. Derivatives. Stability and Reactivity. Quantitative information about the relative stability and reactivity of epimeric derivatives (alcohols, esters, amines, etc.) in fused ring systems is quite limited. Qualitative information, on the other hand, abounds, especially in the steroid and terpene series (Chap. 5). Qualitatively, fused ring systems resemble simple ring systems (Chap. 2) of analogous conformation. In this section we shall examine how far such analogies can be carried on a quantitative basis in the few cases that have been studied.

The saponification rates of the 2-hydrindanyl succinates and the ethanolysis rates of the corresponding p-toluenesulfonates[149b] differ by no more than 33% from the rates of the corresponding cyclopentyl derivatives. (A better comparison might have been with 3-alkylcyclopentyl derivatives, but the necessary rate data are not available.) Thus, taking the saponification rate of cyclopentyl acid succinate as 1.0, that for one of the *meso-cis*-2-hydrindanyl acid succinates is 0.67 and that for the *dl-trans*-2-hydrindanyl succinate is 0.90. Similarly, taking the ethanolysis rate of cyclopentyl tosylate as 1.0, that for *dl-trans*-2-hydrindanyl tosylate is 0.8 and those for the two *meso-cis*-2-hydrindanyl tosylates are 1.25 and 0.9.[149b] Whatever difference exists between the conformation of the five-membered ring in the cyclopentyl derivative on the one hand and the *cis*- and *trans*-hydrindanyl derivatives on the other does not seem to affect the rate greatly, the conclusion being either that there is not much difference in conformation in these particular instances or that the rates are not very sensitive to whatever conformational differences are likely to be found in different cyclopentyl derivatives.

More extensive data are available for the decalin series. *trans*-Decalin-2α-ol* (equatorial hydroxyl) and *trans*-decalin-2β-ol* (axial hydroxyl)

---

[168] J. W. Cook, N. A. McGinnis, and S. Mitchell, *J. Chem. Soc.*, **1944**, 286.

[169] N. S. Crossley and H. B. Henbest, *J. Chem. Soc.*, **1960**, 4413.

[170] J. L. Margrave, M. A. Frisch, R. G. Bautista, R. L. Clarke, and W. S. Johnson, *J. Am. Chem. Soc.*, **85**, 546 (1963).

* For nomenclature, see footnote on p. 89.

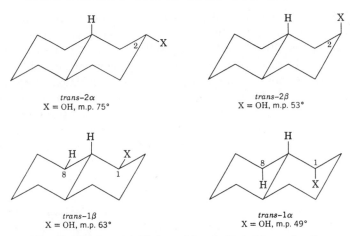

Fig. 26. The four (rigid) 1- and 2-substituted *trans*-decalins.

(Fig. 26, X = OH) have been equilibrated both by means of aluminum isopropoxide and by means of Raney nickel.[171] Aluminum isopropoxide equilibration gives 79% equatorial and 21% axial isomer, in excellent agreement with the findings in the 4-*t*-butylcyclohexanol series (Sec. 2-4c). Equilibrium with Raney nickel at 180° gives 68% equatorial, 32% axial isomer, again in good agreement with what has been reported in the 4-*t*-butylcyclohexanols (Sec. 2-4c).

In discussing the reactivity of the decalins, it is well to start with some *a priori* considerations of their conformation. In the *trans*-decalin series the conformation is, of course, rigidly defined, except in the rare cases where boat forms may play a part. The 2-substituted *trans*-decalins (Fig. 26, top) resemble the corresponding equatorially or axially substituted 4-*t*-butyl substituted cyclohexyl compounds except for the possible disturbing effect of an additional 3-substituent (cf. Fig. 32 in Sec. 2-5b). The situation in the 1-substituted *trans*-decalins (Fig. 26, bottom) is somewhat different, however. The 1β (equatorial) isomer has a "*peri*-interaction" with the equatorial hydrogen at C_8 which resembles a single axial interaction. It may also be affected by the inductive effect of an adjacent alkyl group and, in fact, resembles a *trans*-2-methyl substituted cyclohexyl compound. The 1α (axial) isomer has three *syn*-axial hydrogens instead of the usual two (the third one being the axial hydrogen at C_8) and resembles a *cis*-2-methyl substituted cyclohexyl derivative.

The relative rates of saponification[172] of *trans*-2α-decalyl acid succinate

[171] W. Hückel and D. Rücker, *Ann.*, **666**, 30 (1963).

[172] W. Hückel, *Ber.*, **67A**, 129 (1934); W. Hückel, H. Haveloss, K. K. Kumetat, D. Ullmann, and W. Doll, *Ann.*, **533**, 128 (1937).

and its *trans*-2β epimer (Fig. 26, X $=$ $O_2CCH_2CH_2CO_2H$) are 1.00 and 0.13, a reasonable ratio for equatorial and axial epimers. Unfortunately, comparison data for the 4-*t*-butylcyclohexyl acid succinates are not available, but the reactivity ratio for the corresponding acid phthalates[173] is 1.00:0.11. Moreover, the relative rate of saponification of cyclohexyl acid succinate[172] is 0.81, from which we may calculate a rather reasonable $-\Delta G°$ of 0.8 kcal./mole for the acid succinate group, assuming that the *trans*-2-decalyl derivatives are proper models to determine k_e and k_a (cf. Sec. 2-2c). The relative saponification rate of the equatorial *trans*-1β-decalyl acid succinate (Fig. 26) is 0.18, much less than the corresponding rate for the equatorial 2 derivative but still faster than the axial 2 derivative. Clearly the *peri* effect of the adjacent C_8 carbon is substantial, a bit larger, perhaps, than we might have expected on the basis of the single equatorial hydrogen at C_8 which occupies a *"syn*-axial"-like position. Finally, the rate for *trans*-1α-decalyl acid succinate is very low, only 0.0095. Here, again, there is an unusually large decelerative effect of the third *syn*-axial hydrogen, and the over-all conclusion from the data for the *trans*-1-decalyl series is that the hindrance at the 1-position is larger than anticipated.

The acetolysis rates of the four *p*-toluenesulfonates shown in Fig. 26 (X $=$ $OSO_2C_6H_4CH_3$-*p*) have been determined also.[174] Taking the rate for *trans*-4-*t*-butylcyclohexyl tosylate[173] as 1.00, that for the conformationally analogous 2α derivative is only 0.53; the rate for the axial 2β derivative (1.65) is also less than that of the *cis*-4-*t*-butyl tosylate (2.72). Despite the difference in the actual solvolysis rates, the rate ratio (axial:equatorial) in the *trans*-2-decalyl series (3.1) is very close to that in the 4-*t*-butylcyclohexyl series (2.7) or that in the (very similar) 3-*t*-butylcyclohexyl series[173] (3.3). As might be expected, the axial *trans*-1α derivative (relative rate 14.6) solvolyzes substantially faster than the axial *trans*-2β compound (relative rate 1.65); however, the increase, by a factor of 8.9, seems greater than we would expect from just one extra *syn*-axial hydrogen. A possible complicating factor in this case is hydrogen participation of the tertiary axial hydrogen at C_9, similar to that invoked to explain the high solvolysis rate of neomenthyl tosylate (Sec. 2-5c). Finally, the equatorial *trans*-1β compound (relative rate 0.28) solvolyzes only half as fast as the conformationally analogous *trans*-2α isomer (relative rate 0.53). This is very surprising on the basis of the ground state compression to be expected from the *peri* hydrogen (which should accelerate the reaction). A possible explanation[174] is bond eclipsing of the C_1

[173] S. Winstein and N. J. Holness, *J. Am. Chem. Soc.*, **77**, 5562 (1955).

[174] (a) I. Moritani, S. Nishida, and M. Murakami, *J. Am. Chem. Soc.*, **81**, 3420 (1959); see also (b) H. L. Goering, H. H. Espy, and W. D. Closson, *ibid.*, **81**, 329 (1959), and (c) I. Moritani, S. Nishida, and M. Murakami, *Bull. Chem. Soc. Japan*, **34**, 1334 (1961).

hydrogen in the carbonium ion formed in the rate-determining step with the adjacent (C_9-C_8) carbon-carbon bond which would lead to an increase in free energy of the transition state which more than compensates for the steric compression in the ground state.

Rather similar results are obtained in the ethanolysis of decalyl tosylates.[175] If we again take the rate of *trans*-4-*t*-butylcyclohexyl tosylate[173] as 1.00, that for the conformationally analogous *trans*-2α-decalyl tosylate is only 0.52; the relative rate of the *trans*-2β derivative (2.4) is also less than that of the *cis*-4-*t*-butyl compound (3.9). Here, again, the axial:equatorial rate ratio (4.7) is similar to that in the 4-*t*-butyl compounds (3.9). Corresponding rates in the *trans*-1 series are 0.26 for the (equatorial) 1β isomer and 17.4 for the (axial) 1α. Apart from a somewhat greater relative rate for the axial compounds, the ethanolysis picture is very similar to the acetolysis one. Since cyclohexyl tosylate was included in the ethanolysis study of both the 4-*t*-butylcyclohexyl compounds[173] and the *trans*-2-decalyl derivatives[175] and identical solvolysis rates were obtained, there is no doubt that the differences between the two series are real. An explanation for the difference is not at hand; it is interesting that, whereas the acetolysis rates of the 3-*t*-butylcyclohexyl tosylates are very similar to those of the conformationally analogous 4-*t*-butyl compounds,[173] it appears[175] that *cis*-3-methylcyclohexyl tosylate ethanolyzes at a substantially lesser rate (relative-rate 0.57) than the presumably conformationally analogous *trans*-4-*t*-butylcyclohexyl tosylate, and, in fact, the ethanolysis rate of the *cis*-3-methylcyclohexyl derivative is very close to that of the conformationally analogous *trans*-2-decalyl derivative (which also bears a 3-substituent). Finally, the relative ethanolysis rates of *trans*-2-methylcyclohexyl tosylate (0.30) and *cis*-2-methylcyclohexyl tosylate (27) show the same trends as those of the corresponding *trans*-1-decalyl compounds; the same explanations (see above) for the rate retardation in the equatorial and rate acceleration in the axial 2-isomer may well apply.

The situation in the *cis*-decalyl derivatives is more complex. As shown in Fig. 27, all four isomers may, in principle, exist in two conformations each. In all cases, however, the conformation with the equatorial substituent should predominate. In particular, it might be noted that the alternative conformations for the 1α and 2α compounds have 1,3-diaxial methylene-X interactions.

The ethanolysis rates of the *cis*-decalyl tosylates (Fig. 27, X = OTs) have been determined,[175] and their values at 50°, relative to *trans*-4-*t*-butylcyclohexyl tosylate as unity, are: 2α, 2.7; 2β, 1.96; 1β, 0.77; 1α, 34.

[175] W. Hückel and D. Rücker, *Ann.*, **666**, 30 (1963); W. Hückel and H. Feltkamp, *ibid.*, **649**, 21 (1961); W. Hückel et al., *ibid.*, **624**, 142 (1959); **645**, 115 (1961); W. Hückel, *Bull. Soc. Chim. France*, **1960**, 1369.

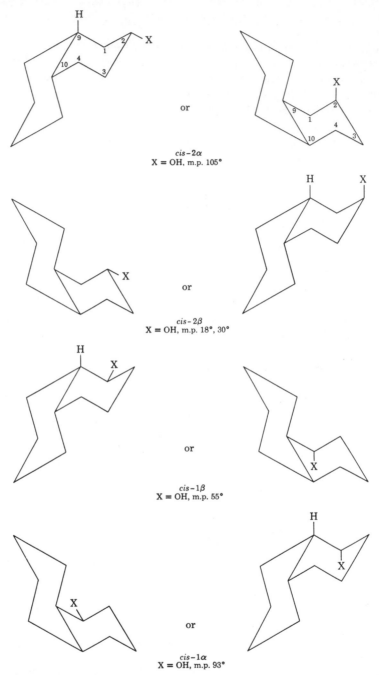

cis–2α
X = OH, m.p. 105°

or

cis–2β
X = OH, m.p. 18°, 30°

or

cis–1β
X = OH, m.p. 55°

or

cis–1α
X = OH, m.p. 93°

Fig. 27. The four (mobile) 1- and 2-substituted *cis*-decalins. (For the sake of convenience in representation, the four structures on the right are conformational isomers *not* of the structures on the left but of the corresponding mirror images.)

The relatively high solvolysis rate of the 2β isomer (equatorial) may be explained on the basis of a steric decompression which occurs in going from the tosylate to the corresponding carbonium ion, since the axial hydrogen at C_2 moves away from the axial methylene at C_9. This situation is similar to a 3-alkylketone effect (Sec. 2-6c). The same effect may explain why the cis-1β isomer (relative rate 0.77) solvolyzes faster than the trans-1β (rate 0.26). A somewhat different decompression occurs in the solvolysis of the 2α-tosylate which is almost entirely equatorial; formation of the carbonium ion relieves interaction of the axial hydrogens at C_1 and C_3 with the axial methylene at C_{10}. The situation resembles the 4-alkylketone effect defined in Sec. 2-6c and is best recognized through manipulation of Dreiding models. The very high solvolysis rate of the cis-1α isomer may signify that the transition state in this isomer corresponds to the axial conformation (whose unfavorable steric compression is, of course, relieved when the tosylate group has departed) and that its energy level is substantially lowered by anchimeric assistance of the adjacent axial hydrogen at C_9 (see, however, below).

The acetolysis rates[174c] of the cis-decalyl tosylates are similar, in relative order, to the ethanolysis rates: 2α, 2.0; 2β, 2.4; 1β, 1.1; 1α, 20 (taking the rate for the trans-4-t-butylcyclohexyl compound[173] as unity). The Japanese authors[174c] explain the high solvolysis rate of the 1α isomer not in terms of anchimeric assistance (see previous paragraph) but in terms of a decompression which occurs when the axial ring juncture in the tosylate becomes pseudoaxial in the olefin product ($\Delta^{1,9}$-octalin) and the corresponding transition state.

The saponification rates of the three cis-decalyl succinates which have been studied,[172] relative to that of trans-2α-decalyl acid succinate (equatorial) taken as 1.00, are: cis-2β, 0.58; cis-2α, 0.95; cis-1α, 0.37. These rates are qualitatively reasonable, considering that the cis-decalyl isomers, because of their mobility, will have the functional group largely (though not exclusively) in the equatorial position so that their rate of saponification should be somewhat less than that of a purely equatorial ester; in addition, the 1 isomer encounters some steric hindrance from the adjacent methylene substituent. Of the two cis-2 isomers, the 2α which is more completely equatorial (since the axial conformation would have a syn-axial methylene interaction) reacts faster.

The course of the reaction of the conformationally homogeneous decalylamines (Fig. 26, $X = NH_2$) with nitrous acid has already been discussed in Sec. 2-5c: the equatorial amines (trans-1β, trans-2α) give equatorial alcohols, and the axial amines (trans-1α, trans-2β) give much elimination product along with an alcohol mixture which is rich in the (inverted) equatorial isomer.[176] In the case of the conformationally mobile isomers (Fig. 27, $X = NH_2$), the cis-1α- and cis-2α-amines which are highly conformationally biased toward the equatorial conformation (because the axial conformation would have a syn-axial NH_2-methylene interaction) give equatorial isomers, but the cis-1β- and cis-2β-amines give mixtures, because these amines are conformationally heterogeneous.

Nevertheless, the major product in these two cases is the alcohol of retained configuration, as would be expected since the starting amines exist predominantly in the equatorial conformation and equatorial amines give rise to equatorial alcohols. The case of the 4- and 5-amino-cis-hydrindanes[177] is similar to that of the amino-cis-decalins.

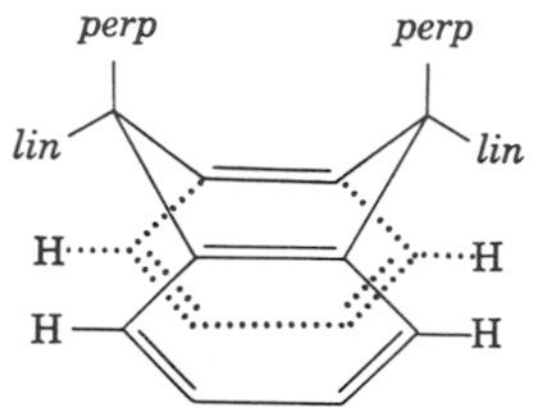

Fig. 28. 9,10-Disubstituted 9,10-dihydroanthracenes.

We shall conclude this section with a consideration of 9,10-disubstituted 9,10-dihydroanthracenes. As shown in Fig. 28, these compounds exist as (readily interconvertible) boat forms. Two types of substituents may be discerned which have been called lin (linear) and perp (perpendicular).[178] Unlike the cyclohexane case (where an axial substituent is more hindered than an equatorial), in the 9,10-dihydroanthracene series the lin substituent is more hindered than the perp, for the lin substituent is quite badly crowded by the peri hydrogens whereas the perp substituents diverge somewhat and are therefore farther apart than syn-axial substituents in cyclohexane. Three consequences follow and have been observed experimentally: (1) A cis-9,10-disubstituted 9,10-dihydroanthracene will be in the perp-perp rather than in the alternative lin-lin conformation, the two conformations being readily interconvertible by ring inversion. The dipole moment of dimethyl cis-9,10-dihydroanthracene-9,10-dicarboxylate in fact confirms that the molecule exists in the perp-perp conformation.[178] (2) A perp substituent is less hindered than a lin substituent. This may be seen in the epimeric methyl esters— morpholides of the 9,10-dihydroanthracene-9,10-dicarboxylic acids. The cis (perp-perp) isomer is readily reduced by lithium aluminum hydride at

[176] W. Hückel, Ann., **533**, 1 (1938); W. G. Dauben, R. C. Tweit, and C. Mannerskantz, J. Am. Chem. Soc., **76**, 4420 (1954). Recent work by W. Hückel and K. Heyder, Chem. Ber., **96**, 220 (1963), throws some doubt on the quantitative accuracy of the earlier results (which were obtained at a time when accurate analytical techniques were not available) but does not affect the qualitative conclusions.

[177] W. G. Dauben and J. Jiu, J. Am. Chem. Soc., **76**, 4426 (1954).

[178] A. H. Beckett·and B. A. Mulley, Chem. & Ind. (London), **1955**, 146; J. Chem. Soc., **1955**, 4159.

both carboxyl functions to *cis*-9-hydroxymethyl-10-morpholinomethyl-9,10-dihydroanthracene, whereas in the *trans* (*lin-perp*) isomer only the ester group and not the morpholide group is reduced so that the product is *trans*-9-hydroxymethyl-10-morpholinocarbonyl-9,10-dihydroanthracene.[178] (3) Finally, one might also expect[178] that the *cis* (*perp-perp*) isomer would be more stable than the *trans* (*perp-lin*) in 9,10-disubstituted-9,10-dihydroanthracenes. This has been confirmed in the case of the 9,10-dimethyl-9,10-dihydroanthracenes in which the *cis* is more stable than the *trans* by 1.6 kcal./mole.[179] This order may, however, be reversed when the substituents have strong dipoles, because then the *cis* (*perp-perp*) isomer is affected by dipolar repulsion more than is the *trans* (*perp-lin*). This situation is observed in the 9,10-dicarboxy-9,10-dihydroanthracenes in which the *trans* isomer is more stable than the *cis*.[178]

4–6. Heterocyclic Rings

The replacement of a methylene group in an alicyclic ring by a hetero-atom usually does not have a profound effect on the conformational properties of the system. Certainly there are quantitative differences between the homocyclic and heterocyclic rings, but in general the similarities, rather than the differences, are the striking features.

a. **Small Rings.** Little is known about the conformations of heterocyclic rings smaller than six-membered. A careful microwave study has been made on trimethylene oxide[180] which has led to the conclusion that the geometry of the ground state of the molecule is best described as planar, but having a puckering vibration of large (14°) amplitude. An attempt was also made to specify the ground state geometry by a comparison of the simple vector addition of bond moments and the experimental dipole moment of the compound.[181] Such calculations do not, however, adequately account for either hybridization changes or lone pair moments, and lead to invalid conclusions in this case. Cyclobutane is non-planar (Sec. 4-2), but the energy difference between the planar and non-planar forms is small. There is less torsional strain in the planar form of trimethylene oxide than in cyclobutane, and the zero point energy of the puckering vibration of the oxide is sufficient to carry it through the planar

[179] M. L. Caspar, J. N. Seiber, and K. Matsumoto, Abstracts, Meeting of American Chemical Society, Denver, Colorado, 1964, p. 30C.

[180] S. I. Chan, J. Zinn, and W. D. Gwinn, *J. Chem. Phys.*, **34**, 1319 (1961).

[181] B. A. Arbousow, *Bull. Soc. Chim. France*, **1960**, 1311.

form.* Theoretical considerations supporting a preferred half-chair form for tetrahydrofuran, tetrahydrothiophene, and pyrrolidine have already been presented in Sec. 4-3.[22] Evaluation of the thermodynamic properties of tetrahydrothiophene[182] and pyrrolidine[183] supports a pseudo-rotating puckered ring for these molecules.

b. Six-Membered Rings Containing One Hetero-atom.

A consideration of the magnitude of the rotational barriers about carbon-carbon, carbon-oxygen, carbon-nitrogen, and carbon-sulfur bonds allows some predictions to be made regarding the preferred geometry of six-membered heterocyclic ring systems. The rotational barriers for ethane, methylamine, methanol, and methyl mercaptan are, respectively, 2.9, 1.9, 1.1, and 1.3 kcal./mole (Sec. 3-3). Although we do not understand in detail the nature of these barriers, it is apparent that for this group of compounds the height of the barrier to rotation of the methyl group is increased by about 1 kcal./mole for each substituent (hydrogen) attached to the hetero-atom. As long as the substituents attached to an ethane are non-polar and not too large, the barrier about the central bond does not vary greatly with the nature of the substituents. It is reasonable to expect that this will also be true for the various carbon to hetero-atom bonds being discussed here. Just as a knowledge of the rotational barrier in ethane enabled a rather good prediction of the energy for the chair → boat conversion in cyclohexane to be made, it can be anticipated that similarly good predictions can be made for at least the relatively simple six-membered heterocyclic ring systems. Thus in piperidine and in tetrahydropyran, respectively, the torsional energies of the illustrated forms (Fig. 29) can be calculated from the data cited above to be 4.7 and 3.9 kcal./mole above those of the corresponding chair forms. The interactions between the hydrogens at the

[182] W. N. Hubbard, H. L. Finke, D. W. Scott, J. P. McCullough, C. Katz, M. E. Gross, J. F. Messerly, R. E. Pennington, and G. Waddington, *J. Am. Chem. Soc.*, **74**, 6025 (1952).

[183] Unpublished observations cited by J. P. McCullough, *J. Chem. Phys.*, **29**, 966 (1958).

* The potential energy diagram of trimethylene oxide (solid curve) as compared to cyclobutane (dotted) will make this point clear.

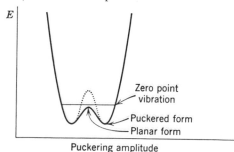

Puckering amplitude

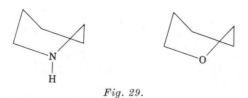

Fig. 29.

prow and stern of the classical boats can be relieved by slight pseudo-rotations, and it seems likely that the numerical values given will be close to the actual energy differences between the chair and flexible forms.[27]

X-ray studies have established that the chair form is the preferred conformation for the piperidine ring in a number of compounds,[184] and the question then arises as to whether the hydrogen on nitrogen is oriented axially or equatorially. It has been suggested[185] on the basis of Kerr constant measurements that the hydrogen is oriented axially in about 80% of the molecules in piperidine and in nearly all the molecules in morpholine, indicating that the free electron pair is more "space-consuming" than a hydrogen atom. However, in the light of the previous discussion of results based on Kerr constants (Sec. 3-8) this conclusion must be viewed with considerable reserve. More recent work based on dipole moments in fact indicates the reverse "sizes," but the difference is not large.[186]

For N-methylpiperidine it was suggested[185] that the methyl group is axial in about half the molecules and equatorial in the other half. This seems, however, to be incorrect as evidenced by the previously discussed results with N-methylpiperidine (Sec. 3-8) and work on N,N'-dimethyl-piperazine to be discussed below (Sec. 4-6c). The indications are that the N-methyl group is similar to the methyl group in methylcyclohexane and prefers the equatorial position and that, in fact, the pair on the hetero-atom may well be disregarded unless there is a polar substituent on a neighboring carbon, as in the halodioxanes (Sec. 4-6c) or the pyranose sugars (Sec. 6-4).

At one time it was suggested[187] that the conformation of an N-alkyl substituent on nitrogen could be inferred from the conformation of the quaternary salt formed on treatment with a different alkyl halide. Clearly, however, this cannot be done. Because the N-alkyl group moves readily

[184] (a) M. Przybylska and W. H. Barnes, *Acta Cryst.*, **6**, 377 (1953); (b) J. W. Visser, J. Manassen, and J. L. DeVries, *Acta Cryst.*, **7**, 288 (1954).

[185] M. Aroney and R. J. W. Le Fèvre, *Proc. Chem. Soc.*, **1958**, 82; *J. Chem. Soc.*, **1958**, 3002.

[186] N. L. Allinger, J. Tai, J. G. D. Carpenter, F. Karkowski, and S. P. Jindal, Presented at the National Meeting of the American Chemical Society, April, 1963.

[187] G. Fodor, J. Toth, and I. Vincze, *J. Chem. Soc.*, **1955**, 3504.

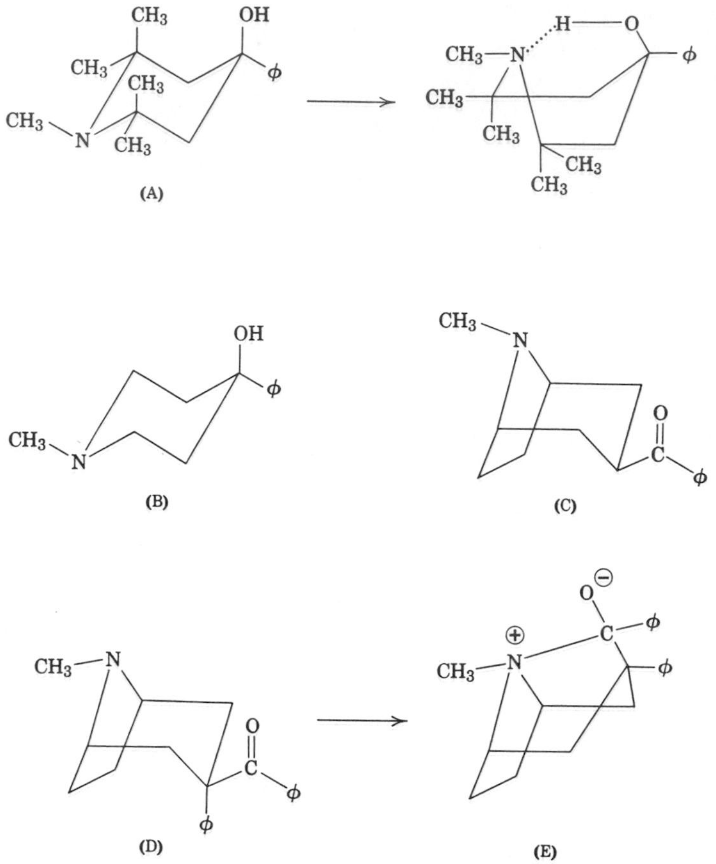

Fig. 30.

from the equatorial to the axial position and vice versa by nitrogen inversion, it may be predicted from the Curtin-Hammett principle (Sec. 1-5) that the product of quaternization does not depend at all on the preferred conformation of the starting material. For example, the fact that 1-ethyl-2-methylpiperidine on treatment with methyl iodide gives 1-ethyl-1,2-dimethylpiperidinium iodide with equatorial ethyl and axial methyl on nitrogen[188] does *not necessarily* mean that the ethyl group in 1-ethyl-2-methylpiperidine is equatorial; it merely indicates that the incoming methyl group approaches more readily from the axial side than from the equatorial, or that the ethyl group in the transition state for quaternization prefers the equatorial to the axial position.

[188] J. McKenna, J. White, and (in part) A. Tulley, *Tetrahedron Letters*, **1962**, 1097; J. E. Becconsall and R. A. Y. Jones, *ibid.*, **1962**, 1103.

As with the cyclohexane system, if sufficient constraint is present a piperidine ring may also preferentially assume a boat form. A well-documented example of such a situation is that reported by Lyle.[189] The infrared spectrum of A (Fig. 30) shows a strongly intramolecularly hydrogen bonded hydroxyl, while that of B shows a free (unbonded) hydroxyl. The conclusion is that in B the chair form is perfectly stable, but in A the *syn*-triaxial interaction of the two methyls and the hydroxyl is sufficient to push the compound over into a boat form. Another interesting example of a boat form is furnished by C and D (Fig. 30). The former shows (from infrared and ultraviolet spectra) a normal benzoyl group, while the latter shows no benzoyl absorption in the ultraviolet, and no C=O stretching frequency in the infrared in methanol solvent. The interaction between the phenyl and the ethylene bridge in D appears to be sufficient to put the piperidine ring into a boat form, whereupon the nitrogen more or less adds to the carbonyl to give E as indicated.[190]

Another interesting example is afforded by the comparison of the dipole moments of *N*-methyl-4-piperidone (A, Fig. 31) with those of the corresponding bridged compounds B and C. The former has a dipole moment of 2.91 D,[191] which is consistent with the methyl group being mainly in the equatorial position. In B, however, it is clear that the dimethylene bridge will pull together the two axial bonds of the carbons which it joins. As a result of this, the five-membered ring will flatten out, in turn pushing up the C=O dipole so that it more nearly opposes the lone pair dipole (the reverse of the reflex effect[192]). Thus it can be predicted that the resultant dipole moment will decrease, if it is assumed that the methyl group is in the equatorial position. This is what is found experimentally, the observed moment of B being 2.66 D.[191] If the positions

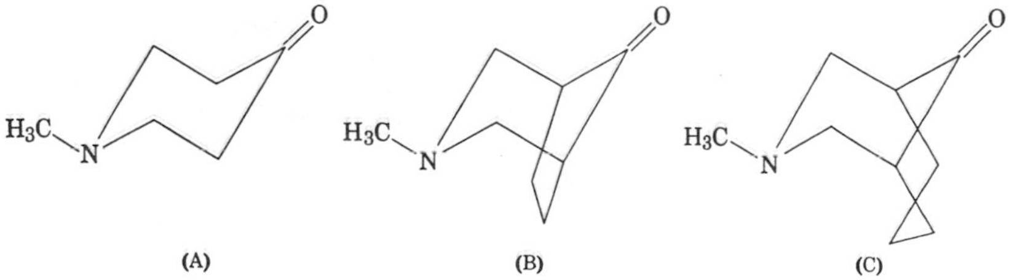

(A)   (B)   (C)

Fig. 31.

[189] R. E. Lyle, *J. Org. Chem.*, **22**, 1280 (1957).

[190] M. R. Bell and S. Archer, *J. Am. Chem. Soc.*, **82**, 151 (1960).

[191] Unpublished measurements by M. A. DaRooge and C. L. Neumann, quoted by H. O. House, P. P. Wickham, and H. C. Muller, *J. Am. Chem. Soc.*, **84**, 3139 (1962).

[192] C. Sandris and G. Ourisson, *Bull. Soc. Chim. France*, **1958**, 1524. See also Sec. 2-7, p. 126, and B. Waegell and G. Ourisson, *Bull. Soc. Chim. France*, **1963**, 495, 496.

adjacent to the carbonyl are joined by a trimethylene bridge, rather than by a dimethylene bridge, no such twisting would be predicted, and the dipole moment should and does remain similar to that of the unbridged compound (2.89 D).[191,*]

The steric course of reduction of variously substituted 4-piperidones[193] is similar to that of analogous cyclohexanones.

c. Six-Membered Rings Containing More Than One Hetero-atom.

We now discuss systems with two or more hetero-atoms. p-Dioxane can be predicted (Sec. 4-6b) to have its flexible form 2.2 kcal./mole above the chair form (twice the methanol barrier). Experimentally, the existence of the substance in the chair form is supported by electron diffraction,[194-196] by infrared and Raman spectra,[197-198] by variation of total polarization with temperature[199] as well as by Kerr constant measurements.[200] NMR data are in accord with the chair conformation and suggest a rather low barrier to chair inversion in a cis-2,3-disubstituted 1,4-dioxane,[201,202] since no splitting of the axial and equatorial proton signals occurs at temperatures as low as $-104°$.[201] It is interesting that dipole, infrared, and Raman measurements of $trans$-2,3- and $trans$-2,5-dihalo-1,4-dioxanes,[203] besides confirming the chair structure of these molecules, also indicate that the greatly preferred conformation is that with the halogens diaxial, the preference being greater in the dihalodioxanes than in corresponding dihalocyclohexanes. This appears to be due to an interaction of the carbon-halogen dipoles with the atomic dipole due to the p electrons on oxygen, an effect similar to the anomeric effect in the sugars (Sec. 6-4).

In analogy with the corresponding situation in cyclohexane, 1,4-dioxane-cis-2,5-dicarboxylic acid (e,a) is less stable than the corresponding $trans$

[193] M. Balasubramanian and N. Padma, *Tetrahedron*, **19**, 2135 (1963).

[194] L. E. Sutton and L. O. Brockway, *J. Am. Chem. Soc.*, **57**, 473 (1935).

[195] O. Hassel and H. Viervoll, *Acta Chem. Scand.*, **1**, 149 (1947).

[196] M. Davis and O. Hassel, *Acta Chem. Scand.*, **17**, 1181 (1963).

[197] D. A. Ramsay, *Proc. Roy. Soc. (London)*, **A190**, 562 (1947).

[198] F. E. Malherbe and H. J. Bernstein, *J. Am. Chem. Soc.*, **74**, 4408 (1952).

[199] J. H. Gibbs, *Discussions Faraday Soc.*, **10**, 122 (1951).

[200] R. J. W. Le Fèvre, A. Sundaram, and R. K. Pierens, *J. Chem. Soc.*, **1963**, 479.

[201] E. Caspi, T. A. Wittstruck, and D. M. Piatak, *J. Org. Chem.*, **27**, 3183 (1962).

[202] C. Altona and C. Romers, *Acta Cryst.*, **16**, 1225 (1963).

[203] C. Altona, C. Romers, and E. Havinga, *Tetrahedron Letters*, **No. 10**, 16 (1959); C. Altona, C. Knobler, and C. Romers, *Acta Cryst.*, **16**, 1217 (1963); *Rec. Trav. Chim.*, **82**, 1089 (1963); C. Altona and C. Romers, *ibid.*, **82**, 1080 (1963).

* This interpretation may be oversimplified, since in the trimethylene compound in the double chair form one hydrogen comes quite close to nitrogen. An appreciable bending of the offending C—H bond, coupled with other minor deformations, can reduce the van der Waals energy to a rather small value, however (Sec. 7-2).

isomer (e,e), whereas in the 2,6-diacid the *cis* isomer (e,e) is more stable then the *trans* (e,a).[204]

m-Dioxane (the formal of 1,3-propanediol) has been studied through dipole measurements[181,205] which are consistent with this molecule existing predominantly in a chair form. The latter is calculated to have an enthalpy only 2.2 kcal. below that of the flexible form. This enthalpy difference is sufficiently small that substituted derivatives might well have stable flexible forms. NMR measurements[206] are in agreement with the chair conformation and suggest that *cis*-2,5-disubstituted 1,3-dioxanes exist as readily interconvertible e,a-a,e isomers; the barrier to interconversion has not, however, been determined. The chair form also appears, from NMR measurements,[207] to be the preferred conformation of 1,2-dioxanes; the barrier to chair inversion in 3,3,6,6-tetramethyl-1,2-dioxane is unusually high ($\Delta G^{\ddagger} = 14.5$ kcal./mole, $\Delta H^{\ddagger} = 18.5$ kcal./mole).[207]

A similarly high barrier is found in 1,2-dithianes,[207,208] $\Delta G^{\ddagger}$ for variously substituted 1,2-dithianes being 11.6–13.9 kcal./mole and $\Delta H^{\ddagger}$ 12.0–16.1 kcal./mole. This is perhaps not so surprising in view of the high barrier found from thermodynamic measurements in simple acyclic disulfides, such as dimethyl disulfide[209] (9.5 kcal./mole) and diethyl disulfide[210] (13.2 kcal./mole). It is also of interest that the preferred torsional angle in dimethyl disulfide is 90° rather than 60°[209] and that, as a result, 1,2-dithiane is quite appreciably strained.[211] Nevertheless, calculation[211] and dipole moment data[212] support the chair structure for 1,2-dithiane.

Dipole moment data also suggest the chair conformation for 1,3-dithiane.[212] In the case of 1,4-dithiane, existence of the molecule in the chair form is exhaustively documented by electron diffraction,[195] dipole,[212,213] infrared-Raman,[214] and (in the solid state) X-ray[215,216] data.

[204] R. K. Summerbell and J. R. Stephens, *J. Am. Chem. Soc.*, **76**, 731, 6401 (1954).

[205] R. Walker and D. W. Davidson, *Can. J. Chem.*, **37**, 492 (1959).

[206] N. Baggett, B. Dobinson, A. B. Foster, J. Homer, and L. F. Thomas, *Chem. & Ind. (London)*, **1961**, 106.

[207] G. Claeson, G. Androes, and M. Calvin, *J. Am. Chem. Soc.*, **83**, 4357 (1961).

[208] A. Lüttringhaus, S. Kabuss, W. Maier, and H. Friebolin, *Z. Naturforsch.*, **16b**, 761 (1961).

[209] D. W. Scott, H. L. Finke, M. E. Gross, G. B. Guthrie, and H. M. Huffman, *J. Am. Chem. Soc.*, **72**, 2424 (1950).

[210] D. W. Scott, H. L. Finke, J. P. McCullough, M. E. Gross, R. E. Pennington, and G. Waddington, *J. Am. Chem. Soc.*, **74**, 2478 (1952).

[211] G. Bergson and L. Schotte, *Arkiv Kemi*, **13**, 43 (1958).

[212] H. T. Kalff and E. Havinga, *Rec. Trav. Chim.*, **81**, 282 (1962).

[213] K. E. Calderbank and R. J. W. Le Fèvre, *J. Chem. Soc.*, **1949**, 199.

[214] K. Hayasaki, *J. Chem. Soc. Japan*, **81**, 1645 (1960). The non-coincidence of infrared and Raman lines indicates that the molecule has a center of symmetry.

[215] H. J. Dothie, *Acta Cryst.*, **6**, 804 (1953).

[216] R. E. Marsh, *Acta Cryst.*, **8**, 91 (1955).

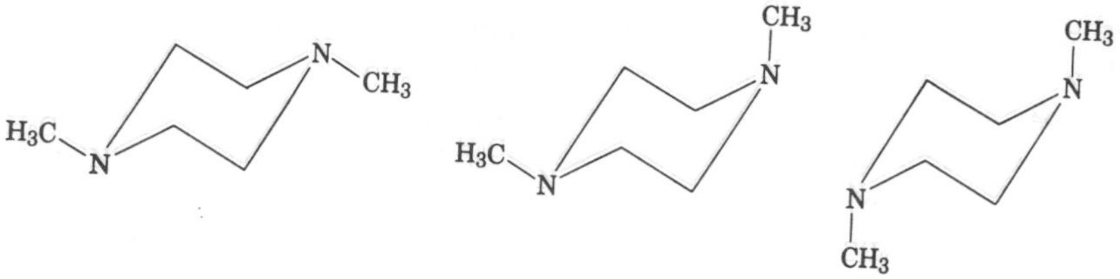

Fig. 32.

The barrier to chair interconversion does not seem to be known, but, since the NMR shows that the protons are still equivalent at −120°, it must be rather low.[217] 1,4-Diselenane also exists in the chair form, at least in the solid state, as indicated by X-ray data;[218] the carbon-selenium bond is unusually long.

The flexible form of piperazine is calculated to be 3.8 kcal./mole less stable than the chair form. Electron diffraction measurements on N,N'-dichloropiperazine[219] indicate that this molecule is chair-shaped, the chlorine atoms occupying equatorial positions. Electron diffraction measurements also support a chair conformation for N,N'-dimethylpiperazine[196] and piperazine itself; moreover, the chair conformation for piperazine is in agreement with the infrared spectrum.[220] The barrier to chair inversion has been determined for N,N'-dimethylpiperazine by low temperature NMR measurements, and it has the value 13.3 ± 0.3 kcal./mole.[221]

Knowing the conformational energy of a methyl group on nitrogen (p. 179), we can calculate the energies of the three chair conformations for N,N'-dimethylpiperazine (Fig. 32). From these energies and the temperature, we can calculate the concentration of these various conformations at equilibrium, and, since the dipole moment of each conformation can be calculated, the dipole moment of the mixture can be calculated. The agreement between this calculated dipole moment and the experimental value is excellent.[186]

In opposition to the other available evidence, Kerr constant measurements on piperazine and on N,N'-dimethylpiperazine have been interpreted as indicating that these compounds exist predominantly in a boat form but contain some chair form also.[222] The difficulties involved in the

[217] F. Lautenschlaeger and G. F. Wright, *Can. J. Chem.*, **41,** 1972 (1963).

[218] R. E. Marsh and J. D. McCullough, *J. Am. Chem. Soc.*, **73,** 1106 (1951).

[219] P. Andersen and O. Hassel, *Acta Chem. Scand.*, **3,** 1180 (1949).

[220] P. J. Hendra and D. B. Powell, *J. Chem. Soc.*, **1960,** 5105; *Spectrochim. Acta,* **18,** 299 (1962).

[221] L. W. Reeves and K. O. Strømme, *J. Chem. Phys.*, **34,** 1711 (1961).

[222] M. Aroney and R. J. W. Le Fèvre, *J. Chem. Soc.*, **1960,** 2161.

calculation of Kerr constants and examples of erroneous conclusions reached by this method were discussed in Sec. 3-8, and the same considerations apply here. The available data, other than the Kerr constants, are all interpretable in terms of chair forms for these molecules. Since the differences between the Kerr constants of the various possible forms are small, conclusions based on them are regarded by the present authors as unacceptable. Not surprisingly, perhaps, in the palladium chloride chelate of N,N'-dimethylpiperazine, the heterocyclic ring assumes the shape of a boat so as to accommodate a bidentate complex.[223]

1,3,5-Trioxane (formaldehyde trimer) has been studied through dipole measurements[181,213,224] and infrared spectral measurements[225] which agree in suggesting that the molecule exists very predominantly in the chair conformation, the calculated dipole moment for the chair, 2.24 D, agreeing well with the experimental value of 2.18 D. X-ray data confirm the chair conformation for the solid.[226]

1,3,5-Trithiane also appears to exist in the chair conformation. This has been deduced (somewhat tenuously) from electron diffraction[195] and dipole[213] data; somewhat more convincing (though, perhaps, not conclusive) evidence comes from NMR measurements of substituted trithianes.[227] Thus the trimer of thioacetaldehyde (2,4,6-trimethyltrithiane) exists in two geometrical isomers, one of which (*cis*) shows a single methyl and a single ring-proton signal in the NMR whereas the other (*trans*) shows two different methyl and ring-proton signals in a ratio of 2:1. A slight anomaly exists in that the axial ring protons appear at lower field than the equatorial (unlike the cyclohexanes; cf. Sec. 3-4d), but this is ascribed to the shielding properties of the carbon-sulfur as distinct from carbon-carbon bonds in the ring rather than to deviation from chair geometry. The two chair forms of 1,3,5-trithiane are rapidly interconvertible; the barrier has not yet been determined.

The chair form also seems to represent the conformation of 2,4,6-trimethyl-1,3,5-triazacyclohexane, at least of the hydrate in the solid state, as evidenced by X-ray measurements.[228]

d. Larger Systems. Conformational studies on molecules which contain piperidine rings incorporated into larger fused systems have been sparse, although some information of this kind is available from work on alkaloids

[223] O. Hassel and B. F. Pedersen, *Proc. Chem. Soc.*, **1959**, 394.

[224] A. A. Maryott and S. F. Acree, *J. Res. Natl. Bur. Std.*, **33**, 71 (1944).

[225] D. A. Ramsay, *Trans. Faraday Soc.*, **44**, 289 (1948).

[226] N. F. Moerman, *Rec. Trav. Chim.*, **56**, 161 (1937).

[227] E. Campaigne, N. F. Chamberlain, and B. E. Edwards, *J. Org. Chem.*, **27**, 135 (1962).

[228] E. W. Lund, *Acta Chem. Scand.*, **5**, 678 (1951).

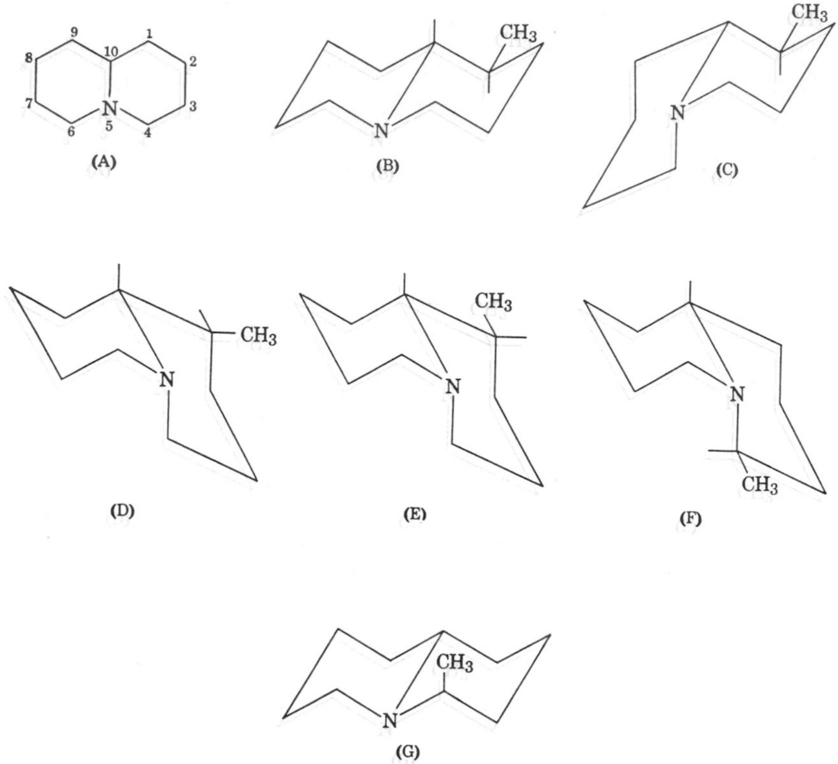

Fig. 33.

containing such systems (Chap. 5). Some additional studies have been carried out on the quinolizidine system (A, Fig. 33).[229] The basic ring system may in principle exist in either *cis* or *trans* forms, comparable to those found in the decalins, except that in this case the interconversion of the two isomers involves only an inversion of configuration at the nitrogen atom. The *cis* and *trans* forms of quinolizidine are therefore not separately isolable compounds but are in rapid equilibrium in solution at room temperature. From the preceding discussion of piperidines, we can estimate that the equilibrium with quinolizidine itself will favor the *trans* isomer by about 2.6 kcal./mole, although there is no experimental information on this point. The series of quinolizidines in which a methyl is substituted in turn at positions 1, 2, 3, and 4 has been studied in some detail,[229] in both the *cis* and *trans* series, relative to the hydrogen at C-10 (Fig. 33). If we take the C-1 methylated quinolizidines as examples,

    [229] T. M. Moynehan, K. Schofield, R. A. Y. Jones, and A. R. Katritzky, *J. Chem. Soc.*, **1962**, 2637.

the compound in which the hydrogens at C-1 and C-10 are *trans* to one another has the methyl group located comfortably in an equatorial position when the ring juncture is *trans* (B), and the two forms with *cis* ring junctures have the methyl equatorial and axial, respectively (C and D). There is one *gauche* interaction when the ring juncture is *trans*, and when it is *cis* there are either one (C) or two (D) *gauche* interactions involving the methyl besides the three *gauche* interactions between the rings. Thus it is clear that the *trans* ring juncture will be stable by approximately 2.6 kcal./mole, and only negligible amounts of the alternative structures are expected in the equilibrium mixture. If, on the other hand, the hydrogens at C-1 and C-10 are *cis* to one another, then the methyl group is axial when the ring juncture is *trans*, but it can become equatorial when the ring juncture is *cis*. If the nitrogen and its lone pair may be considered to be similar to a CH or CH_2 group with respect to the *gauche* interaction energy, then we can see that the *trans* ring juncture for this compound will lead to a system containing three *gauche* interactions, while with a *cis* ring juncture a conformation is possible in which there are only four *gauche* interactions (E). The latter is therefore expected to be a minor but substantial component of the equilibrium mixture. A similar analysis can be carried through for the other methylated compounds, and for each of them it is predicted that, if the methyl can locate itself in an equatorial position when the ring juncture is *trans*, that conformation will predominate, essentially to the exclusion of other possibilities. If the methyl finds itself in an axial position when the ring juncture is *trans*, it is predicted that in each case that conformation will still predominate, but substantial amounts of material containing a *cis* ring juncture and an equatorial methyl group will be found in equilibrium with the preferred conformation. The proton salts of the quinolizidine systems would, by this analysis, be expected to be conformationally quite similar to the free bases, and this appears to be the case.[229] There are a number of methods that might be employed in studying such equilibria. Proton magnetic resonance spectra were studied but were not definitive.[229] Chemical reactions were also studied, but they give information regarding transition states rather than the ground states and are of limited help in this problem. The most useful information would appear to be that from infrared spectra, but it is unfortunately mutually inconsistent.[230] Bohlmann, in a series of studies,[230] noted that a quinolizidine in which there are at least two hydrogens on carbons attached to the nitrogen, both of which are oriented axially and *trans* to the lone pair, will show absorption in the infrared in the 2700–2800 cm.$^{-1}$ region which is absent in quinolizidines

---

[230] F. Bohlmann, *Angew. Chem.*, **69**, 641 (1957); *Ber.* **91**, 2157 (1958). See also S. V. Kessar, *Experientia*, **18**, 56 (1962).

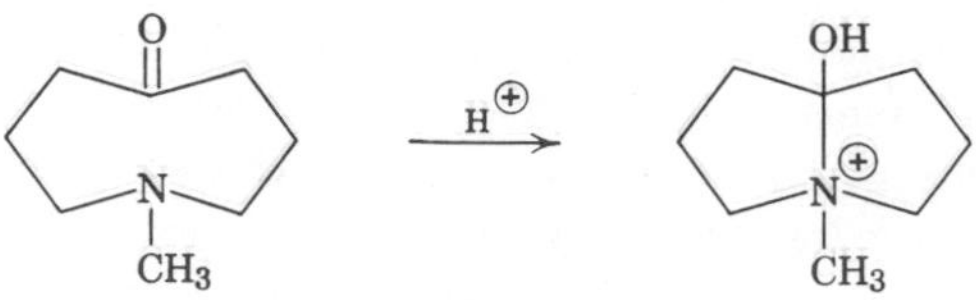

Fig. 34.

not meeting the structural requirement. It is presumed that the absorption is due to a coupling between the CH stretching frequencies. This criterion has been widely applied (Chap. 5), and there appear to be no known exceptions. 1-Methylquinolizidine (B, Fig. 33) in which the hydrogen at C-1 is *trans* to that at C-10 shows this absorption as expected, as does the corresponding *cis* isomer. In the isomer in which the two hydrogens at C-1 and C-10 are *cis*, the methyl is in the axial position if the ring juncture is *trans* but may assume an equatorial position when the ring juncture is *cis* (E). If the lone pair on nitrogen were effectively as large as a methyl group, the arrangement with the *trans* ring juncture would give a serious repulsion from the equivalent of a *syn*-diaxial dimethyl repulsion, and it can be calculated that in this case conformation (E) would predominate over the conformation with the *trans* ring juncture by some 2.6 kcal./mole. Since the Bohlmann bands are seen in the infrared, it would seem that the lone pair is effectively somewhat smaller than methyl.

The situation with the 4-methylquinolizidines is not consistent with the foregoing, unfortunately. One of these isomers shows the Bohlmann bands, the other does not. Certainly if the methyl is equatorial when the ring juncture is *trans*, that conformation will be the stable one. If the methyl is axial when the ring juncture is *trans* (G), it might be anticipated that the compound would go over to the *cis* juncture with an equatorial methyl (F). If the lone pair is the size of a hydrogen atom or smaller, it can be calculated that the *trans* ring juncture is the more favorable. The two structures become of equal energy only when the lone pair has the same effective size as the methyl group, and even in this case the Bohlmann bands should be detectable. That they are not seen is therefore unexpected, and, if the conformation (F) in fact predominates in the equilibrium as indicated, the predominance remains unexplained. It does not appear possible to draw unequivocal conclusions from these conflicting data.

Little definitive information is available on the conformations of heterocyclic rings containing more than six ring members. It has been assumed that such rings would, except in special cases, be conformationally similar to the corresponding cycloalkanes, and the little information available appears consistent with this idea.[12] Special cases, such as the salt of N-methyl-5-azacyclooctanone[231] (Fig. 34), are of course to be excluded from such a generalization.

General References

W. G. Dauben and K. S. Pitzer, "Conformational Analysis," in M. S. Newman, ed., *Steric Effects in Organic Chemistry*, John Wiley and Sons, New York, 1956.

J. D. Dunitz and V. Prelog, "Röntgenographisch bestimmte Konformationen und Reaktivität mittlerer Ringe," *Angew. Chem.*, **72,** 896 (1960).

A. R. Katritzky, ed., *Physical Methods in Heterocyclic Chemistry*, Vols. 1 and 2, Academic Press, New York, 1963.

R. A. Raphael, "Recent Studies on Many-Membered Rings," *Proc. Chem. Soc.*. **1962,** 97.

J. Sicher, "The Stereochemistry of Many-Membered Rings," *Progr. Stereochem.*, **3,** 202 (1962).

[231] N. J. Leonard, R. C. Fox, and M. Oki, *J. Am. Chem. Soc.*, **76,** 5708 (1954).

Chapter 5

Conformational Analysis of Steroids, Triterpenoids, and Alkaloids

5–1. Introduction

Since the principles of conformational analysis, as applied to cyclohexane derivatives, were first clearly enunciated by Barton[1] in 1950 they have been brought to bear on almost every problem of relative stereochemistry in the field of natural products. In this chapter the great value of conformational analysis in natural product chemistry will be illustrated by reference to some of the more important applications of this method to the chemistry of steroids, triterpenoids, and alkaloids.

5–2. Conformation of Steroids and Triterpenoids

a. The Nucleus. Steroids occupy an important place in the development of conformational analysis. They constitute a class of complex natural products whose structure and stereochemistry were elucidated by classical methods. A great quantity of facts concerning the reactions of steroids was accumulated in the five or six decades prior to 1950, and consideration of these in the light of the concept of axial and equatorial bonds revealed many of the generalizations of which the principles of conformational analysis are comprised.

5α-Cholestane (1; rings A and B *trans* fused) and 5β-cholestane (1; rings A and B *cis* fused) exemplify the two types of nuclei which are found in "natural" steroids.* It has been pointed out in earlier chapters that, except in certain special circumstances (see Sec. 7-4), a cyclohexane derivative is most stable in the chair conformation. Assuming that the

---

[1] D. H. R. Barton, *Experientia*, **6**, 316 (1950).

* Throughout this chapter methyl groups are indicated in formulas only by the bonds linking them to the remainder of the molecule, as exemplified by the methyl groups attached at C_{10} and C_{13} in (1).

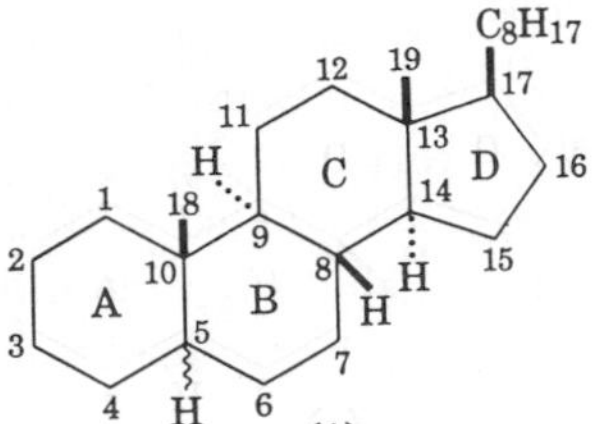

(1)

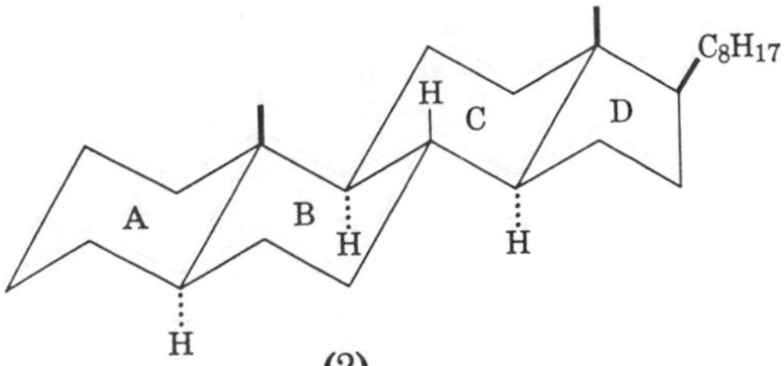

(2)

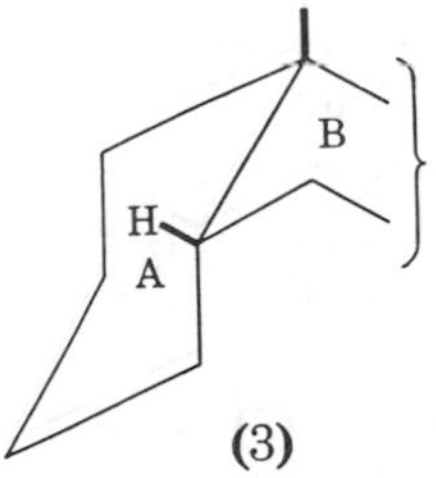

(3)

preferred conformation of a system of fused cyclohexane rings will be that which contains the maximum number of chairs, then 5α-cholestane may be represented by (2) and 5β-cholestane by (3). X-ray analysis of cholesteryl iodide[2] has established that it adopts an all-chair and half-chair conformation in the crystal lattice.

Because of the impossibility of fusing one five- or six-membered ring to another through *trans* diaxial linkages, neither (2) nor (3) can undergo a conformational "flip" to an alternative all-chair system, in the way in which cyclohexane and *cis*-decalin can. They are, like *trans*-decalin, rigid conformations.* A very important consequence of this is that there is a fixed relationship between the configuration and the conformation of any substituent attached to a steroid nucleus. The conformation of a β-substituent at each position in the 5α series is indicated in (4). The

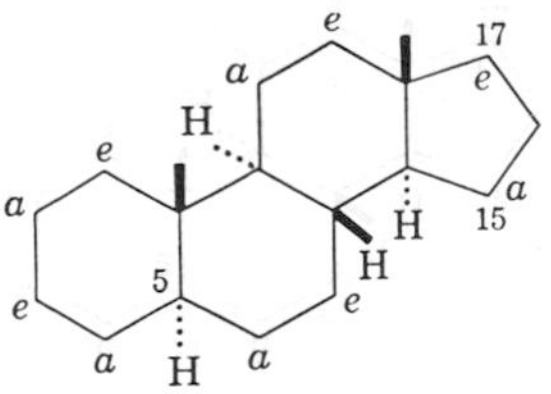

(4)

[2] C. H. Carlisle and D. Crowfoot, *Proc. Roy. Soc. (London), Ser. A*, **184**, 64 (1945).

* Conformations in which one or more of the rings of the steroid nucleus exist in the flexible form are also possible, but since these are generally of much higher energy than the all-chair conformation they are usually unimportant. Some exceptions to this rule will be discussed in Chap. 7.

situation in the 5β series differs only with regard to ring A substituents, the conformations of which are the opposite of those obtaining in *trans* A/B steroids. Conformations at C_{15} and C_{17} are assigned relative to ring C.[3]

The exact shape of ring D of the steroids has been the subject of much recent work.[4-8] It is known that the introduction of a halogen atom into the α-position of a cyclohexanone produces characteristic changes in its spectral properties, depending on whether the halogen atom is in the equatorial or the axial conformation (see Secs. 3-4, 7-3). In the studies referred to above, the infrared and ultraviolet spectra and the optical rotatory dispersion of steroidal ring D ketones were compared with those of their α-halo derivatives of known configuration. The axial or equatorial characters of the halogen atoms were inferred from these data by applying the relationships established in the cyclohexanone series. The results which have been obtained in the 14α series indicate that in 15-halo and 17-halo derivatives of 16-oxosteroids[4] ring D adopts what is termed the half-chair conformation (5)[9] (cf. Sec. 4-3). In this conformation C_{13} lies above the plane defined by C_{15}, C_{16}, and C_{17}, and C_{14} is situated at the same distance below this plane; 15α- and 17β-substituents possess some

(5) (6)

[3] D. H. R. Barton, *J. Chem. Soc.*, **1953**, 1027.

[4] J. Fajkos, *Collection Czech. Chem. Commun.*, **20**, 312 (1955); *J. Chem. Soc.*, **1959**, 3966; J. Fajkos, and J. Joska, *Chem. Ind. (London)*, **1960**, 872; *ibid.*, 1162; *Collection Czech. Chem. Commun.*, **25**, 2863 (1960); **26**, 1118 (1961); **27**, 1849 (1962); J. Fajkos, J. Joska, and F. Šorm, *ibid.*, **27**, 64 (1962); C. W. Shoppee, R. H. Jenkins, and G. H. R. Summers, *J. Chem. Soc.*, **1958**, 3048; J. Fishman, *J. Org. Chem.*, **27**, 1745 (1962).

[5] J. Fishman and C. Djerassi, *Experientia*, **16**, 138 (1960); G. P. Mueller and W. F. Johns, *J. Org. Chem.*, **26**, 2403 (1961); J. Fajkos and V. Sanda, *Collection Czech. Chem. Commun.*, **27**, 355 (1962).

[6] J. Fishman, *J. Am. Chem. Soc.*, **82**, 6143 (1960).

[7] T. Nambara and J. Fishman, *J. Org. Chem.*, **26**, 4569 (1961). See also C. Djerassi, J. Fishman, and T. Nambara, *Experientia*, **17**, 565 (1961).

[8] F. V. Brutcher, Jr., and W. Bauer, Jr., *J. Am. Chem. Soc.*, **84**, 2236 (1962); See also *ibid.*, 2233.

[9] F. V. Brutcher, Jr., T. Roberts, S. J. Barr, and N. Pearson, *J. Am. Chem. Soc.*, **81**, 4915 (1959).

equatorial character and are termed pseudoequatorial (e'), while their epimers are pseudoaxial (a'). In 16-halo-17-oxosteroids[5,6] of the 14α series the spectral data and also the results of dipole moment measurements[8] suggest that ring D is in the "envelope" conformation (6).[9] In the envelope conformation, C_{13}, C_{15}, C_{16}, and C_{17} are coplanar, and 15α- and 15β-substituents are again pseudoequatorial and pseudoaxial, respectively. The 16α and 16β bonds, in contrast, are conformationally equivalent, since the angle between them is bisected by the plane defined by C_{13}, C_{15}, C_{16}, and C_{17}. They are designated as bisectional bonds (b).*

A note of caution must be sounded on the danger of extrapolating from the conformation of ring D ketones to the conformation adopted by ring D when it does not contain a trigonal carbon atom, since, in these cases, there are different steric and electrostatic interactions to be considered. Direct investigation of the conformation of a steroidal ring D which does not carry a functional group is made very difficult by the absence of a suitable parameter, such as carbonyl absorption, or strength of hydrogen bonding in vic-glycols, as judged by the O—H stretching frequency. However, Brutcher and his co-workers have calculated[8] that the preferred conformation of a 17β-substituted steroid, with a $trans$ C/D junction, e.g., cholesterol, is best represented as the envelope form shown in formula (8) with C_{13} situated above the plane defined by C_{14}, C_{15}, C_{16}, and C_{17}, and the 17β-substituent in a pseudoequatorial conformation.

As will be evident in the sequel (see Sec. 5-7), conformational analysis was extensively employed in the elucidation of the stereochemistry of the

---

[10] T. Nambara and J. Fishman, *J. Org. Chem.*, **27**, 2131 (1962); cf. ref. 6.

* A somewhat different situation obtains in 14β-steroids. A study of the epimeric 3β-acetoxy-16-bromo-5α,14β-androstan-17-ones (7) has revealed that in these compounds the 16α and 16β bonds are not conformationally equivalent. On the basis

(7)

of the spectral changes observed, the 16α-bromo substituent is classed as pseudo-equatorial, and its epimer as pseudoaxial.[7]

The results obtained in a determination of the relative stabilities of the four isomeric C_{16}, C_{17} ring D ketols in the 14β series are also anomalous.[10] Further work is required to establish the conformation adopted by the five-membered ring in this series.

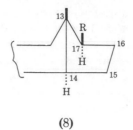

(8)

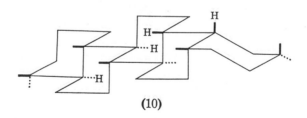

(9)

(10)

(11)   (12)

triterpenoids. For the present, it is sufficient to note that, assuming that wherever it is configurationally possible, an assembly of fused six-membered carbocycles will exist in an all-chair conformation, oleanane (9) and ursane (11) may be represented by the perspective formulas (10) and (12), respectively. It is possible to write preferred all-chair and half-chair conformations for most other triterpenoids.

b. Steroidal Side Chains.   The early X-ray crystallographic studies of cholesteryl iodide[2] and of the related lanost-8-en-3β-ol iodoacetate,[11] a triterpenoid derivative which possesses the same side chain as cholesterol, revealed that in these compounds the saturated side chain adopts a fully staggered, zigzag conformation in the crystal. Such an arrangement would, of course, be predicted by conformational analysis (Sec. 1-3).

Recent work on the conformation of steroid side chains has dealt mainly with the 20-oxo side chain of C_{21}-steroids (i.e., steroids possessing only an acetyl substituent at C_{17}) and has been the subject of a review by Rakhit and Engel.[12]

Lithium aluminum hydride reduction of steroids possessing a 17β-acetyl side chain, but otherwise unsubstituted in rings C and D, affords predominantly the corresponding 20β-hydroxy compound (13).* If it is assumed that the hydride attack takes place on an intermediate complex having a conformation similar to that of the original 20-oxo side chain, then it follows from Cram's rule (Sec. 1-5) that this may be represented by (15),[12] in which the 18- and 21-methyl groups are situated on the same side of the plane defined by C_{20}, C_{17}, and the 17α-hydrogen atom.

[11] J. Fridrichsons and A. McL. Mathieson, *J. Chem. Soc.*, **1953**, 2159.

[12] S. Rakhit and Ch. R. Engel, *Can. J. Chem.*, **40**, 2163 (1962).

[13] L. F. Fieser and M. Fieser, *Steroids*, Reinhold Publishing Corp., New York, 1959, p. 337.

* The stereochemistry at C_{20} of a steroid is designated here and in later sections by employment of the convention proposed by Fieser and Fieser.[13] A model of the side chain is oriented with the longest carbon chain projecting to the rear, and from this a Fischer projection formula is constructed; thus (13) is the Fischer projection corresponding to (14). According to the convention, the configuration of the group

(14)

situated on the left, i.e., the hydroxyl group of (13) is arbitrarily defined as 20β, and that of the group on the right is designated 20α.

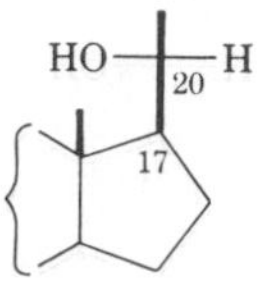

(13) 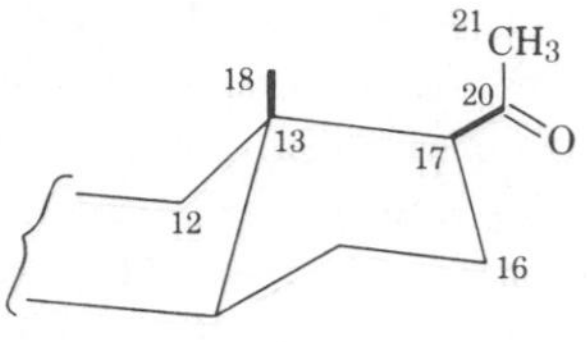   (15)

The same conclusion had previously been reached by Allinger and Da Rooge[14] by means of dipole moment measurements; and application of the axial haloketone rule (Sec. 3-6) to the optical rotatory dispersion of certain 17α-halo-20-oxosteroids[15] had indicated that in these compounds, also, the conformation of the side chain is as shown in (15).

The steric arrangement depicted in general terms in (15) may be represented more precisely by either of the detailed conformations illustrated in the Newman projection formulas (16) and (17). Allinger and Da Rooge rejected (17) as a possibility,[14] since it involves a non-bonded interaction between the 18- and 21-methyl groups similar in degree to that between 1,3-diaxial methyl groups on a cyclohexane ring. It has been suggested[12] that support for this view is provided by the observation that the frequency of carbonyl absorption in the infrared of the 20-keto group is almost unaltered by introduction of a 17α-halo substituent. In

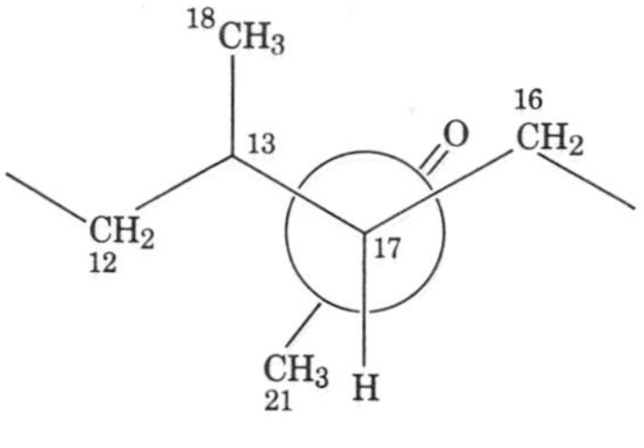

(16) 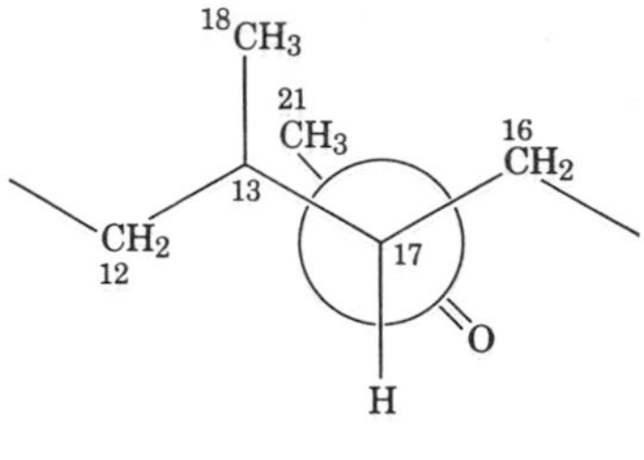   (17)

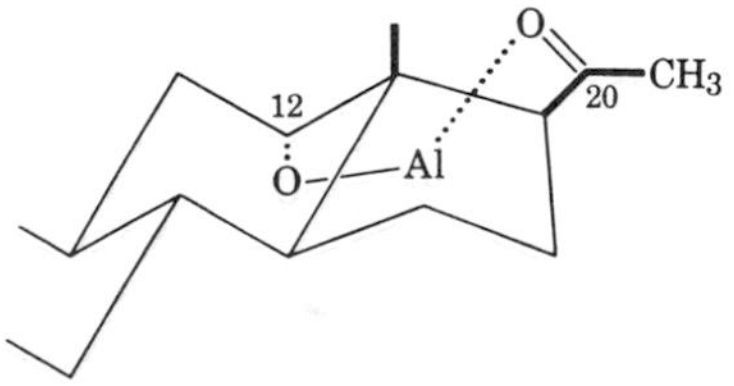

(18)

[14] N. L. Allinger and M. A. Da Rooge, *J. Am. Chem. Soc.*, **83**, 4256 (1961).
[15] C. Djerassi, I. Fornaguera, and O. Mancera, *J. Am. Chem. Soc.*, **81**, 2383 (1959).

conformation (17) the 17α-halogen atom would bear the same relationship to the carbonyl group as it does in a cyclic equatorial α-haloketone and would result in a hypsochromic shift of the carbonyl absorption (Sec. 7-3).

The presence of a hydroxyl group in the neighborhood of the 20-carbonyl group sometimes results in the 20α-hydroxy compound being formed as the major product of lithium aluminum hydride reduction. 3α,12α-Dihydroxy-5β-pregnan-20-one undergoes reduction in that way,[16] and this has been attributed to the side chain adopting a conformation such that the intermediate aluminum complex can link the two oxygen functions (18). Other examples of hydroxyl participation have been similarly explained.[17]

5-3. The Importance of Steric Hindrance and Steric Compression

Much of the chemical behavior of steroids and triterpenoids is determined by the degree of steric hindrance which the molecule exhibits toward particular reagents, and by the strength of the non-bonded interactions to which functional substituents are subjected. Both these effects are directly related to the conformation adopted by the molecule. It is the purpose of this section to trace this relationship and to note the perturbations which may be introduced in certain cases by the operation of hydrogen bonding or electrostatic effects.

a. Axial and Equatorial Substituents

Relative Stability. It has already been pointed out in Chap. 2 that, in the great majority of cases, a substituent on a cyclohexane ring is more stable in the equatorial conformation, since non-bonded interactions are then at a minimum. The application of this generalization to rigid systems is of great importance, since the conformation, *and hence the configuration,* of a substituent may be determined by observing whether or not it is epimerized under equilibrating conditions.

An interesting example of this method is provided by its employment in establishing the configuration of the C_{20} epimers of the 3α,12β-dihydroxy-bisnorcholanic acids, (19; R = H) and (20; R = H).[18] Each of the epimeric hydroxyacids was heated in toluene at 190°C. to bring about lactonization. From the acid (19; R = H) there was obtained,

---

[16] G. Just and R. Nagarajan, *Can. J. Chem.*, **39**, 548 (1961).

[17] K. Heusler and A. Wettstein, *Helv. Chim. Acta*, **45**, 347 (1962), and references cited therein.

[18] D. Arigoni, B. Riniker, and O. Jeger, *Helv. Chim. Acta*, **37**, 878 (1954); M. Sorkin and T. Reichstein, *ibid.*, **27**, 1631 (1944).

(19)   (20)

(21)   (22)

(21a)   (22a)

together with other products, a mixture of two lactones, (21; R = H) and (22; R = H), isolated as the corresponding acetates. The acetoxy-lactone (21; R = Ac) had the same configuration at C_{20} as the starting acid, since it was reconverted to (19; R = H) upon mild hydrolysis. Saponification of the second lactone (22; R = H) afforded the 20-epimeric acid (20; R = H), from which it could be obtained as the sole product of lactonization. When the acetoxylactone (21; R = Ac) was treated under the conditions of lactonization, it was transformed into (22; R = Ac). Each of the lactones (21) and (22) can be represented by only one all-chair conformation. Since (21) is the less stable of the two, it is to be represented by perspective formula (21a), in which the 20-methyl group is axial and interacts strongly with the 13-methyl group. In lactone (22) the 20-methyl group is in the favoured equatorial conformation, as shown in formula (22a). It follows then that the carboxylic acid (19; R = H) has the 20S configuration, while (20; R = H) has the 20R configuration.[19]

Some of the more subtle aspects of steroid and triterpenoid behavior find a ready explanation in terms of the relative stabilities of axial and equatorial substituents. Thus the base-catalyzed conversion of veracevine (23; $R_1 = R_2 = R_3 = H$, $R_4 = OH$) to cevine (25; $R_1 = R_2 = R_3 = H$,

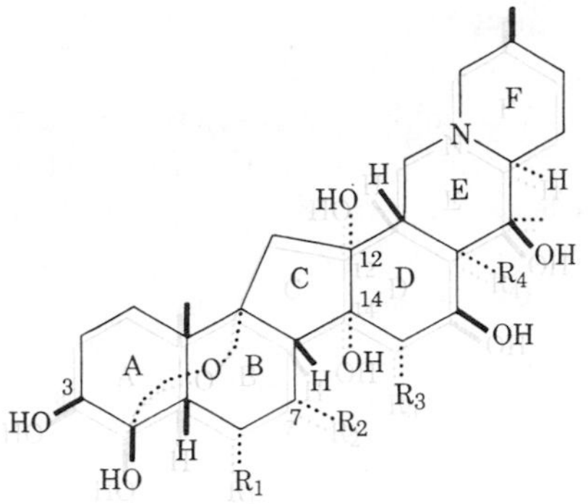

(23)

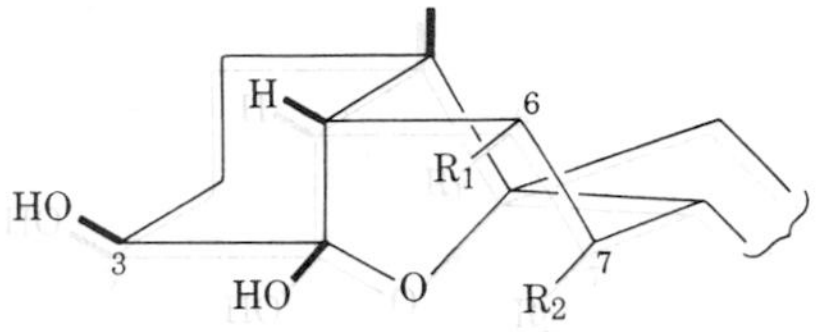

(23a)

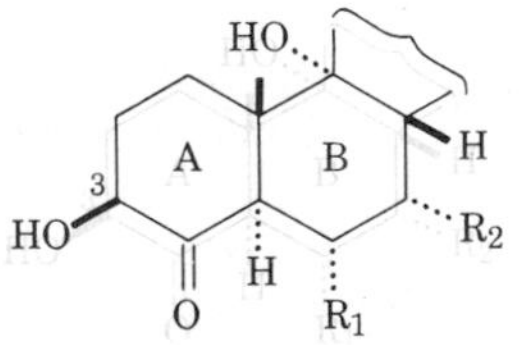

(24)

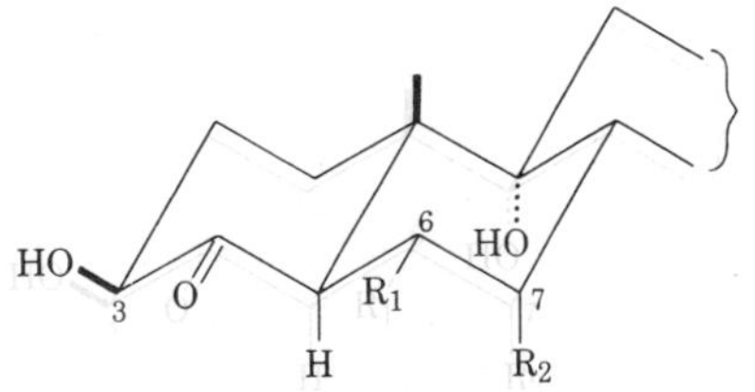

(24a)

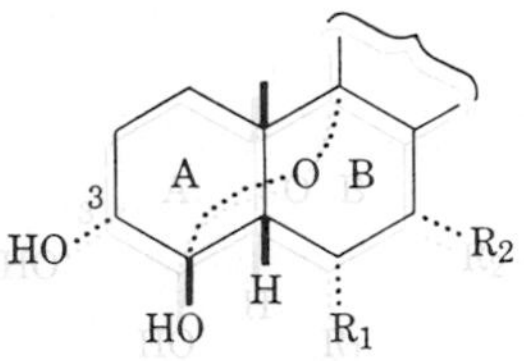

(25)

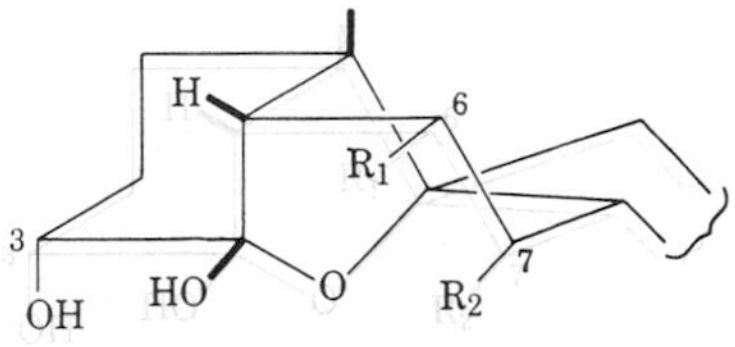

(25a)

[19] For a description of the R and S designations of absolute configurations, see R. S. Cahn, C. K. Ingold, and V. Prelog, *Experientia*, **12**, 81 (1956).

$R_4 = OH$) through a third isomer, cevagenin (24; $R_1 = R_2 = R_3 = H$, $R_4 = OH$), of intermediate stability, is a reflection of the preference of the 3-hydroxyl group to be situated in an equatorial conformation.[20] The steric situation in rings A and B in each compound is shown in formulas (23a), (25a), and (24a), respectively. Similar stability sequences are encountered in the corresponding triads based on zygadenine (23; $R_1 = R_2 = R_4 = H$, $R_3 = OH$, H in place of OH on C_{12})[21] and protoverine (23; $R_1 = R_2 = R_3 = OH$, $R_4 = H$, H in place of OH on C_{12}).[22]

In the protoverine (23; $R_1 = R_2 = R_3 = OH$, $R_4 = H$, H in place of OH on C_{12}) series, (23a) and (24a) are stabilized by intramolecular hydrogen bonding, between the ethereal oxygen atom and the 7α-hydroxyl group, and between the carbonyl group and the 6α-hydroxyl group, respectively. In the triad based on germine (23; $R_1 = R_4 = H$, $R_2 = R_3 = OH$, H in place of OH on C_{12}) only (23a) is stabilized in this way, and this has been suggested as a rationale for the observation that in this series the usual stability order of the first two members of the triad is reversed.[23]

Despite the close similarity in their structures, the chemistry of α-amyrin (26; $R = CH_3$) and its derivatives is different in a number of important respects from that of β-amyrin (27; $R = CH_3$) and related compounds. Some of these differences may be rationalized on the basis of the differing stability of axial and equatorial substituents. Thus the observation that, while the D/E ring junction of the 11-oxo derivative of methyl oleanolate (27; $R = CO_2CH_3$) is isomerized from *cis* to *trans* by base treatment, the corresponding compound in the α-amyrin series is stable in the *cis* configuration[24] may be explained as a consequence of the fact that epimerization at C_{18} of an ursane derivative would involve changing the conformation of the methyl groups attached at C_{19} and C_{20} from diequatorial to the less stable diaxial arrangement, with the 17- and 19-methyl groups participating in a 1,3-diaxial interaction.[25]

Another notable difference between the two series is that the double bond in β-amyrin (27; $R = CH_3$) is readily isomerized from the Δ^{12}-position to the fully substituted $\Delta^{13(18)}$-position. α-Amyrin (26; $R = CH_3$) cannot be similarly isomerized. Again, this is readily explicable, since, if

[20] D. H. R. Barton, O. Jeger, V. Prelog, and R. B. Woodward, *Experientia*, **10**, 81 (1954).

[21] S. M. Kupchan, *J. Am. Chem. Soc.*, **81**, 1925 (1959).

[22] S. M. Kupchan, C. I. Ayres, M. Neeman, R. H. Hensler, T. Masamune, and S. Rajagopalan, *J. Am. Chem. Soc.*, **82**, 2242 (1960).

[23] S. M. Kupchan and C. R. Narayanan, *J. Am. Chem. Soc.*, **81**, 1913 (1959); S. M. Kupchan, S. McLean, G. W. A. Milne, and P. Slade, *J. Org. Chem.*, **27**, 147 (1962).

[24] D. H. R. Barton and N. J. Holness, *J. Chem. Soc.*, **1952**, 78.

[25] E. J. Corey, and J. J. Ursprung, *J. Am. Chem. Soc.*, **78**, 183 (1956).

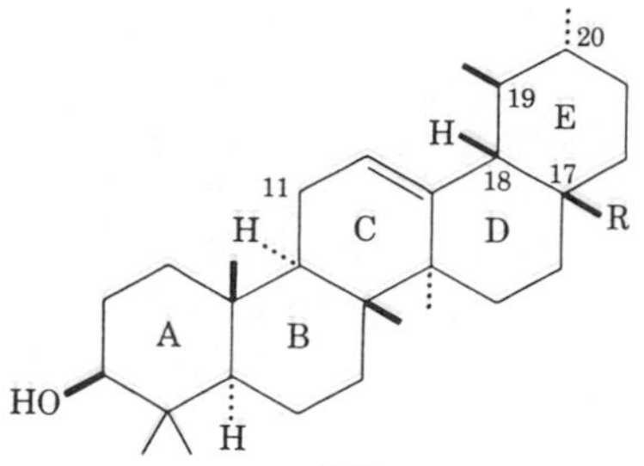

(26)

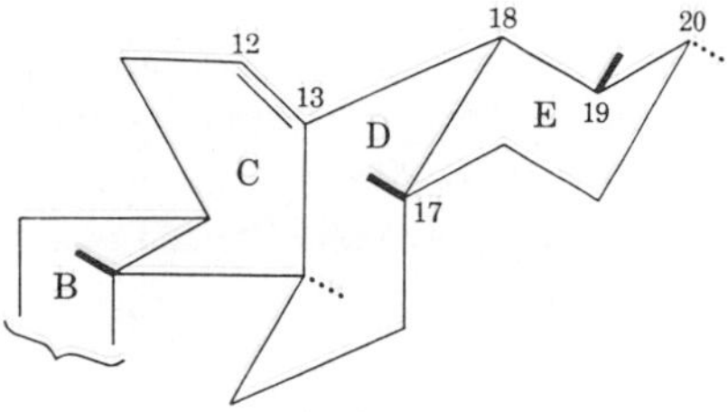

(26a)

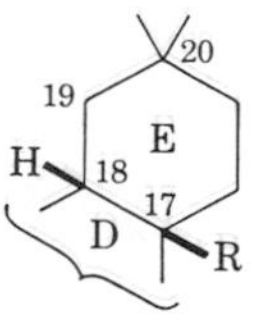

(27)

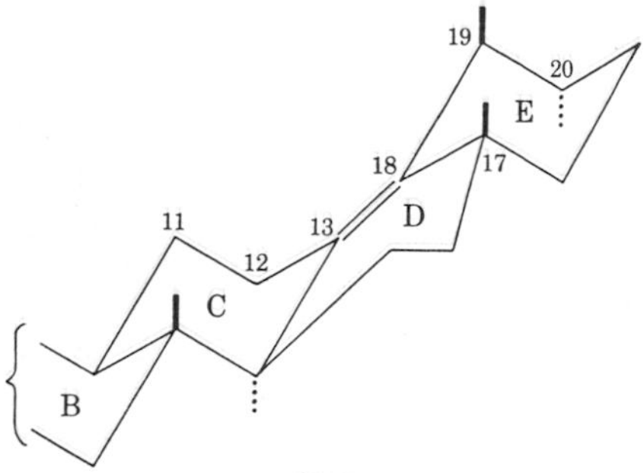

(28)

the double bond of α-amyrin were to migrate, the 19- and 20-methyl groups would be forced to change from the diequatorial conformation (26a) to the diaxial conformation (28), in which the 17- and 19-methyl groups would again be in a 1,3-diaxial relationship.

While in the great majority of cases the simple rule holds that a substituent is more stable in an equatorial conformation, some exceptions are known. Those which may be explained in terms of hydrogen bonding or dipole interaction have already been noted in Chaps. 2 and 3.

In the more complex systems represented by steroids and triterpenoids it sometimes happens that a substituent interacts more strongly with other groups in the molecule when it is equatorial than when it is axial. In such cases there is a reversal of the usual stability relationship. This phenomenon has been observed where a bulky substituent (carboxyl or methoxycarbonyl) is flanked by other bulky groups, so that in the equatorial conformation the substituent is subjected to unusually severe 1,2-interactions. Such a situation obtains in (29)[26] and (30),[27] which are degradation products of triterpenoids of the oleanane series. In each case the C_9-methoxycarbonyl group is in its more stable (axial) conformation. Similarly, the preferred pseudoaxial conformation of the carboxyl group

[26] L. Ruzicka and K. Hofmann, *Helv. Chim. Acta*, **19**, 114 (1936).
[27] G. Brownlie and F. S. Spring, *J. Chem. Soc.*, **1956**, 1949.

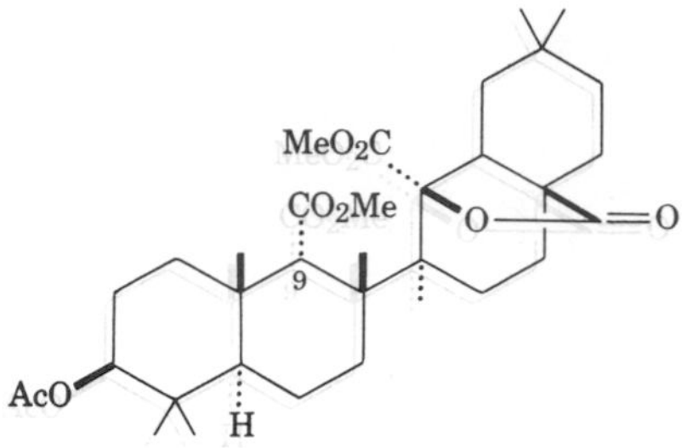

(29)

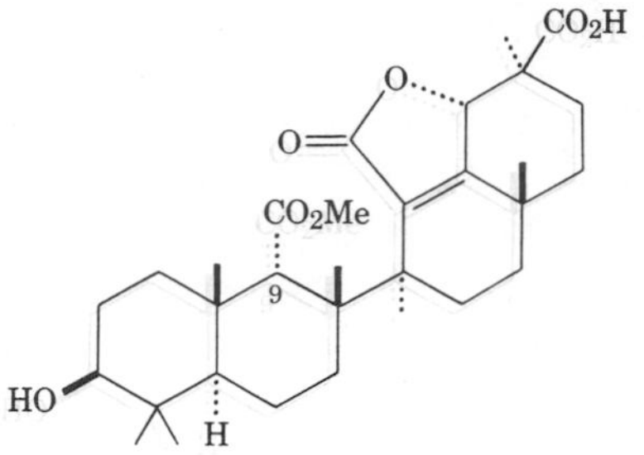

(30)

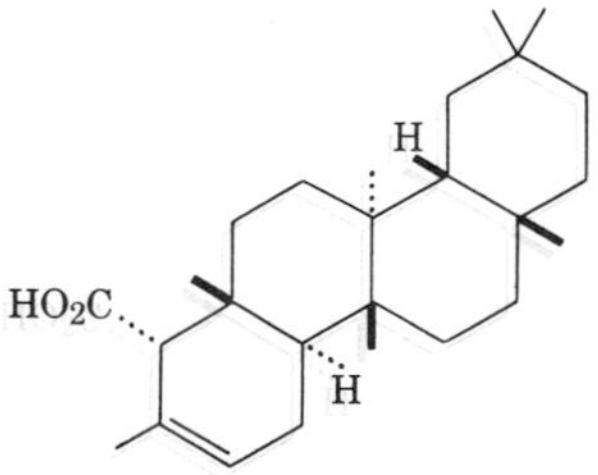

(31)

of (31), which is obtained by degradation of friedelin, is a consequence of the severe 1,2-interactions to which the equatorial epimer would be subjected.[28]

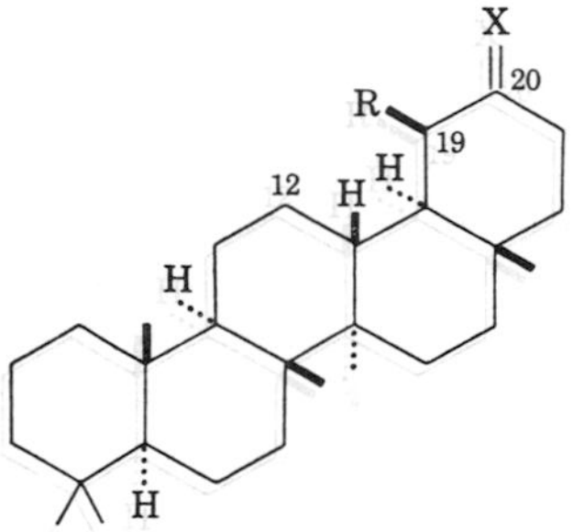

(32)

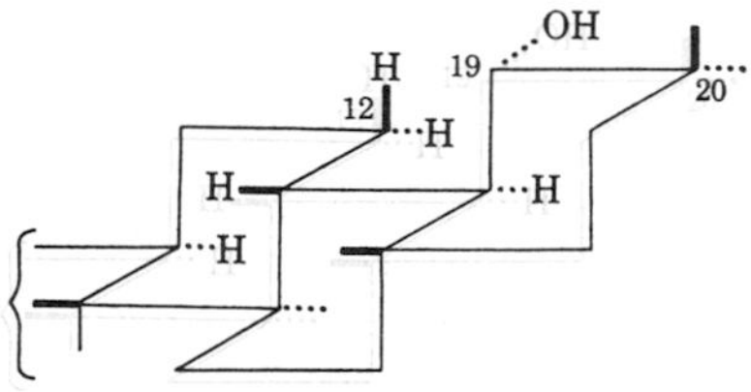

(33)

The preferred axial conformation of the hydroxyl group of 18α-oleanan-19β-ol [32; R = OH, X = $(CH_3)_2$][29] may also be noted here. It has been

[28] E. J. Corey and J. J. Ursprung, *J. Am. Chem. Soc.*, **78**, 5041 (1956); see also R. L. Autrey, D. H. R. Barton, A. K. Ganguly, and W. H. Reusch, *J. Chem. Soc.*, **1961**, 3313.

[29] T. R. Ames, J. L. Beton, A. Bowers, T. G. Halsall, and E. R. H. Jones, *J. Chem. Soc.*, **1954**, 1905.

suggested that the epimeric equatorial conformer is destabilized by virtue of the non-bonded interactions which it enters into with the C_{20}-*gem*-dimethyl group and the C_{12}-methylene group (see 33), the latter being equivalent to a 1:3-diaxial interaction.

The stable β-configuration of the 19-methyl group of (32; $R = CH_3$, $X = O$)[29] implies a preferred axial conformation for the 19-substituent in this compound also, provided that the molecule exists in the all-chair arrangement.

Esterification, and Hydrolysis of Esters.   It follows, from the fact that an axially disposed group is subject to greater steric hindrance than its equatorial epimer, that an equatorial hydroxyl or carboxyl group should be more easily esterified than an axial one, and that of a pair of epimeric esters the axial one should be the more difficult to hydrolyze (Sec. 2-5a). Such has indeed proved to be the case in steroid chemistry,[3] and this difference in reactivity is of value in assigning configurations.

This is well illustrated by the procedure adopted by Beaton and Spring[30] to determine the configuration of the carboxyl group in glycyrrhetic acid (34; $R = H$, 18β-H), a triterpenoid of the oleanane series. Base treatment of methyl glycyrrhetate (34; $R = CH_3$, 18β-H) caused isomerization of the D/E ring junction from *cis* to *trans*, and as a consequence inverted the conformations of the substituents on C_{20}. Comparative hydrolyses revealed that the methoxycarbonyl group of methyl glycyrrhetate (34; $R = CH_3$,

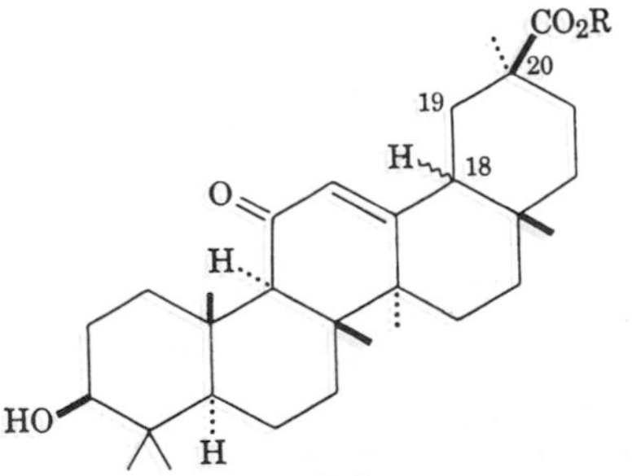

(34)

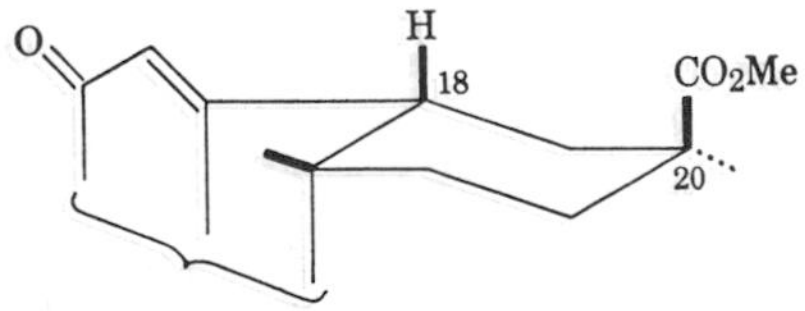

(34a)

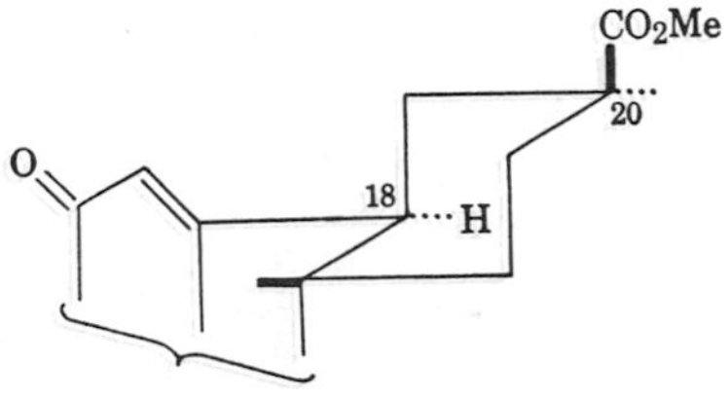

(34b)

[30] J. M. Beaton and F. S. Spring, *J. Chem. Soc.*, **1955**, 3126.

18β-H) is much more hindered than that of methyl 18α-glycyrrhetate (34; $R = CH_3$, 18α-H), and from this it was deduced that the ester group in these compounds was, respectively, axial (34a) and equatorial (34b) and therefore had the β-configuration.

More recently Allen has investigated[31] the conformations of some of the indole alkaloids by measuring the rates of hydrolysis of the methoxy-carbonyl groups which they contain. The interpretation of these results is complicated by the possibility of participation of suitably disposed hydroxyl groups.

Exceptions to the rule that an equatorial acetate is more readily hydro-lyzed than its epimer occur when there is an axial hydroxyl group in a 1,3-relationship with the axial acetate. In such cases hydrolysis of the ester is facilitated by hydrogen bonding to the hydroxyl group. Henbest and Lovell have investigated[32] this phenomenon in the isomeric 3,5-dihydroxysteroids and have proposed a transition state of type (35; n = nucleophile). On the basis of kinetic measurements, it has been suggested[33] that the solvolysis of 1,3-diaxial hydroxyacetates is an instance of concerted general base-general acid catalysis; and the inter-esting proposal has been made[34] that the facile methanolyses of C_{16}-acetoxy

(35)   (36)

[31] M. J. Allen, *J. Chem. Soc.*, **1960**, 4904; **1961**, 4252.

[32] H. B. Henbest and B. J. Lovell, *J. Chem. Soc.*, **1957**, 1965; see also W. S. Johnson, J. J. Korst, R. A. Clement, and J. Dutta, *J. Am. Chem. Soc.*, **82**, 614 (1960); R. West, J. J. Korst, and W. S. Johnson, *J. Org. Chem.*, **25**, 1976 (1960); S. M. Kupchan, P. Slade, and R. J. Young, *Tetrahedron Letters*, No. 24, 22 (1960); T. C. Bruice and T. H. Fife, *ibid.*, 263 (1961); *J. Am. Chem. Soc.*, **84**, 1973 (1962); S. M. Kupchan, P. Slade, R. J. Young, and G. W. A. Milne, *Tetrahedron*, **18**, 499 (1962).

[33] S. M. Kupchan, S. P. Eriksen, and M. Friedman, *J. Am. Chem. Soc.*, **84**, 4159 (1962).

[34] S. M. Kupchan, A. Afonso, and P. Slade, Abstracts of Papers, 88Q, Division of Organic Chemistry, 140th Meeting, American Chemical Society, Chicago, Ill., September, 1961. Evidence for such a mechanism has been obtained from kinetic studies of the methanolysis of cevadine D-orthoacetate diacetate; see S. M. Kupchan, S. P. Eriksen, and Y. -T. Shen, *J. Am. Chem. Soc.*, **85**, 350 (1963).

derivatives of the *Veratrum* alkaloids, such as cevine (25; $R_1 = R_2 = R_3 = H$, $R_4 = OH$) are examples of intramolecular bifunctional general acid-general base catalysis in which both the 20-hydroxyl group and the tertiary nitrogen atom participate (see 36).

Recognition of the operation of the neighboring hydroxyl group effect in some reactions of the *Veratrum* alkaloids was of importance in elucidating the stereochemistry of these complex molecules. Thus an axial 7α-acetoxyl group in the germine series (see 23; $R_1 = R_4 = H$, $R_2 = R_3 = OH$, H in place of OH on C_{12}) is very readily methanolyzed, owing to participation by the axial hydroxyl group at C_{14}.[23] Other examples are known.[22,23,35]

The facilitation of ester hydrolysis by the presence of an axial hydroxyl group is only one example of a range of reactions between 1,3-diaxial groups. More instances are quoted in a later section (Sec. 5-3d).

Oxidation of Secondary Alcohols. Oxidation of a secondary hydroxyl group with chromic acid is thought to proceed through the formation of an intermediate chromate ester. Generally, however, the rate-determining step in the reaction is not the formation of the ester, but its decomposition, involving breaking of the bond to the carbinol hydrogen atom. Evidence in support of such a mechanism is provided by the data reported for 5α-cholestan-3α-ol by Schreiber and Eschenmoser.[36a] This compound is oxidized with chromic acid some seven times faster than its 3β-deuterio derivative. A similar clear isotope effect was observed for 5α-pregnan-11β-ol;[36a] hence, even in the case of the strongly hindered 11β-hydroxysteroids, formation of the chromate ester is not the rate-determining step.*

The rates of chromic acid oxidation of a series of epimeric pairs of steroidal alcohols have been measured.[36c] From the results, as listed in Table 1, it may be seen that in each case the axial epimer is oxidized more rapidly. This is thought to be a reflection of the greater compression to

[35] S. M. Kupchan, W. S. Johnson, and S. Rajagopalan, *Tetrahedron*, **7**, 47 (1959).

[36a] We thank Dr. Eschenmoser for communicating these results to us in advance of publication.

[36b] J. Roček, F. H. Westheimer, A. Eschenmoser, L. Moldoványi, and J. Schreiber, *Helv. Chim. Acta*, **45**, 2554 (1962).

[36c] J. Schreiber and A. Eschenmoser, *Helv. Chim. Acta*, **38**, 1529 (1955).

* It should be noted, however, that the effect of steric hindrance to esterification becomes dominant in the case of 3,28-diacetoxy-6β-hydroxy-18β-olean-12-ene, which exhibits no isotope effect when oxidized with chromic acid.[36b] The hydroxyl group of this compound is subject to three 1,3-diaxial interactions with methyl groups, and formation of the chromate ester is hindered sufficiently to render it the slow, rate-determining step. Consequently, the usual relationship between the steric compression of a hydroxyl group and its rate of oxidation with chromic acid (see Table 1) does not hold for this compound, and it is oxidized some seven times more slowly than is 5α-pregnan-11β-ol.[36b]

which an axial substituent is subjected, resulting in its energy being closer to that of the transition state of the reaction.

Chromic acid oxidation of secondary alcohols has already been discussed in Sec. 2-5c.

In the oxidation of steroidal alcohols to the corresponding ketones with N-bromosuccinimide—a reaction in which free bromine is the active

Table I. Relative Rates of Oxidation of Steroidal Alcohols with Chromic Acid[36c]

Axial alcohols	k	Equatorial Alcohols	k
5α-Cholestan-1α-ol	13.0	5α-Cholestan-1β-ol	9.7
5α-Cholestan-2β-ol	20	5α-Cholestan-2α-ol	1.3
5α-Cholestan-3α-ol	3.0	5α-Cholestan-3β-ol	1.0
5α-Cholestan-4β-ol	35	5α-Cholestan-4α-ol	2.0
5α-Cholestan-6β-ol	36	5α-Cholestan-6α-ol	2.0
5α-Cholestan-7α-ol	12.3	5α-Cholestan-7β-ol	3.3
5α-Pregnan-11β-ol[36a]	>900	5α-Pregnan-11α-ol	14
5α-Pregnan-11β-ol-3,20-dione	>60	5α-Pregnan-11α-ol-3,20-dione	7.0

oxidizing agent—the rate-determining step appears to be attack of bromine on the hydrogen attached to the hydroxylated carbon atom.[37] There is evidence that the accessibility of the hydrogen atom is an important factor in determining the ease of oxidation,[37] and in this reaction also axial hydroxyl groups are more readily oxidized than their equatorial epimers.[38,*]

In contrast, evidence has accumulated that, when a pair of epimeric allylic or benzylic hydroxyl groups is oxidized to the corresponding ketones, it is the pseudoequatorial epimer which reacts more rapidly.[38,42a,43]

[37] G. Langbein and B. Steinert, *Chem. Ber.*, **95**, 1873 (1962).

[38] R. Filler, *Chem. Rev.*, **63**, 21 (1963).

[39] F. Mareš and J. Roček, *Collection Czech. Chem. Commun.*, **26**, 2370 (1961).

[40] J. Roček, *Tetrahedron Letters*, **1962**, 135.

[41] Cf. J. Iriarte, J. N. Shoolery, and C. Djerassi, *J. Org. Chem.*, **27**, 1139 (1962).

[42] (a) G. Stork, in R. H. F. Manske, ed., *The Alkaloids*, Vol. VI, Academic Press, New York, 1960; (b) cf. H. L. Goering and R. R. Josephson, *J. Am. Chem. Soc.*, **84**, 2779 (1962).

[43] H. Rapoport and S. Masamune, *J. Am. Chem. Soc.*, **77**, 4330 (1955); H. L. Goering, R. W. Greiner, and M. F. Sloan, *ibid.*, **83**, 1391 (1961).

* It is interesting to note that recently it has been demonstrated,[39] by experiments on the isomeric 1,4-dimethylcyclohexanes, that an equatorial C—H bond is oxidized with chromium trioxide more than four times as fast as an axial bond. The natures of the intermediates involved in this reaction are still uncertain,[40] and it is impossible to decide from the available data whether the difference in rates is due to the greater accessibility of an equatorial hydrogen atom or to the effect of steric acceleration in the case of the compound with an axial methyl group.

(37)   (38)

Thus it has been found[44] that cholest-5-en-7β-ol, with a pseudoequatorial hydroxyl group, is more rapidly oxidized with manganese dioxide than is its 7-epimer; similarly the Amaryllidaceae alkaloid, buphanamine (37),[45] which has a pseudoaxial hydroxyl group, is unaffected by manganese dioxide under conditions which oxidize 1-epibuphanamine to the corresponding ketone.*,[46]

Stork[42] has suggested that these results might be due to the fact that overlap between the π electrons of the allylic double bond and the electrons of the C—H bond which must be broken in the oxidation is at a maximum when these are parallel, a situation which exists when the C—H bond is pseudoaxial, i.e., in the pseudoequatorial alcohol (see 38).

Solvolysis. As has been discussed in Chap. 2, an axial substituent on a cyclohexane ring is more readily replaced, in both S_N1 and S_N2 processes, than is its equatorial epimer.

The rates of acetolyses, by S_N1 mechanisms, of a number of decalyl[47] and cholestanyl[48] tosylates have been correlated with the relief of steric compression associated with the departure of the leaving group in each case. This effect is important in the related $E1$ eliminations also.[49]

[44] A. Nickon and J. F. Bagli, J. Am. Chem. Soc., **83**, 1498 (1961).

[45] H. A. Lloyd, E. A. Kielar, R. J. Highet, S. Uyeo, H. M. Fales, and W. C. Wildman, J. Org. Chem., **27**, 373 (1962).

[46] H. M. Fales and W. C. Wildman, J. Org. Chem., **26**, 881 (1961).

[47] I. Moritani, S. Nishida, and M. Murakami, J. Am. Chem. Soc., **81**, 3420 (1959); Bull. Chem. Soc. Japan, **34**, 1334 (1961).

[48] S. Nishida, J. Am. Chem. Soc., **82**, 4290 (1960).

[49] G. Just and Ch. R. Engel, J. Org. Chem., **23**, 12 (1958); Ch. R. Engel and S. F. Papadopoulos, ibid., **26**, 2868 (1961).

* Measurements of the rates of oxidation of allylic alcohols must be interpreted with caution, since under certain conditions chromic acid oxidation of the pseudoaxial epimer may follow an unusual course.[36a] Thus, while oxidation of cholest-4-en-6α-ol with chromium trioxide in acetic acid affords the corresponding $\alpha\beta$-unsaturated ketone in good yield, with consumption of 1 mole of oxidant, the 6β-epimer reacts faster under the same conditions and consumes 2 moles of oxidant. The reaction mixture contains no detectable amount of the expected enone, but it is possible to isolate from it some 4,5-epoxy-5β-cholestan-6-one.[36a,41]

Physical Properties. It is only reasonable that the different environments experienced by a functional group in the axial and the equatorial conformations should be reflected in its physical properties as well as in its reactivity.

For example, an equatorial group, being more accessible than its axial epimer, should be more strongly adsorbed on a chromatographic column. In many instances this simple relationship does indeed hold.[1,3,50,51,51a] However, the conformation adopted by a molecule adsorbed on a solid surface is not necessarily its preferred conformation in solution; it is not surprising, therefore, that exceptions to the generalization above have been noted.[52] Intramolecular hydrogen bonding between 1,3-diaxial groups may also reduce their polarity and result in their being eluted before their (*e,a*) isomers.[53]

In partition chromatography[54] and in gas-liquid chromatography[55] of steroids it has similarly been found that a compound with an axial polar substituent is more mobile than its equatorial epimer. The conformation adopted by the molecule during these techniques should be closer to that predicted by conformational analysis. Nevertheless, exceptions to the general rule have been reported;[56] they may be due to preferential solvation of the less hindered equatorial epimer with a consequent reduction in its polarity.[57]

The degree of ionization of a carboxyl group is determined, at least partly, by the ease with which the corresponding ion can be solvated.

[50] D. H. R. Barton and R. C. Cookson, *Quart. Rev. (London)*, **10**, 44 (1956).

[51] C. S. Barnes and A. Palmer, *Australian J. Chem.*, **9**, 105 (1956).

[51a] The same relationship applies in thin-layer chromatography also; see, for example, L. Labler and V. Černý, *Collection Czech. Chem. Commun.*, **28**, 2932 (1963), and also ref. 57.

[52] R. V. Brooks, W. Klyne, and E. Miller, *Biochem. J.*, **54**, 212 (1953); C. W. Shoppee, D. N. Jones, and G. H. R. Summers, *J. Chem. Soc.*, **1957**, 3100; E. J. Corey, *J. Am. Chem. Soc.*, **75**, 4832 (1953).

[53] H. B. Henbest and J. McEntee, *J. Chem. Soc.*, **1961**, 4478.

[54] K. Savard, *J. Biol. Chem.*, **202**, 457 (1953); *Recent Progr. Hormone Res.*, **9**, 185 (1954); R. Hirschmann, G. A. Bailey, R. Walker, and J. M. Chemerda, *J. Am. Chem. Soc.*, **81**, 2822 (1959).

[55] R. B. Clayton, *Nature*, **190**, 1071 (1961); W. J. A. van den Heuvel, E. O. A. Haahti, and E. C. Horning, Abstracts of Papers, 68Q, Division of Organic Chemistry, 140th Meeting, American Chemical Society, Chicago, Ill., September, 1961.

[56] C. Sannié and H. Lapin, *Bull. Soc. Chim. France*, **1952**, 1080; W. J. McAleer and M. A. Kozlowski, *Arch. Biochem. Biophys.*, **66**, 120 (1957); W. J. McAleer, M. A. Kozlowski, T. H. Stoudt, and J. M. Chemerda, *J. Org. Chem.*, **23**, 958 (1958); S. G. Brooks, J. S. Hunt, A. G. Long, and B. Mooney, *J. Chem. Soc.*, **1957**, 1175.

[57] Cf. H.-J. Petrowitz, *Angew. Chem.*, **72**, 921 (1960); see also M. Takeuchi, *Chem. Pharm. Bull. (Tokyo)*, **11**, 1183 (1963).

Equatorial carboxyl groups, being less hindered than their epimers, are therefore more strongly acidic (Sec. 3-11).

A semi-empirical method for calculating the apparent pK_a of an alicyclic carboxylic acid in 80% methyl Cellosolve-20% water (w:w) as solvent has recently been described;[58] it is based on data obtained from twenty-one carboxylic acids. An allowance of 0.25 pK unit is made for each 1,3-diaxial interaction involving the carboxyl group (including $H:CO_2H$ interactions), and 0.22 pK unit for an α-substituent. Hence $pK_a = 7.44 + 0.25a + 0.22b$, where a = number of 1,3-diaxial interactions, and $b = 1$, if there is an α-substituent, 0 if there is an α-hydrogen atom. The potential application of this approach to the determination of the configuration of carboxyl groups in naturally occurring compounds is obvious.

The pK_a values of a range of steroidal amines have been measured by Bird and Cookson,[59] and it would appear that, in general, of a pair of epimeric amines the one with the functional group equatorial is the stronger base. Again, this is best explained as being a result of the greater ease of solvation of an equatorial ion. Equatorial amino groups have also been found to form N-oxides, and to undergo the Menshutkin reaction more rapidly than their axial epimers.[60,60a]

A few exceptions to the generalization concerning the relative basic strengths of epimeric aminosteroids have been noted.[59] The most striking one is 7β-dimethylamino-5α-cholestane, which is some 0.8 pK unit weaker than its axial epimer. It has been suggested that the reason is that the most stable conformation of an equatorial cyclohexyldimethylammonium ion (i.e., that in which the N—H bond is parallel to the axial bonds of the ring, with the hydrogen atom strongly bonded to the solvent) is denied to the ion derived from 7β-dimethylamino-5α-cholestane, since this would involve an extremely severe interaction between one of the N-methyl groups and the 15-methylene group (39). In accord with this explanation, the epimeric 7-amino-5α-cholestanes display the usual order of basic strength. The epimeric 7-dimethylamino-5α-cholestanes are also anomalous in their relative rates of N-oxidation[60a] and of quaternization.[60b]

[58] P. F. Sommer, V. P. Arya, and W. Simon, *Tetrahedron Letters*, No. 20, 18 (1960); P. F. Sommer, C. Pascual, V. P. Arya, and W. Simon, *Helv. Chim. Acta*, **46**, 1734 (1963).

[59] C. W. Bird and R. C. Cookson, *Chem. Ind. (London)*, **1955**, 1479; *J. Chem. Soc.*, **1960**, 2343.

[60] B. B. Gent and J. McKenna, *J. Chem. Soc.*, **1956**, 573.

[60a] R. Ledger, J. McKenna, and P. B. Smith, *Tetrahedron Letters*, **1963**, 1433.

[60b] R. Ledger, J. McKenna, and P. B. Smith, *Chem. Ind. (London)*, **1963**, 863.

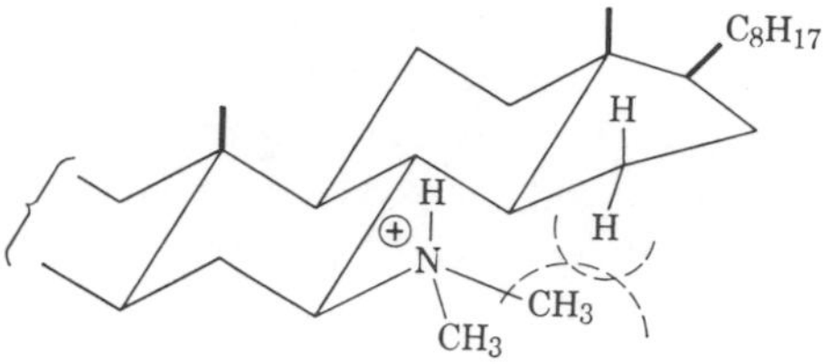

(39)

In the natural product field the correlations established between the spectral properties of a molecule and its conformation have proved to be very valuable.[61] The rapid development of optical rotatory dispersion[61a] as a powerful and versatile technique in conformational studies is particularly noteworthy. An account of the various spectroscopic correlations has already been given in Chap. 3, and the investigations into the conformation of ring D of the steroids (Sec. 5-2) provide some recent examples of their application.

A major advantage of spectral methods is that the information which is obtained relates directly to the preferred conformation of the molecule under investigation. In drawing conclusions from chemical evidence alone it must always be borne in mind that a molecule may react in one of its less stable conformations.[62] Thus it is likely that the abnormally fast hydrolysis of methyl machaerate (40)[63] and of methyl treleasegenate (41),[64] which are both triterpenoids of the oleanane series, is due to participation of the C_{21} oxygen function; this implies that during the reaction both rings D and E of these compounds adopt boat conformations.

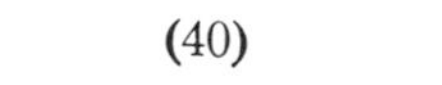

(40)

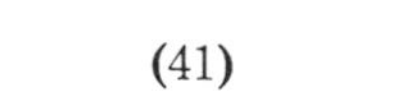

(41)

[61] M. Legrand and J. Mathieu, *Bull. Soc. Chim. France*, **1961**, 1679.

[61a] C. Djerassi, *Optical Rotatory Dispersion. Applications to Organic Chemistry*, McGraw-Hill Book Co., New York, 1960; *Tetrahedron*, **13**, 13 (1961); W. Klyne, *ibid.*, 29.

[62] J. Sicher, F. Šipoš, and M. Tichý, *Collection Czech. Chem. Commun.*, **26**, 847 (1961); J. Sicher, M. Tichý, F. Šipoš, and M. Pánková, *ibid.*, 2418.

[63] C. Djerassi and A. E. Lippman, *J. Am. Chem. Soc.*, **77**, 1825 (1955).

[64] C. Djerassi and J. S. Mills, *J. Am. Chem. Soc.*, **80**, 1236 (1958).

It is virtually certain, however, that the preferred conformations of these molecules involve only chair and half-chair forms.[65]

b. Relative Hindrance of Substituents at Different Positions on the Steroid Nucleus.   The differing reactivities of epimeric functional substituents described in the preceding section arise because, at a given position on a steroid nucleus, an axial substituent is subject to greater steric compression and is more hindered than an equatorial one.   By considering the different non-bonded interactions to which substituents at *different* positions on a steroid nucleus are subjected, it is possible to estimate their relative hindrance.[3]

If this procedure is applied to the isomeric 5α-cholestan-2- and 3-ols, it is seen that in the 2β-alcohol (42) the hydroxyl group, which interacts strongly with the axial methyl group at C_{10} and the axial hydrogen atom at C_4, should be considerably hindered while the 3α-alcohol (43), in which the hydroxyl group has axial interactions only with hydrogen atoms, should be somewhat less hindered; and the 2α-(44) and 3β-(45) isomers should be relatively unhindered, since the functional groups in these compounds experience only 1,2-interactions with hydrogen atoms.   In agreement with this reasoning, the relative rates of oxidation of these alcohols with chromium trioxide are 20, 3, 1.3, and 1, respectively,[36c] and the percentage hydrolyses of the corresponding acetates under standard conditions are 11, 34, 87, and 92 respectively.[66]

A similar analysis may be carried out for each of the other positions on a steroid molecule.   It is then found that, owing to the severe 1,3-diaxial interactions which it enters into with the two angular methyl

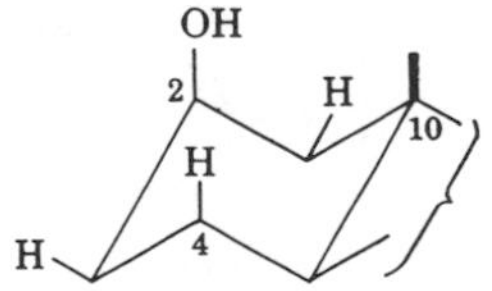

(42) 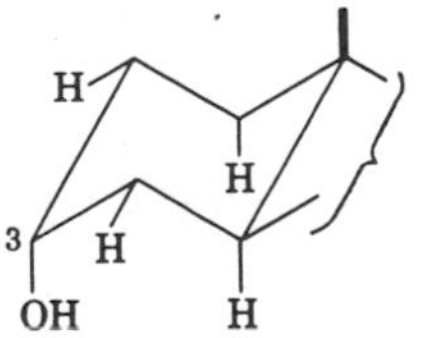   (43)

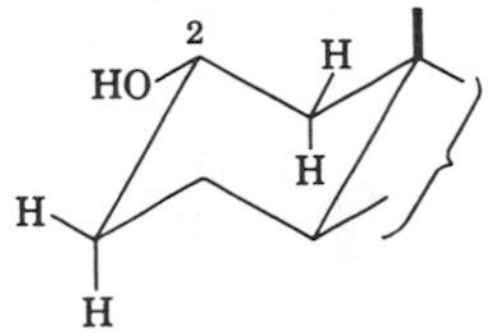

(44) 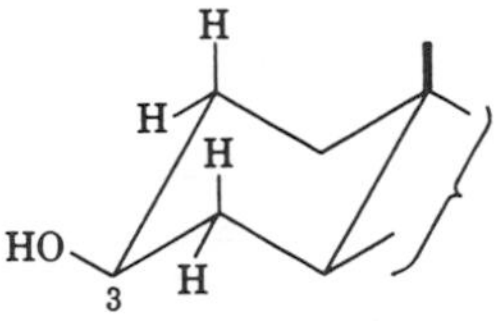   (45)

[65] A. M. Abd El Rahim and C. H. Carlisle, *Chem. Ind. (London)*, **1954,** 279.
[66] A. Fürst and Pl. A. Plattner, *Helv. Chim. Acta*, **32,** 275 (1949).

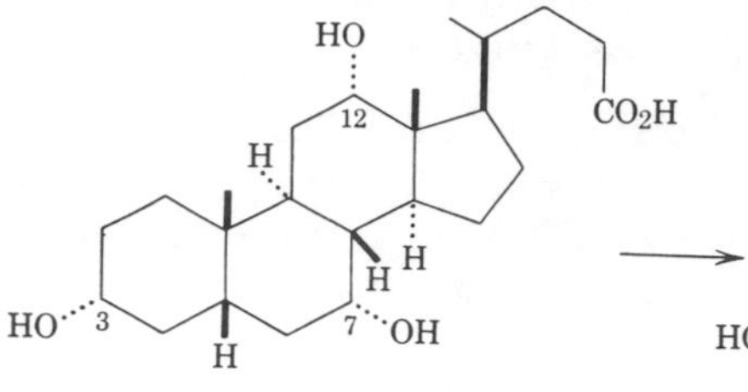

(46)

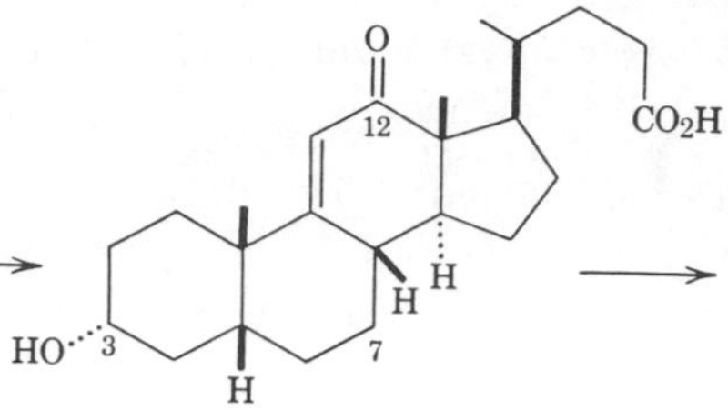

(47)

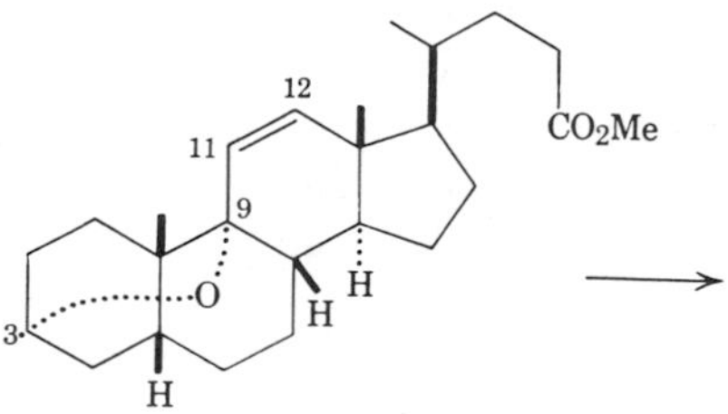

(48)

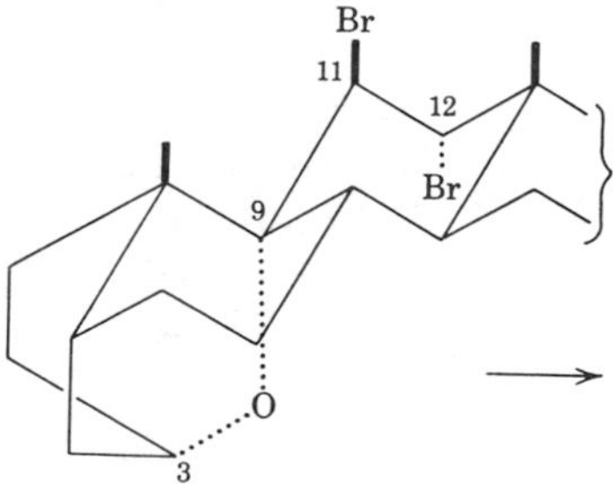

(49)

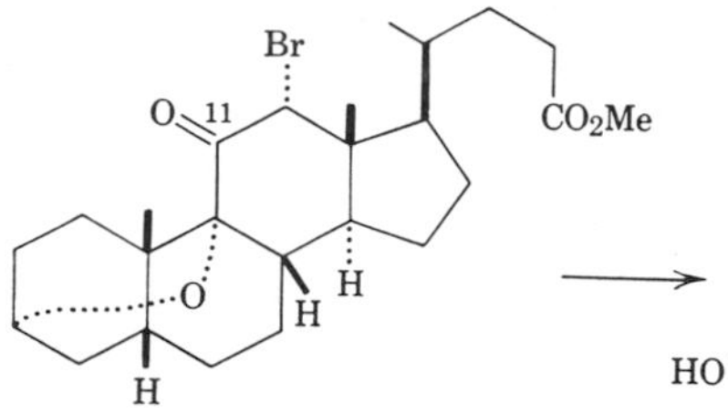

(50)

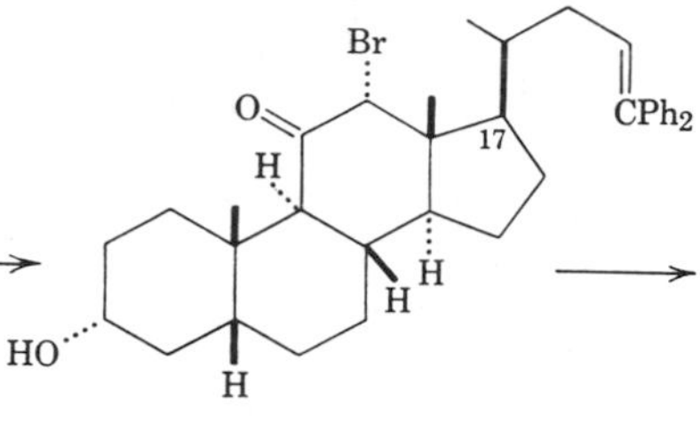

(51)

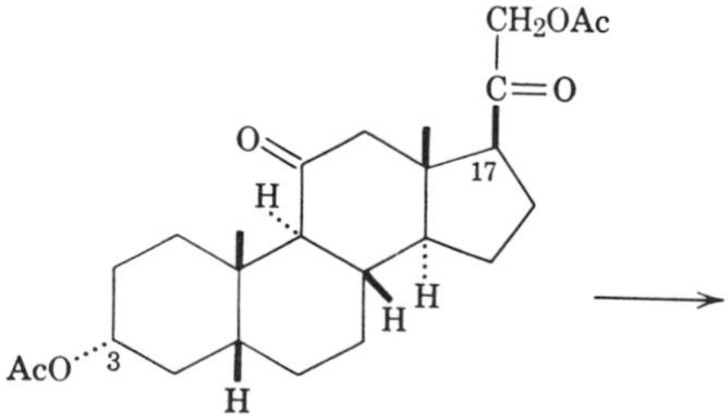

(52)

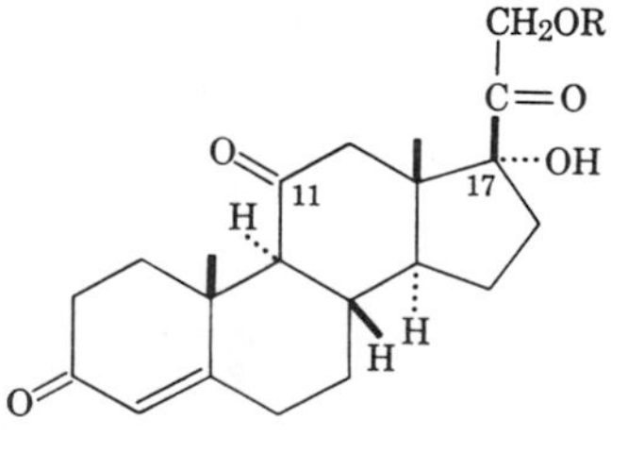

(53)

groups, an 11β-hydroxyl group is the most sterically hindered. Grimmer has developed an analytical procedure for determining the position and configuration of hydroxyl groups attached to the steroid nucleus, based on their differing reactivities toward chromium trioxide.[67]

Since the formation of most carbonyl derivatives involves an intermediate of the type

$$\begin{array}{c} \text{OH} \\ \diagdown \diagup \\ \text{C} \\ \diagup \diagdown \\ \text{NH— (or —O—)} \end{array}$$

in which the grouping —NH— (or —O—) has steric requirements similar to those of an —OH group, the relative hindrance of a ketone at any position may be assessed by summing the non-bonded interactions in which each of the epimeric alcohols at that position participates. When this is done,[3] it is seen that, in the natural steroids, an 11-oxo group is the most hindered. The same conclusion is reached by considering the observed rates of reduction of various steroidal ketones with sodium borohydride.[68] The relative ease with which steroidal ketones form mercaptol derivatives[69] and the extent to which they exist as the derived (hemi)ketals in alcoholic solution[70]—as judged by the diminution in the amplitude of their Cotton effects—are also in reasonable agreement with the relative degrees of steric hindrance predicted by conformational analysis.

The variation in steric hindrance of substituents or carbonyl groups around the steroid nucleus underlies many of the selective reactions[71] which have made possible the spectacular advances since about 1940 in the synthesis of steroids. The synthesis of cortisone (53; R = H) and its 11β-hydroxy analog, cortisol, using cholic acid (46) as starting material, provides several examples of such selective reactions.

The first stage in the synthesis involved the conversion of cholic acid (46) to deoxycholic acid by replacement of the 7-hydroxyl group with a hydrogen atom. Since this is the most sterically compressed of the three

[67] G. Grimmer, *Ann.*, **636**, 42 (1960).

[68] O. H. Wheeler and J. L. Mateos, *Can. J. Chem.*, **36**, 1049 (1958); J. L. Mateos, *J. Org. Chem.*, **24**, 2034 (1959).

[69] H. Hauptmann. *J. Am. Chem. Soc.*, **69**, 562 (1947); H. Hauptmann and M. M. Campos, *ibid.*, **74**, 3179 (1952); H. Hauptmann and F. O. Bobbio, *Chem. Ber.*, **93**, 2187 (1960).

[70] C. Djerassi, L. A. Mitscher, and B. J. Mitscher, *J. Am. Chem. Soc.*, **81**, 947 (1959). Ketals rather than hemiketals may be involved; cf. D. G. Kubler and L. E. Sweeney, *J. Org. Chem.*, **25**, 1437 (1960).

[71] H. J. E. Loewenthal, *Tetrahedron*, **6**, 269 (1959).

hydroxyl groups,* its removal could be accomplished by selective oxidation[74] followed by Wolff-Kishner reduction of the 7-monoketone. The
second objective—introduction of a $\Delta^{9(11)}$-double bond—was achieved by
selective benzoylation of the relatively unhindered equatorial 3α-hydroxyl
group,[75a] oxidation of the 12-hydroxyl to a carbonyl group, and dehydrogenation with selenium dioxide.[75a,75b] Alkaline saponification gave (47)
which, after esterification of the carboxyl group, was converted to a
mixture of 12-epimeric allylic alcohols by catalytic hydrogenation.[75a]
The 12-oxygen function was then converted to a methoxyl group, which
with hydrochloric acid afforded the corresponding chloro compound. As
a consequence of the *cis* A/B ring fusion, treatment of the allylic 12-chloro

---

[72] L. F. Fieser, S. Rajagopalan, E. Wilson, and M. Tishler, *J. Am. Chem. Soc.*,
73, 4133 (1951).

[73] A. Lardon, *Helv. Chim. Acta*, **30**, 597 (1947).

[74] G. A. D. Haslewood, *Biochem. J.*, **37**, 109 (1943); T. F. Gallagher, and W. P.
Long, *J. Biol. Chem.*, **147**, 131 (1943); W. M. Hoehn and J. Linsk, *J. Am. Chem. Soc.*,
67, 312 (1945); L. F. Fieser, and S. Rajagopalan, *ibid.*, **71**, 3935 (1949).

[75a] B. F. McKenzie, V. R. Mattox, L. L. Engel, and E. C. Kendall, *J. Biol. Chem.*,
173, 271 (1948).

[75b] E. Schwenk and E. Stahl, *Arch. Biochem. Biophys.*, **14**, 125 (1947).

* It may be seen from the perspective formula of cholic acid (54) that, while the
7α-hydroxyl group participates in two (1:3-H) interactions and a strong 1,3-diaxial
interaction with the 4-methylene group, the 12α-hydroxyl group is subject to only
three (1:3-H) interactions. The observed relative ease of oxidation of these groups
is therefore in agreement with expectation. It is to be noted, however, that the
7α-hydroxyl group is more readily acylated than the 12α-hydroxyl group.[72] The
explanation of this result may be that in cholic acid, although the 7α-hydroxyl
group is subjected to the greatest destabilizing non-bonded interactions, it is the
12α-hydroxyl group which, because of the conformation adopted by the side chain,
is the most difficultly accessible. This view is supported by the observation[73] that
acetylation of methyl 3α,7α,12α-trihydroxy-5β-androstan-17β-carboxylate (55)
affords chiefly the 3,12-diacetate. The relative ease of oxidation of the hydroxyl
groups is, however, the same as in cholic acid.[73]

(54)   (55)

compound with sodium bicarbonate resulted in formation of the $3\alpha,9\alpha$-oxide (48).[75a]

Bromination of (48) gave the expected diaxial dibromide (49)[75c] (cf. Sec. 5-4b) as the preponderant product. Since, of the two bromine atoms in this compound, that attached to C_{11} is the more strongly destabilized (by virtue of its two 1,3-diaxial interactions with the angular methyl groups), it is more readily solvolyzed (cf. Sec. 5-3a). Advantage of this fact was taken to introduce the 11-oxygen function which is present in cortisone and in cortisol; oxidative hydrolysis of (50) with silver chromate selectively attacked the 11β-bromine atom to afford the bromoketone (50).[75d]

The further conversion of (50) to cortisone (53; R = H) proceeded through the bromoketone (51), which was obtained by treatment of (50) with phenylmagnesium bromide (during which reaction the 12α-bromine atom was lost), dehydration of the resultant diphenylcarbinol, and cleavage of the oxide ring with hydrogen bromide (with reintroduction of the 12α-bromine).[76] The bromine was reduced out with zinc, and the C_{17} side chain was degraded by a combination of allylic bromination, dehydrobromination, acetolysis, and oxidation to afford (52).[77] Introduction of the $\Delta^{4(5)}$-double bond was achieved by a process of bromination and dehydrobromination of the corresponding 3-ketone.[78] A variety of procedures have been developed for construction of the dihydroxyacetone side chain of cortisone (53; R = H).[79] One of these consists of hydroxylation of a $\Delta^{17(20)}$-20-cyano compound, obtained from the corresponding 20-ketone by dehydration of its cyanohydrin, followed by regeneration of the 20-carbonyl group by base treatment of the initially formed cyanohydrin.[80]

Cortisone acetate (53; R = Ac) may be converted to cortisol (53; R = H; 11β-OH instead of carbonyl group) by forming the 3,20-bis-semicarbazone under conditions which do not affect the severely hindered 11-carbonyl group, and reducing the latter with sodium borohydride,

[75c] V. R. Mattox, R. B. Turner, L. L. Engel, B. F. McKenzie, W. F. McGuckin, and E. C. Kendall, *J. Biol. Chem.* **164**, 569 (1946).

[75d] R. B. Turner, V. R. Mattox, L. L. Engel, B. F. McKenzie, and E. C. Kendall, *J. Biol. Chem.*, **166**, 345 (1946).

[76] V. R. Mattox and E. C. Kendall, *J. Biol. Chem.*, **185**, 589 (1950).

[77] L. F. Fieser and M. Fieser, *Steroids*, Reinhold Publishing Corp., New York, 1959, p. 648; cf. Ch. Meystre and A. Wettstein, *Helv. Chim. Acta*, **30**, 1037, 1256 (1947).

[78] V. R. Mattox and E. C. Kendall, *J. Am. Chem. Soc.*, **72**, 2290 (1950); W. F. McGuckin and E. C. Kendall, *ibid.*, **74**, 3951 (1952).

[79] L. F. Fieser and M. Fieser, *Steroids*, Reinhold Publishing Corp., New York, 1959, pp. 652–659.

[80] J. Heer and K. Miescher, *Helv. Chim. Acta*, **34**, 359 (1951); R. Tull, R. E. Jones, S. A. Robinson, and M. Tishler, *J. Am. Chem. Soc.*, **77**, 196 (1955).

when the corresponding 11β-alcohol is produced in good yield, and can be hydrolyzed to afford cortisol.[81]

The formation of the axial 11β-alcohol rather than its more stable epimer is a consequence of the severely hindered character of the 11-carbonyl group. Hydride reduction of such a carbonyl group is subject to "steric approach control" (Chap. 2);[82,83] this implies that the major product is the one arising by attack of the reagent on the less hindered side of the carbonyl group. In the steroid series this generally results in the formation of an axial hydroxyl group.

In contrast, lithium aluminum hydride reduction of 5α-cholestan-3-one affords the corresponding equatorial alcohol in 90% yield.[84] In this reaction the carbonyl group is not significantly hindered, and the reagent can attack with equal ease from either side of the molecule. The product composition is therefore determined by "product development control" (Chap. 2),[82,83] and the equilibrium mixture of the two possible intermediate complexes is formed. Hydrolysis of this mixture results in a preponderance of the more stable equatorial alcohol.

Hydride reduction of a carbonyl group at any other position on the steroid nucleus is subject to both types of control, "steric approach control" becoming progressively more important with increasing steric hindrance of the ketonic function. In the reduction of 3β-acetoxy-5α-cholestan-7-one, for example, approach of the reagent to the β-face is slightly impeded by the 18- and 19-methyl groups, while the axial hydrogen atoms at C_5, C_9, and C_{14} seriously hinder any attack on the carbonyl group from the α side of the molecule. The influence of steric approach control is reflected in the composition of the mixture of epimeric alcohols which is obtained. This mixture contains 55% of the axial 7α-hydroxy compound,[82] as against the 20% which is present at equilibrium.[85]

The stereochemical course of the reduction of many other steroidal ketones may be similarly explained.[86] Introduction of unsaturation in

[81] R. E. Jones, and S. A. Robinson, *J. Org. Chem.*, **21**, 586 (1956); E. P. Oliveto, R. Rausser, L. Weber, E. Shapiro, D. Gould, and E. B. Hershberg, *J. Am. Chem. Soc.*, **78**, 1736 (1956); S. G. Brooks, R. M. Evans, G. F. H. Green, J. S. Hunt, A. G. Long, B. Mooney, and L. J. Wyman, *J. Chem. Soc.*, **1958**, 4614.

[82] W. G. Dauben, E. J. Blanz, Jr., J. Jiu, and R. A. Micheli, *J. Am. Chem. Soc.*, **78**, 3752 (1956).

[83] W. G. Dauben, G. J. Fonken, and D. S. Noyce, *J. Am. Chem. Soc.*, **78**, 2579 (1956).

[84] W. G. Dauben, R. A. Micheli, and J. F. Eastham, *J. Am. Chem. Soc.*, **74**, 3852 (1952).

[85] D. H. R. Barton and W. J. Rosenfelder, *J. Chem. Soc.*, **1951**, 1048.

[86] O. H. Wheeler and J. L. Mateos, *Can. J. Chem.*, **36**, 1431 (1958); D. N. Jones and G. H. R. Summers, *J. Chem. Soc.*, **1959**, 2594; J.-C. Richer and E. L. Eliel, *J. Org. Chem.*, **26**, 972 (1961); P. W. Schiess, D. M. Bailey, and W. S. Johnson, *Tetrahedron Letters*, **1963**, 549.

the vicinity of a carbonyl group often has a profound effect on the steric outcome of its reduction with hydride.[87],[88] For example, while the reduction of a 6-oxosteroid with lithium aluminum hydride generally affords the axial 6β-alcohol,[89] similar treatment of cholest-4-en-6-one gives the corresponding equatorial 6α-alcohol.[87] This unusual reaction course must be at least partly due to the flattening of rings A and B caused by the additional trigonal carbon atoms at positions 4 and 5, with consequent easier access to the β face of the carbonyl group.

A factor in hydride reduction which has yet to be fully evaluated is the role of the solvent. It has, however, been demonstrated[90] that the stereochemical outcome of a reduction can be altered by changing the solvent. One possible explanation for this is that the effective size of the attacking species might depend on its degree of solvation and might therefore vary from one solvent to another.

Recently it has been shown that a higher proportion of axial alcohol is obtained by hydride reduction of a cyclohexanone derivative when there is an electron-attracting substituent on the ring.[91] This has been explained[91] by assuming that in the transition state the carbonyl carbon atom has acquired considerable tetrahedral character, with the greatest negative charge residing on the oxygen atom. The transition state (56) for formation of an axial hydroxyl would then be of lower energy than that (57) leading to an equatorial alcohol, since the distance between the carbon atom of the $C \xrightarrow{++} X$ dipole and the oxygen atom of the incipient hydroxyl group is less in the former. Wheeler and Wheeler,[92] however,

(56)   (57)

[87] E. J. Becker and E. S. Wallis, *J. Org. Chem.*, **20**, 353 (1955).

[88] E. J. Bailey, J. Elks, J. F. Oughton, and L. Stephenson, *J. Chem. Soc.*, **1961**, 4535.

[89] C. W. Shoppee and G. H. R. Summers, *J. Chem. Soc.*, **1952**, 3361.

[90] A. H. Beckett, N. J. Harper, A. D. J. Balon, and T. H. E. Watts, *Tetrahedron*, **6**, 319 (1959); H. Haubenstock and E. L. Eliel, *J. Am. Chem. Soc.*, **84**, 2363, 2368 (1962). See also C. D. Ritchie, *Tetrahedron Letters*, **1963**, 2145; C. D. Ritchie and A. L. Pratt, *J. Am. Chem. Soc.*, **86**, 1571 (1964).

[91] M. G. Combe and H. B. Henbest, *Tetrahedron Letters*, **1961**, 404.

[92] D. M. S. Wheeler and M. M. Wheeler, *J. Org. Chem.*, **27**, 3796 (1962).

have invoked the Hammond principle[93a] to support their view that the geometry of the transition state in ketone-borohydride reductions resembles that of the starting material more closely than that of the product. Their explanation of the observed stereoselectivity is that the hydride attacks the carbonyl group on the side remote from the electronegative substituent so as to minimize electrostatic repulsion in the transition state.

In the steroid series it has been demonstrated that this effect can be transmitted across three saturated rings. The proportion of 12α(axial)-alcohol obtained when a 12-oxotigogenin is reduced with borohydride in tetrahydrofuran solution is increased from the 22% formed when position 3 is unsubstituted to 29% in the case of the corresponding 3β-chloro derivative, and to 33% in the case of the 3α-chloro derivative.[93b]

c. *cis* Additions to Double Bonds. The stereochemistry of a product obtained by *cis* addition to a double bond is determined by the fact that the reagent attacks predominantly from the less hindered side of the molecule. The steric course of hydroxylation, hydroboration, and epoxidation of a double bond is therefore controlled to a large extent by the conformation adopted by the unsaturated molecule. This is well illustrated by the results obtained when these reactions have been applied to steroids.*

In the natural steroids the axial 10β- and 13β-methyl groups seriously hinder the β face of the molecule, and there is a general tendency for reagents to attack from the α side.[96] It appears, however, that this "rule of α attack" is infringed in the hydroxylation or epoxidation of the 5,6-double bond of steroids containing the partial structure (58).[97] This is a consequence of the conformation imposed on rings A and B by the unsaturated linkage; the 10β-methyl group projects away from the double bond, so rendering it more accessible to β attack, while the axial

---

[93a] G. S. Hammond, *J. Am. Chem. Soc.*, **77,** 334 (1955).

[93b] H. B. Henbest, *Proc. Chem. Soc.*, **1963,** 159.

[94] H. I. Hadler, *Experientia*, **11,** 175 (1955); J. R. Lewis and C. W. Shoppee, *J. Chem. Soc.*, **1955,** 1365; C. W. Shoppee, B. D. Agashe, and G. H. R. Summers, *ibid.*, **1957,** 3107; M. J. T. Robinson, *Tetrahedron*, **1,** 49 (1957).

[95] E. Caspi, *J. Org. Chem.*, **24,** 669 (1959); M. C. Dart and H. B. Henbest, *J. Chem. Soc.*, **1960,** 3563.

[96] L. F. Fieser, *Experientia*, **6,** 312 (1950); T. F. Gallagher and T. H. Kritchevsky, *J. Am. Chem. Soc.*, **72,** 882 (1950).

[97] S. Bernstein, W. S. Allen, C. E. Linden, and J. Clemente, *J. Am. Chem. Soc.*, **77,** 6612 (1955); S. Bernstein and R. Littell, *J. Org. Chem.*, **26,** 3610 (1961).

* Conformational analysis has also been employed[94] to explain the stereochemistry of catalytic hydrogenation of steroids. Owing to uncertainty concerning the actual conformation adopted by a molecule adsorbed on a catalyst surface, however, the products obtained are not always those predicted by theory.[95]

(58)   (59)

3α-oxygen function is responsible for a measure of steric hindrance normally absent on the α face of a steroid of natural configuration.

A 5β-steroid containing a 2,3-double bond has also been found to react with peracids and with osmium tetroxide to afford products arising from β attack of the reagent.[98] It may be seen from perspective formula (59) that, in this case, owing to the "folding back" of ring A, the β face of the double bond is indeed the less hindered.

This folding back of ring A impedes the approach of reagents to the α side of a *cis* A/B steroid, with the result that a 9,11-double bond in such a compound is unusually inert.[99] Thus, of the pair of epimers represented by (60), only the 5α-isomer undergoes hydroboration under standard conditions.[100] Similarly, using perbenzoic acid, it is possible to epoxidize selectively the 17,20-double bond of the dienol diacetate (61); this is a key reaction in one partial synthesis of cortisone.[101] Under the same conditions, both double bonds in the corresponding 5α compound afford α-epoxides,[102] and during a recent steroid synthesis it was found necessary to protect the 9,11-double bond of (62) by converting it to the corresponding dichloride before reacting the side chain double bond with peracid.[103]

The influence of molecular conformation in determining the stereo-chemistry of *cis* addition to double bonds is strikingly illustrated by some of the reactions which have been carried out on steroids of unnatural

[98] C. Djerassi and J. Fishman, *J. Am. Chem. Soc.*, **77**, 4291 (1955).

[99] Ch. R. Engel, S. Rakhit, and W. W. Huculak, *Can. J. Chem.*, **40**, 921 (1962).

[100] W. J. Wechter, *Chem. Ind. (London)*, **1959**, 294. Cf. S. Wolfe, M. Nussim, Y. Mazur, and F. Sondheimer, *J. Org. Chem.*, **24**, 1034 (1959); F. Sondheimer and M. Nussim, *ibid.*, **26**, 630 (1961); A. Hassner, and C. Pillar, *ibid.*, **27**, 2914 (1962).

[101] T. H. Kritchevsky, D. L. Garmaise, and T. F. Gallagher, *J. Am. Chem. Soc.*, **74**, 483 (1952).

[102] D. H. R. Barton, R. M. Evans, J. C. Hamlet, P. G. Jones, and T. Walker, *J. Chem. Soc.*, **1954**, 747.

[103] J. Attenburrow, J. E. Connett, W. Graham, J. F. Oughton, A. C. Ritchie, and P. A. Wilkinson, *J. Chem. Soc.*, **1961**, 4547.

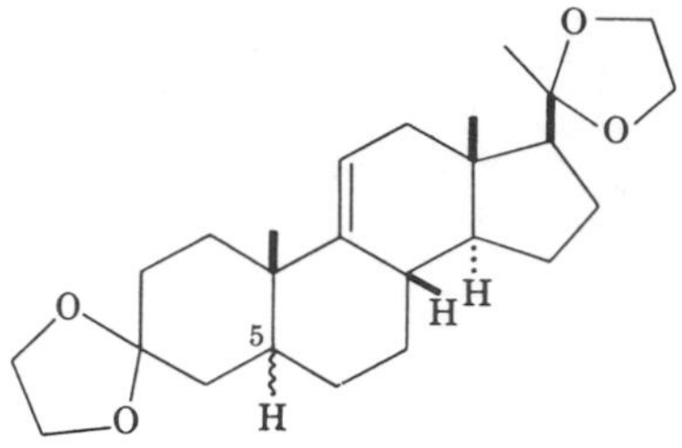

(60)

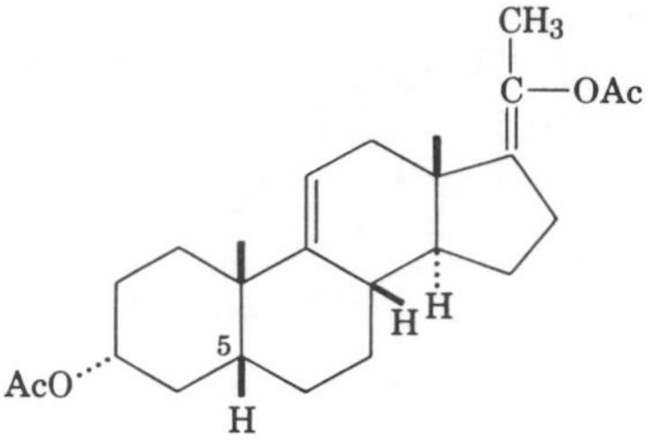

(61)

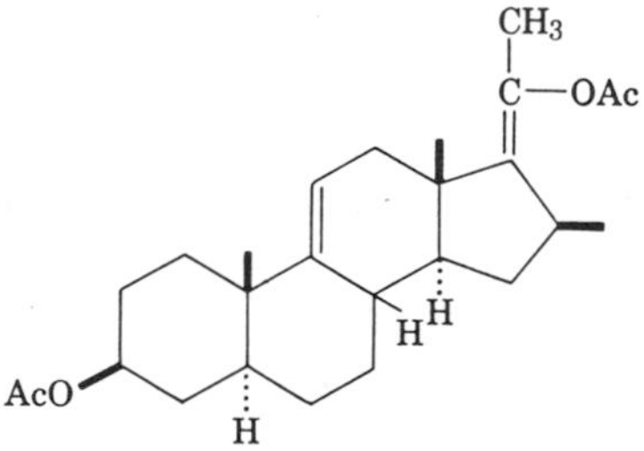

(62)

configuration. Thus, while the unsaturated steroid (63), in conformity with the rule of α attack, affords a 7α,8α-epoxide when treated with perbenzoic acid, its 9β-epimer is converted to a β-oxide under the same

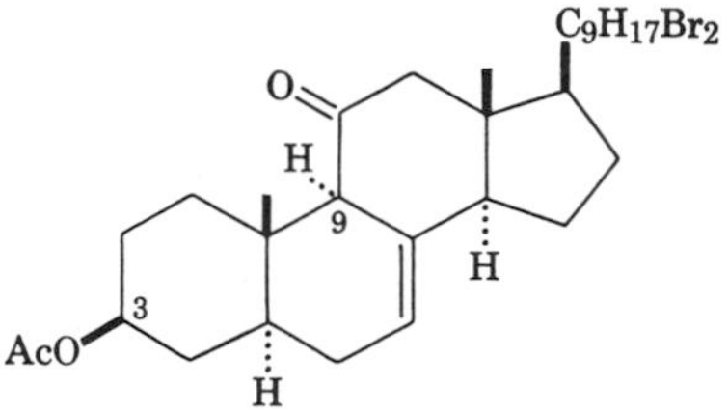

(63)

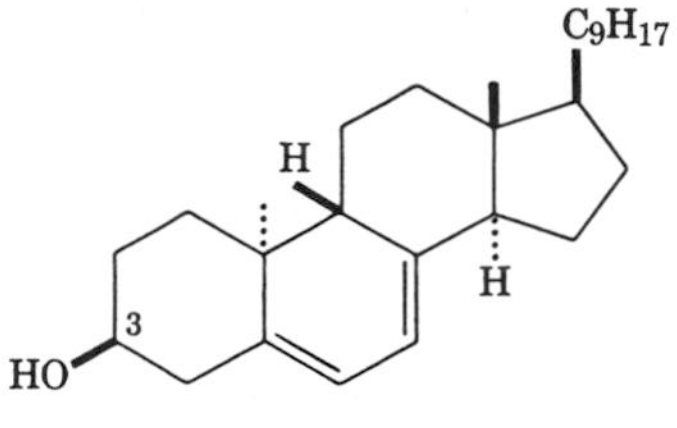

(64)

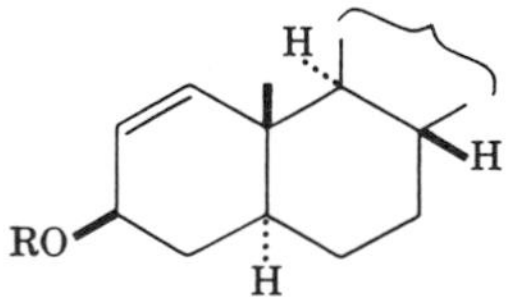

(65)

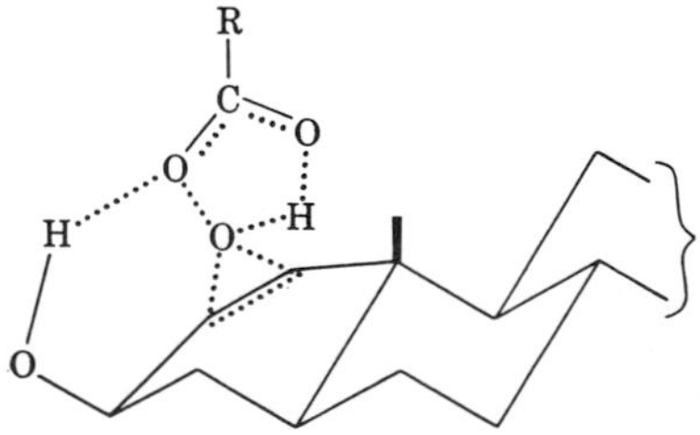

(66)

conditions.[104] The greater accessibility of the β face of the unnatural steroid is due to the fact that ring C is forced to adopt a distorted boat conformation, causing the angular methyl groups to be directed away from the double bond. The 5,6-double bonds in lumisterol (64)[105] and in its 3-epimer, epilumisterol,[106] are also attacked by perbenzoic acid principally on the β face.

It is now known that steric factors are not always the only ones which affect the stereochemistry of epoxidation of double bonds. The operation of another effect is indicated by the observation that, while oxidation of (65; R = Ac) with perbenzoic acid affords the expected $1\alpha,2\alpha$-epoxide, similar treatment of (65; R = H) results in the formation of the corresponding $1\beta,2\beta$-epoxide.[107] This *cis*-directing effect of an allylic hydroxyl group is ascribed[108] to hydrogen bonding with the reagent, as illustrated in (66). The interplay of steric and hydrogen bonding effects in the epoxidation reaction has been studied using a steroid molecule as substrate.[109]

Recent work on alicyclic olefins[93b,110] and on 17-substituted 3-oxo-Δ^4-steroids[93b] has shown that electrostatic effects can also exert some influence on the stereochemistry of epoxidation.

d. Reactions between 1,3-Diaxial Groups.

As a result of the close proximity of 1,3-diaxial groups on a cyclohexane chair, such systems provide a very favorable substrate for various intramolecular reactions. The relative ease of hydrolysis of an axial ester when there is an axial hydroxyl group in the 3-position is one example of this class (Sec. 5-3a). The base-catalyzed cleavage of lumisterol-5β,8β-epidioxide (67) under conditions to which ergosterol-5α,8α-epidioxide is stable[111] is likewise a consequence of the 1,3-diaxial disposition of the substituents on ring A in the former case (see 67a). The conformation adopted by the ergosterol derivative (68) does not permit an intramolecular reaction.

[104] J. Grigor, W. Laird, D. MacLean, G. T. Newbold, and F. S. Spring, *J. Chem. Soc.*, **1954**, 2333.

[105] P. A. Mayor and G. D. Meakins, *J. Chem. Soc.*, **1960**, 2792.

[106] G. D. Meakins and M. W. Pemberton, *J. Chem. Soc.*, **1961**, 4676.

[107] R. Albrecht and Ch. Tamm, *Helv. Chim. Acta*, **40**, 2216 (1957).

[108] H. B. Henbest and R. A. L. Wilson, *J. Chem. Soc.*, **1957**, 1958. For a recent example of the operation of this effect in the aliphatic series, see S. Sasaki, M. Ito, T. Suzukamo, and S. Fujise, *J. Chem. Soc. Japan, Pure Chem. Sect.*, **84**, 351 (1963).

[109] H. B. Henbest, B. Nicholls, W. R. Jackson, R. A. L. Wilson, N. S. Crossley, M. B. Meyers, and R. S. McElhinney, *Bull. Soc. Chim. France*, **1960**, 1365; see also A. C. Darby, H. B. Henbest, and I. McClenaghan, *Chem. Ind. (London)*, **1962**, 462.

[110] N. S. Crossley, A. C. Darby, H. B. Henbest, J. J. McCullough, B. Nicholls, and M. F. Stewart, *Tetrahedron Letters*, **1961**, 398.

[111] P. Bladon, *J. Chem. Soc.*, **1955**, 2176.

(67)

(67a)   (68)

Since about 1960 a good deal of effort has gone into devising methods for carrying out substitution reactions at saturated, unactivated carbon atoms, in a specific manner. In the steroids[112] this work has been stimulated by the desire for a simple synthesis of the rare adrenocortical hormone aldosterone (69) from steroids lacking substitution at C_{18}. Several partial syntheses of aldosterone have recently been reported[113–116] in all of which the key reaction involves intramolecular attack on C_{18} by a suitably oriented functional group. The scheme employed by Barton and Beaton,[116] in which aldosterone 21-acetate 19-oxime (71) was obtained by photolysis of corticosterone 21-acetate 11β-nitrite (70), provides an elegant example of this approach. Products arising by attack on C_{19}, which like C_{18}, is in a 1,3-diaxial relationship with the nitrite group, were also formed in the reaction.

[112] K. Schaffner, D. Arigoni, and O. Jeger, *Experientia*, **16**, 169 (1960).

[113] K. Heusler, J. Kalvoda, Ch. Meystre, P. Wieland, G. Anner, A. Wettstein, G. Cainelli, D. Arigoni, and O. Jeger, *Experientia*, **16**, 21 (1960); *Helv. Chim. Acta*, **44**, 502 (1961).

[114] L. Velluz, G. Muller, R. Bardoneschi, and A. Poittevin, *Compt. Rend.*, **250**, 725 (1960).

[115] M. E. Wolff, J. F. Kerwin, F. F. Owings, B. B. Lewis, B. Blank, A. Magnani, and V. Georgian, *J. Am. Chem. Soc.*, **82**, 4117 (1960).

[116] D. H. R. Barton and J. M. Beaton, *J. Am. Chem. Soc.*, **82**, 2641 (1960); **83**, 4083 (1961); See also M. Akhtar, D. H. R. Barton, J. M. Beaton, and A. G. Hortmann, *ibid.*, **85**, 1512 (1963).

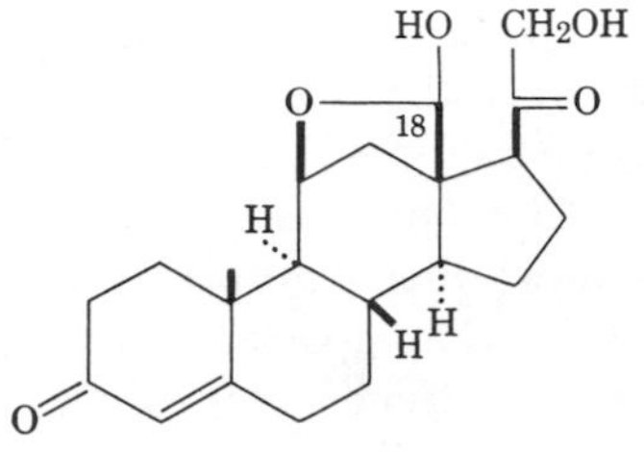

(69)

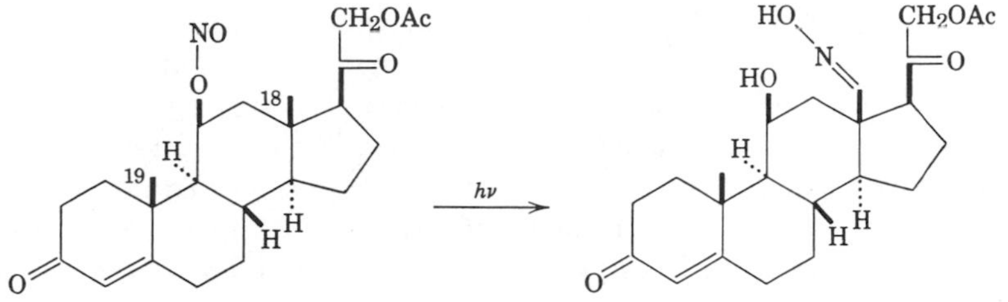

(70) $h\nu$ (71)

It has been proposed[117] that the light-catalyzed nitrite rearrangement proceeds by a free radical process involving a 1,5-hydrogen transfer. In accord with this, it has been found [118] that, while irradiation of pregn-4-en-20α-ol-3-one 20α-nitrite (72; R = NO, X = H$_2$) affords a 60% yield of the 18-oximino compound (72; R = H, X = NOH), similar treatment of the corresponding 20β-nitrite results in only a 15% yield of an 18-oximino-derivative. In the latter case the transition state is destabilized by a non-bonded interaction between C$_{12}$ and C$_{21}$ which are 1,3-diaxial substituents on the transition state chair (73; R$_1$ = H, R$_2$ = CH$_3$).

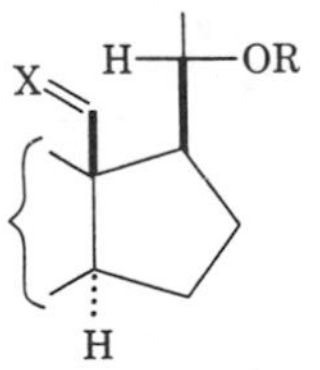

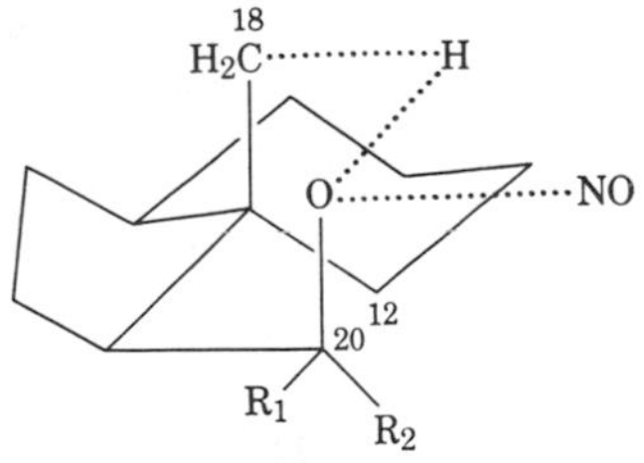

(72)   (73)

[117] D. H. R. Barton, J. M. Beaton, L. E. Geller, and M. M. Pechet, J. Am. Chem. Soc., **83**, 4076 (1961); see also D. H. R. Barton and J. M. Beaton, ibid., **84**, 199 (1962); M. Akhtar and M. M. Pechet, ibid., **86**, 265 (1964).

[118] A. L. Nussbaum, F. E. Carlon, E. P. Oliveto, E. Townley, P. Kabasakalian, and D. H. R. Barton, J. Am. Chem. Soc., **82**, 2973 (1960); Tetrahedron, **18**, 373 (1962).

There is no such unfavorable interaction in the transition state (73; $R_1 =$ CH_3, $R_2 = H$) through which the 20α-nitrite is transformed.[119]

In the lead tetraacetate oxidation of 20-hydroxysteroids to the corresponding 18,20-oxides, which must also involve a 1,5-hydrogen transfer, a parallel relationship between configuration at C_{20} and ease of reaction has been established.[114,120] The importance of the conformation adopted by the molecule in the transition state may also be discerned in the different products obtained by lead tetraacetate oxidation of various 11α-hydroxysteroids.[121] It is probable that related reactions involving chloroamines,[122] hypohalites,[123] azides,[124] ketones,[125] and diazoketones[126] proceed through transition states of similar geometry and are subject to similar steric effects. It has been demonstrated[124a] that the nitrene generated by photolysis of a carbonyl azide may rearrange by 1,6-hydrogen transfer, and so produce ultimately a δ-lactam, if the transition state for 1,5-hydrogen transfer is destabilized by severe non-bonded interactions. However, in the absence of such a constraint, 1,5-hydrogen transfer predominates, as in the other reactions mentioned, and a γ-lactam is the major lactam product.[124b]

[119] A. L. Nussbaum and C. H. Robinson, *Tetrahedron*, **17**, 35 (1962).

[120] G. Cainelli, B. Kamber, J. Keller, M. Lj. Mihailovic, D. Arigoni, and O. Jeger, *Helv. Chim. Acta*, **44**, 518 (1961).

[121] J. Kalvoda, G. Anner, D. Arigoni, K. Heusler, H. Immer, O. Jeger, M. Lj. Mihailovic, K. Schaffner, and A. Wettstein, *Helv. Chim. Acta*, **44**, 186 (1961).

[122] E. J. Corey and W. R. Hertler, *J. Am. Chem. Soc.*, **81**, 5209 (1959); **82**, 1657 (1960); P. Buchschacher, J. Kalvoda, D. Arigoni, and O. Jeger, *ibid.*, **80**, 2905 (1958); J. F. Kerwin, M. E. Wolff, F. F. Owings, B. B. Lewis, B. Blank, A. Magnani, C. Karash, and V. Georgian, *J. Org. Chem.*, **27**, 3628 (1962).

[123] M. Akhtar and D. H. R. Barton, *J. Am. Chem. Soc.*, **86**, 1528 (1964); J. S. Mills and V. Petrow, *Chem. Ind. (London)*, **1961**, 946; Ch. Meystre, K. Heusler, J. Kalvoda, P. Wieland, G. Anner, and A. Wettstein, *Experientia*, **17**, 475 (1961); *Helv. Chim. Acta*, **45**, 1317 (1962); K. Heusler, J. Kalvoda, Ch. Meystre, G. Anner, and A. Wettstein, *ibid.*, 2161; K. Heusler, J. Kalvoda, P. Wieland, G. Anner, and A. Wettstein, *ibid.*, 2575; J. Kalvoda, K. Heusler, G. Anner, and A. Wettstein, *ibid.*, **46**, 618, 1017 (1963); K. Heusler and J. Kalvoda, *ibid.*, **46**, 2020.

[124] D. H. R. Barton and L. R. Morgan, Jr., *Proc. Chem. Soc.*, **1961**, 206; *J. Chem Soc.*, **1962**, 622. But see also D. H. R. Barton and A. N. Starratt, *J. Chem. Soc.*, **1965**, 2444.

[124a] J. W. ApSimon and O. E. Edwards, *Can. J. Chem.*, **40**, 896 (1962).

[124b] W. L. Meyer and A. S. Levinson, *J. Org. Chem.*, **28**, 2859 (1963); cf. R. F. C. Brown, *Australian J. Chem.*, **17**, 47 (1964).

[125] P. Buchschacher, M. Cereghetti, H. Wehrli, K. Schaffner, and O. Jeger, *Helv. Chim. Acta*, **42**, 2122 (1959); M. Cereghetti, H. Wehrli, K. Schaffner, and O. Jeger, *ibid.*, **43**, 354 (1960); M. S. Heller, H. Wehrli, K. Schaffner, and O. Jeger, *ibid.*, **45**, 1261 (1962); N. C. Yang and D. H. Yang, *Tetrahedron Letters*, No. 4, 10 (1960); J. Iriarte, K. Schaffner, and O. Jeger, *Helv. Chim. Acta*, **46**, 1599 (1963); I. Orban, K. Schaffner, and O. Jeger, *J. Am. Chem. Soc.*, **85**, 3033 (1963); N. C. Yang, A. Morduchowitz, and D. H. Yang, *ibid.*, **85**, 1017.

[126] F. Greuter, K. Kalvoda, and O. Jeger, *Proc. Chem. Soc.*, **1958**, 349.

5–4. Reactions with Sterically Demanding Transition States

The stereoelectronic factors which influence the course of many reactions (Chap. 1) assume great importance in conformationally rigid systems, where a small change in stereochemistry may completely inhibit a chemical reaction or cause it to take a quite different course. In this section the operation of these factors in the natural product field is illustrated, mainly by reference to steroid chemistry.

a. Bimolecular Eliminations. Treatment of 5β-pregnane-$3\beta,20\alpha$-diol-3β-acetate-20α-tosylate (74) with pyridine affords the olefin (75)[127], while treatment of a tosylate of the 20β series (77)[128] with collidine yields, as the major isolable product, the olefin (78) in which the 17(20)-double bond is of opposite configuration. The configurations of the olefinic linkages in (75) and (78) were established by allowing each to react with osmium tetroxide when, in accord with the rule of α attack (Sec. 5-3c), the 20β-alcohol (76) and the 20α-alcohol (79), respectively, were obtained. The different reaction courses followed by the two epimeric tosylates is a consequence of the fact that $E2$ eliminations proceed most readily when the four participating centers can achieve an *anti* coplanar conformation in the transition state (Chaps. 1, 2).[1,129,*,†]

This condition is fulfilled for neighboring groups on a cyclohexane chair only when both are axial. In accordance with this, methyl $11\alpha,12\beta$-dibromo-$3\alpha,9\alpha$-epoxy-5β-cholanate, in which both bromine atoms are equatorial, fails to undergo debromination with iodide ion under conditions

---

[127] H. Hirschmann, *J. Biol. Chem.*, **140**, 797 (1941).

[128] L. H. Sarett, *J. Am. Chem. Soc.*, **70**, 1690 (1948).

[129] W. Klyne, *Chem. Ind.* (*London*), **1951**, 426.

[130] C. H. De Puy, R. D. Thurn, and G. F. Morris, *J. Am. Chem. Soc.*, **84**, 1314 (1962).

[130a] J. Levisalles, J. C. N. Ma, and J.-P. Pete, *Bull. Soc. Chim. France*, **1963**, 1126.

[131] D. M. Glick and H. Hirschmann, *J. Org. Chem.*, **27**, 3212 (1962).

* The recent suggestion that a concerted bimolecular elimination may also proceed easily through a *syn* coplanar transition state (Sec. 4-3)[130] has no bearing on the behavior of compounds possessing an all-chair assemblage of fused six-membered rings, and it will not be discussed in this chapter. The mechanism of the *cis* elimination observed when 4,4-dimethyl-5α-cholestan-3β-ol tosylate is treated with boiling pyridine or collidine[130a] has yet to be established. This may be a pyrolytic elimination of the type considered in Sec. 5-4e.

† It should be noted, however, that in a recent study of the pyridine-induced dehydrotosylation of the 20-epimers of 5α-pregnane-$3\beta,20$-diol-3β-acetate-20-tosylate and of $5\alpha,17\alpha$-pregnane-$3\beta,20$-diol-3β-acetate-20-tosylate, Glick and Hirschmann have shown[131] that where an *anti* coplanar conformation of 17-H, C_{17}, C_{20}, and 20-OTs is energetically unfavorable, *cis* elimination may preponderate.

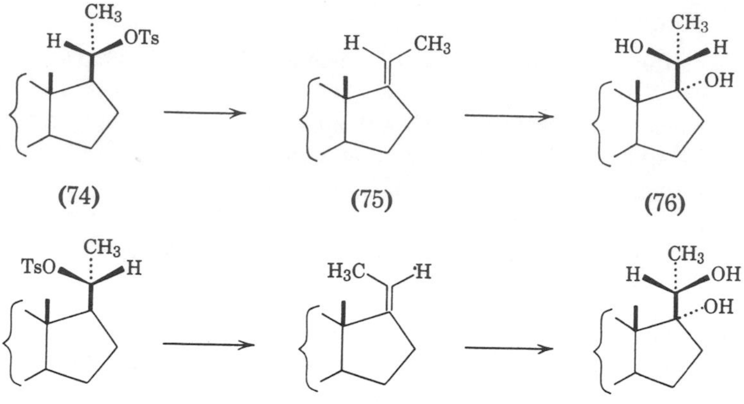

(74)   (75)   (76)

(77)   (78)   (79)

which are sufficient to eliminate bromine from the diaxial $11\beta,12\alpha$-dibromo isomer (49).[85]

The importance of a coplanar transition state is apparent also from the results obtained in the Hofmann degradation of steroidal amines[132]—equatorial quaternary ammonium groups yielding principally the corresponding tertiary amines—and in the dehydration of steroidal alcohols with phosphorus oxychloride.[133,134] For example, 3α-hydroxy-3β-methyl-5α-cholestane gives 3-methyl-5α-cholest-2-ene by *trans* diaxial elimination of a molecule of water. In contrast, the 3-epimer, in which the hydroxyl group is equatorial, affords mainly the corresponding 3-methylene compound, through the *anti* coplanar conformation (80).*

There is some evidence that reduction of an α-acetoxy- or α-hydroxy ketone with zinc proceeds more readily when the leaving group is in an axial conformation.[136] This stereospecificity is explicable on the basis of an E1cB mechanism if the intermediate carbanion adopts a conformation in which the lone pair of electrons is disposed axially: the reaction may

[132] R. D. Haworth, J. McKenna, and R. G. Powell, *J. Chem. Soc.*, **1953**, 1110; B. B. Gent and J. McKenna, *ibid.*, **1959**, 137.

[133] D. H. R. Barton, A. da S. Campos-Neves,· and R. C. Cookson, *J. Chem. Soc.*, **1956**, 3500.

[134] J. L. Beton, T. G. Halsall, E. R. H. Jones, and P. C. Phillips, *J. Chem. Soc.*, **1957**, 753.

[135] S. G. Levine and M. E. Wall, *J. Am. Chem. Soc.*, **82**, 3391 (1960).

[136] R. S. Rosenfeld and T. F. Gallagher, *J. Am. Chem. Soc.*, **77**, 4367 (1955); D. K. Fukushima, S. Dobriner, and R. S. Rosenfeld, *J. Org. Chem.*, **26**, 5025 (1961).

* The proportions of *exo*- and *endo*-cyclic olefin obtained when a tertiary alcohol is dehydrated with thionyl chloride appear to be independent of the conformation of the hydroxyl group.[135] The observed results are consistent with *cis* elimination occurring by a cyclic process.[134]

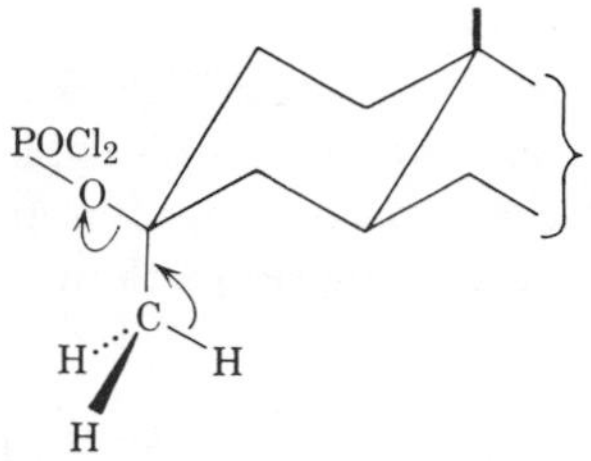

(80) 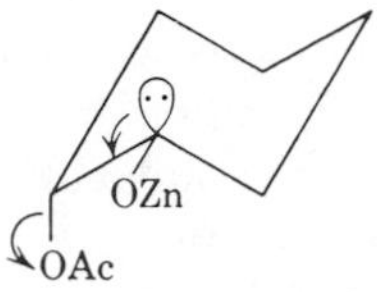   (81)

then be regarded as another example of *trans* diaxial elimination, as illustrated in (81). The observation[137] that there is a higher proportion of olefin among the products of Wolff-Kishner reduction of a steroidal α-ketol when the hydroxyl group is axial than when it is equatorial may be similarly explained. It has been suggested[138] that the reduction of a bromohydrin with zinc also involves an $ElcB$ mechanism; in the steroidal bromohydrins which have been investigated, however, the relative dispositions of the groups eliminated appear to be of only minor importance.[138]

$E2$ eliminations which result in the cleavage of a cyclohexane ring also proceed most readily when the four participating centers are *anti* coplanar: in certain cases this condition is fulfilled by the loss of an *equatorial*

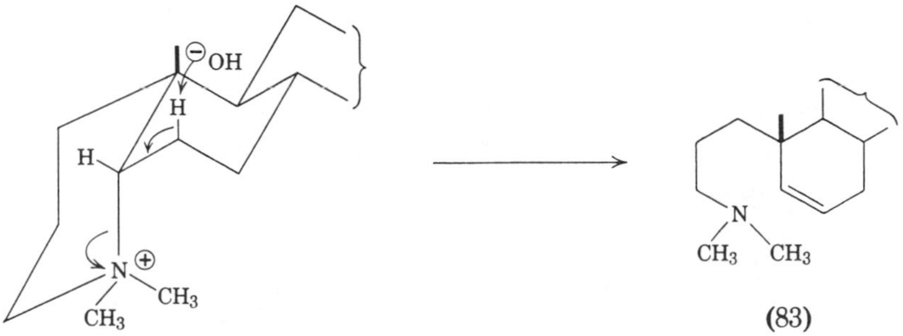

(82)   (83)

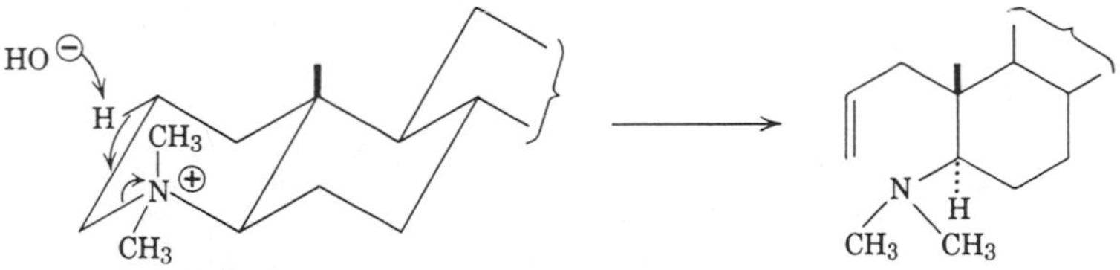

(84)   (85)

[137] R. B. Turner, R. Anliker, R. Helbling, J. Meier, and H. Heusser, *Helv. Chim. Acta*, **38**, 411 (1955).

[138] D. R. James, R. W. Rees, and C. W. Shoppee, *J. Chem. Soc.*, **1955**, 1370.

proton.[139] Thus, while Hofmann degradation of N-methyl-4-aza-5β-cholestane affords the methine (83) by the usual *trans* diaxial elimination (82), the corresponding 5α-azasteroid can achieve an *anti* coplanar transition state for the elimination only if the equatorial proton attached to C_2 is lost (84). In this case, therefore, the isomeric methine (85) is the major product.[140]

b. Halonium Ions and Epoxides. Additions to double bonds which proceed through cyclic halonium ions[141] yield *trans* diaxial products[142] (Sec. 2-5d). Thus addition of hypobromous acid to 21-acetoxypregn-4,9(11)-dien-17-ol-3,20-dione (86) affords the diaxial 9α,11β-bromohydrin (87),[143] presumably through an intermediate 9α,11α-bromonium ion formed by attack on the less hindered α face of the double bond.*

(86)   (87)

Stereospecific *trans* diaxial additions of this type are evidence that in the transition state of the reaction an *anti* coplanar arrangement of the four participating centers is preferred. It is to be noted that the formation

[139] K. Jewers and J. McKenna, *J. Chem. Soc.*, **1960**, 1575; G. H. Whitham, *ibid.*, 2016.

[140] J. McKenna and A. Tulley, *J. Chem. Soc.*, **1960**, 945.

[141] I. Roberts and G. E. Kimball, *J. Am. Chem. Soc.*, **59**, 947 (1937).

[142] G. H. Alt and D. H. R. Barton, *J. Chem. Soc.*, **1954**, 4284.

[143] J. Fried and E. F. Sabo, *J. Am. Chem. Soc.*, **79**, 1130 (1957).

[144] D. H. R. Barton and E. Miller, *J. Am. Chem. Soc.*, **72**, 1066 (1950).

[145] D. H. R. Barton, *Bull. Soc. Chim. France*, **1956**, 973.

[146] C. A. Grob and S. Winstein, *Helv. Chim. Acta*, **35**, 782 (1952); H. Kwart and L. B. Weisfeld, *J. Am. Chem. Soc.*, **78**, 635 (1956).

[147] D. H. R. Barton and J. F. King, *J. Chem. Soc.*, **1958**, 4398.

[147a] J. F. King, R. G. Pews, and R. A. Simmons, *Can. J. Chem.*, **41**, 2187 (1963).

* During bromination under equilibrating conditions the initially formed diaxial dibromide may be converted to its more stable diequatorial isomer.[144] This rearrangement is of general occurrence in the steroid field[145] and is thought to proceed through a dipolar transition state (88), without the intermediacy of free ions.[146] A whole series of such diaxial-diequatorial rearrangements is now known: results obtained in a study of the rearrangements of various 2β-halo-3α-acyloxy-5α-cholestanes may be interpreted in terms of the transition state (89).[147] Evidence has beed adduced[147a] that in the corresponding rearrangement of a diaxial bromohydrin sulfonate ester the bromine atom bears a partial positive charge in the transition state. See p. 295.

of the bromohydrin (87) from (86) conflicts with the generalized Markownikoff rule, which states that, in electrophilic additions to double bonds, the positive fragment of the reagent (in this case Br) adds to the *less* substituted carbon atom (in this case C_{11}). In general, in rigid systems the steric factors which affect the orientation of attack on a halonium ion outweigh any electronic considerations.[133]

In contrast, the addition of hydrogen halides to unsaturated steroids follows the Markownikoff rule, and there is no evidence of diaxial addition.[133] It is probable that such reactions proceed through a classical carbonium ion which is attacked by halide ion from the less hindered side of the molecule.

Steroidal *vic* diaxial halohydrins, in consequence of the *anti* coplanar arrangement of the four centers which are involved in the reaction, are much more rapidly converted to epoxides than are their diequatorial isomers.[148],* Hence, under standard basic conditions, 3α-bromo-5α-cholestan-2β-ol (90) and 2α-bromo-5α-cholestan-3β-ol (91) require reaction

(88)

(89)

[148] D. H. R. Barton, D. A. Lewis, and J. F. McGhie, *J. Chem. Soc.*, **1957**, 2907.

[149] S. Winstein and collaborators, *J. Am. Chem. Soc.*, **64** (1942), *et seq.*; W. Lwowski, *Angew. Chem.*, **70**, 483 (1958).

* Similarly, the geometrical requirement for neighboring group participation[149] in a replacement reaction is that the four participating centers should adopt an *anti* coplanar conformation; consequently, replacement of the hydroxyl group of a steroidal halohydrin with a halogen atom proceeds much faster when both functional groups are axial,[142] as, for example in the conversion of 3α-bromo-5α-cholestan-2β-ol (90) to 3α-bromo-2β-chloro-5α-cholestane by treatment with thionyl chloride, through the bromonium ion (93). Under the same conditions the diequatorial

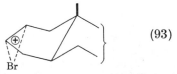

(93)

isomer fails to react. The corresponding diaxial chlorohydrin exhibits neighboring group participation only when the strongly electrophilic reagent, phosphorus pentachloride, is employed; this is consistent with the lesser nucleophilicity of the chlorine atom.

(90)   (92)   (91)

times of less than 30 seconds and 75 hours, respectively, for the conversion of each, to the extent of 70%, to 2β,3β-epoxy-5α-cholestane (92). *cis*-Halohydrins, in which, of necessity, one group must be axial and the other equatorial, undergo dehydrohalogenation with base to yield the corresponding ketone.

The geometry of an epoxide ring is similar to that of a bromonium ion; it is not surprising, therefore, that, in the great majority of cases, nucleophilic or electrophilic attack should result in predominantly diaxial cleavage.[150] The stereospecific nature of epoxide formation and cleavage has made such compounds valuable intermediates in various steroid transformations. Thus, under mild basic conditions, the diaxial bromohydrin (87) affords the 9β,11β-epoxide (94),[143] which, with hydrogen fluoride, undergoes diaxial cleavage to 9α-fluorocortisol 21-acetate (95), which finds application in medicine. The β-epoxide (94) can be obtained from (86) only by the indirect route outlined above; direct oxidation of the 9(11)-double bond of (86) with perbenzoic acid results in formation of the corresponding α-epoxide.[143] The few infringements of the rule of diaxial epoxide cleavage which have been reported in the steroid and

(94)   (95)

[150] A. Fürst and Pl. A. Plattner, Abstr. of Papers, 12th Intern. Congr. Pure Appl. Chem., p. 409, New York, 1951; A. Fürst and R. Scotoni, *Helv. Chim. Acta*, **36**, 1332, 1410 (1953); S. J. Angyal, *Chem. Ind.* (*London*), **1954**, 1230; A. S. Hallsworth and H. B. Henbest, *J. Chem. Soc.*, **1957**, 4604; see also *idem.*, *ibid.*, **1960**, 3571. An ethyleneimine ring fused to a 6-membered ring also appears to undergo acid-catalyzed diaxial cleavage; see A. Hassner and C. Heathcock, *Tetrahedron Letters*, **1963**, 393.

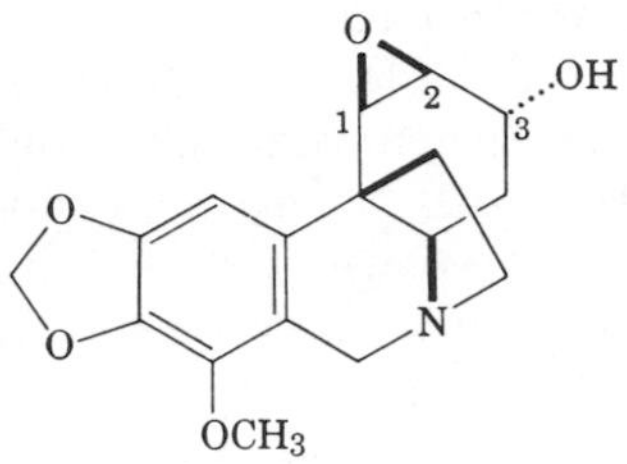

(96)

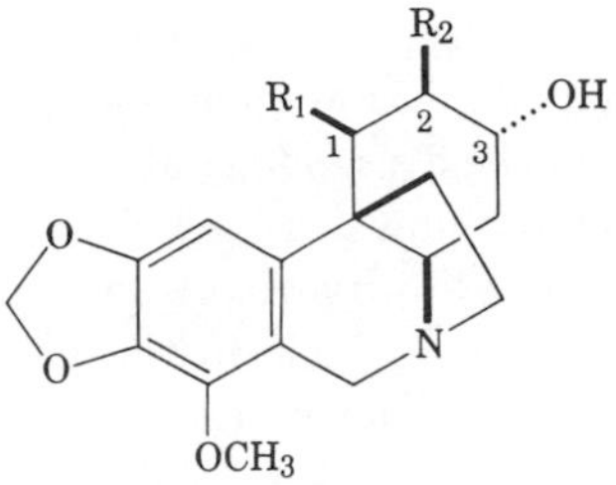

(97)

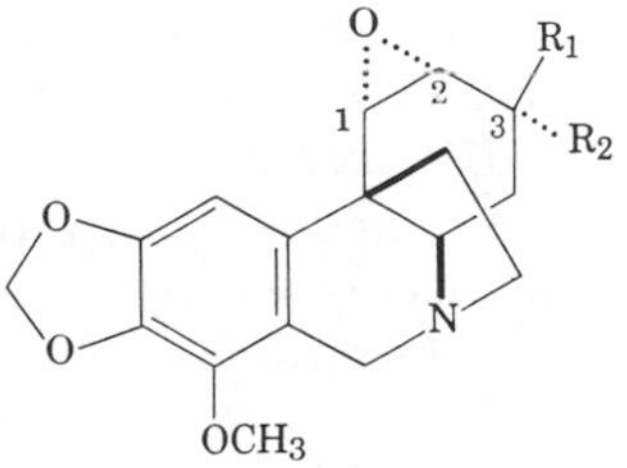

(98)

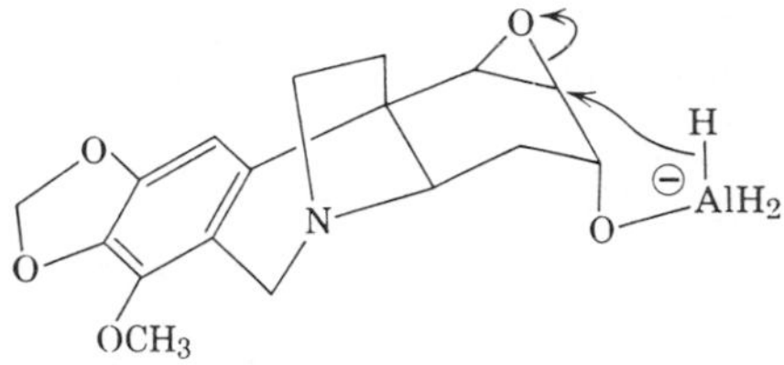

(96a)

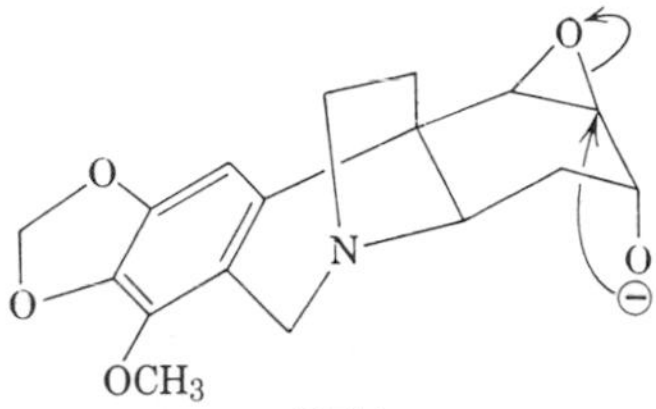

(96b)

triterpenoid literature can mostly be explained either in terms of a two-step reaction involving an intermediate benzylic[151] or allylic[152] carbonium ion, or as being due to the original epoxide existing in a preferred half-boat conformation.[148]

An interesting exception to the generalization that axial alcohols are obtained by hydride reduction of epoxides occurs in the case of the Amaryllidaceae alkaloid crinamidine (96) which is reduced with lithium aluminum hydride to the diol (97; $R_1 = OH$, $R_2 = H$), in which the 1-hydroxyl group is equatorial.[153] The reaction appears to take this sterically anomalous course only when there is a *trans* pseudoaxial hydroxyl group on a neighboring carbon atom. The normal product (97; $R_1 = H$, $R_2 = OH$), arising by diaxial reduction of the epoxide, is obtained if the 3-hydroxyl group of crinamidine is masked during the

151 R. C. Cookson and J. Hudec, *Proc. Chem. Soc.*, **1957**, 24.
152 P. A. Mayor and G. D. Meakins, *J. Chem. Soc.*, **1960**, 2792.
153 H. M. Fales and W. C. Wildman, *J. Org. Chem.*, **26**, 181 (1961).

reaction by conversion to its tetrahydropyranyl ether derivative. Similarly, the oxide rings of epoxypowelline (98; $R_1 = H$, $R_2 = OH$), in which the 3-pseudoaxial hydroxyl group is *cis* to the epoxy grouping, and *epi*-epoxypowelline (98; $R_1 = OH$, $R_2 = H$), with a *trans* pseudoequatorial 3-hydroxyl group, undergo diaxial cleavage with lithium aluminum hydride in the usual way. It is probable that the abnormal opening of the oxirane ring of crinamidine (96) is the result of intramolecular attack by the favorably oriented 3-hydroxyl group,* either in the form of the derived alkoxyaluminohydride (96a)† or as the anion (96b). The $2\alpha,3\alpha$-epoxide which would be the initial product of the process illustrated in formula (96b) would itself undergo diaxial hydride reduction to afford the diol (97; $R_1 = OH$, $R_2 = H$).

Epoxides may be isomerized to ketones by treatment with boron trifluoride etherate; this reaction has been employed[155,156] to prepare 11-oxosteroids from those lacking substitution at C_{11}. This rearrangement involves a stereospecific hydride shift[157a,157b] the transformation of (99) to (100) being a typical example. The mechanism of the reaction doubtless involves coordination of the Lewis acid with the oxide oxygen. Polarization of the bond to the more alkylated carbon atom is enhanced by the strongly electrophilic boron trifluoride, and the reaction is completed by a 1,2-hydride migration to the positive center, (101) → (102).

The isomerization is greatly affected both by conformational and by electronic factors. Thus, whereas the isomerization of a $9\beta,11\beta$-epoxysteroid (103) to the corresponding ketone (104; 9α-H) is complete in 5 minutes, the isomeric $9\alpha,11\alpha$-epoxide (105) requires a reaction time of 75 hours for its

[154] F. Dalton and G. D. Meakins, *J. Chem. Soc.*, **1961**, 1880.

[155] H. Heusser, K. Eichenberger, P. Kurath, H. R. Dällenbach, and O. Jeger, *Helv. Chim. Acta*, **34**, 2106 (1951); C. Djerassi, A. J. Lemin, G. Rosenkranz, and F. Sondheimer, *J. Chem. Soc.*, **1954**, 2346; H. H. Inhoffen and W. Mengel, *Chem. Ber.*, **87**, 146 (1954). See also W. E. Bachmann, J. P. Horwitz, and R. J. Warzynski, *J. Am. Chem. Soc.*, **75**, 3268 (1953).

[156] P. Bladon, H. B. Henbest, E. R. H. Jones, B. J. Lovell, G. W. Wood, G. F. Woods, J. Elks, R. M. Evans, D. E. Hathway, J. F. Oughton, and G. H. Thomas, *J. Chem. Soc.*, **1953**, 2921; K. Heusler and A. Wettstein, *Helv. Chim. Acta*, **36**, 398 (1953).

[157a] H. B. Henbest and T. I. Wrigley, *J. Chem. Soc.*, **1957**, 4596.

[157b] Cf. D. N. Kirk and V. Petrow, *J. Chem. Soc.*, **1960**, 4657; J. P. Dusza, J. P. Joseph, and S. Bernstein, *J. Org. Chem.*, **28**, 92 (1963).

* As may be verified with the aid of models, the pseudoequatorial hydroxyl group in *epi*-epoxypowelline (98; $R_1 = OH$, $R_2 = H$) is also suitably placed for participation in an intramolecular epoxide cleavage. In this case, however, the end product would be the same as that formed by direct diaxial reduction of the oxide ring.

† Similar oxide cleavages by intramolecular attack of alkoxyaluminohydride have been proposed in the cyclohexane[108] and steroid[154] series.

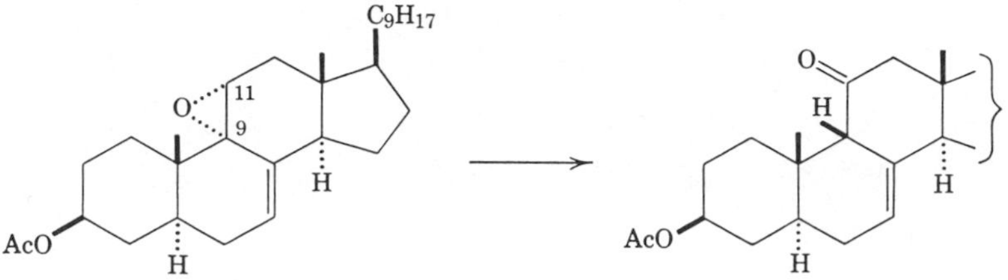

C_9H_{17}

(99)   (100)

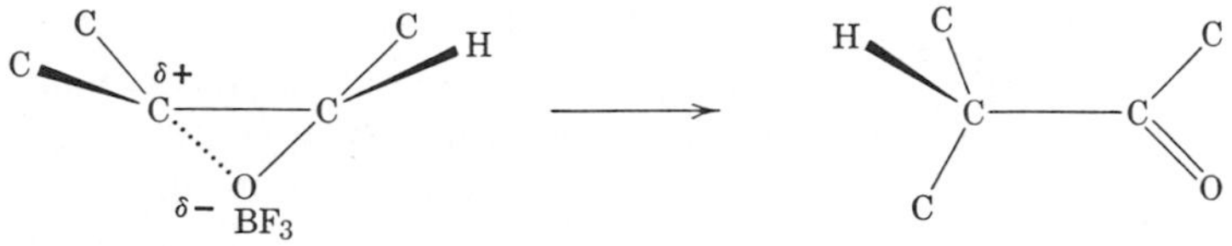

(101)   (102)

conversion to the ketone (104; 9β-H). This great difference in reactivity may be ascribed to the fact that in the latter reaction the 9β-configuration of the product necessitates a boat conformation for either ring B or ring C. In contrast, the 9α,11α-epoxide (99) is wholly transformed into the 9β-steroidal ketone (100) in less than 30 seconds,[157a] the unfavorable conformation of the product being more than compensated by the stabilization afforded to the partial positive charge developed on the tertiary carbon atom during the reaction (see 101) by resonance partici-pation of the 7(8)-double bond.

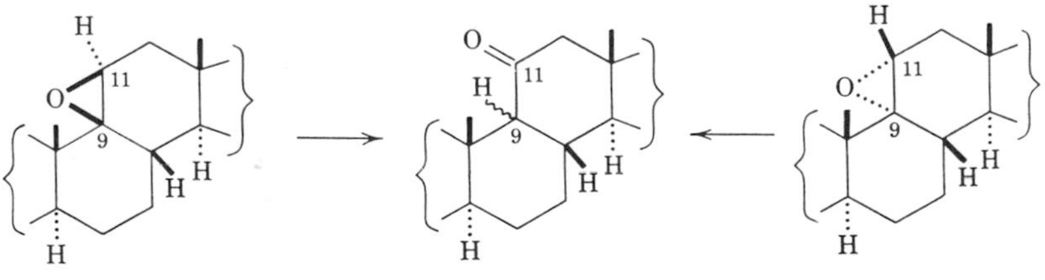

(103)   (104)   (105)

The delicate balance between conformational and electronic factors in the reaction of boron trifluoride with epoxides is further illustrated by the differing reaction courses followed by 3β-acetoxy-5,6-epoxy-5α-cholestane (106; R_1 = AcO, R_2 = H) and its 3-epimer (106; R_1 = H, R_2 = AcO). Boron trifluoride-induced ionization of the C_5—O bond of (106; R_1 = AcO, R_2 = H) is inhibited by the —I effect of the equatorial 3β-acetoxyl group;

(106)   (107)   (108)

consequently, instead of the isomeric ketone (108; R_1 = AcO, R_2 = H), the major product is the diaxial fluorohydrin (107; R_1 = AcO, R_2 = H), formed by dual attack of the Lewis acid and external fluoride ion.[158] The same electronic considerations apply to (106; R_1 = H, R_2 = AcO), but nevertheless it yields the ketone (108; R_1 = H, R_2 = AcO) as the major product.[158] In this case it appears that conformational factors predominate: the diaxial fluorohydrin (107; R_1 = H, R_2 = OAc) is, by virtue of the 1,3-diaxial interaction between the 3α-acetoxyl group and the 5α-hydroxyl group (although this is probably rendered less severe by hydrogen bonding), of higher energy than the ketone (108; R_1 = H, R_2 = OAc), with its equatorial acetoxyl group.

c. 1,2-Diaxial Rearrangements.

In cyclohexane derivatives, 1,2-migrations of the Wagner-Meerwein type occur most readily when both the migrating group and the leaving group are axial (Sec. 2-5d). The Westphalen-Lettre rearrangement, e.g., (109) → (110), is typical of this class of reaction.[159]

It has been suggested[50,160] that such rearrangements may be facilitated if they involve some relief of steric compression. It is probable that the

(109)   (110)

[158] H. B. Henbest and T. I. Wrigley, *J. Chem. Soc.*, **1957**, 4765. See also A. Bowers, L. C. Ibanez, and H. J. Ringold, *Tetrahedron*, **7**, 138 (1959); A. Bowers, E. Denot, R. Urquiza, and L. M. Sanchez-Hidalgo, *ibid.*, **8**, 116 (1960).

[159] H. Aebli, C. A. Grob, and E. Schumacher, *Helv. Chim. Acta*, **41**, 774 (1958), *et loc. cit.* See also, O. R. Rodig, P. Brown, and P. Zaffaroni, *J. Org. Chem.*, **26**, 2431 (1961); J. S. Mihina, *ibid.*, **27**, 2807 (1962).

[160] D. H. R. Barton, J. F. McGhie, M. K. Pradhan, and S. A. Knight, *J. Chem. Soc.*, **1955**, 876.

(111)   (112)

(113)

driving force for the acid-catalyzed rearrangement of α-amyrin (111) to α-amyradiene (112)[161] is at least partly conformational in origin. This reaction, which involves contraction of ring A (Sec. 5-4d), and the concerted migration of three axial methyl groups and one axial hydrogen atom (see 113), appears to be quite general for triterpenoids of similar structure in which rings D and E are locked in a *cis* fusion.[162,163] In all such cases the rearrangement relieves the 1,3-diaxial interaction between the C_{14}-methyl group and the C_{19}-methine group. In triterpenoids with a D/E *trans* fusion the corresponding interaction, between the C_{14}-methyl group and the 18α-hydrogen atom, is considerably less severe; these compounds do not undergo the rearrangement.[162] The failure of β-amyrin to undergo the amyradiene rearrangement is attributed to the fact that this compound differs from α-amyrin in that its D/E *cis* fusion is unstable under the conditions of the reaction[162] (cf. Sec. 5-3a).

d. Ring Contraction and Ring Expansion Reactions. As with the other transformations discussed above, the steric requirement for contraction of a six-membered ring is that the four centers involved should adopt an

[161] M. B. E. Fayez, J. Grigor, F. S. Spring, and R. Stevenson, *J. Chem. Soc.*, **1955**, 3378.

[162] G. G. Allan, M. B. E. Fayez, F. S. Spring, and R. Stevenson, *J. Chem. Soc.*, **1956**, 456.

[163] W. Laird, F. S. Spring, and R. Stevenson, *J. Am. Chem. Soc.*, **82**, 4108 (1960).

(114)   (115)

(116)   (117)   (118)

(119)   (120)   (121)

anti coplanar conformation. This condition is fulfilled in the familiar phosphorus pentachloride-catalyzed ring A contraction of 3β-hydroxy-triterpenoids, e.g., (114), to yield as end products isopropylidene derivatives of partial formula (117). This rearrangement is of diagnostic value, since 3α-hydroxytriterpenoids, when treated under the same conditions, yield principally products with partial structures (118) and (119). By solvolysis of the 3β-sulfonic esters of some triterpenoids and 4,4-dimethyl-steroids in aqueous acetone in the presence of calcium carbonate, Biellmann and Ourisson[164] were able to isolate tertiary alcohols with the stereochemistry shown in (120), thus demonstrating that the rearrangement proceeds with retention of configuration at C_5 and inversion at C_3. This stereochemistry might be explained by postulating a mechanism in which the initial step is formation of the non-classical carbonium ion (115), which subsequently rearranges to the tertiary carbonium ion (116). However, kinetic measurements have revealed that the reaction is not anchimerically assisted;[165-167] thus, if the non-classical carbonium ion

[164] J.-F. Biellman and G. Ourisson, *Bull. Soc. Chim. France*, **1962**, 331.

[165] C. W. Shoppee and G. A. R. Johnston, *J. Chem. Soc.*, **1961**, 3261; cf. *idem.*, *ibid.*, **1962**, 2684.

[166] G. Bancroft, Y. M. Y. Haddad, and G. H. R. Summers, *J. Chem. Soc.*, **1961**, 3295.

[167] J.-F. Biellmann and G. Ourisson, *Bull. Soc. Chim. France*, **1962**, 341.

(115) is formed, its intervention affects only the structural outcome of the reaction, and not the kinetics of the process.[167,168] An alternative explanation[165,166] of the experimental data is that the rate-determining step is the formation of the classical ion (121) which is converted to the thermodynamically more stable bridged ion (115) in a subsequent fast step; in order to account for the observed stereoselectivity it is necessary to postulate that (121) exists as an intimate ion pair.[165]

(122)   (123)

(124)   (125)

(126)   (127)

The rearrangement, (122) → (123), which occurs when a 17α-amino-17aβ-hydroxy-17aα-methyl-D-homosteroid is treated with nitrous acid[169] is another example of a ring contraction involving an *anti* coplanar conformation of the migrating bond and the departing group. The importance of this geometrical preference for four coplanar reaction centers is illustrated by the different reaction courses, (124) → (125) and (126) → (127), which are followed if the stereochemistry of ring D is

[168] Cf. S. Winstein and N. J. Holness, *J. Am. Chem. Soc.*, **77**, 3054 (1955); H. L. Goering and M. F. Sloan, *ibid.*, **83**, 1992 (1961).

[169] R. J. W. Cremlyn, D. L. Garmaise, and C. W. Shoppee, *J. Chem., Soc.*, **1953**, 1847.

altered. In each case the reaction follows the course imposed on it by the geometry of the molecule.[169],*

Many other examples of stereospecific ring contractions could be quoted.[171]

Ring expansion reactions also proceed through a four-center coplanar transition state. Where two such transition states are possible, that which has the less severe non-bonded interactions will give rise to the major product. Thus solvolysis of 5α-pregnane-$3\beta,17,20\alpha$-triol 3β-acetate 20α-tosylate (128) affords the D-homosteroid (129) by migration of the C_{13}—C_{17} bond.[172] Migration of the C_{16}—C_{17} bond cannot compete effectively,

(128)   (129)

(130)   (131)

[170] J. H. Ridd, *Quart. Rev. (London)*, **15**, 418 (1961).

[171] See, for example, J. Elks, G. H. Phillipps, D. A. H. Taylor, and L. J. Wyman, *J. Chem. Soc.*, **1954**, 1739; R. Hirschmann, C. S. Snoddy, Jr., C. F. Hiskey, and N. L. Wendler, *J. Am. Chem. Soc.*, **76**, 4013 (1954); R. Anliker, O. Rohr, and H. Heusser, *Helv. Chim. Acta*, **38**, 1171 (1955); J. M. Coxon, M. P. Hartshorn, and D. N. Kirk, *Tetrahedron Letters*, **1965**, 119; N. L. Wendler, D. Taub, and R. P. Graber, *Tetrahedron*, **7**, 173 (1959); H. Mitsuhashi and Y. Shimizu, *ibid.*, **19**, 1027 (1963); V. Georgian and N. Kundu, *ibid.*, 1037.

[172] K. I. H. Williams, M. Smulowitz, and D. K. Fukushima, *J. Org. Chem.*, **30**, 1447 (1965). See also F. Ramirez, and S. Stafiej, *J. Am. Chem. Soc.*, **78**, 644 (1956).

* It is to be noted that, although treatment of the 17a-epimer of (122) with nitrous acid affords some of the expected ring-contracted product (123), the major product of the reaction is the epoxide (127).[169] It is probable that this arises, not by a concerted displacement of nitrogen from the diazonium ion, for which the molecular geometry is unsuitable, but through an intermediate free carbonium ion. A similar mechanism has been suggested to account for certain results obtained in studies on the deamination of acyclic amines.[170]

because the transition state (130) would involve a serious non-bonded interaction between the methyl groups attached to C_{13} and to C_{20}. In the solvolytic ring expansion of the 20-epimer of (128), on the other hand, the transition state for migration of the C_{16}—C_{17} bond is less compressed than that for migration of the C_{13}—C_{17} bond: consequently, in this case the product is the D-homosteroid (131).[172]

e. Reactions Involving Coplanar Cyclic Transition States.

It has already been noted (p. 77) that less expenditure of energy is required to bring adjacent axial and equatorial bonds on a cyclohexane ring into the same plane than to bring two adjacent equatorial bonds into coplanarity. It would, therefore, be predicted that, in reactions which involve in their transition states a 1,2 fusion of an approximately planar ring, cis (e,a) compounds would be more reactive than trans (e,e) compounds, while trans (a,a) compounds would be extremely unreactive.

Among the most commonly employed of such reactions in the chemistry of natural products are oxidative cleavages of α-glycols with lead tetraacetate or periodic acid. In general, the results obtained accord well with expectation.[20,35,173–177] A few anomalies have been reported, however,[174,176–178] notably the fact that in certain 5α-steroidal 2,3-diols, the diaxial glycol reacts faster with lead tetraacetate than does its diequatorial isomer.[177] This could be due to ring A of the steroid adopting a flexible conformation under the conditions of the reaction: however, this explanation will not suffice for the observation that the diaxial trans-decalin-9,10-diol is also cleaved with lead tetraacetate (but not with periodic acid). It may be that lead tetraacetate can in certain circumstances cleave diols by a mechanism which does not involve a cyclic intermediate.[174]

Evidence for the configurations of various triterpenoid 2,3-diols[179] has been obtained by comparing their rates of oxidation with lead

[173] L. F. Fieser and S. Rajagopalan, J. Am. Chem. Soc., 71, 3938 (1949); G. Roberts, C. W. Shoppee, and R. J. Stephenson, J. Chem. Soc., 1954, 3178; R. P. Graber, C. S. Snoddy, Jr., H. B. Arnold, and N. L. Wendler, J. Org. Chem., 21, 1517 (1956).

[174] S. J. Angyal and R. J. Young, J. Am. Chem. Soc., 81, 5251 (1959).

[175] Md. E. Ali and L. N. Owen, J. Chem. Soc., 1958, 2119.

[176] R. Granger, P. F. G. Nau, and C. François, Bull. Soc. Chim. France, 1962, 1902.

[177] C. Djerassi and R. Ehrlich, J. Org. Chem., 19, 1351 (1954).

[178] S. L. Hsia, J. T. Matschiner, T. A. Mahowald, W. H. Elliott, E. A. Doisy, Jr., S. A. Thayer, and E. A. Doisy, J. Biol. Chem., 226, 667 (1957).

[179] C. Djerassi, D. B. Thomas, A. L. Livingston, and C. R. Thompson, J. Am. Chem. Soc., 79, 5292 (1957); J. Polonsky and M. J. Zylber, Bull. Soc. Chim. France, 1961, 1586; see also F. E. King, T. J. King, and J. D. White, J. Chem. Soc., 1958, 2830.

(132)   (133)

tetraacetate with those of a number of steroidal 2,3-glycols of known stereochemistry.[177,180] A striking feature of these reactions is the great enhancement of the rate of oxidation when the molecule contains a 5(6)-double bond.[177] Thus methyl bassate (132) is cleaved with lead tetraacetate some sixteen times faster than methyl medicagenate (133).[181] It is likely that this difference in rates is due at least partly to conformational transmission (Sec. 5-9).

The thermal decomposition of esters to yield olefins,[182] the decarboxylation of β,γ-unsaturated carboxylic acids,[183,184] and the pyrolysis of amine oxides[185] all take place through cyclic transition states, and in general, therefore, these reactions are cis stereospecific. However, pyrolytic trans elimination of an ester is possible if the proton which is eliminated is sufficiently acidic. Thus it has been shown that pyrolysis of

(134)   (135)

[180] C. Djerassi, T. T. Grossnickle, and L. B. High, J. Am. Chem. Soc., **78**, 3166 (1956).

[181] T. J. King and J. P. Yardley, J. Chem. Soc., **1961**, 4308.

[182] D. H. R. Barton, J. Chem. Soc., **1949**, 2174; D. H. R. Barton and W. J. Rosenfelder, ibid., 2459; ibid., **1951**, 1048; D. H. R. Barton, A. J. Head, and R. J. Williams, ibid., **1953**, 1715; R. A. Benkeser and J. J. Hazdra, J. Am. Chem. Soc., **81**, 228 (1959); D. H. Froemsdorf, C. H. Collins, G. S. Hammond, and C. H. DePuy, ibid., **81**, 643 (1959); C. H. DePuy and R. W. King, Chem. Rev., **60**, 431 (1960).

[183] D. H. R. Barton and C. J. W. Brooks, J. Chem. Soc., **1951**, 257.

[184] R. T. Arnold and M. J. Danzig, J. Am. Chem. Soc., **79**, 892 (1957).

[185] A. C. Cope and E. M. Acton, J. Am. Chem. Soc., **80**, 355 (1958).

(134) results in a *trans* elimination to afford (135) as the major product[186] (cf. Sec. 2-5d).

5–5. Reactions of Enols and Enolates

a. Bromination and Protonation of Enols.   In 1953 it was proposed by Corey that when a cyclohexanone derivative is brominated under conditions of kinetic control the major product has the bromine atom in an axial conformation.   This generalization was based on the results of spectroscopic studies[187,188] and was explained[187] as being due to the bromine atom attacking the enol predominantly in the direction best suited for maximum overlap with the π orbital of the enolic double bond, i.e., perpendicular to the plane of the double bond (136).

Similarly, it was shown by deuterium isotope studies[189,190] that, in the protonation of an unhindered enol, the entering proton assumed the axial conformation predominantly in the product.   However, it is known that in ketonization reactions steric hindrance also is important in determining the direction of attack of the incoming proton;[191] the relative proportions of axial and equatorial protonation of the Δ^6-enol derived from a 7-oxosteroid have been explained in terms of opposing stereoelectronic and steric factors.[189]

As a result of recent studies of the bromination of steroidal ring A

(136)

[186] F. G. Bordwell and P. S. Landis, *J. Am. Chem. Soc.*, **80**, 2450 (1958).

[187] E. J. Corey, *Experientia*, **9**, 329 (1953).

[188] E. J. Corey, *J. Am. Chem. Soc.*, **75**, 2301 (1953); **76**, 175 (1954).

[189] E. J. Corey and R. A. Sneen, *J. Am. Chem. Soc.*, **78**, 6269 (1956).

[190] See also J. Elks, G. H. Phillipps, T. Walker, and L. J. Wyman, *J. Chem. Soc.*, **1956**, 4330.

[191] H. E. Zimmerman, *J. Org. Chem.*, **20**, 549 (1955).

ketones and enol acetates* it has become clear that the simple theory that the major product of kinetically controlled bromination is always an axial α-bromoketone must also be modified in order to take account of the effect of steric hindrance.

When axial bromination is unopposed by any significant steric hindrance, Corey's simple hypothesis correctly predicts the configuration of the kinetically controlled product. Thus 5α-cholestan-2-one (137), or its enol acetate (138), affords the axial bromoketone, 3α-bromo-5α-cholestan-2-one (139), as the major product of kinetically controlled bromination.[194] However, if formation of an axial bromoketone is strongly hindered, the principal product of bromination under kinetic control is the equatorial epimer. This is well illustrated by the bromination of 17β-acetoxy-5α-androstan-3-one (140; R = CH_3) or its enol acetate (141). In these cases, β attack at C_2, which would lead to the formation of an axial bromoketone, is strongly hindered by the axial 10β-methyl group; consequently 2α-bromo-5α-androstan-17β-ol-3-one acetate (142), in which the bromine atom is equatorial, is formed almost exclusively.[192] Similarly, under the

(137)   (139)   (138)

(140)   (142)   (141)

[192] (a) R. Villotti, H. J. Ringold, and C. Djerassi, *J. Am. Chem. Soc.*, **82**, 5693 (1960); (b) see also E. W. Warnhoff, *J. Org. Chem.*, **28**, 887 (1963); A Lablache-Combier, J. Levisalles, J. -P. Pete, and H. Rudler, *Bull. Soc. Chim. France*, **1963**, 1689.

[193] M. P. Hartshorn and E. R. H. Jones, *J. Chem. Soc.*, **1962**, 1312.

[194] C. Djerassi and T. Nakano, *Chem. Ind.* (*London*), **1960**, 1385; T. Nakano, M. Hasegawa, and C. Djerassi, *Chem. Pharm. Bull.* (*Tokyo*), **11**, 465 (1963).

* Djerassi and his co-workers have demonstrated that bromination of the enol acetates derived from steroidal ring A ketones results in virtually the same product mixtures as are obtained from the ketones themselves.[192a] However, it has recently been shown[193] that in other systems ketones and the enol acetates derived from them may react quite differently under the same conditions of bromination.

same conditions, 3-acetoxy-5α-cholest-2-ene (141) affords almost pure 2α-bromo-5α-cholestan-3-one (142).[195] The importance of the part played by the angular methyl group is apparent from the fact that 19-nor-5α-androstan-3-one (140; R = H) gives only its 2β(a)- and 4β(a)-bromo derivatives (in yields of 63% and 37%, respectively) as products of kinetically controlled bromination. The same mixture is obtained also from the corresponding enol acetate, which is known to be a 2:1 mixture of Δ²- and Δ³-isomers.[192a]

In an important review of the stereochemistry of addition to cyclo-hexenic double bonds, Valls and Toromanoff[196] have sought to reconcile the results described above with the principle of perpendicular attack on the enolic double bond. They have pointed out that perpendicular attack, with consequent maximum orbital overlap, may involve either *parallel attack*, in which the bromine approaches the double bond from the same side as the adjacent pseudoaxial bond (see 143), or *antiparallel attack* (146) which results when the bromine approaches from the opposite side. In the latter case the product is an axial α-bromoketone, with the six-membered ring in a chair conformation (147), and in the absence of opposing steric effects this is the usual mode of attack. When antiparallel attack is inhibited by steric hindrance as in the bromination of (140; R = CH₃) and (141), parallel addition supervenes. The initial product of this

(143)   (144)   (145)

(146)   (147)

[195] C. Djerassi, N. Finch, R. C. Cookson, and C. W. Bird, *J. Am. Chem. Soc.*, **82**, 5488 (1960).

[196] J. Valls and E. Toromanoff, *Bull. Soc. Chim. France*, **1961**, 758.

(148)   (149)

reaction is a cyclohexanone boat, with the bromine in a boat-axial conformation (144); in general, the ring immediately undergoes a conformational flip to the chair conformation, and the isolated product is an equatorial α-bromoketone (145).*

Bromination of 5α-chlorocholestan-3-one (148; R = Cl) or of the Δ²-enol acetates derived from (148; R = Cl) or (148; R = Me) in which attack from either side of the molecule is seriously hindered results in formation of the corresponding 2β(a)-bromo derivative (part structure 149)[197] Presumably antiparallel attack is favored because it does not involve an energetically unfavorable boat intermediate, as would the alternative parallel attack.

Several examples of bromination are known[195,198–200] in which the boat conformation initially formed by parallel attack of bromine on an enol or enol acetate is preferred to the chair form which would arise by a subsequent conformational inversion. For example, by a combination of chemical and physical techniques it has been shown that kinetically controlled bromination of 2α-methyl-5α-androstan-17β-ol-3-one (150) or its enol acetate yields 2α-bromo-2β-methyl-5α-androstan-17β-ol-3-one (151), ring A of which exists in a preferred flexible conformation (151a).[199] Doubtless, the energy of the alternative chair conformation (151b) is greater by virtue of the unfavorable 1,3-diaxial interaction between the 2β- and 10β-methyl groups and the repulsive interaction of the $C\overset{+-}{=}O$ and $C\overset{+-}{=}Br$ dipoles.

Treatment of (151) with hydrogen bromide causes epimerization at C_2, with formation of the thermodynamically more stable 2β-bromo isomer, in which ring A adopts the chair conformation and the bromine atom is axial (152).[199]

[197] J.-C. Jacquesy and J. Levisalles, *Bull. Soc. Chim. France*, **1962**, 1866.

[198] C. Djerassi, N. Finch, and R. Mauli, *J. Am. Chem. Soc.*, **81**, 4997 (1959).

[199] R. Mauli, H. J. Ringold, and C. Djerassi, *J. Am. Chem. Soc.*, **82**, 5494 (1960).

[200] D. T. Cropp, B. B. Dewhurst, and J. S. E. Holker, *Chem. Ind. (London)*, **1961**, 209. See also Sec. 7-4.

* Formulas (143) to (147) are intended to illustrate the general case of bromination of a 6-membered cyclic ketone, when formulas (145) and (147) represent different compounds, and not—as in the case of bromocyclohexanone itself—different conformations of an enantiomeric pair.

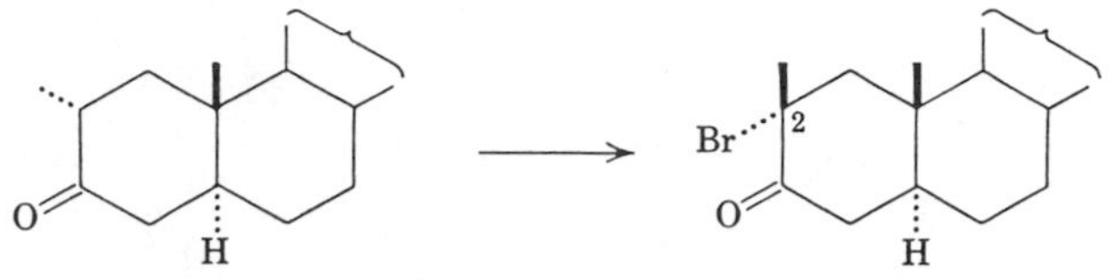

(150) (151)

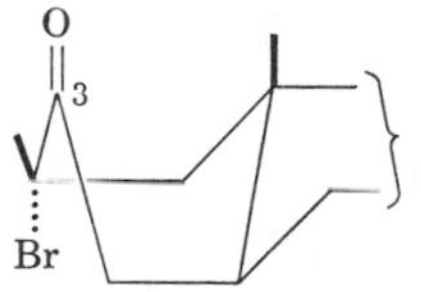

(151a) 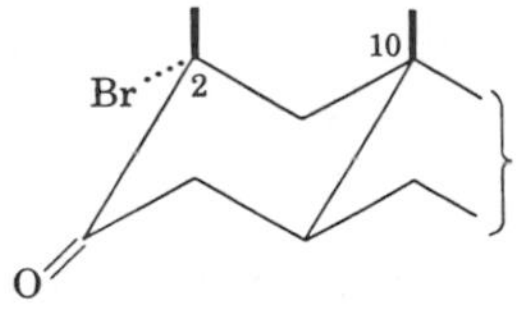 (151b) 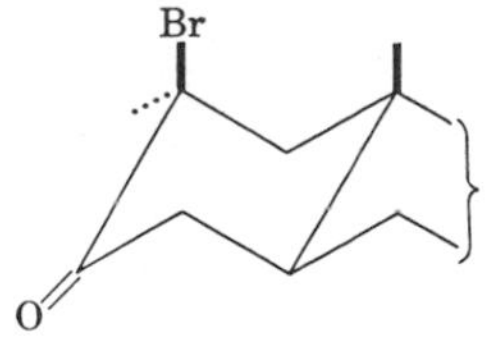 (152)

b. Alkylation of Enolates. Total syntheses of steroids often involve alkylation of an appropriate ketone as a step in a process of homo-annulation, or as a means of introducing an angular methyl group. Such reactions are usually carrried out with base catalysis, and an understanding of the factors which determine the stereochemistry of the products obtained is obviously a matter of some importance.

On the assumption that, in the transition state of an alkylation reaction, the stereochemistry of the enolate ion is very similar to that of the corresponding enol, Valls and Toromanoff[196] have suggested that, as in the bromination of enols, the methylating agent will attack either in a *parallel* or an *antiparallel* fashion, depending on the relative degrees of steric hindrance to be overcome. By analogy with the bromination reactions discussed above (Sec. 5-5a), in the alkylation of an unhindered enolate the incoming alkyl group should assume the axial conformation predominantly.[201]

Experimental support for this hypothesis is provided by the careful studies on the methylation of fused ring ketones carried out by Johnson and his colleagues.[202,203] Methylation of 2-furfurylidene *trans*-1-decalone (153; R = H) affords the *cis*- (154) and *trans*- (153; R = CH$_3$) methyl derivatives in a ratio of 3:1.[204]

The preponderance of the *cis*- isomer is explicable on the basis of an enol-type transition state (155);[203] antiparallel attack of the methylating

[201] W. S. Johnson, *Chem. Ind.* (*London*), **1956**, 167; see also J. -M. Conia and F. Rouessac, *Bull. Soc. Chim. France*, **1963**, 1930.

[202] W. S. Johnson and D. S. Allen, Jr., *J. Am. Chem. Soc.*, **79**, 1261 (1957).

[203] W. S. Johnson, D. S. Allen, Jr., R. R. Hindersinn, G. N. Sausen, and R. Pappo, *J. Am. Chem. Soc.*, **84**, 2181 (1962).

[204] W. S. Johnson, B. Bannister, and R. Pappo, *J. Am. Chem. Soc.*, **78**, 6331 (1956).

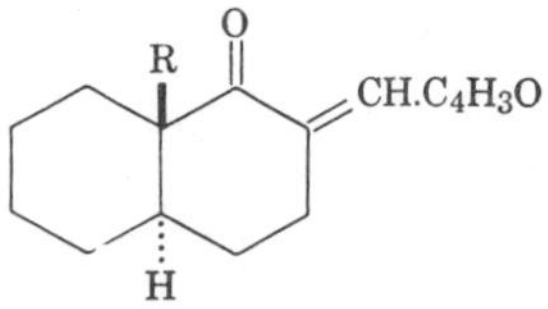

(153)

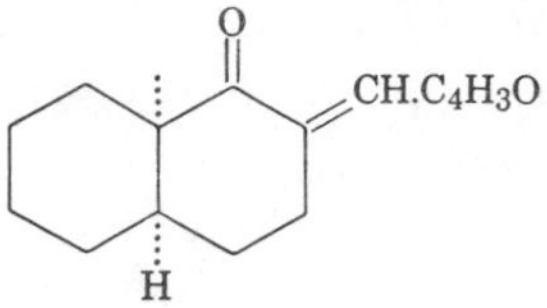

(154)

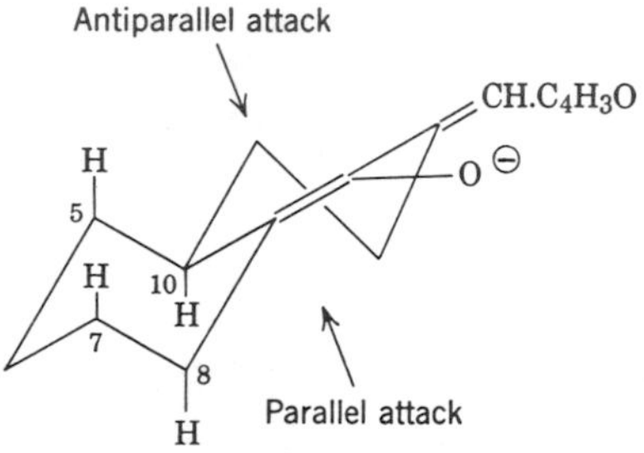

(155)

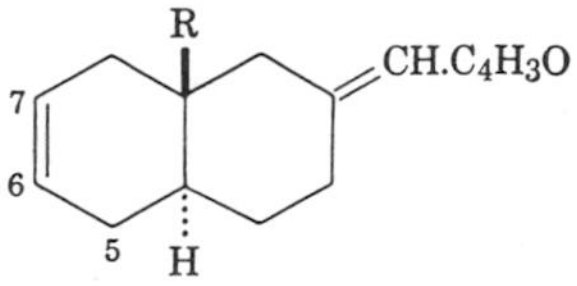

(156)

agent is subject to steric hindrance by the 5- and 7-axial hydrogen atoms situated in a 1,3-relationship to the approaching methyl group, while *cis* methylation, resulting from parallel attack, is unimpeded by any serious steric hindrance.

This interpretation of the reaction is quite consistent with the observation[203] that the *trans* fused octalone derivative (156; R = CH$_3$), isolated in 56% yield, is the major product of methylation of (156; R = H), for in the latter compound there is no axial hydrogen atom at C$_7$ to interfere with antiparallel methylation, and because of the changed geometry of the molecule the hindrance caused by the pseudoaxial hydrogen atom at C$_5$ is much less severe than in the saturated analog (153; R = H).*

The effect of a suitably placed double bond in promoting antiparallel attack has been exploited in a stereoselective total synthesis of oestrone (159).[206] Introduction of the 18-methyl group in the required configuration was achieved by methylation of the intermediate (157), which gave a 56% yield of the C/D *trans* compound (158), convertible to oestrone by

[205] R. E. Ireland and J. A. Marshall, *J. Org. Chem.*, **27**, 1620 (1962).

[206] J. E. Cole, Jr., W. S. Johnson, P. A. Robins, and J. Walker, *J. Chem. Soc.*, **1962**, 244.

* It is to be noted that the nature of the blocking group employed also appears to have some effect on the ratio of *cis* and *trans* methylated products. Thus, if the furfurylidene group of (156; R = H) is replaced by the *n*-butylthiomethylene group, methylation affords a mixture of *trans*- and *cis*-9-methyl derivatives in a ratio of 3 : 1.[205]

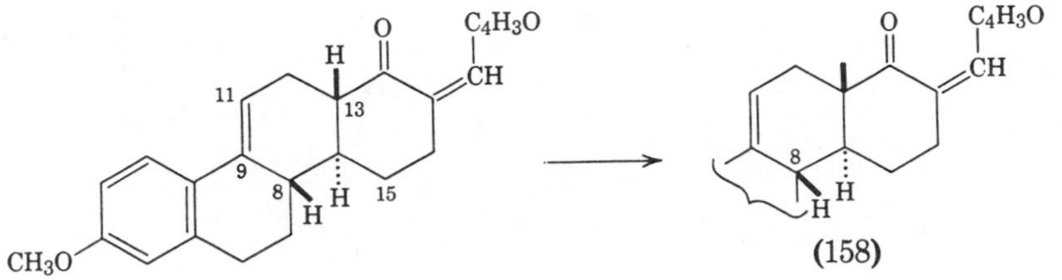

(157)

(158)

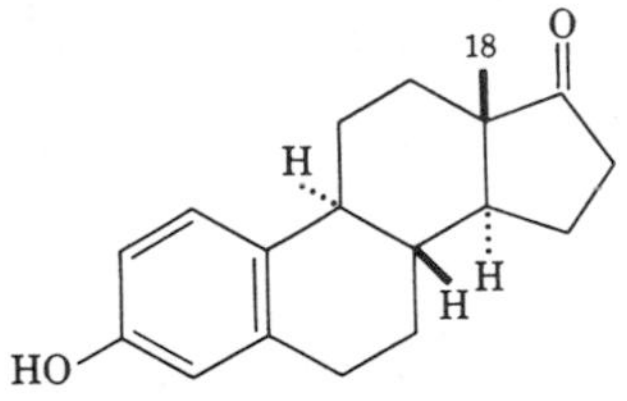

(159)

standard procedures.* Previous work had shown that, in the absence of a $\Delta^{9(11)}$-double bond, methylation affords a considerable preponderance of unwanted C/D *cis* product.[207]

The stereochemical courses of a great many alkylation reactions are explicable in the light of the concept of perpendicular attack and the effect of steric hindrance.[196,208–210] It is interesting that, while alkylation of the enolate of the tricyclic ketone (160) with ethyl bromoacetate affords (161) as the major product[210] (antiparallel attack being inhibited because of the steric hindrance of the angular methyl group), methylation of (162) leads to the formation of (163) in very good yield.[211] It has been suggested[210] that in the latter case the major product arises by antiparallel attack, despite the hindrance of the angular methyl group, because

[207] W. S. Johnson, I. A. David, H. C. Dehm, R. J. Highet, E. W. Warnhoff, W. D. Wood, and E. T. Jones, *J. Am. Chem. Soc.*, **80**, 661 (1958).

[208] G. I. Poos, G. E. Arth, R. E. Beyler, and L. H. Sarett, *J. Am. Chem. Soc.*, **75**, 422 (1953); L. H. Sarett, W. F. Johns, R. E. Beyler, R. M. Lukes, G. I. Poos, and G. E. Arth, *ibid.*, 2112; J. H. Fried, G. E. Arth, and L. H. Sarett, *ibid.*, **82**, 1684 (1960).

[209] G. Stork, P. Rosen, and N. L. Goldman, *J. Am. Chem. Soc.*, **83**, 2965 (1961).

[210] G. Stork and J. W. Schulenberg, *J. Am. Chem. Soc.*, **84**, 284 (1962).

[211] M. E. Kuehne, *J. Am. Chem. Soc.*, **83**, 1492 (1961).

* It might have been expected that methylation of the Δ^8-isomer of (157) would afford the Δ^8-isomer of (158) as the major product, since in this case the absence of the axial hydrogen atom at C_8 should facilitate antiparallel attack on the derived enolate. In fact, a high yield of the corresponding C/D *cis* compound is obtained.[206] The reason for this anomaly is not clear: it might, however, be due to the $15\beta(a)$-hydrogen atom being brought into closer proximity with C_{13}, as a result of the introduction of a Δ^8-double bond, and so effectively blocking alkylation on that side of the molecule.[206]

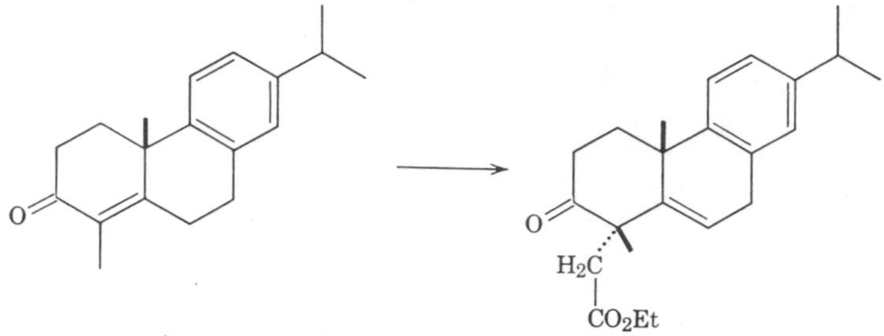

(160)

(161)

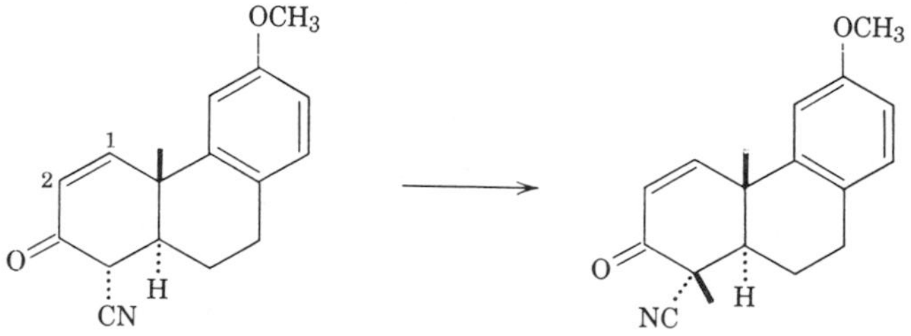

(162)

(163)

parallel attack would lead to an intermediate half-boat conformation for ring A (see Sec. 5-5a) in which there would be incomplete overlap of the orbitals of the carbonyl group and the C_1—C_2 double bond.

5–6. Reactions of $\alpha\beta$-Unsaturated Ketones

a. Conjugate Addition. During the course of early work on the synthesis of aldosterone from steroids lacking substitution at C_{18}, it was found that when (164) was caused to undergo an intramolecular Michael reaction the product, after acetylation, was 3β-acetoxy-5α,13α,17α-pregnane-11,20-dione (165).[212] The establishment of the unnatural configuration at C_{13} is best explained as a consequence of perpendicular attack on the eneone system by the enolate anion derived from the side chain carbonyl group. The formation of the 13-epimer of (165) would require equatorial attack by this anion.

Similarly, the formation of the tricyclic ketone (167) by base treatment of (166)[213] may be interpreted as involving in the initial step an intramolecular Michael addition proceeding by perpendicular attack on

[212] D. H. R. Barton, A. da S. Campos-Neves, and A. I. Scott, J. Chem. Soc., 1957, 2698.

[213] W. S. Johnson, S. Shulman, K. L. Williamson, and R. Pappo, J. Org. Chem., 27, 2015 (1962).

the eneone system by the enolate derived from the methyl ketone (see 166a).

Recently, the hypothesis has been put forward that in the addition of anions to cyclohexenone derivatives, perpendicular attack, leading initially to axially substituted compounds, is a general phenomenon.[214] According to this theory, the positively charged component of the attacking species (e.g., the metal cation associated with an enolate anion) promotes polarization of the eneone system, and the negative component provides a pair of electrons to fill the vacant orbital which consequently develops at the β-carbon atom. The process, which is probably concerted, is indicated schematically in (168). It follows that the anion approaches the β-carbon atom at right angles to the plane defined by the coordinates of the atoms which form the eneone system.

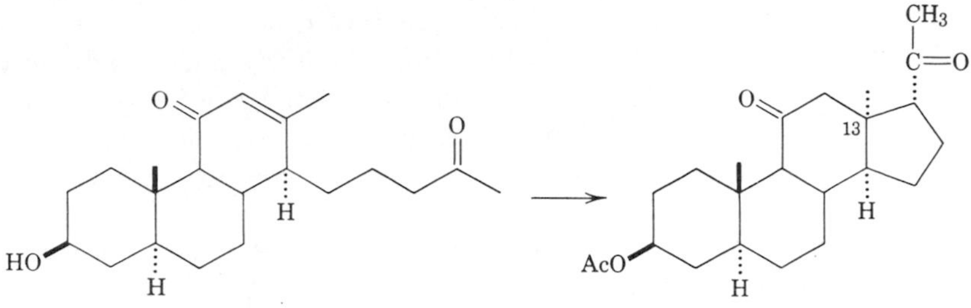

(164)

(165)

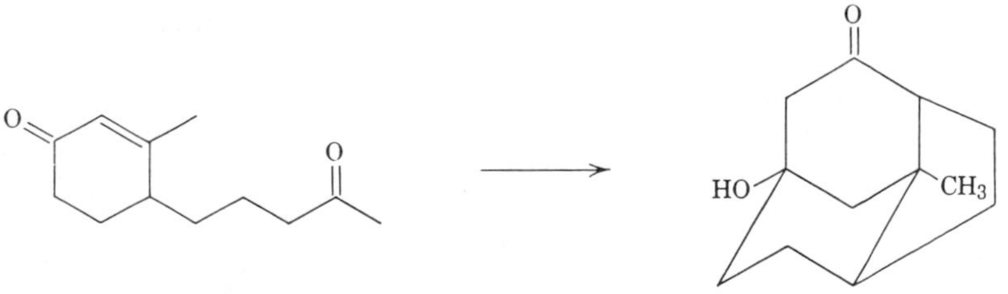

(166)

(167)

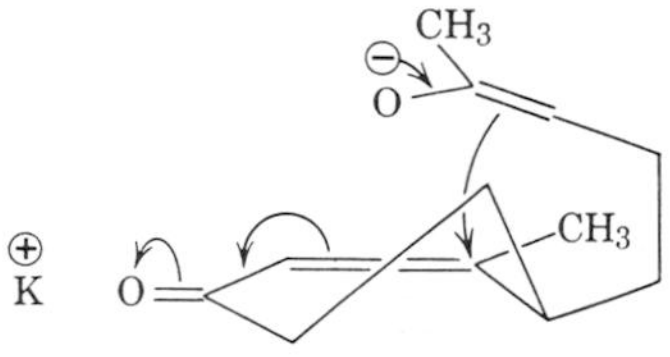

(166a)

[214] E. Toromanoff, *Bull. Soc. Chim. France*, **1962,** 708; see also H. O. House and H. W. Thompson, *J. Org. Chem.*, **28,** 360 (1963).

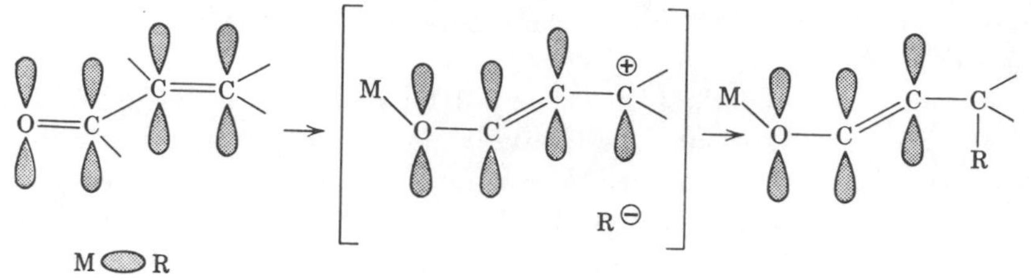

(168)

In general, the anion may approach from either side of the molecule, and the major product will arise by attack on the less hindered face. However, in intramolecular Michael reactions, such as those illustrated above, the configuration of the molecule may be such that perpendicular attack is possible on only one side of the conjugated system.

The concept of perpendicular attack from the less hindered side of a conjugated eneone (or dieneone) accounts in a satisfactory way for the steric course of 1,4- or 1,6-addition of Grignard reagents. Thus treatment of (169) with ethylmagnesium iodide affords as major product the *cis* C/D compound (170).[214] In this case β approach, which would have afforded the corresponding *trans* C/D compound, is inhibited by the 1,3-diaxial interactions with the β-hydrogen atoms at C_9, C_{11}, and C_{15} to which the incoming ethyl group would be subjected.

Anomalous results may sometimes be obtained in 1,4- or 1,6-additions of Grignard reagents when there is a suitably disposed hydroxyl or amino group, which, by complexing with the organomagnesium reagent, might

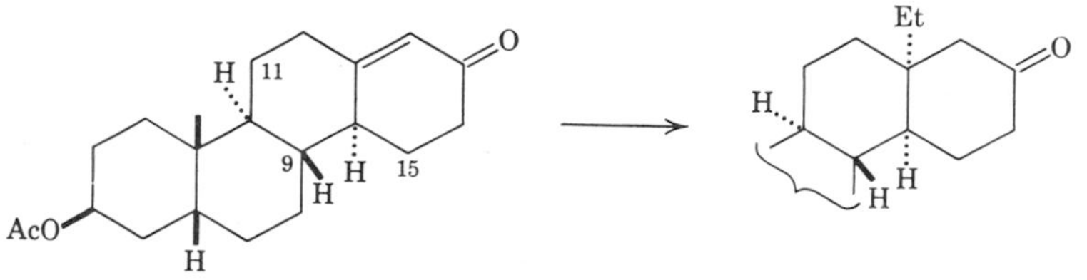

(169)

(170)

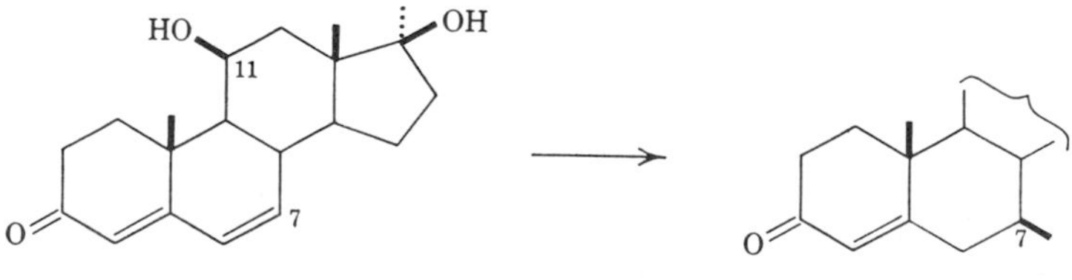

(171)

(172)

direct attack to one side or the other of the unsaturated system. It has been suggested[215] that the formation of the 7β-methyl derivative (172) as the principal product of the reaction between (171) and methyl-magnesium bromide is a consequence of such an effect, involving the 11β-hydroxyl group. In the absence of the hydroxyl group the major product has a $7\alpha(a)$-methyl group.

When the product of a Michael addition is a 1,5-dicarbonyl compound or a β-cyanoketone, the reaction is, of course, reversible and under the experimental conditions it is possible that equilibration might take place, resulting in the isolation of the thermodynamically more stable equatorial epimer.[214] The same result would also arise if the initially formed axial epimer were to undergo a bimolecular nucleophilic displacement.

b. Metal-Ammonia Reduction. As a result of a study of the stereo-chemical course of the reduction of various multiple bond systems, Barton and Robinson proposed that protonation of a carbanion always affords the more stable epimer.[216] Thus, for example, lithium-ammonia reduction of a 6-membered cyclic ketone in general yields an equatorial hydroxyl group,* and similar treatment of an oxime gives an equatorial amino group. It was suggested[133] that these data could most easily be explained if a carbanion possessed a definite, though readily inverted, tetrahedral configuration, with an electron pair having steric requirements intermediate between those of a carbon-hydrogen bond and of a carbon-carbon bond.

215 J. A. Campbell and J. C. Babcock, *J. Am. Chem. Soc.*, **81,** 4069 (1959).

216 D. H. R. Barton and C. H. Robinson, *J. Chem. Soc.*, **1954,** 3045.

217 J. W. Huffman, D. M. Alabran, and T. W. Bethea, *J. Org. Chem.*, **27,** 3381 (1962); see also M. Alauddin and M. Martin-Smith, *ibid.*, **28,** 886 (1963).

218 G. Ourisson and A. Rassat, *Tetrahedron Letters*, No. 21, 16 (1960).

219 D. H. R. Barton and G. A. Morrison, *Fortschr. Chem. Org. Naturstoffe*, **19,** 165 (1961).

* More recent work has revealed a number of exceptions to this rule (see Sec. 2-6c). For example, 12-oxo-5β-cholanic acid gives the corresponding $12\alpha(a)$-hydroxy compound as the major product upon reduction with lithium and liquid ammonia.[217] Since hecogenin, a sapogenin of the 5α series, gives rockogenin, the corresponding $12\beta(e)$-hydroxy compound, as the only isolable product under the same conditions, it has been suggested that the obtainment of the thermodynamically less stable alcohol from 12-oxo-5β-cholanic acid might result from shielding of the reaction site due to the *cis* A/B junction and to the side chain.

It has also been established that the steric courses of various reductions with dissolving metals are to some extent dependent on the nature of the metals employed (see Sec. 2-6). Thus, in the alkali metal-ammonia reduction of camphor, the pro-portion of the less stable product, isoborneol, increases with increasing size of the metal atoms.[218] A possible explanation for this is that these reactions involve intermediates containing carbon-metal bonds,[219] and that the metal atoms undergo electrophilic substitution by protons, with retention of configuration.

If the reduction of an $\alpha\beta$-unsaturated ketone by a metal-ammonia solution is admitted to involve an intermediate of type (173),[216,220] then,

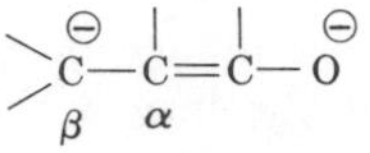

$$\overset{\ominus}{\underset{\beta}{C}}-\overset{|}{\underset{\alpha}{C}}=\overset{|}{C}-\overset{\ominus}{O}$$

(173)

on the above view, subsequent protonation should afford the more stable configuration at the β-carbon atom. In the great majority of cases this is found to be true, and the reduction of $\alpha\beta$-unsaturated ketones and other conjugated systems with dissolving metals has been widely employed as a means of obtaining the reduced material in its most stable configuration. A good example of this technique is provided by the stereospecific conversion of the tetracyclic ketone (174) to (175) in the total synthesis of dl-epiandrosterone.[204] Reduction of (174) with lithium and alcohol in ammonia gave a mixture of (175) and (176). The latter was further reduced by catalytic hydrogenation, and the crude dihydro product initially formed was treated with alkali to cause epimerization at C_{13} with production of (175). The two-step reduction sequence therefore produced six centers of asymmetry in a stereospecific fashion.

It is important to note that, assuming (173) to be an intermediate in the metal-ammonia reduction of conjugated ketones, the process would be expected to ensure the more stable configuration only at the β-carbon

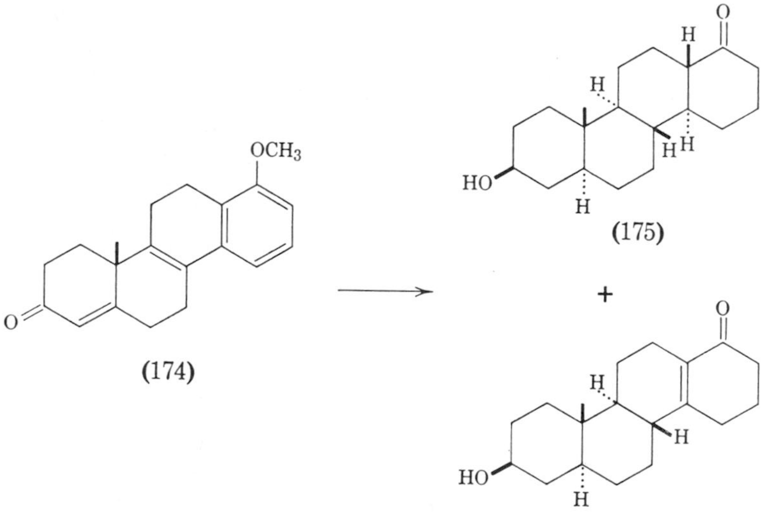

(174)

(175)

+

(176)

[220] A. J. Birch and H. Smith, Quart. Rev. (London), 12, 17 (1958).

(177)   (178)   (179)

atom, that at the α-position being determined by the stereochemistry of
ketonization of the derived enol and consequently subject to the usual
steric and stereoelectronic control (Sec. 5-5a). It is reasonable, therefore,
that by employing non-epimerizing conditions during the working-up
procedure, it is sometimes possible to obtain the less stable configuration
at the α-position. Thus (178) may be prepared by reduction of (177) with
lithium and alcohol in ammonia; subsequent acid-catalyzed isomerization
converts the product to the more stable *trans* fused compound.[221]

Recently, on the basis of studies of the metal-ammonia reduction of
certain conjugated octalones of type (179), it has been proposed that the
steric course of the reaction is controlled by the necessity for the incoming
proton to affix itself axially at the β-position of the enone system, so as
to ensure maximum orbital overlap in the transition state[222] (see Sec. 2-6).
According to this hypothesis, the major product should be the more stable
isomer (*cis* or *trans*) arising by perpendicular attack on the enolate anion.

Although the theory of axial protonation accounts in a satisfactory
manner for results obtained in the octalone series,[222] its applicability to
more complex systems remains to be established.[219] Thus lithium-ammonia
reduction of (180) followed by re-acetylation affords the *cis* B/C steroid
(181) as the major product.[223] If an all-chair conformation is assumed for
the molecule, the newly introduced hydrogen must occupy an equatorial
conformation with respect to ring B. For this to arise by axial protonation
it is necessary to postulate an intermediate (182) in which both rings B
and C adopt a flexible conformation. Axial protonation leading to the
8α-epimer of (181) would, on the other hand, involve a boat conformation
for ring C only (183). The fact that, despite this, protonation on the α face
of the molecule to give (183) is at most a minor process might be an
indication that steric hindrance has a role to play, since the incoming
proton would enter into a 1,3-diaxial interaction with the 10α-methyl

[221] A. J. Birch, H. Smith, and R. E. Thornton, *J. Chem. Soc.*, **1957**, 1339.
[222] G. Stork and S. D. Darling, *J. Am. Chem. Soc.*, **82,** 1512 (1960).
[222a] G. Stork and S. D. Darling, *J. Am. Chem. Soc.*, **86,** 1761 (1964).
[223] P. A. Mayor and G. D. Meakins, *J. Chem. Soc.*, **1960,** 2800.

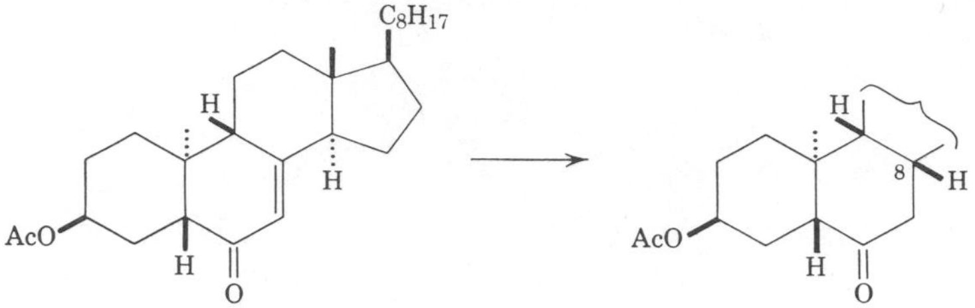

(180)　　　　　　　　　　　　　(181)

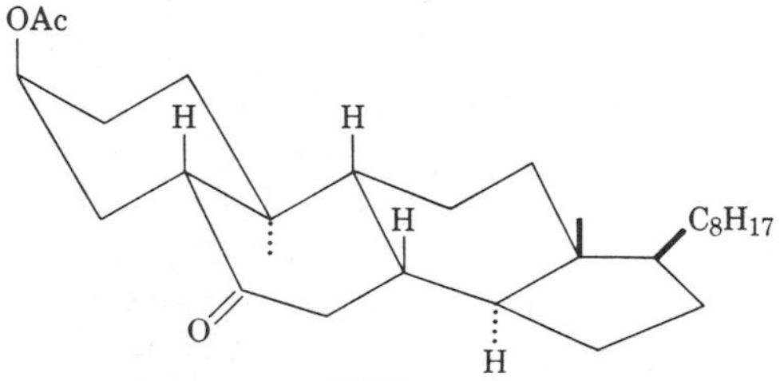

(182)

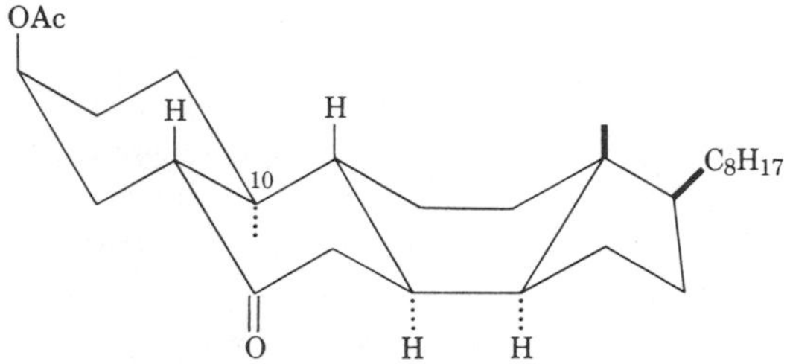

(183)

group. Alternatively, since (183) cannot undergo a conformational inversion to an all-chair conformation, it might be argued that the obtainment of (181) as the major product is due to the protonation of an intermediate carbanion in its most stable conformation (i.e., with the electron pair equatorially situated). (See, however, ref. 222a.)

5–7. Configurations of the Triterpenoids

Among the earliest and most important applications of conformational analysis in structural organic chemistry was its employment as an aid in elucidating the stereochemistry of the triterpenoids.

In this field the determination of the stereochemistry of the oleanane group of triterpenoids, all of which have structures based on the parent

(184)   (185)   (186)

hydrocarbon oleanane (184), was an outstanding achievement.[24] The *trans* fusion of rings A and B followed from the smooth phosphorus pentachloride-catalyzed ring A contraction of various 3β-hydroxy derivatives (see Sec. 5-5d); and from the fact that acid-catalyzed cleavage of the lactone (185) to give oleanolic acid (186) proceeded more rapidly than the alternative reaction to give the thermodynamically more stable[224] $\Delta^{13(18)}$-isomer of (186) it was concluded that the 18-hydrogen atom and the 13-oxygen function of (185) bore a *cis* relationship and therefore were not favorably disposed for ionic elimination (cf. Sec. 5-4a). Consequently, a *cis* fusion of rings D and E was assigned to the oleanane group.

The configuration at C_{13} of the oleanane group was related to that at C_{17} by conversion of siaresinolic acid (187) to (189) through a reaction

(187)   (188)   (189)

(190)

[224] G. S. Davy, T. G. Halsall, and E. R. H. Jones, *J. Chem. Soc.*, **1951**, 458.

sequence involving pyrolysis of the $\beta\gamma$-unsaturated carboxylic acid (188).[183] From the mode of formation of (189) it followed that its C_{13}-hydrogen atom was situated on the same side of the ring system as the original 17-carboxyl group of siaresinolic acid (cf. Sec. 5-4e). Furthermore, since it was known[183] that the conjugated ketone (189) possessed the same configuration at C_{13} as the naturally occurring oleanane triterpenoids, it followed that throughout the group the 13-hydrogen atom and the 17-substituent were *cis* to each other.

Methyl 12-oxo-oleananolate acetate (190; $R_1 = Ac$, $R_2 = R_3 = H$, $R_4 = CO_2CH_3$, $X = O$) is not epimerized by treatment with base. Since this implies that the 13-hydrogen atom is in the thermodynamically stable configuration, it was concluded that oleanane and its derivatives must possess a *trans* C/D ring junction. If it were *cis*, rings C, D, and E would constitute a *cis-syn-cis*-perhydrophenanthrene system, and the 13-epimer ought to be thermodynamically more stable[225] (cf. Sec. 4-5a).

Reduction of methyl 11-oxo-oleananolate acetate (190; $R_1 = Ac$, $R_2R_3 = O$, $R_4 = CO_2CH_3$, $X = H_2$) with lithium aluminum hydride, followed by reacetylation, affords 3,28-diacetoxyoleanan-11β-ol (190; $R_1 = Ac$, $R_2 = OH$, $R_3 = H$, $R_4 = CH_2OAc$, $X = H_2$). The hydroxyl group in this compound was assigned an axial conformation (from which it followed, on the basis of an all-chair arrangement of the molecule, that it was situated on the same side of the molecule as the 13-hydrogen atom), since it was formed by hydride reduction of a hindered ketone (cf. Sec. 5-3b) and since it was resistant to acetylation (cf. Sec. 5-3a). This assignment was supported by the observation that the alcohol was easily dehydrated with phosphorus oxychloride and pyridine to yield the corresponding $\Delta^{9(11)}$-compound. The ease with which the dehydration reaction occurred indicated an axial conformation for the 9-hydrogen atom also, and on the assumption of an all-chair conformation, this implied that it was *trans* to the 13-hydrogen atom and *cis* to the 14-methyl group. In agreement with these conclusions the configuration[226] at C_9 in the oleanane series, as well as that at C_{13} (see above), is the thermodynamically more stable one. These data require[3] that the B-C-D ring system must be based either on *trans-anti-trans*-perhydrophenanthrene or on *cis-syn-trans*-perhydrophenanthrene (Sec. 4-5a).[225]

Although β-amyrin was known to have rings A and B *trans* fused, the conformational arguments did not relate the configurations at C_5 and C_{10} to the stereochemistry of the remainder of the molecule. However, since ring C could be linked to ring B only through two equatorial bonds or

[225] W. S. Johnson, *Experientia*, **7**, 315 (1951).

[226] R. Budziarek, J. D. Johnston, W. Manson, and F. S. Spring, *J. Chem. Soc.*, **1951**, 3019.

(191)

through an equatorial and an axial bond, and since it had been demonstrated that an axial 11-hydroxyl group bore a *trans* relationship to the 9-hydrogen atom, only two possible representations of the relative stereochemistry of oleanane remained for consideration. These were (184) and (191).

Employing generalized molecular rotation arguments, Klyne[227] was able to reach a decision in favor of (184), which was also shown to represent the correct *absolute* stereochemistry of oleanane. Subsequently, structure (184) (excluding the configuration at C_{13}) was confirmed by the results of an X-ray investigation of methyl oleanolate iodoacetate.[65]

Much of the work carried out in the oleanane series was highly relevant to the configurational problem posed by the lupane group of triterpenoids. Lupeol (192) is the principal member of this group which is named after the corresponding saturated hydrocarbon. Numerous interrelationships exist between the oleanane and lupane series, and the configuration assigned to lupeol (192) rests on these and on conformational arguments similar to those already described.[24,228–231] The α-configuration of the isopropenyl side chain of lupeol follows from the re-formation of the triterpenoid by treatment of its ring-expanded hydrochloride (193) with ethanolic silver nitrate.[231] This reaction, which is presumably initiated by attack of Ag^+ on the chlorine atom, requires an *anti* coplanar arrangement of the participating centers (see 193a) (cf. Sec. 5-4d).

The configurations at C_3, C_5, C_8, and C_{10} of α-amyrin (194) were shown to be identical with those of the corresponding centers of β-amyrin by conversion of both triterpenoids to common degradation products.[232] The remaining stereochemistry of the ursane group was deduced largely by

[227] W. Klyne, *J. Chem. Soc.*, **1952**, 2916.
[228] T. R. Ames, T. G. Halsall, and E. R. H. Jones, *J. Chem. Soc.*, **1951**, 450.
[229] G. S. Davy, T. G. Halsall, and E. R. H. Jones, *J. Chem. Soc.*, **1951**, 2696.
[230] G. S. Davy, T. G. Halsall, E. R. H. Jones, and G. D. Meakins, *J. Chem. Soc.*, **1951**, 2702.
[231] T. G. Halsall, E. R. H. Jones, and G. D. Meakins, *J. Chem. Soc.*, **1952**, 2862.
[232] A. Meisels, O. Jeger, and L. Ruzicka, *Helv. Chim. Acta*, **33**, 700 (1950).

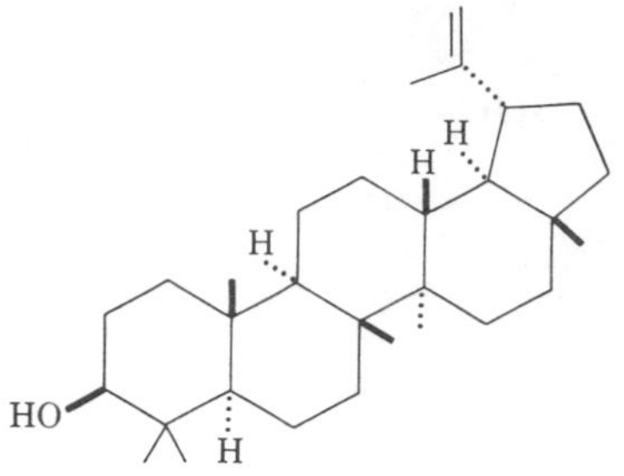

(192)

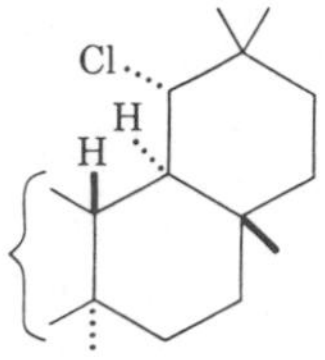

(193)

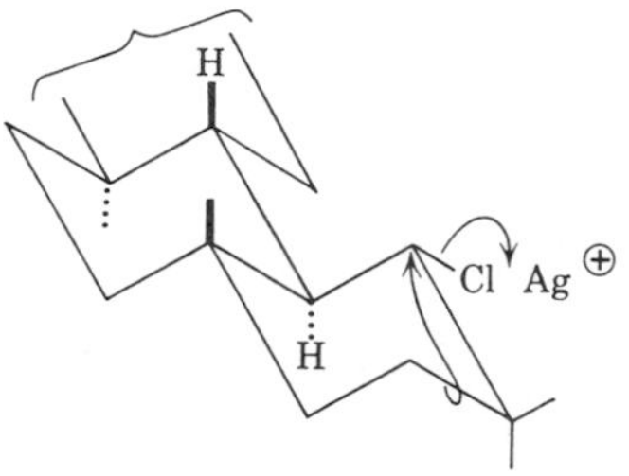

(193a)

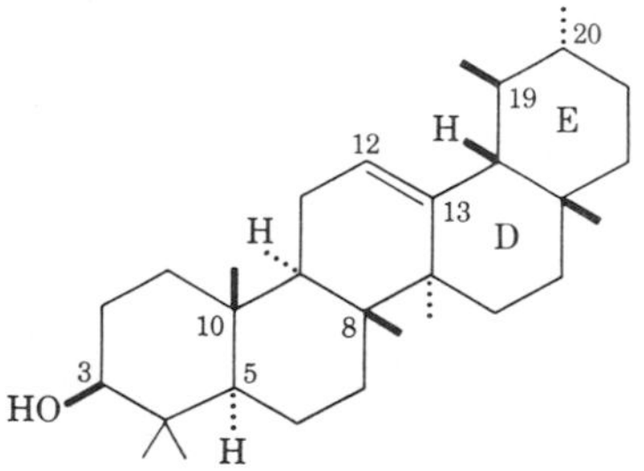

(194)

application of the principles of conformational analysis.[25,233,234] The *cis* fusion of rings D and E results in considerable hindrance of the Δ^{12}-double bond (cf. Sec. 5-3c), and additional hindrance due to the equatorial 19β-methyl group results in the double bond of α-amyrin being less reactive than that of β-amyrin.[25] Other differences in the chemistry of α- and β-amyrin resulting from the equatorial 19β- and 20α-methyl groups present in the former have already been discussed (see Sec. 5-3a). The structure and stereochemistry of α-amyrin have been confirmed by a partial synthesis of its acetate from glycyrrhetic acid, a triterpenoid of the oleanane group.[235]

The principal members of the tetracyclic lanostane and euphane groups of triterpenoids are, respectively, lanosterol (195) and euphol, which is epimeric with lanosterol at C_{13}, C_{14}, and C_{17}. The stereochemistry of lanosterol was deduced by Barton and his co-workers,[236] employing conformational arguments similar in the main to some of those already described in this section. The structure and stereochemistry depicted in (195)

[233] G. G. Allan and F. S. Spring, *J. Chem. Soc.*, **1955**, 2125.

[234] J. M. Beaton, F. S. Spring, R. Stevenson, and W. S. Strachan, *J. Chem. Soc.*, **1955**, 2610.

[235] E. J. Corey and E. W. Cantrall, *J. Am. Chem. Soc.*, **81**, 1745 (1959).

[236] C. S. Barnes, D. H. R. Barton, J. S. Fawcett, and B. R. Thomas, *J. Chem. Soc.*, **1953**, 576.

(195)

(196)   (197)

were also established by an X-ray investigation of lanost-8-en-3β-ol iodoacetate.[11] The close relationship existing between the lanostane group of triterpenoids and the steroids is underlined by the achievement of a partial synthesis of lanosterol from cholesterol.[237]

The configurations of the methyl groups attached to C_{13} and to C_{14} in euphol were assigned[160,238] largely on the basis of the acid-catalyzed rearrangement of euphenol (196) to isoeuphenol (197). No comparable rearrangement takes place when lanostenol (195; saturated side chain) is treated under the same conditions. The structures assigned to lanosterol and to euphol account for this difference in behavior in a highly satis-factory manner. Thus, whereas lanosterol (195) can adopt an all-chair and half-chair conformation, the configuration of (196) is such that ring C is forced to adopt a conformation which is close to a half-boat. There is therefore a conformational driving force (cf. Sec. 5-4c) for the rearrange-ment of euphenol (196) into isoeuphenol (197), which can readily assume an all-chair conformation. No such driving force is operative in the case of lanostenol; if lanostenol were to undergo a similar rearrangement, the product would possess an energetically unfavorable *cis* B/C junction.

The α-configuration of the side chain of euphol was first put forward on the assumption that the rearrangement of euphenol to isoeuphenol

[237] R. B. Woodward, A. A. Patchett, D. H. R. Barton, D. A. J. Ives, and R. B. Kelly, *J. Chem. Soc.*, **1957**, 1131.

[238] D. Arigoni, R. Viterbo, M. Dünnenberger, O. Jeger, and L. Ruzicka, *Helv. Chim. Acta*, **37**, 2306 (1954).

(198)   (199)   (200)

is a fully concerted process.[238] This stereochemical assignment was later confirmed by conversion of elemadienolic acid, a triterpenoid of the 20-isoeuphane series, and lanosterol (195) to enantiomeric degradation products by reactions which did not affect the configuration at C_{17}.[239]

Rearrangements analogous to the euphenol-isoeuphenol change are characteristic of compounds of the euphane and 20-isoeuphane series which possess a Δ^8-double bond.[240] A similar conformational driving force is no doubt responsible for the acid-catalyzed isomerization of bauerenol acetate (198), a triterpenoid possessing a modified ursane skeleton, first into isobauerenol acetate (199), then, under more vigorous conditions, into a mixture of α-amyrin acetate and 3β-acetoxyurs-13(18)-ene (200).[241]

5–8. Application of Conformational Analysis to Alkaloid Chemistry

For the most part this chapter has so far been concerned only with the application of conformational analysis to the chemistry of carbocyclic compounds. However, as has been indicated in Sec. 4-6, the same principles may also be extended to heterocycles; this approach has been very fruitful when applied to stereochemical problems encountered in studies of alkaloids. It has been applied with special success to the morphine alkaloids,[42a] the indole alkaloids,[242] and the Amaryllidaceae alkaloids.[243]

[239] E. Ménard, H. Wyler, A. Hiestand, D. Arigoni, O. Jeger, and L. Ruzicka, Helv. Chim. Acta, 38, 1517 (1955).

[240] D. Arigoni, O. Jeger, and L. Ruzicka, Helv. Chim. Acta, 38, 222 (1955); J. B. Barbour, W. A. Lourens, F. L. Warren, and K. H. Watling, J. Chem. Soc., 1955, 2194.

[241] F. N. Lahey and M. V. Leeding, Proc. Chem. Soc., 1958, 342; cf. P. Sengupta, and H. N. Khastgir, Tetrahedron, 19, 123 (1963).

[242] J. E. Saxton, "The Indole Alkaloids," in R. H. F. Manske, ed., The Alkaloids, Vol. VII, Academic Press, New York, 1960.

[243] W. C. Wildman, "Alkaloids of the Amaryllidaceae," in R. H. F. Manske, ed., The Alkaloids, Vol. VI, Academic Press, New York, 1960; Y. Inubushi, H. M. Fales, E. W. Warnhoff, and W. C. Wildman, J. Org. Chem., 25, 2153 (1960).

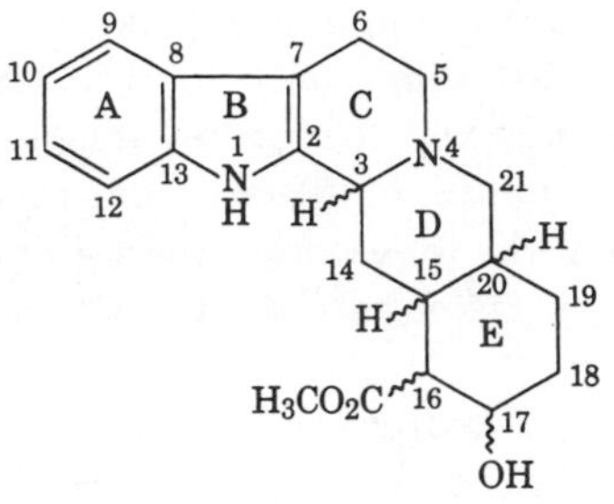

(201) 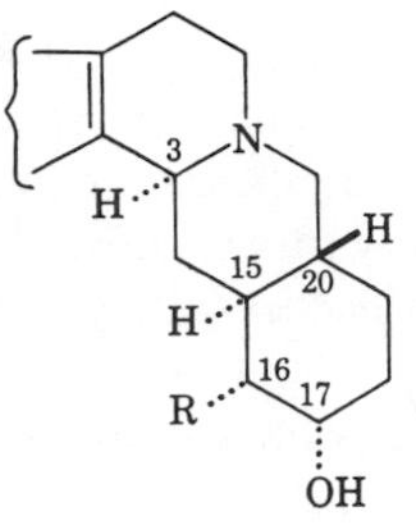 (202) 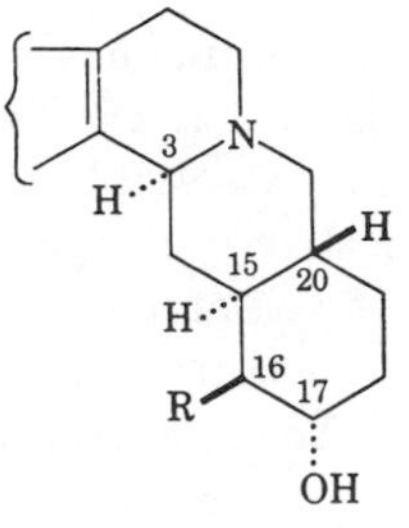 (203)

In the following pages the arguments which have been employed are illustrated by reference to a few selected topics drawn from the chemistry of the *Yohimbehe* and *Rauwolfia* alkaloids.

a. Yohimbine and Its Stereoisomers. At least seven stereoisomers of yohimbine occur in nature, and several others have been obtained by isomerizations carried out in the laboratory.[244] All are to be represented by the gross structure (201). The configurations of two of these isomers, yohimbine (202; $R = CO_2CH_3$) and coryanthine (203; $R = CO_2CH_3$), were deduced independently by Cookson[245] and by Janot, Goutarel, Le Hir, Amin, and Prelog,[246] mainly by application of the principles of conformational analysis.

The *trans* D/E ring fusion of yohimbine, shown in (202; $R = CO_2CH_3$) follows from the formation of $(+)$-N-methyl-*trans*-decahydroisoquinoline by successive catalytic hydrogenation and Hofmann degradation of (204), a compound obtained by distillation of yohimbic acid (202; $R = CO_2H$) from thallous oxide.[247]

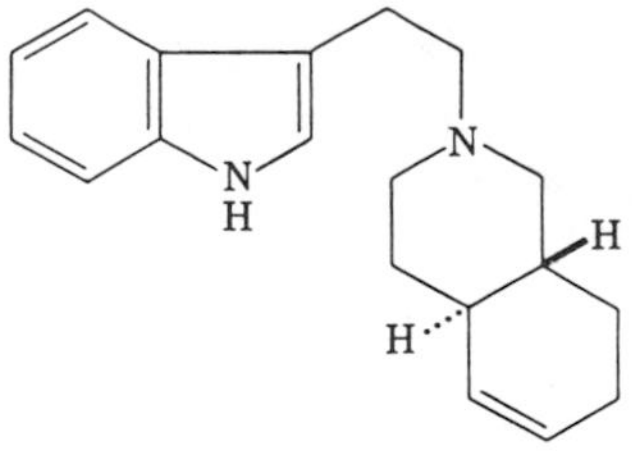

(204) 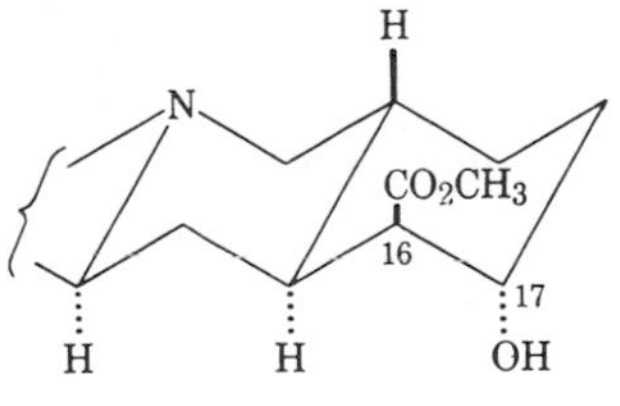 (205)

[244] M.-M. Janot, R. Goutarel, E. W. Warnhoff, and A. Le Hir, *Bull. Soc. Chim. France*, **1961**, 637.

[245] R. C. Cookson, *Chem. Ind. (London)*, **1953**, 337.

[246] M.-M. Janot, R. Goutarel, A. Le Hir, M. Amin, and V. Prelog, *Bull. Soc. Chim. France*, **1952**, 1085.

[247] B. Witkop, *J. Am. Chem. Soc.*, **71**, 2559 (1949).

Since both yohimbine and corynanthine afford yohimbic acid (202; R = CO_2H) upon alkaline hydrolysis,[248] they must differ only in the configuration of their 16-methoxycarbonyl group, which may be assigned the less stable (axial) conformation in corynanthine. This is indicated in (205). In agreement with this formulation, the (equatorial) ester group of yohimbine is saponified at an appreciably faster rate than is that of corynanthine.[246]

The α-(axial)-configuration assigned to the 17-hydroxyl group of yohimbine and corynanthine may be inferred from the fact that the mixture of epimeric alcohols obtained by Meerwein-Ponndorf reduction of yohimbone, the ketone derived from yohimbol (202; R = H), becomes progressively richer in *epi*-yohimbol as the reaction time is increased;[249] *epi*-yohimbol must therefore be the thermodynamically more stable (i.e., equatorial) epimer, which of course implies an axial conformation for the hydroxyl group originally present in yohimbine. The more recent reduction of yohimbone to *epi*-yohimbol with lithium aluminum hydride supports these assignments.[250]

The *trans* diaxial arrangement of the ring E substituents of corynanthine is confirmed by the course of the reaction of corynanthine sulfuric ester with dilute alkali, elimination of the sulfuric ester group occurring with concomitant decarboxylation (206). The product is *apo*-corynanthol (207).[248] Similar treatment of the sulfuric ester of yohimbine (or yohimbic acid itself), which does not have an *anti* coplanar disposition of ring E

(206)   (207).

(208)   (209)

[248] M.-M. Janot and R. Goutarel, *Bull. Soc. Chim. France*, **1949**, 509.
[249] B. Witkop, *Ann.*, **554**, 83 (1943).
[250] E. Wenkert and D. K. Roychaudhuri, *J. Am. Chem. Soc.*, **80**, 1613 (1958).

(210)   (211)

substituents, results only in the elimination of sulfuric acid (208) with formation of *apo*-yohimbine (209).[251]

The configuration of yohimbine at C_3 is known to be the more stable of the two possible arrangements, (210) and (211).[244,245] Since the indole ring is attached to ring D equatorially in the former, and axially in the latter, it was concluded[244,245] that (210) correctly represents the stereochemistry of yohimbine. A total synthesis of this alkaloid was reported in 1958.[252]

Both yohimbine and corynanthine are based on the "normal" yohimbane skeleton, characterized by the $3\alpha,15\alpha\text{-}20\beta$-configuration of the angular hydrogen atoms.* A third member of this group is β-yohimbine (212), the 16-methoxycarbonyl group of which was assigned an equatorial conformation (and hence the α-configuration) by comparison of its rate

(212)   (213)   (214)

[251] G. Barger and E. Field, *J. Chem. Soc.*, **1923**, 1038.

[252] E. E. van Tamelen, M. Shamma, A. W. Burgstahler, J. Wolinsky, R. Tamm, and P. E. Aldrich, *J. Am. Chem. Soc.*, **80**, 5006 (1958).

[253] W. Klyne, *Chem. Ind. (London)*, **1953**, 1032.

[254] C. Djerassi, R. Riniker, and B. Riniker, *J. Am. Chem. Soc.*, **78**, 6362 (1956).

[255] Y. Ban and O. Yonemitsu, *Chem. Ind. (London)*, **1961**, 948.

* While conformational analysis yields information concerning only relative configurations, all the formulas reproduced here represent the correct absolute configurations. These were originally proposed on the basis of molecular rotation differences[253] and optical rotatory dispersion studies.[254] More recently, the result of an asymmetric synthesis with yohimbine as one of the participants has provided supporting chemical evidence.[255]

(215)   (216)

of saponification and the pK value of the resultant β-yohimbic acid with the corresponding figures for yohimbine and corynanthine.[256] In support of such an assignment, β-yohimbine does not undergo epimerization with base. β-Yohimbine is therefore the 17-epimer of yohimbine, and it affords yohimbone by Oppenauer oxidation.[256]

Those isomers of yohimbine possessing the $3\beta,15\alpha,20\beta$-configuration are designated members of the "pseudo" series. The only naturally occurring representative of this group is pseudoyohimbine (213), which was related to yohimbine by lead tetraacetate oxidation, when tetra-dehydroyohimbine (214) was obtained.[246]

Two other basic skeletons are known among the *Yohimbehe* alkaloids. These differ from compounds of the normal and pseudo series in that they have a *cis* D/E ring junction. They are based, respectively, on alloyo-himbane (215) ($3\alpha,15\alpha,20\alpha$-configuration) and on 3-epialloyohimbane (216) ($3\beta,15\alpha,20\alpha$-configuration).

A notable difference between compounds of the normal and pseudo series on the one hand and those of the allo and 3-epiallo series on the other is that, while rings D and E of the former can exist in only one all-chair conformation, i.e. (210) in the case of normal compounds, and (211) for compounds of the pseudo series, those of both alloyohimbane (215) and 3-epialloyohimbane (216) can each exist in two all-chair confor-mations. For alloyohimbane these are (215a) and (215b); for 3-epialloyo-himbane, (216a) and (216b).

Because of the axial disposition of the indole ring system in (211), a compound of the pseudo series is always less stable than its 3-epimer of the normal series, in which C_2 is equatorially disposed.

The situation regarding compounds of either of the two *cis* D/E series is less clear cut. An analysis of conformations (215a)–(216b) will reveal that the preferred conformations of alloyohimbane and 3-epialloyohimbane are probably (215a) and (216b), respectively. It is not easy to predict, *a priori*, which of these two preferred conformations will be the more

[256] A. Le Hir and R. Goutarel, *Bull. Soc. Chim. France*, **1953**, 1023.

stable. In fact, when either *dl*-alloyimbane (215) or *dl*-epialloyimbane (216) was equilibrated with acid, a $74 \pm 5\%$ yield of a mixture of allo and epiallo isomers was obtained, the latter predominating in a ratio of 3.6:1.[257] However, since when (216b) undergoes epimerization to (215a), or vice versa, all the bonds attached to rings D or E which were originally equatorial become axial, and those which were axial become equatorial, it is obvious that the presence of substituents in either of these rings might result in a stability relationship between the 3-epimers very different from that obtaining in the parent hydrocarbons.[250] This point is very relevant to the discussion of the configurations of reserpine and deserpidine (Sec. 5-8b).

Infrared spectroscopy has proved to be a useful technique for investigating the steric arrangement at C_3 of the yohimbines and related indole alkaloids. It was suggested by Wenkert and Roychaudhuri[250] that those alkaloids with an α-hydrogen atom at C_3 exhibit two or more peaks between 2700 and 2900 cm.$^{-1}$ on the low wave number side of the major band (at about 2900 cm.$^{-1}$) while those with a 3β-configuration do not. Two exceptions to this rule, 3-epialloyohimbone[250] and 3-epialloyohimbine,[244] have, however, been reported, and it has been suggested[258] that the differences in the infrared spectra reflect differences in the *conformation*

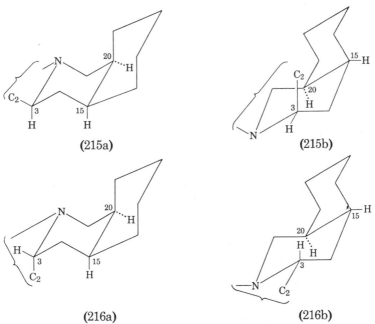

(215a)   (215b)

(216a)   (216b)

[257] E. Wenkert and L. H. Liu, *Experientia*, **11**, 302 (1955).
[258] W. E. Rosen, *Tetrahedron Letters*, **1961**, 481.

of the 3-hydrogen atom and only incidentally in its configuration. According to this view, additional peaks in the infrared are exhibited only by those alkaloids which possess an axial hydrogen atom at C_3 in their stable conformations. As regards the normal and the pseudo series, both rules predict the same results, as may be verified by reference to formulas (210) and (211). In the allo and 3-epiallo series, however, the bands exhibited in the infrared spectrum will, according to the modified generalization, depend on which of the possible all-chair conformations is adopted by the molecule. The exceptions to the original generalization concerning C_3 configurations are readily accommodated by the modified rule, if the molecules in question adopt conformations similar to (216b). The revised generalization as quoted here is simply a special case of a more general correlation concerning the infrared spectra of quinolizidines and their preferred conformations.[259]

Nuclear magnetic resonance has also been employed[260,261] to make conformational assignments at C_3 of the indole alkaloids. It appears that only an equatorial 3-hydrogen atom gives rise to a signal downfield from the region of saturated C—H absorption[260,261] (ca. 6.88 τ). The signal given by a 3-equatorial hydrogen atom is usually observed near 5.55 τ.

It has been suggested that the mercuric acetate oxidation of tertiary amines, e.g., quinolizidine, involves an *anti* coplanar elimination from an N-mercurated complex (217).[262,263] The possible diagnostic value of this reaction in investigating alkaloids of the yohimbine type has been tested by a number of workers.[264-266] In agreement with expectation, compounds of the normal series, such as yohimbine, and those of the allo series, e.g., α-yohimbine and 3-isoreserpine, all of which might be expected to adopt a preferred conformation with the 3-hydrogen atom axial, and *trans* to the p-electrons of N_4, readily gave the corresponding 3-dehydro derivatives, e.g. (218), while pseudoyohimbine (213), for which such a conformation is not possible, was not oxidized under the conditions of the reaction.[264]

The situation concerning compounds of the 3-epiallo series is, however, anomalous. Such compounds can in principle adopt either of the conformations (216a) and (216b), only the latter possessing the required *anti*

[259] F. Bohlmann, *Chem. Ber.*, **91**, 2157 (1958).

[260] E. Wenkert and B. Wickberg, *J. Am. Chem. Soc.*, **84**, 4914 (1962).

[261] W. E. Rosen and J. N. Shoolery, *J. Am. Chem. Soc.*, **83**, 4816 (1961); E. Wenkert, B. Wickberg, and C. L. Leicht, *ibid.*, 5037; *Tetrahedron Letters*, **1961**, 822.

[262] N. J. Leonard, A. S. Hay, R. W. Fulmer, and V. W. Gash, *J. Am. Chem. Soc.*, **77**, 439 (1955).

[263] N. J. Leonard and D. F. Morrow, *J. Am. Chem. Soc.*, **80**, 371 (1958).

[264] F. L. Weisenborn and P. A. Diassi, *J. Am. Chem. Soc.*, **78**, 2022 (1956).

[265] E. Wenkert and D. K. Roychaudhuri, *J. Org. Chem.*, **21**, 1315 (1956).

[266] E.Farkas, E. R. Lavagnino, and R. T. Rapala, *J. Org. Chem.*, **22**, 1261 (1957).

(217)   (218)

coplanar arrangement of the 3-hydrogen atom and the p-electrons of N_4. Under the usual conditions of the oxidation, reserpine (219; $R = OCH_3$; $R' = $ 3,4,5-trimethoxyphenyl), deserpidine (219; $R = H$; $R' = $ 3,4,5-trimethoxyphenyl), and methyl reserpate (219; $R = OCH_3$, OH in place of $O.CO.R'$), all of which are based on the 3-epialloyohimbane skeleton (216), failed to react.[264],* While these results might be attributed to the alkaloids existing predominantly in conformation (216a), so as to achieve an equatorial orientation for their ring E substituents (see Sec. 5-8b), a similar explanation will not suffice for the observation that dl-3-epialloyohimbane itself is not readily oxidized by mercuric acetate.[265] The preferred conformation of this compound is almost certainly (216b), in which the C/D ring junction is correctly disposed for a process of the type illustrated in (217).

b. Reserpine and Deserpidine. Conformational analysis figured largely in the elucidation of the stereochemistry of the pharmacologically important *Rauwolfia* alkaloids reserpine (219; $R = OCH_3$, $R' = $ 3,4,5-trimethoxyphenyl) and deserpidine (219; $R = H$, $R' = $ 3,4,5-trimethoxyphenyl).[267]

In both alkaloids the configuration at C_3 is the thermodynamically unstable one.[268-270] By conversion of deserpidine to α-yohimbine (220) under

[267] P. E. Aldrich, P. A. Diassi, D. F. Dickel, C. M. Dylion, P. D. Hance, C. F. Huebner, B. Korzun, M. E. Kuehne, L. H. Liu, H. B. MacPhillamy, E. W. Robb, D. K. Roychaudhuri, E. Schlittler, A. F. St. André; E. E. van Tamelen, F. L. Weisenborn, E. Wenkert, and O. Wintersteiner, *J. Am. Chem. Soc.*, **81**, 2481 (1959).

[268] H. B. MacPhillamy, L. Dorfman, C. F. Huebner, E. Schlittler, and A. F. St. André, *J. Am. Chem. Soc.*, **77**, 1071 (1955).

[269] H. B. MacPhillamy, C. F. Huebner, E. Schlittler, A. F. St. André, and P. R. Ulshafer, *J. Am. Chem. Soc.*, **77**, 4335 (1955).

[270] C. F. Huebner, A. F. St. André, E. Schlittler, and A. Uffer, *J. Am. Chem. Soc.*, **77**, 5725 (1955).

* Dehydrogenation may be achieved by carrying out the oxidation in refluxing 10% acetic acid.[266] It is possible that under these conditions epimerization occurs at C_3, and that the compounds actually react in a conformation of the same type as (215a).

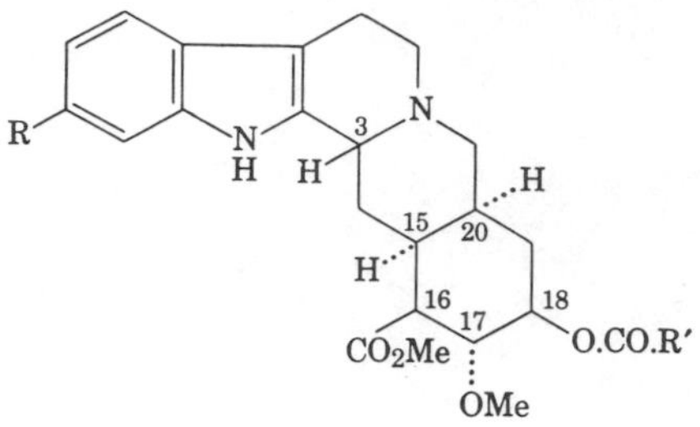

(219) 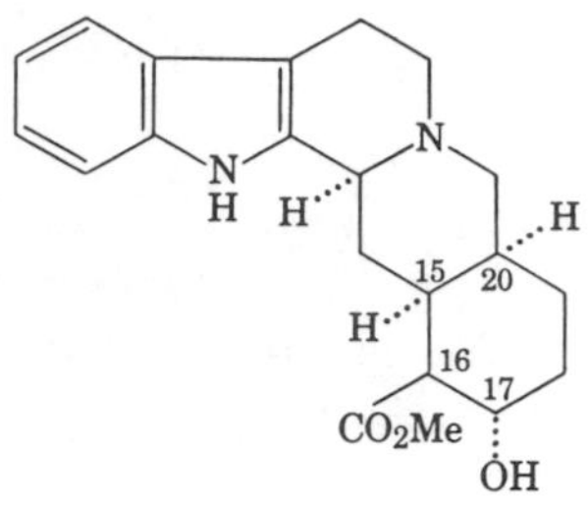 (220)

conditions known to cause epimerization at C_3, it was shown that the former must possess the same skeleton as 3-epi-α-yohimbine, i.e., the 3-epialloyohimbane skeleton (216).[269,270] The *cis* D/E configuration of the latter was based on firm evidence, including the achievement of several stereospecific syntheses of *dl*-epialloyohimbane and its diastereomers,[271] and the configuration at C_3 was inferred from the finding that 3-epialloyohimbane is a stronger base than alloyohimbane, and that it affords an

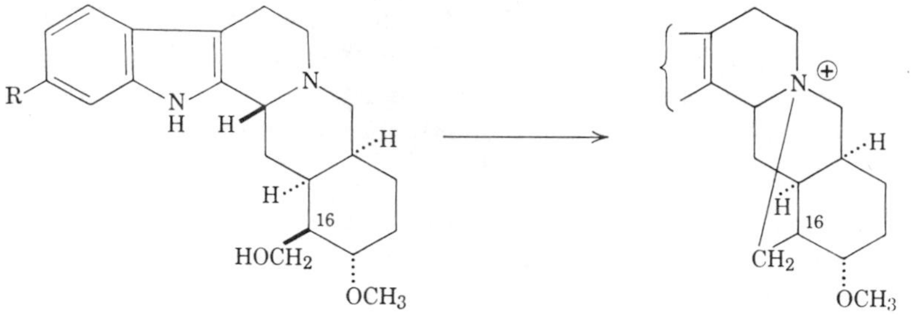

(221) (222)

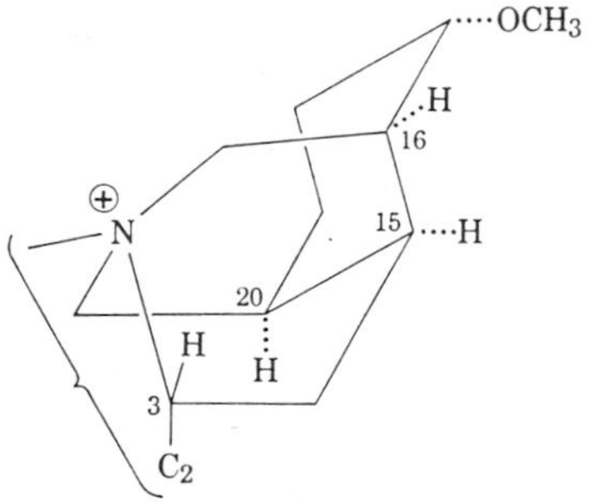

(222a)

[271] E. E. van Tamelen, M. Shamma, and P. Aldrich, *J. Am. Chem. Soc.*, **78**, 4628 (1956); G. Stork and R. K. Hill, *ibid.*, **79**, 495 (1957); R. T. Rapala, E. R. Lavagnino, E. R. Shepard, and E. Farkas, *ibid.*, 3770.

N-oxide under conditions which do not affect its 3-epimer.[257] From a consideration of the more stable conformations of the $3\alpha,15\alpha,20\alpha$- and the $3\beta,15\alpha,20\alpha$-isomers, (215a) and (216b), respectively, it was concluded that the quinolizidine nitrogen atom was less hindered in the latter. Accordingly, 3-epialloyohimbane (216), and consequently also both 3-epi-α-yohimbine and deserpidine (219; R = H, R' = 3,4,5-trimethoxyphenyl), could be assigned the $3\beta,15\alpha,20\alpha$-configuration; as a corollary, alloyohimbane could be represented as the $3\alpha,15\alpha,20\alpha$-isomer (215). The fact that 3-epialloyohimbane (216) undergoes ring C dehydrogenation with palladium and maleic acid significantly faster than does alloyohimbane (215) was construed as further evidence in support of these formulations.[257]

The configuration of deserpidine at C_{16} follows from the conversion of deserpidinol (221; R = H) to the quaternary ammonium tosylate (222; R = H) upon treatment with p-toluenesulfonyl chloride and pyridine.[267] The formation of such a compound necessarily implies a *cis* relationship for the hydrogen atoms attached to C_{15}, C_{16}, and C_{20}, as is evident from the partial formula (222a).

The ready formation of deserpidic acid lactone (224; R = H) by treatment of deserpidic acid (223; R = H) with acetic anhydride[269] indicates a *cis* relationship of the C_{16} and C_{18} substituents of deserpidine and leaves only the configuration of the 17-methoxyl group to be derived.

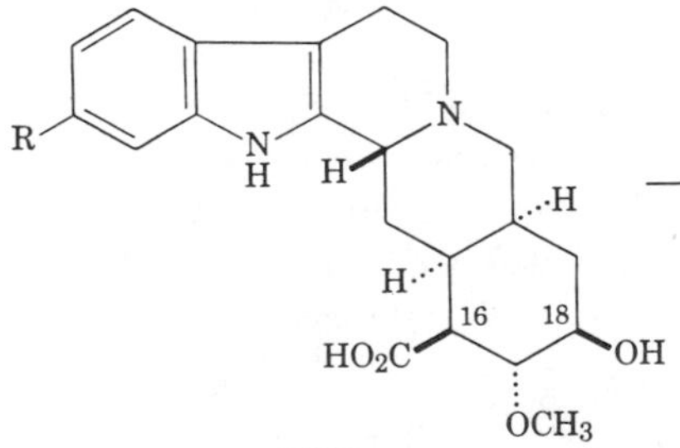

(223) 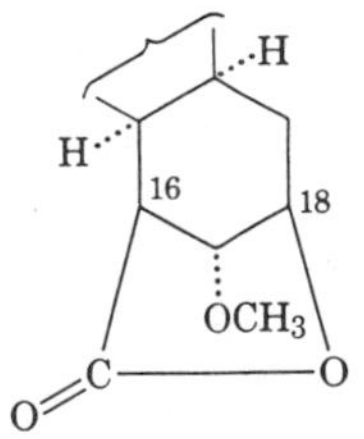   (224)

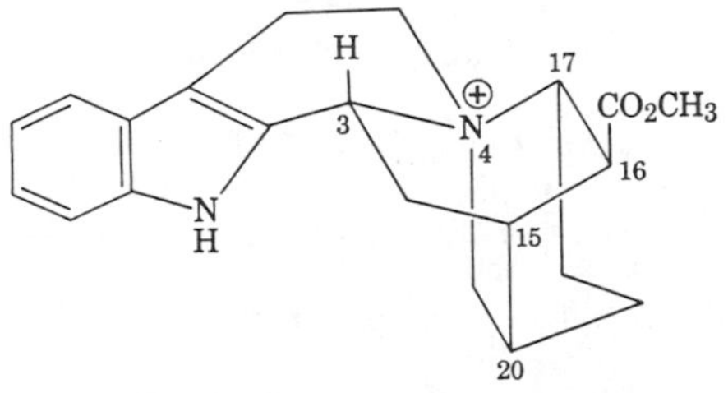

(225) 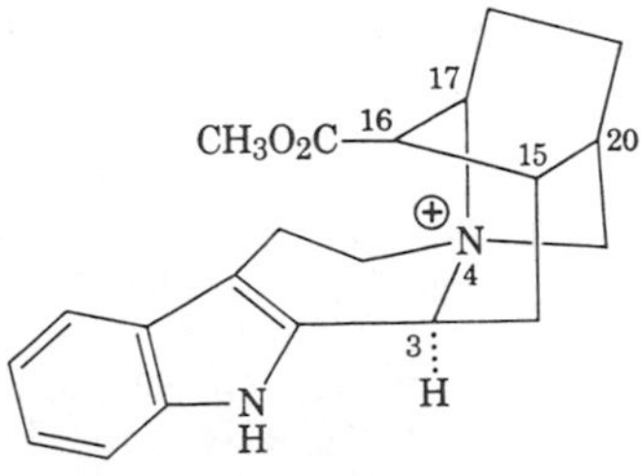   (226)

The last stereochemical point was resolved by comparing the action of p-toluenesulfonyl chloride and pyridine on α-yohimbine (220) and on 3-epi-α-yohimbine, both of these alkaloids being known to possess the same configurations as deserpidine at C_{15}, C_{16}, C_{17}, and C_{20}.[268,269,272] The 17-tosylate obtained from α-yohimbine was quite stable, but the 3-epi-α-yohimbine derivative could be isolated only if special precautions were taken.[273] Under very mild conditions it was transformed in good yield into the quaternary ammonium tosylate (225). Such ready formation of a quaternary ammonium salt under essentially non-ionizing conditions is indicative of a direct intramolecular displacement of the covalent tosylate by N_4, and it follows, therefore, that the 17-substituent in 3-epi-α-yohimbine, and hence also in α-yohimbine and in deserpidine, is α-oriented. The complete stereoformulas of α-yohimbine and of deserpidine are thus established as (220) and (219; R = H, R' = 3,4,5-trimethoxyphenyl), respectively. The failure of α-yohimbine tosylate to undergo a similar isomerization is probably a consequence of the prohibitive non-bonded interaction which would exist between the 16-carbomethoxy group and the indole ring in the analogous structure (226).

Much of the stereochemistry of reserpine (219; R = OCH_3, R' = 3,4,5-trimethoxyphenyl) was deduced by the application of arguments analogous to those already described for deserpidine.[267] Thus the cis relationship of the C_{15}, C_{16}, and C_{20} hydrogen atoms was inferred from the formation of a quaternary ammonium tosylate (222; R = OCH_3) by treatment of reserpinol (221; R = OCH_3) with p-toluenesulfonyl chloride and pyridine,[274-276] while the obtainment of reserpic acid lactone (224; R = OCH_3) from reserpic acid (223; R = OCH_3)[277] established that the C_{16} and C_{18} substituents were also cis to each other. The relationship existing between the configurations at C_3 and C_{18} was revealed when the absolute configuration at each of these centers was inferred from molecular rotation data.[278,279]

[272] F. E. Bader, D. F. Dickel, C. F. Huebner, R. A. Lucas, and E. Schlittler, J. Am. Chem. Soc., 77, 3547 (1955).

[273] C. F. Huebner and D. F. Dickel, Experientia, 12, 250 (1956).

[274] C. F. Huebner and E. Wenkert, J. Am. Chem. Soc., 77, 4180 (1955).

[275] P. A. Diassi, F. L. Weisenborn, C. M. Dylion, and O. Wintersteiner, J. Am. Chem. Soc., 77, 4687 (1955).

[276] E. E. van Tamelen and P. D. Hance, J. Am. Chem. Soc., 77, 4692 (1955).

[277] L. Dorfman, A. Furlenmeier, C. F. Huebner, R. Lucas, H. B. MacPhillamy, J. M. Mueller, E. Schlittler, R. Schwyzer, and A. F. St. André, Helv. Chim. Acta, 37, 59 (1954).

[278] P. A. Diassi, F. L. Weisenborn, C. M. Dylion, and O. Wintersteiner, J. Am. Chem. Soc., 77, 2028 (1955).

[279] C. F. Huebner, H. B. MacPhillamy, E. Schlittler, and A. F. St. André, Experientia, 11, 303 (1955).

As with deserpidine, the configuration at C_{17} of reserpine was assigned on the basis of a quaternization reaction. Treatment of methyl reserpate tosylate (223; R = OCH_3; CH_3O_2C in place of HO_2C; OTs in place of OH) with collidine afforded a mixture of methyl anhydroreserpate (227) and the quaternary ammonium tosylate (228).[278] Since the tosylate group of methyl reserpate tosylate has a β-orientation, a direct intramolecular displacement mechanism for the formation of (228) is excluded, and it is necessary to postulate participation of the 17-methoxyl group in some manner.[257,276] The simplest possibility would involve the intermediacy of the cyclic oxonium ion (229)[276], but this was rejected by Aldrich and co-workers since no compounds possessing structure (230) or (231) were isolated from the reaction.[267] They prefer to regard both the elimination reaction—leading, by a subsequent prototropic rearrangement to methyl anhydroreserpate (227)—and the quaternization reaction as proceeding through an intermediate of type (232). However, whatever may be the

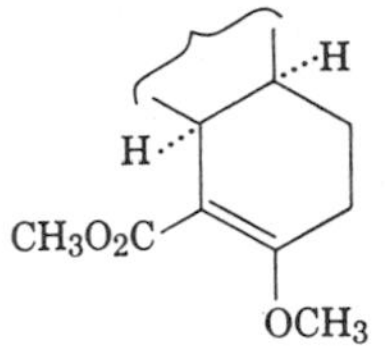

(227)

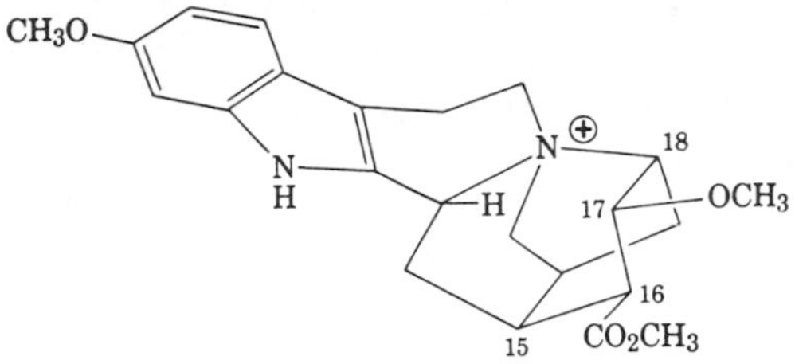

(228)

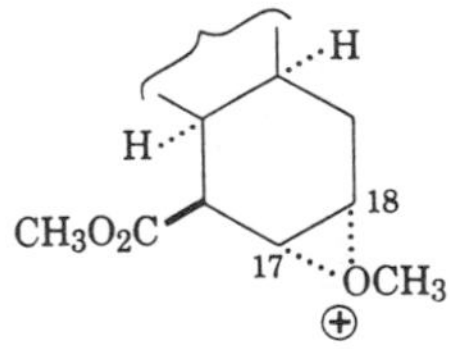

(229)

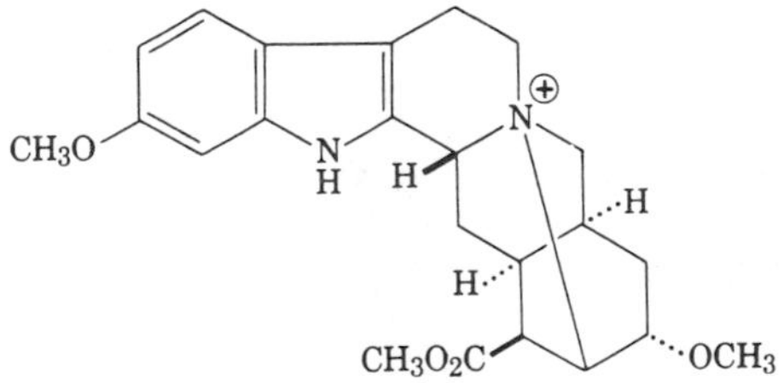

(230)

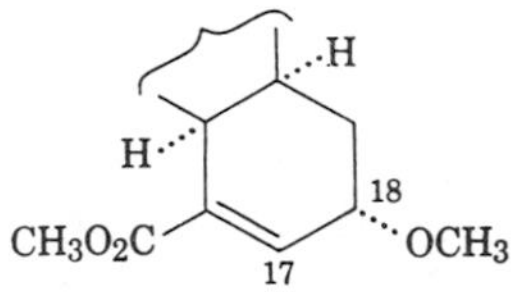

(231)

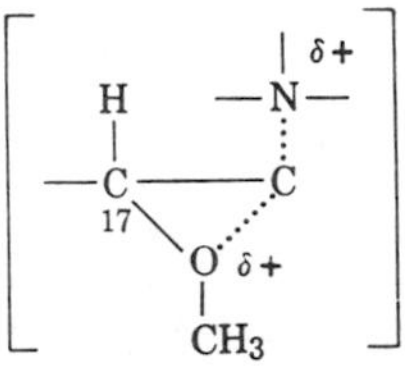

(232)

true mechanism of the process, participation of the 17-methoxyl group in the reaction requires that it be situated *trans* with respect to the 18-tosylate.* The configuration of reserpine is therefore correctly represented by (219; $R = OCH_3$, $R' = 3,4,5$-trimethoxyphenyl).

With the configurations of reserpine and deserpidine firmly established, several features of the chemistry of these compounds find ready explanations in terms of conformational analysis. It has already been noted that, while 3-epialloyohimbane (216) is rather more stable than alloyohimbane (215), deserpidine and reserpine, which are based on the same skeleton as (216), can be isomerized to their more stable 3-iso derivatives. The greater stability of 3-isoreserpine and 3-isodeserpidine is doubtless due to the fact that rings D and E of these compounds can adopt a di-chair conformation (234) in which the indole ring and all three ring E substituents are equatorially situated; of the two di-chair conformations of rings D and E which are possible for the alkaloids themselves, one of them (235) has the indole ring in an unfavorable axial conformation with respect to ring D, while the other (236) requires the C_{16}, C_{17}, and C_{18} substituents to adopt axial conformations.

Deserpidic acid (223; $R = H$) and reserpic acid (223; $R = OCH_3$) must react in conformations similar to (236) when they undergo lactonization upon treatment with acetic anhydride. The corresponding conformation of 3-isoreserpic acid (237) is of much higher energy, since it involves an extremely severe non-bonded interaction between C_2 and the C_{16}-carboxyl group. Consequently, reaction conditions which bring about lactonization of reserpic acid result only in acetylation of the 18-hydroxyl group when applied to 3-isoreserpic acid.[279] However, 3-isoreserpic acid has been converted to the corresponding lactone under

---

[280] W. E. Rosen and J. M. O'Connor, *J. Org. Chem.*, **26**, 3051 (1961).

* No corresponding internal quaternization occurs when the tosylate of methyl neoreserpate (233) is heated with collidine under reflux,[280] for in this compound,

(233)

which is obtained by treatment of methyl reserpate with methanolic sodium methoxide, the 17- and 18-oxygen substituents bear a *cis* relationship.

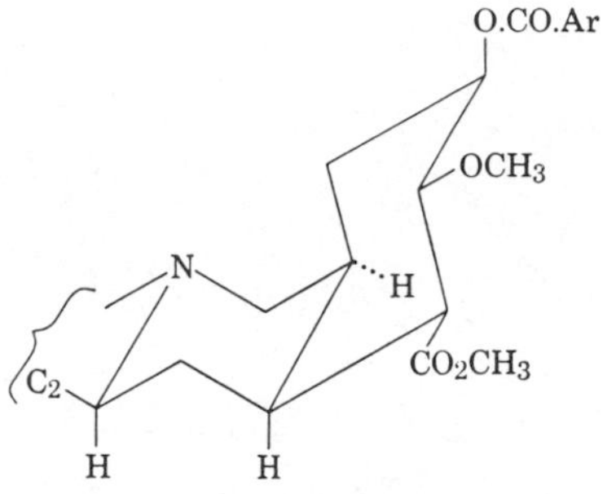

(234) 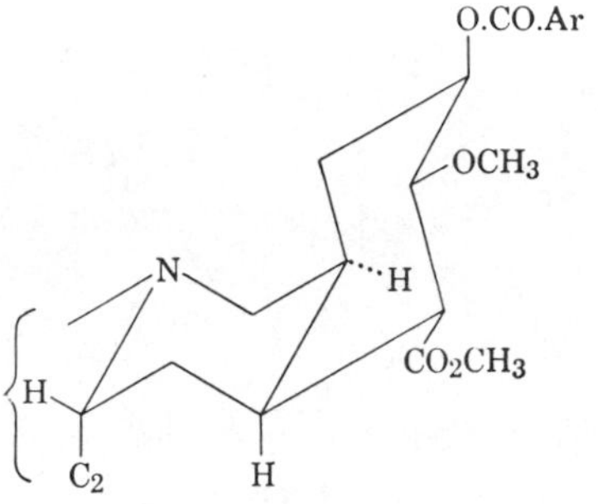   (235)

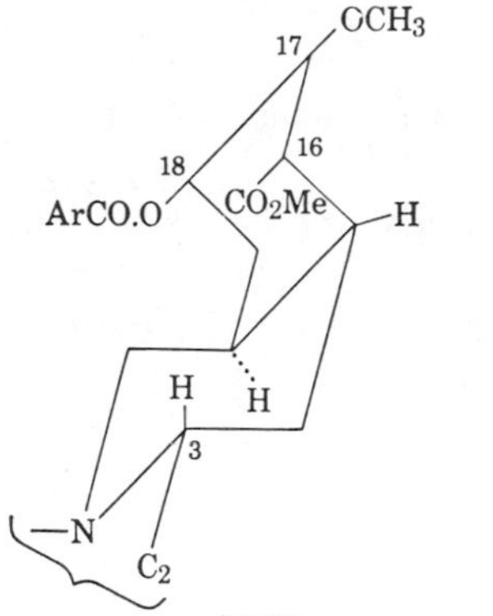

(236) 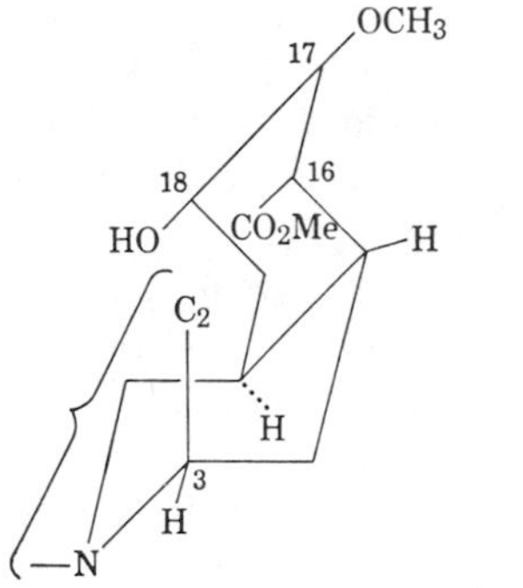   (237)

forcing conditions, e.g., by treatment with N,N'-dicyclohexylcarbodi-imide.[281] It is possible that rings D and E of this lactone may adopt flexible conformations in order to relieve the non-bonded interaction between the indole ring and the lactone bridge.

In 1958, a stereospecific total synthesis of reserpine was reported by Woodward and his colleagues.[281] This brilliant achievement, which has been described[281] as "an exercise in stereochemistry," provides some excellent examples of how the principles of conformational analysis can be employed to guide the synthetic organic chemist.

Of the six centers of asymmetry in reserpine (219; $R = OCH_3$, $R' =$ 3,4,5-trimethoxyphenyl), five (C_{15}, C_{16}, C_{17}, C_{18}, and C_{20}) are situated in ring E. Accordingly, the first objective in the synthesis was the construction of a ring E progenitor of type (238) containing all the ring E substituents, and incipient D/E junction, all in the correct stereochemical relationship.

[281] R. B. Woodward, F. E. Bader, H. Bickel, A. J. Frey, and R. W. Kierstead, *Tetrahedron*, **2**, 1 (1958).

(238)

In the first stage of the synthesis, quinone and vinylacrylic acid were condensed together to give a product which, from the known stereochemistry of the Diels-Alder reaction (see 239), could reasonably be assigned the configuration shown in (240; R = H). This structure was confirmed by the preparation of a lactone (242) from the ketol (241; R = H), obtained by reduction of the Diels-Alder adduct (240; R = H) with sodium borohydride.*

C_9, C_1, and C_{10} of (241; R = H) correspond in stereochemistry and in function to C_{15}, C_{16}, and C_{20}, respectively, of reserpine. The first step in the introduction of substituents corresponding to those at C_{17} and C_{18} of reserpine was the epoxidation of the 2,3-double bond of (241; R = H) with perbenzoic acid. The product obtained, (243), was that arising by attack of the reagent on the less hindered face of (241; R = H) in its more stable conformation (241a).

Since, like (241; R = H), the epoxide (243) will adopt a preferred conformation in which the carboxyl and hydroxyl groups are pseudo-equatorially disposed, it was predictable that hydrolytic cleavage of the oxirane ring would afford, following the usual diaxial course of such reactions (Sec. 5-4b), the triol (244), in which the hydroxyl groups at C_2 and C_3 would have configurations which are opposite those of the corresponding functions in reserpine. In order to obtain the desired stereochemistry at these centers, the epoxide (243) was converted to the derived lactone (245) in which the C_1 and C_5 substituents are obliged to assume pseudoaxial conformations. Diaxial cleavage of the oxide ring of this lactone, brought about by the action of boiling acetic acid, afforded the diol monoacetate (246) the configuration of which corresponded exactly to the stereochemistry of ring E of reserpine.

Further transformation of (246) was not pursued, when it was found that parallel investigations had opened up a synthetic route to reserpine

[282] For the expression of a contrary view, see H. O. House, H. Babad, R. B. Toothill, and A. W. Noltes, *J. Org. Chem.*, **27**, 4141 (1962).

* The preferential reduction of the C_5-carbonyl group of (240; R = H) has been attributed[281] to "electrostatic screening" of the C_8-carbonyl function by the proximate carboxyl group attached to C_1. Other examples of the same phenomenon have since been observed.[92,282]

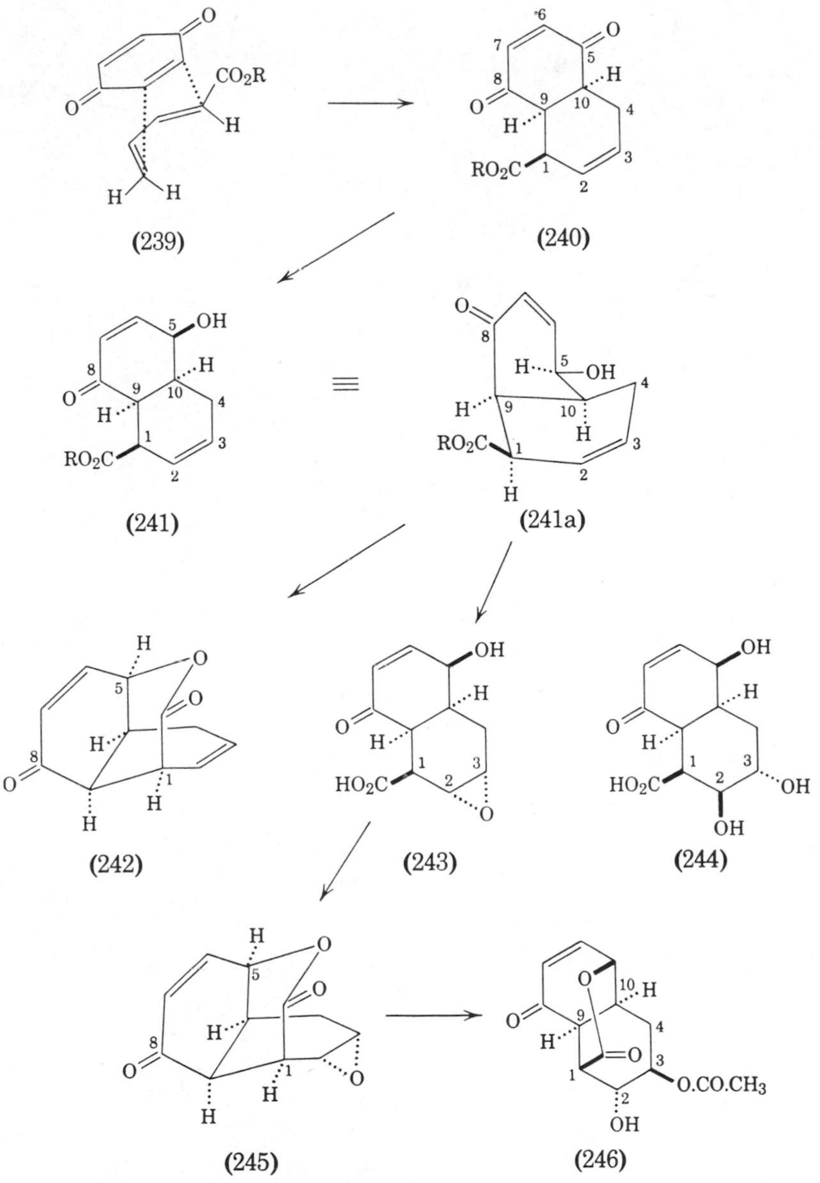

(239)

(240)

(241)   ≡   (241a)

(242)   (243)   (244)

(245)   (246)

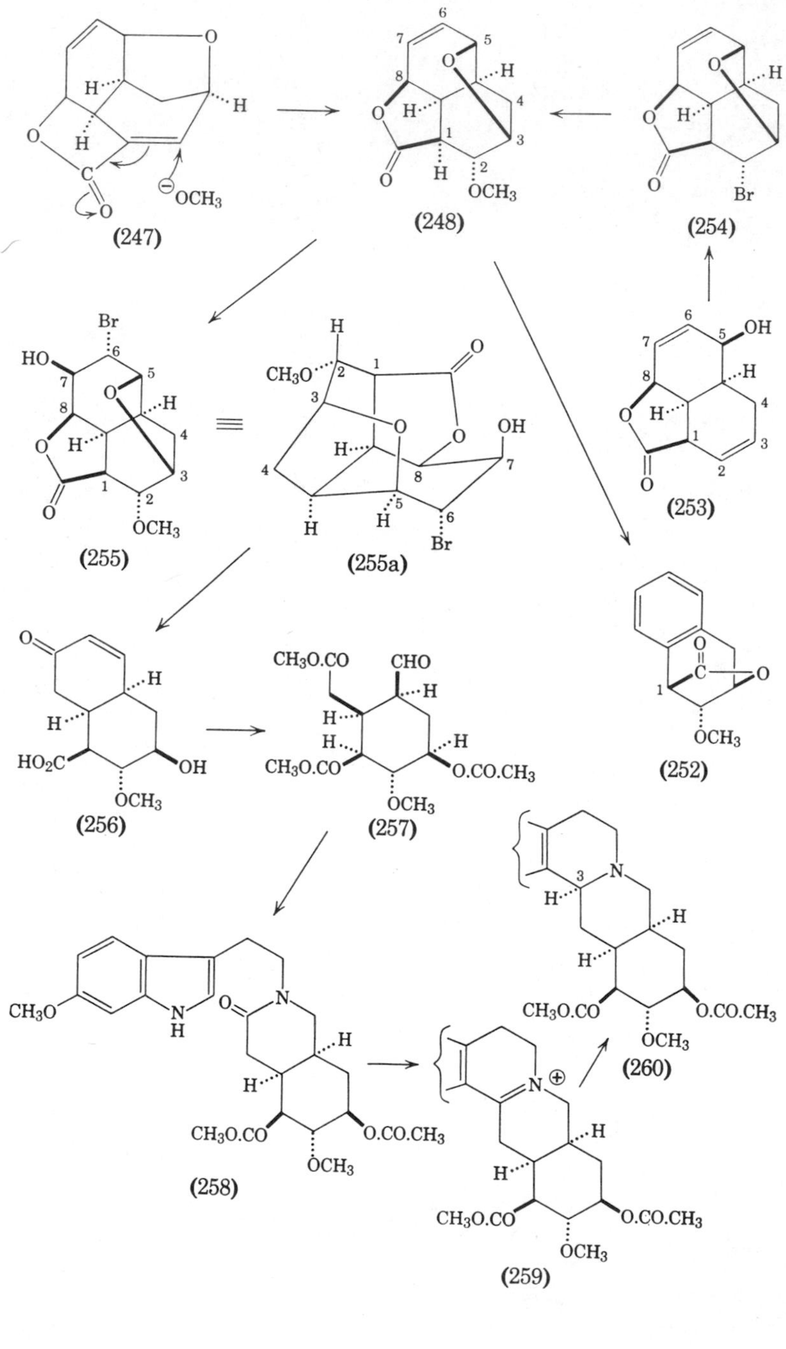

(247)
OCH3
(248)
OCH3
(254)
Br
(255)
HO
Br
OCH3
(255a)
CH3O
OH
Br
(253)
OH
(252)
OCH3
(256)
HO2C
OH
OCH3
(257)
CH3O.CO
CHO
CH3O.CO
O.CO.CH3
OCH3
(258)
CH3O
CH3O.CO
O.CO.CH3
OCH3
(259)
CH3O.CO
O.CO.CH3
OCH3
(260)
CH3O.CO
O.CO.CH3
OCH3

in which the key intermediate was the ether lactone (248). This intermediate could be prepared by a number of routes. It was originally obtained by Michael addition of methoxide ion to (247), which was itself the product of Meerwein-Ponndorf reduction of (245).* The α-configuration of the methoxyl group in (248) is a consequence of perpendicular attack by the anion on the less hindered side of the eneone system (Sec. 5-6a) (see 247). The configuration at C_1 is, of course, the thermodynamically more stable one, since the conditions of the addition reaction were sufficiently basic to cause equilibration at that center. An α-orientation for the C_1 hydrogen was established by conversion of (248) to the aromatic lactone (252) by treatment with stannic chloride.

The epoxide (243) could also be converted to the intermediate (248) by Meerwein-Ponndorf reduction of its methyl ester to afford (247) to which methoxide ion could be added as described above.

Finally, (248) could also be obtained from the lactone (253), itself readily available by reduction with aluminum isopropoxide of (241; R = CH_3), (242), or (240; R = CH_3). Bromination of (253) left the 6,7-double bond untouched and gave (254) as the product, formed by intramolecular attack of the 5-hydroxyl group on an intermediate bromonium ion. Conversion of (254) to (248) was accomplished by the action of methanolic sodium methoxide.

Although the double bond of (248) proved to be inert to the action of bromine at room temperature, it could be induced to add the elements of hypobromous acid by treatment with N-bromosuccinimide in sulfuric acid at 80°. The product was the expected diaxial bromohydrin (255 ≡ 255a), which, by oxidation to the corresponding ketone and subsequent reduction with zinc in glacial acetic acid, gave the $\alpha\beta$-unsaturated ketone (256).

With the preparation of (256), the primary object of the synthesis, i.e., the construction of an intermediate of type (238), was accomplished.

---

* The course of this reduction is thought to be as follows. Addition of hydride ion to the less hindered side of the C_8-carbonyl group of (245) with concomitant

(245) →

(249)   (250)   (251)

alcoholysis of the δ-lactone yields (249) which is then successively converted to (250), (251), and finally (247).

Conversion of (256) to the lactam (258) was unexceptional, proceeding through the aldehyde (257), and the product obtained by condensing it with 6-methoxytryptamine. The lactam (258) was transformed into the quaternary cation (259) by treatment with boiling phosphorus oxychloride.

Reduction of (259) with aqueous methanolic sodium borohydride led to *dl*-methyl-*O*-acetyl isoreserpate (260). The 3α-compound was the expected product, whether the reduction were subject to thermodynamic or steric control, since, in addition to being the more stable isomer, it is the one which would be formed by addition of hydride from the less hindered side of the molecule.

The last stereochemical problem to be surmounted in the synthesis of reserpine was the inversion of the configuration at C_3 of (260). Direct isomerization with acid could not be employed, since (260) has the stable configuration at C_3. The method adopted was to saponify the methoxy-carbonyl and acetoxy groups of (260), and, by treatment with *N*,*N'*-dicyclohexylcarbodiimide in pyridine, to remove the elements of water from the 3-isoreserpic acid thus obtained, with formation of the hitherto unknown 3-isoreserpic acid lactone. The only di-chair conformation possible for rings D and E of this compound (261) involves an extremely

(261)

severe non-bonded interaction between the lactone bridge and the indole ring, attached axially at C_3 to ring D. Because of this factor, 3-isoreserpic acid lactone, unlike 3-isoreserpic acid itself, possesses the unstable con-figuration at C_3. Accordingly, it was found possible to isomerize the lactone with pivalic acid in boiling xylene. The product of the isomeri-zation, *dl*-reserpic acid lactone (224; R = OCH_3), was readily transformed by standard procedures into *dl*-reserpine (219; R = OCH$_3$, R' = 3,4,5-trimethoxyphenyl). The total synthesis of reserpine was completed by resolution of the racemic material, with production of *l*-reserpine.

5-9. Conformational Transmission

It has already been pointed out (Sec. 2-7) that non-bonded interactions between substituents on a cyclohexane ring are often relieved by distortion of bond angles with consequent deformation of the cyclohexane chair. As a result of such deformation the properties of a molecule at a reactive site relatively remote from the origin of the angular distortion may be substantially altered. The term "conformational transmission" has been coined[283] to describe this type of phenomenon.

Ourisson's "reflex effect" (Sec. 2-7) may be regarded as an example of short range conformational transmission. It has been invoked[284] to explain the observation that, while friedelanone (262) contains none of its 4α-(a)-epimer at equilibrium, the D-homosteroid (263) is converted to

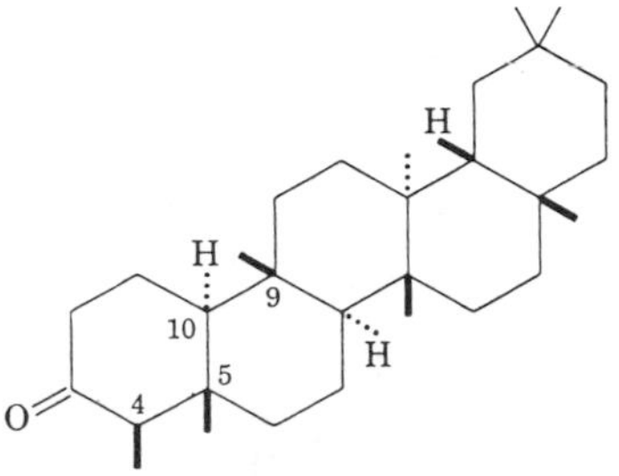

(262) 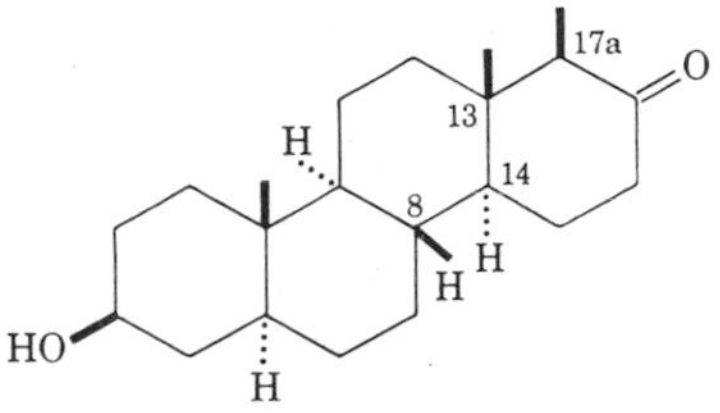   (263)

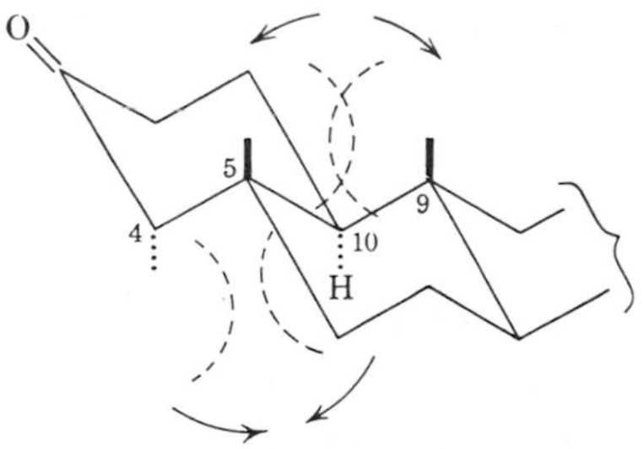

(264) 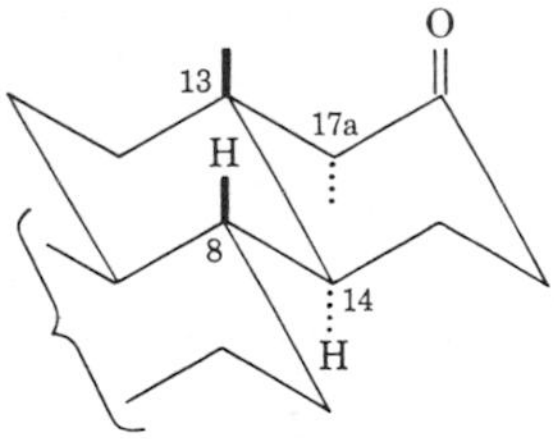   (265)

283 D. H. R. Barton, A. J. Head, and P. J. May, *J. Chem. Soc.*, **1957**, 935.
284 C. Sandris and G. Ourisson, *Bull. Soc. Chim. France*, **1958**, 1524.

its 17aα-epimer to the extent of about 30% under equilibrating conditions. The 1,3-diaxial interaction between the 5β- and 9β-methyl groups of 4-epifriedelanone produces a reflex effect which enhances the destabilizing

Table 2. Relative Rates of Condensation with Benzaldehyde of Steroid and Triterpenoid 3-Ketones

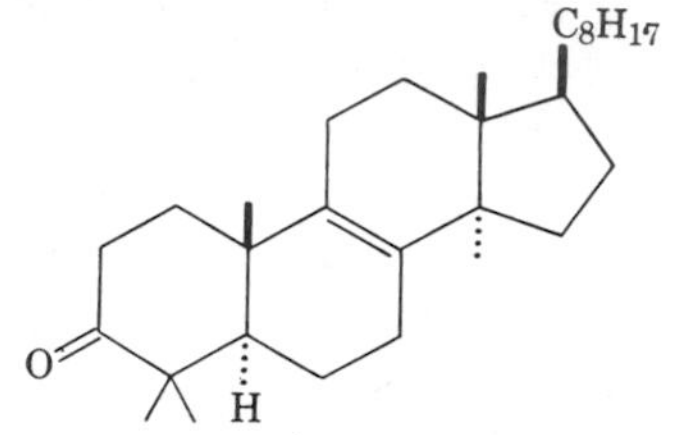

C_8H_{17}

($r = 100$)

(266)

X_2
H

C_8H_{17}

X_1

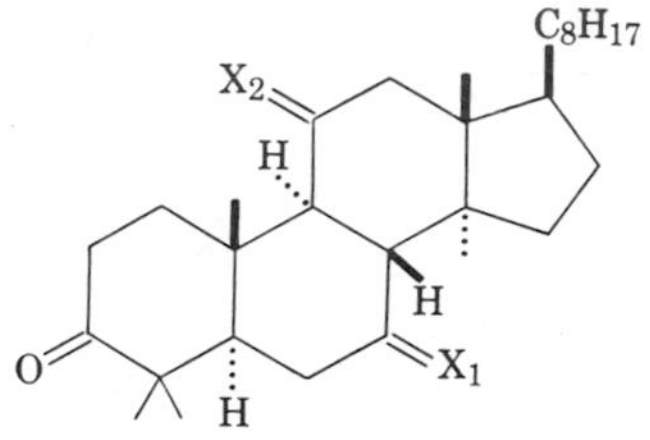

(a) $X_1 = H_2$, $X_2 = H_2$ ($r = 55$)
(b) $X_1 = O$, $X_2 = H_2$ ($r = 200$)
(c) $X_1 = H_2$, $X_2 = O$ ($r = 43$)
(d) $X_1 = O$, $X_2 = O$ ($r = 92$)

(267)

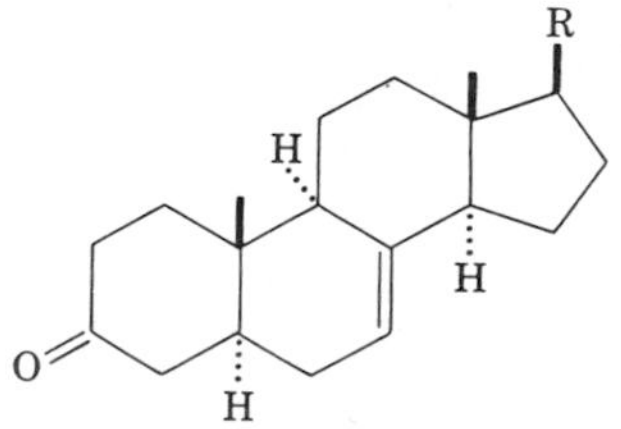

R

H
H
H

(a) $R = C_9H_{17}$ ($r = 43$)
(b) $R = C_9H_{19}$ ($r = 47$)

(268)

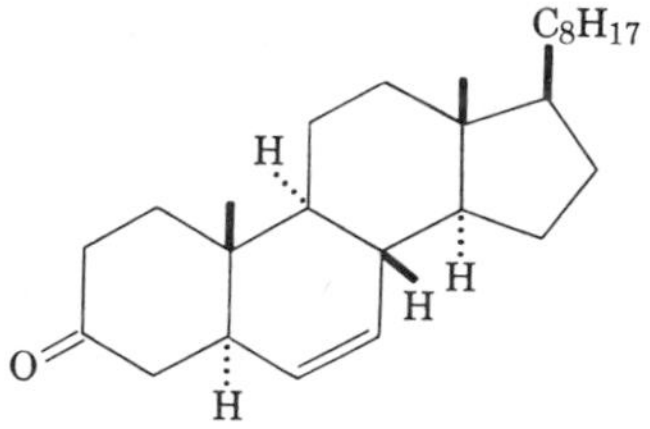

C_8H_{17}

H
H
H
H

($r = 645$)

(269)

R

H
H
H
H

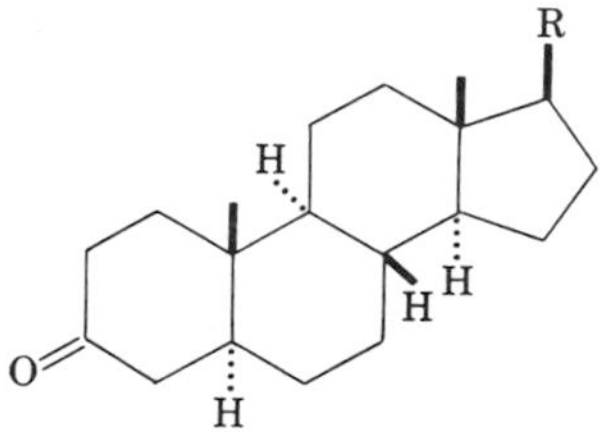

(a) $R = C_{10}H_{21}$ ($r = 180$)
(b) $R = C_8H_{17}$ ($r = 182$)
(c) $R = C_9H_{19}$ ($r = 188$)
(d) $R = C_9H_{17}$ ($r = 188$)

(270)

non-bonded interaction between the 4α-methyl group and the 10α-hydrogen atom (see 264). In the 17a-epimer of (263) the 1,3-diaxial ($CH_3 : CH_3$) interaction on the β face of the molecule is replaced by a 1,3-diaxial ($CH_3 : H$) interaction (between the 13β-methyl group and the 8β-hydrogen atom), and the reflex effect is therefore correspondingly less severe (see 265).

In recent years, a considerable amount of research has been devoted to examining long range conformational transmission in triterpenoids and steroids.[283,285-288] The most detailed of these studies have dealt with the influence of structural modifications on the rates of condensation of 3-ketones with benzaldehyde under base catalysis.[283,286,287]

Table 3.[219] Group Rate Factors, f, for the Condensation of 3-Oxo Derivatives of 5α-Steroids, Triterpenoids, and trans-Decalin with Benzaldehyde

Group	Group Rate Factor		
	5α-Steroid	Triterpenoid	trans-Decalin
7-Ketone	3.38	3.66	—
11-Ketone	0.62	0.78	—
12-Ketone	1.89	1.93	—
11α-Hydroxyl	0.67	0.64	—
11β-Hydroxyl	0.36	0.31	—
10-Methyl	—	3.40	4.20
4,4-Dimethyl	0.31	—	0.23
5(6)-Ethylenic linkage	1.0	1.0	—
6(7)-Ethylenic linkage	3.55	—	—
7(8)-Ethylenic linkage	0.24	0.31	0.23
8(9)-Ethylenic linkage	1.49	1.82	—
8(14)-Ethylenic linkage	0.50	—	—
9(11)-Ethylenic linkage	1.27	1.33	—
11(12)-Ethylenic linkage	2.18	—	—
14(15)-Ethylenic linkage	1.19	—	—
16(17)-Ethylenic linkage	1.14	—	—
7(8),9(11)-Diene	0.53	0.80	—

The condensation reaction is kinetically first order in benzaldehyde, alkali, and ketone. However, the alkali is not consumed, and since the technique adopted[283,286] was to carry out the reactions in a large excess of benzaldehyde, the condensation was then first order in ketone. The progress of the reaction was followed by observing the development of an absorption band at 292 mμ in the ultraviolet spectrum, typical of benzylidene ketones. Some representative results are listed in Table 2, in which the rates of condensation, r (relative to an arbitrary value of 100

285 D. H. R. Barton and A. J. Head, *J. Chem. Soc.*, **1956**, 932.

286 D. H. R. Barton, F. McCapra, P. J. May, and F. Thudium, *J. Chem. Soc.*, **1960**, 1297.

287 D. H. R. Barton, "Some Recent Progress in Conformational Analysis," in *Theoretical Organic Chemistry*, Butterworth's Sci. Publ., London, 1959, p. 127.

288 J. Mathieu, M. Legrand, and J. Valls, *Bull. Soc. Chim. France*, **1960**, 549.

for lanost-8-en-3-one), are placed in parentheses after the appropriate formulas. Molecules possessing an unsubstituted 4-position, or a second carbonyl group, were shown to undergo condensation with benzaldehyde only at the 2-position, under the experimental conditions.*

The results obtained cannot be explained simply on the basis of electrostatic effects or of bond induction; nor, since there is a wide variation in the rates of steroidal 3-ketones, is the presence of an axial 4β-methyl group essential for the operation of these long range effects. Consequently, buttressing is not an important contributing factor.

The likeliest explanation[286,292] is that long range effects of the type uncovered in the benzaldehyde condensation studies originate in bond

[289] For some recent examples in the yohimbanone series, and for leading references to investigations of 5α-steroidal 3-ketones, see J. D. Albright, L. A. Mitscher, and L. Goldman, *J. Org. Chem.*, **28**, 38 (1963).

[290] For a summary of this work, see L. F. Fieser and M. Fieser, *Steroids*, Reinhold Publishing Corp., New York, 1959, pp. 276–279.

[291] E. J. Corey and R. A. Sneen, *J. Am. Chem. Soc.*, **77**, 2505 (1955).

[292] See also R. Bucourt, *Bull. Soc. Chim. France*, **1962**, 1983; *ibid.*, **1963**, 1262.

[293] B. Berkoz, E. P. Chavez, and C. Djerassi, *J. Chem. Soc.*, **1962**, 1323.

[294] A. S. Dreiding, *Chem. Ind. (London)*, **1954**, 1419; see also T. G. Halsall and D. B. Thomas, *J. Chem. Soc.*, **1956**, 2431.

[295] H. H. Inhoffen, G. Kölling, G. Koch, and I. Nebel, *Chem. Ber.*, **84**, 361 (1951); R. O. Clinton, R. L. Clarke, F. W. Stonner, D. K. Phillips, K. F. Jennings, and A. J. Manson, *Chem. Ind. (London)*, **1961**, 2099.

* In most of the cases investigated, 3-oxo derivatives of *trans*-decalins enolize predominantly towards the 2-position,[289] so that this is the preferred site for bromination, aldol condensation, etc. In line with this, 5α-cholest-2-ene has been shown to be more stable than 5α-cholest-3-ene both by infrared studies and by measurements of their heats of hydrogenation.[290] One possible explanation for this difference which has been put forward on the basis of vector analysis calculations in the octalin series[291,292] is that the 6β-(a)-hydrogen atom and the 10β-(a)-methyl group of *trans* A/B steroids approach more closely in compounds containing a Δ^3-double bond than than they do in the Δ^2-isomers. Support for this view is provided by the recent observation that the proportion of Δ^3-enol acetate obtained from a 3-oxo-5α-steroid is greater when the 10β-methyl group is absent.[293] That this increase is further augmented if the molecule contains a 4α-methyl group, but disappears completely in the 2α-methyl series,[293] suggests that hyperconjugative stabilization of the enolic double bond is also an important factor in determining the direction of enolization.

It was formerly thought that 3-oxo-5β-steroids invariably enolized toward the 4-position, and it was suggested by Dreiding[294] that this could be explained if the assumption were made that a cyclohexane system with a trigonal carbon atom in an axial position has a lower energy content than the corresponding one with a tetragonal carbon atom in the same position. However, it is now known that reactions which involve the formation of an enolic intermediate may result in substitution of a 3-oxo-5β-steroid at either or both of the positions α to the carbonyl group.[295] The factors which determine the direction of enolization in these cases have yet to be established.

angle distortion produced by the introduction of a substituent, or more especially of an unsaturated linkage, into the molecule. In some cases the variation in rate produced by a small structural alteration is very substantial. Thus 5α-cholest-6-en-3-one (269) reacts some fifteen times faster than the related compounds (268a) and (268b) which differ significantly only in having a 7(8)-double bond instead of a 6(7)-double bond. Recently, Robinson and Whalley[295a] have elaborated upon the concept of conformational transmission to account in a semiquantitative manner for the results obtained by Barton and his collaborators.

The effect of introducing more than one structural modification into a steroid or triterpenoid molecule has been found to be cumulative. All the data so far gathered[283,286] have been correlated by the assignment of group rate factors to the various structural modifications investigated. The rate of condensation, R_s, of a 5α-steroid or triterpenoid 3-ketone with benzaldehyde may be calculated from the relationship

$$R_s = R_0 \prod_1^i f_i$$

where R_0 is the rate of condensation of the parent ketone, and $f_1, f_2, \ldots, f_i$ are the group rate factors for the various additional functional groups in the ketone under consideration. The group rate factors listed in Table 3 were calculated[287] from measurements made on monosubstituted ketones. Table 4 presents a comparison between the observed and calculated rates

Table 4. Observed and Calculated Rates of Condensation of 5α-3-Oxo Steroids with Benzaldehyde (Relative to lanost-8-en-3-one = 100)

Parent Compound and Its Rate	Derivative	Appropriate Group Rate Factors	Calculated Rate	Observed Rate
5α-Ergost-22-en-3-one (270d) ($r = 188$)	7,11-Dioxo-	7-(C=O); 3.38 11-(C=O); 0.62	393	360
5α-Cholestan-3-one (270b) ($r = 182$)	4,4-Dimethyl-7(8)-dehydro-	4,4-Dimethyl; 0.31 7(8)-(C=C); 0.24	13.5	17
5α-Ergost-22-en-3-one (270d) ($r = 188$)	7(8),14(15)-Dehydro-	7(8)-(C=C); 0.24 14(15)-(C=C); 1.19	54	62

[295a] M. J. T. Robinson and W. B. Whalley, *Tetrahedron*, **19**, 2123 (1963).

of condensation of three typical steroidal ketones. The agreement obtained
is satisfactory; it would not be expected to be perfect, since the functional
groups must interact with each other as well as with C_2.

The concept of conformational transmission has been invoked also to
account for the alterations in the rates of solvolysis of steroidal 17β-
tosylates produced by structural modifications in ring A.[288] The dependence of the pK_a values of A-phenolic steroids on the nature of ring A and
of the side chain,[296] and the influence of 3β- and 17β-substituents on the
rate of addition of bromine to steroidal 5,6-double bonds,[297] are thought
to be due to the operation of long range electrical effects and not to
conformational transmission of bond angle distortion.

Other long range effects which have been investigated in the steroid
field are the variation in the rates of saponification of 17α-hydroxy-20-
oxo-21-acetoxysteroids resulting from structural changes in rings A and
C,[298] and the influence of the structures of rings A and C on the rates of
oxidation of the side chains of 17α,21-dihydroxy-20-oxosteroids.[298] As
yet, these effects have not been satisfactorily explained.

[296] M. Legrand, V. Delaroff, and J. Mathieu, *Bull. Soc. Chim. France*, **1961**, 1346.
[297] V. Schwarz, S. Hermanek, and J. Trojanek, *Chem. Ind. (London)*, **1960**, 1212;
Collection Czech. Chem. Commun., **26**, 1438 (1961); **27**, 2778 (1962); V. Schwarz and
S. Hermanek, *Tetrahedron Letters*, **1962**, 809; see also P. E. Petersen, *ibid.*, **1963**, 181.
[298] M. Pesez and J. Bartos, *Bull. Soc. Chim. France*, **1962**, 1928.

Chapter 6

Conformational Analysis
in Carbohydrate Chemistry

6-1. Introduction

The application of conformational analysis is of great actual or potential usefulness in carbohydrate chemistry for two reasons. Firstly, all the sugars belong to a few families of diastereomers; within these families they differ only in their steric arrangement, and these differences can therefore be accounted for by conformational factors. If these conformational factors could be exactly evaluated, the different chemical and physical properties of the various sugars could be explained and predicted.

Secondly, a reducing sugar in solution represents an equilibrium mixture of one open chain and at least four cyclic isomers. The physical and chemical properties of sugars in solution will depend on the composition of this equilibrium mixture which, in turn, is a function of the free energy content of each isomer. If all the relevant conformational factors were exactly known, the composition of the equilibrium mixture could be calculated.

Despite the importance of this subject, much less work has been done on the conformational analysis of carbohydrates than, for example, on that of the cyclohexanes; in particular, little is known yet about its quantitative aspects, and most of the discussions in the literature are restricted to qualitative considerations. Even these discussions have often been fruitful and provide a substantial part of this chapter. Recent attempts at quantitative applications have given, despite all the approximations involved, results in reasonable agreement with observed data (which themselves are not so abundant as one would wish). In this chapter the quantitative aspect will be developed as far as possible; inevitably, owing to lack of sufficient information, this approach will involve more speculations than the preceding chapters have contained. It is hoped that a stimulus will thereby be provided for further work on

351

the conformations of carbohydrates: rapid progress could be made in the next few years, particularly by the exploitation of nuclear magnetic resonance spectroscopy.

Conformational analysis of sugars in their pyranose forms differs from that of simple cyclohexane systems in two respects: in the accumulation of hydroxyl groups and in the presence of an oxygen atom in the ring. The effect of the first factor can be studied in the chemistry of the cyclitols without the disturbing effects—steric and electronic—of a ring oxygen atom and without the possibility of ring opening. It is convenient, therefore, that the discussion of the carbohydrates be preceded by that of the cyclitols.

6–2. The Cyclitols

a. The Conformation of Cyclitols.   The term "cyclitol" is used to describe the polyhydroxycyclohexanes, of which the inositols form the most important group. The inositols are particularly suitable model compounds for the study of stereochemistry and of conformational effects: they constitute the only group of cyclohexanes substituted on each carbon atom in which every possible diastereomer is known; moreover, the hydroxyl groups, being reactive, allow a great variety of reactions. The more stable conformation for each isomer is readily predicted from the tenets of conformational analysis: it is the chair form which has fewer axial hydroxyl groups. These conformations are shown, together with the distinguishing prefixes of the inositols, for each isomer in Fig. 1. Similar considerations apply to the conformations of other cyclitols.

These conformations have not been proved by direct methods such as X-ray crystallography or electron diffraction. They are, however, in accordance with certain reactions of the cyclitols, such as their dehydrogenation, complex formation with borate, and ease of esterification. Physical evidence also substantiates these assigned conformations. Thus the absorption band at 873 ± 11 cm.$^{-1}$ in the infrared spectra, attributed to deformation of equatorial C—H bonds, is shown by all the cyclitols investigated except *scyllo*-inositol and *scyllo*-quercitol.[1] The NMR spectra[2] of the inositols and their acetates can also be interpreted on the basis of the conformations in Fig. 1, and they agree closely with those of the corresponding hexachlorocyclohexanes[3] (whose conformations have been

[1] S. A. Barker, E. J. Bourne, R. Stephens, and D. H. Whiffen, *J. Chem. Soc.*, **1954**, 4211.

[2] R. U. Lemieux, R. K. Kullnig, H. J. Bernstein, and W. G. Schneider, *J. Am. Chem. Soc.*, **79**, 1005 (1957).

[3] S. Brownstein, *J. Am. Chem. Soc.*, **81**, 1606 (1959).

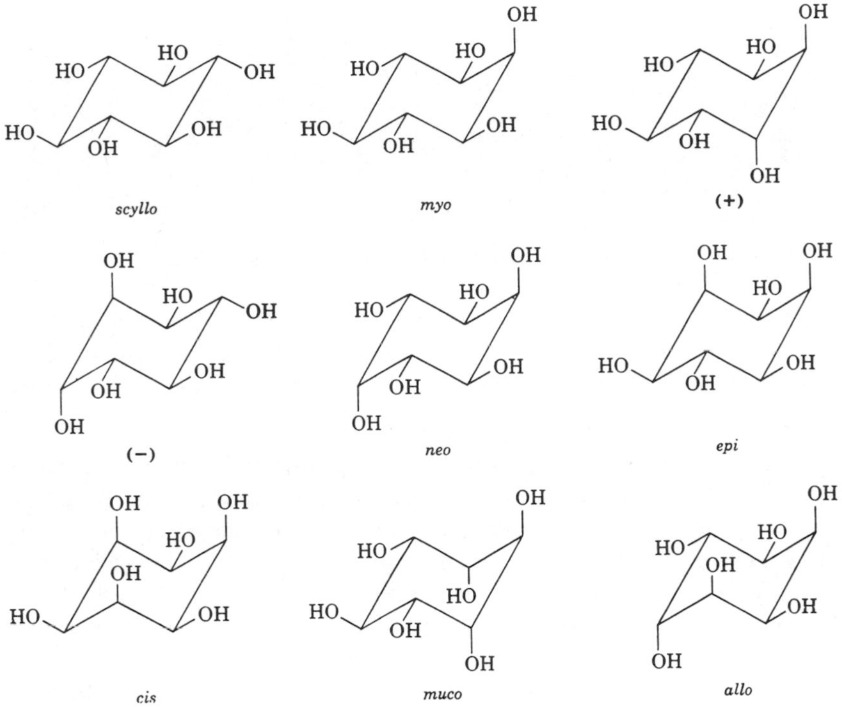

Fig. 1. The nine inositols.

ascertained by other methods). More modern investigations, including evaluation of spin-spin coupling, have been carried out on bromo, nitro, and amino derivatives of inositols, but not on the inositols themselves (see Sec. 6-5c).

It is to be noted that, in those inositols which have *three* axial hydroxyl groups, the two possible chair forms are equally stable. In (unsubstituted) *cis*- and *muco*-inositol the two chair conformations of each are superimposable and are therefore indistinguishable. In *allo*-inositol they are mirror images of each other: *allo*-inositol is, therefore, an inseparable racemic mixture of rotational isomers. Every hydroxyl group in *cis*-, *muco*-, and *allo*-inositol has an equal chance of being axial.

Reeves[4] suggested that *muco*-inositol may be in a flexible conformation. The calculated total of the interaction energies for the chair form, 5.1 kcal./mole (see Sec. 6-2c), however, is not sufficiently large to cause a flip-over into a skew form for which the excess conformational energy would be considerably greater (allowing, in addition to the 3.8 kcal./mole of the flexible form, for at least two eclipsed or semi-eclipsed interactions

[4] R. E. Reeves, *Ann. Rev. Biochem.*, **27**, 15 (1958).

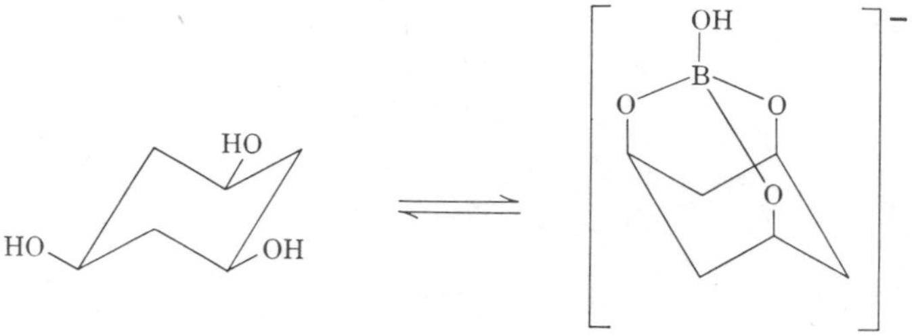

Fig. 2. Formation of tridentate borate.

between neighboring hydroxyl groups). It is relevant that γ-hexachloro-cyclohexane, which has the same configuration as *muco*-inositol, has been shown[5] by X-ray crystallography to be in the chair form (at least in the solid state).

b. Tridentate Borate Complexes. The accumulation of hydroxyl groups on a cyclohexane ring in cyclitols led to the discovery[6] of tridentate borate complexes, i.e., borate anions in which three oxygen atoms of the same molecule are bound to the boron atom. Polyols of the cyclohexane series form such anions in borate solutions if they have three hydroxyl groups in *cis*-1,3,5 relationship (this arrangement is not possible in the pyranose form of sugars). Formation of the tridentate complex involves an inversion from the more stable chair form, in which the three hydroxyl groups are equatorial, to the other chair form, in which they are axial (Fig. 2). An interesting example is the double borate dianion[7] (Fig. 3) formed from *scyllo*-inositol. Tridentate borates of non-cyclic compounds have recently been described,[8] but from the viewpoint of conformational analysis those derived from substituted cyclohexanes are of interest because they allow the quantitative evaluation of conformational interactions.

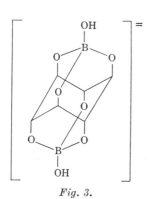

Fig. 3.

From the change of pH caused by successive additions of cyclitols to a borate solution, the equilibrium constants of complex formation have been calculated;[6] they show that the complexes are formed from cyclitol and borate in a 1:1 ratio. The stability of the tridentate borates depends

[5] G. W. van Vloten, C. A. Kruissink, B. Strijk, and J. M. Bijvoet, *Acta Cryst.*, **3**, 139 (1950).

[6] S. J. Angyal and D. J. McHugh, *J. Chem. Soc.*, **1957**, 1423.

[7] A. Weissbach, *J. Org. Chem.*, **23**, 329 (1958).

[8] J. Dale, *J. Chem. Soc.*, **1961**, 922.

scyllo *epi* *cis*

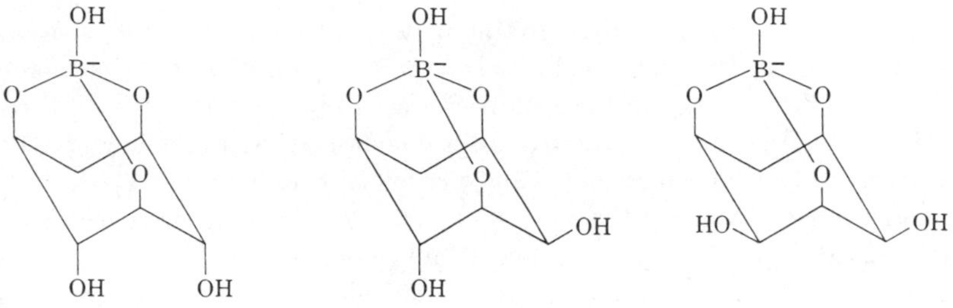

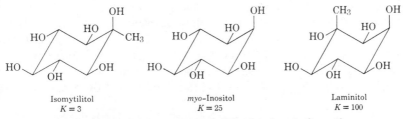

Fig. 4. Borate complexes of some quercitols.

on the steric disposition of the *free* hydroxyl groups in the complex: the more of these are in *axial* positions, the less stable the complex. This is illustrated by the values of the equilibrium constant in the following two series, which are arranged by decreasing numbers of free axial hydroxyl groups in the complex: *scyllo*-quercitol, 5.0; *epi*-quercitol, 310; and *cis*-quercitol, 7900 (Fig. 4); *myo*-inositol, 25; *epi*-inositol, 700; and *cis*-inositol, 1.1×10^6. When the constitution of a cyclitol allows the formation of both the tridentate and the classical *cis*-1,2 type of complex, the former predominates.

A particularly striking example[9] of conformational effects is provided by the comparison of the three compounds shown in Fig. 5. The configuration of all hydroxyl groups is identical in the three compounds; they differ only in the presence or absence, and in the configuration, of a methyl group. In the first case the methyl group is conformationally unfavorable to the formation of a tridentate borate because it is equatorial in the free cyclitol but axial in the borate complex. The converse is true for the third compound, whereas the second forms an intermediate case.

c. Non-bonded Interaction Energies. Since the equilibrium constants of tridentate borate formation depend on the non-bonded interactions

Isomytilitol $K = 3$ *myo*-Inositol $K = 25$ Laminitol $K = 100$

Fig. 5. Equilibrium constants of tridentate borate formation.

[9] S. J. Angyal and J. E. Klavins, unpublished data.

existing in the molecules, they enable us to calculate the values of inter-action energies. In order to do so, two assumptions have to be made: (1) that the free energy of formation of a tridentate anion from axial *cis*-1,3,5-hydroxyl groups and borate ion is constant, that is, independent of the nature of other substituents on the cyclohexane ring; (2) that the free energies of conformational isomers are additive functions of energy terms associated with the presence of non-bonded interactions, that is, the occurrence of one interaction in a molecule does not affect the mag-nitude of another one. Strictly speaking, neither of these statements is true, but their assumption appears to introduce only small errors.

Table I. Energies (in kcal./mole) of Interactions be-tween Substituents in Cyclitols (aqueous soln., 25°)

Between	ax. OH	ax. CH$_3$	eq. OH
ax. H[a]	0.45	0.9	(0)
ax. OH	1.9	1.6	0.35
eq. OH	0.35	0.35	0.35
eq. CH$_3$	0.35		0.35

[a] Since in a monosubstituted cyclohexane an axial substituent X interacts with two axial hydrogen atoms, the interaction energy between *one* hydrogen atom and the group X equals $\frac{1}{2}\Delta G_X^\circ$ (as defined in Chap. 2).

The interaction energies are calculated in the following way. The energies of each non-bonded interaction in each cyclitol, and in their tridentate borates, are listed and separately added up; for the borate a term is added to represent the free energy change on formation of the tridentate borate from the three axial hydroxyl groups; and the differ-ence between the totals for each cyclitol and for its borate is equated with the experimentally determined free energy difference (calculated from $\Delta G^\circ = -RT \ln K$). A series of equations results from which all the unknown quantities can be calculated. Working with cyclitols and with *C*-methylinositols (like those in Fig. 5), Angyal and McHugh[10] and Angyal and Klavins[9] obtained the values listed in Table 1. The inter-actions between axial groups refer to 1:3-interactions,* those of equatorial

[10] S. J. Angyal and D. J. McHugh, *Chem. Ind. (London)*, **1956,** 1147.

* What is described as "interaction between two axial groups" tacitly includes the interaction between each axial group and the carbon atom to which the other axial group is attached.

groups to the interaction with a group *gauche* on an adjacent carbon atom (interaction with an equatorial and with an axial group being taken as equal). All these interactions are repulsive.*

All other interactions (e.g., between two axial hydrogen atoms, or between a hydrogen atom on one carbon and any group on the adjacent carbon) are believed to be negligibly small (less than 0.1 kcal./mole). The interaction shown in the table of an axial methyl group with an axial hydrogen atom agrees with the accepted value (Sec. 2-2b); that of an axial hydroxyl group with an axial hydrogen atom is in good agreement with other values obtained in hydroxylic solvents (though it is larger than values obtained in other solvents). The interaction energy between an axial hydroxyl and an axial methyl group was found to be somewhat smaller than expected (cf. Sec. 2-3a).

It was not possible to determine the interaction energies of the hydroxymethyl group, required for calculations in carbohydrate chemistry, by this method because the primary hydroxyl group itself took part in the reaction with borate. Preliminary experiments[9] indicated that the interactions of the methoxymethyl group are approximately of the same value as those of the methyl group. One can safely assume that this would also apply to the hydroxymethyl group; only the two hydrogen atoms on the carbon interact, the oxygen atom being turned away (cf. Sec. 7-2).

By using the figures in Table 1, the following values can be calculated for the relative free energy content† of the various inositols at 25°: *myo* 3.0, *scyllo* 3.2, (±) 3.55, (+) 4.0, *neo* 4.3, *epi* 4.9, *allo* 5.05, *muco* 5.1, *cis* 8.05. In calculating these values the interaction energies in the more stable chair form have been totaled; entropy of mixing, $RT \ln 2 = 0.4$ kcal./mole, has been deducted for *allo*-inositol, because its two chair forms are mirror images, and for (±)-inositol, because of the existence of two enantiomers; and 1.1 kcal./mole has been added for *scyllo*-, 0.65 for *cis*-, and 0.4 for (±)- and *neo*-inositol, because these molecules have symmetry numbers of 6, 3, and 2, respectively (cf. Sec. 2-3b). Values in good agreement with these calculated ones have been obtained, in several cases, from the equilibrium constants of the acid-catalyzed epimerization of isomeric inositols.[11]

[11] S. J. Angyal, P. A. J. Gorin, and M. Pitman, *Proc. Chem. Soc.*, **1962**, 337.

* Owing to hydrogen bonding, the interaction between hydroxyl groups in aprotic solvents may be attractive; for example, it has been estimated that in dilute tetrachloroethylene solution the attraction between two hydroxyl groups in adjacent equatorial positions has the value 0.8 kcal./mole (J. Pitha, J. Sicher, F. Šipoš, M. Tichý, and S. Vašíčková, *Proc. Chem. Soc.*, **1963**, 301). See also Sec. 6-5b.

† These are relative to an imaginary inositol in which there are no non-bonded interactions.

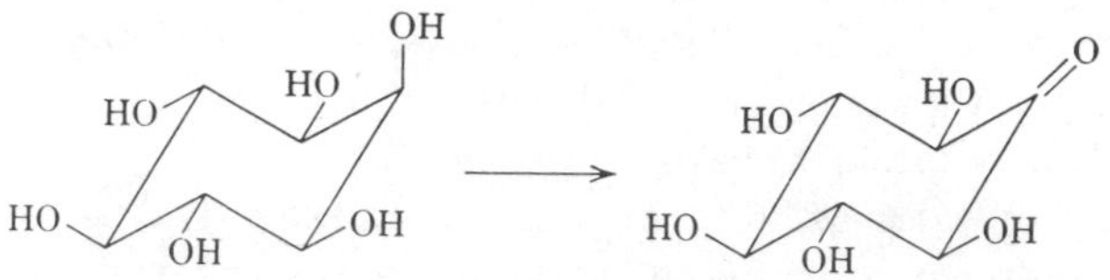

Fig. 6.

d. Dehydrogenation. A reaction showing high conformational specificity is the dehydrogenation of cyclitols which can be carried out enzymically by *Acetobacter suboxydans*, or catalytically by oxygen over platinum. In both reactions, *only axial* hydroxyl groups are dehydrogenated to yield inososes. The reaction differs from the dehydrogenation of steroids by chromic acid (Sec. 5-3a) in which an axial hydroxyl group reacts at a faster rate than an equatorial one; in this reaction of cyclitols, *only* axial groups react at all.[12] *myo*-Inositol, for example, is converted in high yield to *scyllo*-inosose (Fig. 6).

Catalytic dehydrogenation has been applied[13] also in carbohydrate chemistry where it generally attacks an axial hydroxyl group in the more stable pyranose form; dehydrogenation by *Acetobacter* is not applicable to sugars in cyclic forms.

e. Epoxide Opening and Migration. Opening of epoxides in the cyclitol series proceeds according to the Fürst-Plattner rule (Sec. 5-4b). Either of the two half-chair forms may react, but the predominant product of the reaction will be the one derived by diaxial ring opening from the predominant half-chair form of the epoxide.[14] The immediate result may be a chair containing more axial than equatorial hydroxyl groups which then undergoes chair inversion. The over-all result is ring opening in such a way that the predominant product has the larger number of axial groups in its stable conformation.*

As an example, the behavior of 1,2-anhydro-*myo*-inositol will be considered (Fig. 7). The more stable half-chair form has four equatorial (or quasi-equatorial) hydroxyl groups; diaxial opening of the epoxide, either by acids or by bases, gives ($\pm$)-inositol which is the main product

---

[12] S. J. Angyal and L. Anderson, *Advan. Carbohydrate Chem.*, **14,** 135 (1959).

[13] K. Heyns and J. Lenz, *Angew. Chem.*, **73,** 299 (1961); K. Heyns, J. Lenz, and H. Paulsen, *Chem. Ber.*, **95,** 2964 (1962); K. Heyns and H. Paulsen, *Advan. Carbohydrate Chem.*, **17,** 169 (1962).

[14] S. J. Angyal, *Chem. Ind.* (*London*), **1954,** 1230.

* Surprisingly, the recently described opening of epoxides with sodium borohydride in methanol, which gives a methyl ether, occurs contrary to the Fürst-Plattner rule: M. Nakajima, N. Kurihara, and T. Ogino, *Chem. Ber.*, **96,** 619 (1963). The reaction is not clearly understood, and its further investigation might be rewarding.

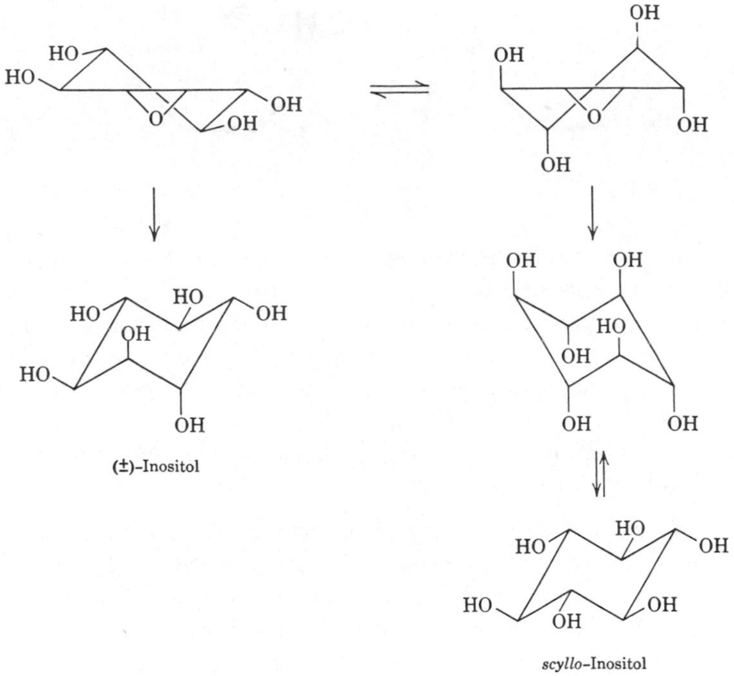

Fig. 7. Hydrolysis of 1,2-anhydro-myo-inositol.

of the reaction. The other half-chair form, with four axial (or quasi-axial) hydroxyl groups, would be present in very small amounts; its diaxial ring opening yields scyllo-inositol (initially in its less stable chair conformation) which is, accordingly, a minor product of the reaction.*

The anhydroinositols have provided an opportunity for studying[15] the phenomenon of "epoxide migration," i.e., opening of an epoxide ring by rearward attack from an adjacent trans situated hydroxyl group, with formation of another epoxide ring. Occurrence of this rearrangement has often been postulated in carbohydrate chemistry,[16] but no clear-cut example had previously been described. All anhydroinositols in which there is a trans-hydroxyl group adjacent to the epoxide ring undergo epoxide migration in alkaline solution at room temperature. For example, 1,2-anhydro-allo-inositol rearranges to 1,2-anhydro-neo-inositol (Fig. 8).

[15] S. J. Angyal and P. T. Gilham, J. Chem. Soc., 1957, 3691.

[16] F. H. Newth, Quart. Rev. (London), 13, 30 (1959).

* Strictly speaking, the argument applies, not to the two ground states, but to the two transition states; in the latter, however, the same arrangement of axial and equatorial substituents is found, and the energy of the transition state leading to (±)-inositol is considerably lower than that of the other one.

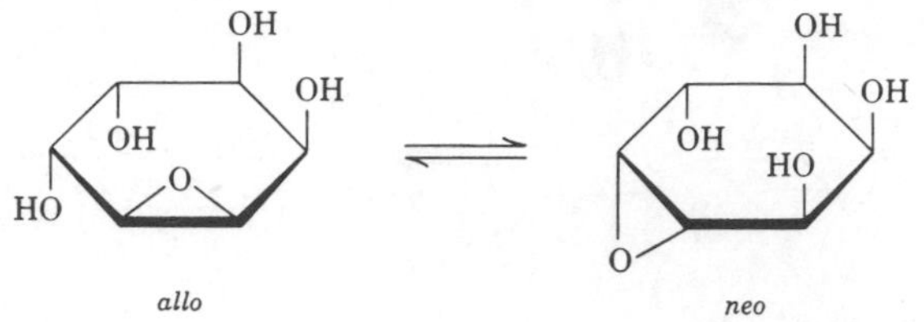

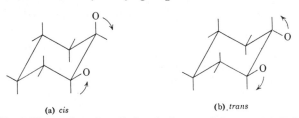

Fig. 8.

The reaction is reversible, and the position of the equilibrium is in accordance with conformational considerations, the isomer with the fewer axial hydroxyl groups being the more stable. Of the two anhydrides, the *allo* isomer has two axial (or quasi-axial) hydroxyl groups, whereas the *neo* isomer has only one; in the equilibrium mixture they were found in the ratio of 1:9.

Similar considerations apply to epoxide opening and migration in carbohydrate chemistry (Sec. 6-8d).

f. Cyclic Ketals and Acetals. Cyclitols, like other glycols, form cyclic ketals when treated with ketones in the presence of acids and dehydrating agents. Acetone and zinc chloride is the most frequently used combination.[17] Under these conditions vicinal *cis-* (but not *trans-*) hydroxyl groups react to form 1,3-dioxolane rings. (The formation of six-membered cyclic ketals from cyclitols has never been observed; the ring of such a ketal, which incorporates two axial oxygen atoms of the cyclitol, would be strained.[16a]) The distance in an undistorted chair between vicinal equatorial hydroxyl groups (*trans*) is the same as between an axial and an equatorial hydroxyl group (*cis*). However, formation of the dioxolane ring, which cannot readily incorporate a dihedral angle of 60°, requires distortion of the chair (Fig. 9), and, as shown in Fig. 33 of Chap. 2, the energy required for such distortion is less in the case of *cis-* than in that of *trans*-hydroxyl groups.[17]

(a) *cis*   (b) *trans*

Fig. 9. Formation of cyclic ketals from cyclohexane-1,2-diols.

[16a] W. A. C. Brown, G. Eglinton, J. Martin, W. Parker, and G. A. Sim, *Proc. Chem. Soc.*, **1964**, 57.

[17] S. J. Angyal and C. G. Macdonald, *J. Chem. Soc.*, **1952**, 686.

Surprisingly, it was found that a vicinal *trans* pair reacts readily with acetone when the cyclitol already has two *cis*-ketals attached to it. 1,2:5,6-Di-*O*-isopropylidene-(−)-inositol (Fig. 10) and the corresponding *epi* isomer are relevant examples. Originally, Angyal and Macdonald explained[17] the reactivity of the *trans* pair in terms of the distortion of the chair conformation: the formation of each *cis*-ketal has moved two axial groups away from their axial neighbors, and thus the way had been opened for the remaining axial groups to undergo the movement indicated in Fig. 9b, necessary for ketal formation on the *trans*-hydroxyl groups.

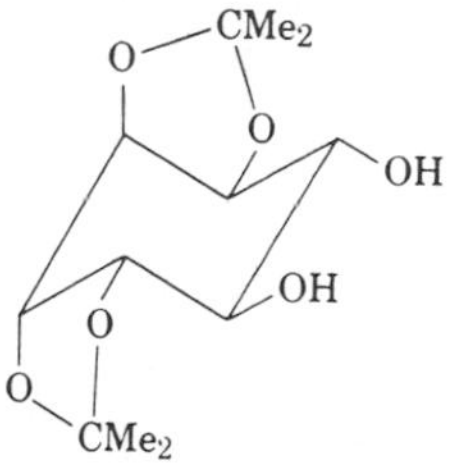

It appears now, however, that ketal formation of the *trans*-hydroxyl groups can be better explained in terms of skew forms. From the study of hydrogen bonding by Kuhn's method (Sec. 3-4a), evidence has recently been adduced[18] that the *cis*-diketals of inositols are in a skew-boat conformation, at least in the very dilute solution in carbon tetra-

Fig. 10.

chloride in which the infrared spectra were taken. In such a skew-boat form the dihedral angle between the free vicinal *trans*-hydroxyl groups is 49° as contrasted to the angle of 60° in an undistorted chair; hence ketal formation is facilitated.* These considerations would apply to the formation of other five-membered cyclic derivatives from cyclitols. Thus 1,2:5,6-di-*O*-isopropylidene-(−)-inositol (Fig. 10) gives a cyclic phosphate with diphenyl phosphorochloridate,[19] just like inositols with vicinal *cis*-hydroxyl groups,[20] whereas a *trans* pair in the undistorted chair conformation yields a diphosphate.[21]

It is probable that 1,2:3,4-diacetals of pyranoid sugars also take up a skew-boat conformation, but the question does not appear to have been investigated.

It is interesting to note that the *cis*-1,2:3,4-diketal system is more stable than the *cis*-1,2:4,5 arrangement; the latter would tend to twist toward the energetically unfavorable boat conformation. Whenever either diketal can be formed (*epi*-inositol, *cis*-inositol), the former is the

[18] S. J. Angyal and R. M. Hoskinson, *J. Chem. Soc.*, **1962**, 2991.

[19] G. L. Kilgour and C. E. Ballou, *J. Am. Chem. Soc.*, **80**, 3956 (1958).

[20] F. L. Pizer and C. E. Ballou, *J. Am. Chem. Soc.*, **81**, 915 (1959); T. Posternak, *Helv. Chim. Acta*, **42**, 390 (1959).

[21] S. J. Angyal and M. E. Tate, *J. Chem. Soc.*, **1961**, 4122.

* The dihedral angle between the two vicinal carbon atoms in 2,2-dimethyl-dioxolane was found (by NMR) to be 41°: R. U. Lemieux, J. D. Stevens, and R. R. Fraser, *Can. J. Chem.*, **40**, 1955 (1962).

major product.* (The 1,2:4,5-diketal arrangement is not possible, however, for pyranoid sugars.)

Under more vigorous conditions—reaction with cyclohexanone while water is azeotropically removed, or ketal interchange with acetone diethyl ketal[23]—*trans*-ketals can be formed from cyclitols which have only one *cis*-ketal group. For example, a tricyclohexylidene derivative of *myo*-inositol can be made.[24] In all cases, however, the *trans* linked five-membered rings are less stable than those formed by *cis* fusion and can be preferentially removed by hydrolysis.

6–3. Determination of the Conformations of Sugars

a. Historical Notes. The term "conformation" was introduced into organic chemistry by Haworth in his book on the constitution of sugars.[25] There had been some previous discussion on the chair and boat forms of pyranoid sugars by Sponsler and Dore[26] who applied the concept of puckered rings to the interpretation of the X-ray diagram of cellulose. They found that the chair conformation of glucopyranose units gave a satisfactory explanation of the spacings in the X-ray diagram of cellulose, whereas boat conformations failed to do so.

Haworth recognized the importance of this new concept in carbohydrate chemistry and, with the aid of models, considered the various possible shapes of sugar molecules. There were, at that time, no experimental data available to guide his speculation, but Haworth predicted that "these considerations open up a large field of inquiry into the conformation of groups as distinct from structure or configuration."

An application of conformational analysis (though not yet so called) was the realization of the physical and chemical similarities between "homomorphous" sugars, i.e., sugars which have the same configurations of the atoms which compose the pyranose ring, and therefore the same conformation of the ring. One such homomorphous series includes, for

[22] F. W. Lichtenthaler and H. O. L. Fischer, *J. Am. Chem. Soc.*, **83**, 2005 (1961).

[23] S. J. Angyal and R. M. Hoskinson, *J. Chem. Soc.*, **1962**, 2985.

[24] M. E. Tate, unpublished observation.

[25] W. N. Haworth, *The Constitution of Sugars*, Edward Arnold & Co., London, 1929, p. 90.

[26] O. L. Sponsler and W. H. Dore, *Colloid Symposium Monograph*, **4**, 174 (1926).

* A model shows that twisting together a *cis* pair of hydroxyl groups on C-1 and C-2 also twists together *cis*-3,4 substituents but twists apart *cis*-4,5 groups. Probably for the same reason 1,4-dideoxy-1,4-dinitro-*neo*-inositol appears to yield only a monoketal with acetone.[22]

example, β-L-arabinose, α-D-fucose, α-D-galactose, D-*glycero*-α-D-*galacto*-heptose, and L-*glycero*-α-D-*galacto*-heptose,[27] all represented by the formula in Fig. 11. This relationship was first recognized by Hann, Merrill, and Hudson[28] and was further elaborated by Isbell[29] who explained it in terms of conformations; although Isbell was hampered by lack of knowledge of the principles of conformational analysis, he recognized clearly, and probably for the first time, the effect of conformation on reactions rates and attempted to classify sugar derivatives according to the presence of what are now known as axial and equatorial substituents.

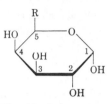

Fig. 11. A homomorphous series of sugars (R = H, CH₃, CH₂OH, or CHOH—CH₂OH).

In their paper which introduced conformational analysis Hassel and Ottar[30] suggested that the facts then recently established for the cyclohexane system could also be applied to sugars in their pyranose forms. They discussed the relative stabilities of some sugar derivatives and predicted the prevalent conformations of many sugars; though some of their conclusions were incorrect, they established conformational analysis as a method useful in carbohydrate chemistry.

The pioneering work on the conformations of sugars was carried out by Reeves.[31] By the study of the complexes formed from sugars and their derivatives in cuprammonium solution, he proved that the pyranoid sugars exist in chair conformations, and that in most cases one of the chair forms (designated C1 by Reeves for the D series) is prevalent. In order to explain his results, Reeves introduced "instability factors" for which values were suggested on a tentative basis. Reeves's assignments of conformations were generally accepted, and they formed the basis of all subsequent work and speculation in this field. After a discussion on nomenclature, Reeves's work will be considered in detail.

b. Nomenclature. It is not customary to designate the different conformations of cyclohexane by special symbols. In carbohydrate chemistry, however, the pyranose ring lends itself to such treatment owing to the presence of one hetero-atom in the ring, and of numerous substituents.

[27] For nomenclature, see W. W. Pigman, *The Carbohydrates*, Academic Press, New York, 1957.

[28] R. M. Hann, A. T. Merrill, and C. S. Hudson, *J. Am. Chem. Soc.*, **57**, 2100 (1935); R. M. Hann and C. S. Hudson, *ibid.*, **59**, 548 (1937).

[29] H. S. Isbell, *J. Res. Natl. Bur. Std.*, **18**, 505 (1937); **20**, 97 (1938).

[30] O. Hassel and B. Ottar, *Acta Chem. Scand.*, **1**, 929 (1947).

[31] R. E. Reeves, (a) *J. Am. Chem. Soc.*, **71**, 215 (1949); (b) **72**, 1499 (1950); *Advan. Carbohydrate Chem.*, **6**, 107 (1951).

Reeves introduced the symbols C1 and 1C to describe the two possible chair forms of pyranoses and B1 to B6 for the boat forms. The system was extended by Bentley[32] to include half-chair and skew forms. These symbols have found fairly extensive usage. The system suffers, however, from the grave disadvantage that it assigns a different symbol to enantiomers having the same equatorial and axial arrangement of groups, e.g., the stable conformation of D-glucose is described as C1 but that of L-glucose is 1C. Confusion can easily result.

Isbell[33] suggested the use of C1 for the stable form of glucose, in both the D and the L series, and C2 for the other chair form. A different designation was proposed by Guthrie:[34] Ca and Ce, according to whether the anomeric hydroxyl group is axial or equatorial, respectively; a similar but more complex nomenclature was suggested for the boat forms. In later papers Isbell and Tipson[35] proposed a similar system, in which the two chairs were described as CA and CE, and which included a detailed nomenclature for all the possible boat and skew forms.

There appears to be little need for a complex system to describe the pyranose conformations; the flexible forms are of little importance and therefore seldom require description by a code. In this chapter a simple expedient will be used to define the chair forms. The conformation which is prevalent in all the common aldohexoses will be described as the *normal* conformation (designated by N), the other as the *alternative* (A) (Fig. 12). For aldohexoses the normal conformation is the one in which (a) C-6 is equatorial, and (b) the anomeric hydroxyl group in the β form is equatorial.† The latter definition is applicable to aldopentoses and

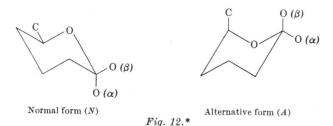

Normal form (N)   Alternative form (A)

*Fig. 12.**

[32] R. Bentley, *J. Am. Chem. Soc.*, **82**, 2811 (1960).

[33] H. S. Isbell, *J. Res. Natl. Bur. Std.*, **57**, 171 (1956).

[34] R. D. Guthrie, *Chem. Ind.* (*London*), **1958**, 1593.

[35] H. S. Isbell and R. S. Tipson, *Science*, **130**, 793 (1959); *J. Res. Natl. Bur. Std.*, **64A**, 171 (1960).

* In Figs. 12, 13, and 14 each formula shows the α- and the β-anomers simultaneously.

† Shafizadeh first used this simple nomenclature: he described the two chairs as "normal" and "reverse": F. Shafizadeh, *Advan. Carbohydrate Chem.*, **13**, 9 (1958).

ketohexoses, whereas the former, of course, is not. Glucose, mannose, and galactose, whether D or L, occur mainly in the normal chair conformation. In the D series, N corresponds to Reeves's C1, A to 1C.

The nomenclature of carbohydrates is not always well adapted for conformational analysis. For example, the fact that the α-anomer (rather than the β) of arabinose shows great similarities to the β form of galactose does not indicate that arabinose is in any way anomalous; the apparent anomaly is caused merely by nomenclature. The α and β designations, according to Hudson's rule, are tied to the D-L system of nomenclature.* Figure 11 represents D-galactose ($R = CH_2OH$) and L-arabinose ($R = H$) because the prefix is governed by the configuration of C-5 in the former but of C-4 in the latter. Consequently the anomeric hydroxyl group is described as α in the galactose and as β in the arabinose, although it has the same configuration in each case. Fortunately the same difficulty no longer exists in the nomenclature of sugars with more than six carbon atoms, e.g., heptoses—where it was particularly troublesome at one time—because the present rules of nomenclature make C-5 the reference point in names like D-*glycero*-α-D-*galacto*-heptose.

c. Cuprammonium Complexes. Reeves determined[31] the prevalent conformations of many methyl glycosides in aqueous solution by the study of the complexes formed with cuprammonium reagent. This method, like several others which have been used for the same purpose, determines the relative positions of hydroxyl groups; but, as in other chemical methods, the reaction itself may change the prevalent conformations. In most cases, however, the sugars can form complexes with cuprammonium in more than one of the possible conformations, and the reaction

---

* Anomers (i.e., sugars differing only in the configuration of the hemiacetal carbon atom) are named according to the relationship of the configuration at the anomeric carbon atom to that of a reference carbon atom of the chain of the acyclic monosaccharide. If the chain has four or less asymmetric carbon atoms, the reference atom is the highest-numbered asymmetric carbon atom; if the chain has more than four asymmetric carbon atoms, the reference atom is the highest-numbered asymmetric carbon atom in the group of four asymmetric carbon atoms next to the functional group. If the reference atom has the same configuration, in Fischer projection, as D-glyceraldehyde, the sugar (or if there are more than four asymmetrical carbon atoms, that part of the name of the sugar which describes the configuration next to the functional group) is designated by D; otherwise it is designated by L. The anomeric center is given the symbols α or β, according to the following convention: For a D-sugar, when the hemiacetal ring is so oriented that the ring oxygen is at the rear and the anomeric carbon atom on the right side (as in Fig. 11), the α form is the one which has the anomeric hydroxyl (or other functional) group below the plane of the ring, and the β form that which has it above that plane. In the L series of sugars, the opposite applies. By this definition, the mirror image of the α-D-isomer is the α-L-isomer, *not* the β-L-isomer.

does not necessarily favor one conformation; moreover the evidence is based not only on formation or non-formation, but also on the nature, of the complex. Hence the method appears, in the majority of cases, to give reliable information about the nature of the most stable conformation.

Conformational analysis of the pyranoid sugars is based on the assumption that the geometry of the pyranose ring is the same as that of cyclohexane. This is not strictly true because, the carbon-oxygen bonds being slightly shorter than the carbon-carbon bonds, a small deviation from the regular cyclohexane structure will exist.[30] Reeves suggested that in a chair form this deviation would bring adjacent *cis* groups a bit closer to each other than adjacent *trans* groups, and this effect would account for the greater reactivity of *cis*-hydroxyl groups in ketal formation, glycol fission, and complexing with cuprammonium. It has been shown, however, in Sec. 6-2f that *cis*-hydroxyl groups are more reactive in cyclitols too, where an oxygen atom is not present in the ring, and that the greater reactivity is connected with the ease of deformation of the chair form. For most purposes the difference between the geometry of cyclohexane and of tetrahydropyran can be neglected.

Cuprammonium solution contains mainly $Cu(NH_3)_4^{++}$ ions. The exact structure of the complexes it not known, but it is certain that two hydroxyl groups, mostly on adjacent carbon atoms, are involved. By studying complex formation with compounds of known rigid conformation, Reeves found that a very small dihedral angle (*cis*-2,3-dihydroxy-tetrahydrofuran) between two hydroxyl groups resulted in very strong complexing, a compound with an angle of 60° (cellulose) was reactive, but with angles of 120° (2,5-anhydrosorbitol) and 180° (methyl 4,6-*O*-benzylidene-α-D-altropyranoside) no complexing occurred. Occasionally, complex formation was observed between two axial hydroxyl groups (as in 1,6-anhydro-3-*O*-methyl-β-D-glucose). It was concluded that the distance between the oxygen atoms must be less than 3.45 Å for complexing to occur.

Complex formation is observed as a decrease in the conductivity of the cuprammonium solution. It also causes a change in the optical rotation which is particularly useful because it yields further information on the relative positions of the oxygen atoms. If the hydroxyl groups are located in true *cis* positions, or in a 1,3-diaxial arrangement, the rotational change is small because the conformation of the molecule is not substantially altered on complex formation. But, if the dihedral angle is about 60°, formation of the copper-containing ring deforms the molecule, causing a large rotational change the sign of which is determined by the sign of the dihedral angle. Examples of this behavior are shown by the glucose derivatives listed in Table 2. When the formation of both dextro- and

Table 2. The Reaction of Cuprammonium with Various
O-Methyl-β-D-glucopyranosides[31a]

Methyl β-D-Glucopyranoside	Shift in Molecular Rotation	Increase in Specific Resistance	Nature of Complex	Location of Complex
2-O-Methyl	+2190°	67	Dextro	C-3:C-4
4,6-Di-O-methyl	−1990	52	Levo	C-2:C-3
3-O-Methyl	−83	16	No complex	—
6-O-Methyl	+435	70	Dextro and levo	C-2:C-3 and C-3:C-4

levorotatory complexes is possible in the same molecule, as in the case of methyl 6-O-methyl-β-D-glucoside, the change in rotation is often small, but the increase in specific resistance indicates the extent of complex formation.

Complexing of glucose at the 2,3-positions causes a levorotatory shift, at the 3,4-positions a dextrorotatory shift. Absence of complex formation in the 3-O-methyl derivative indicates that cuprammonium does not bridge the 4,6-positions. The behavior of these compounds clearly indicates that the β-D-glucopyranosides are in the normal chair conformation (Fig. 13).

Careful study of glucopyranosides, galactopyranosides, and 1,6-anhydrohexopyranosides convinced Reeves that the boat forms play no significant role in the equilibrium of sugar conformations. On the assumption that only chair forms need to be considered, the prevalent chair forms of many methyl glycopyranosides were determined, and most of them were found to be present in the normal chair form. Only methyl α-D-idopyranoside (and some of its derivatives) and methyl 4,6-O-benzylidene-β-D-idopyranoside were found in the alternative form. The lyxo- and altropyranosides showed a behavior intermediate between those expected for the normal and the alternative chair forms, and it was concluded that both conformations contribute to the equilibrium to a significant extent.

An inspection of the two chair forms of the methyl D-glucopyranosides

Normal form · Alternative form

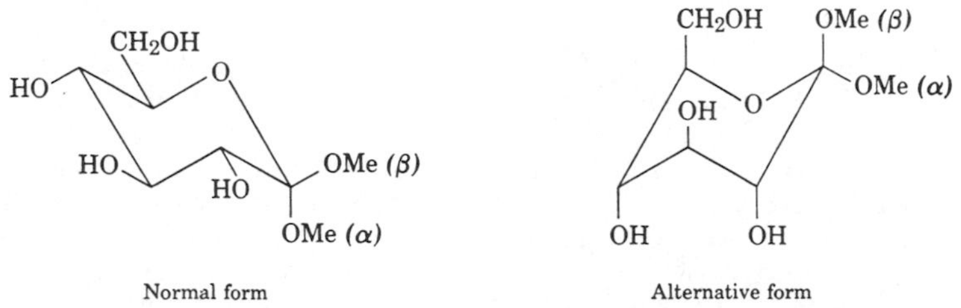

Fig. 13. Conformations of the methyl D-glucopyranosides.

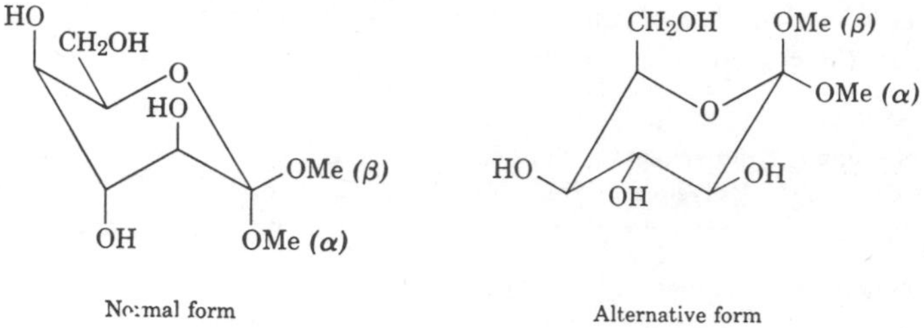

Fig. 14. Conformations of the methyl D-idopyranosides.

(Fig. 13) clearly shows why the normal chair is predominant; in this form all the substituent groups are equatorial in the β-isomer and only one hydroxyl group is axial in the α-isomer. The other extreme is shown by the methyl D-idopyranosides (Fig. 14); in the α-anomer the alternative chair appears more stable, whereas the β-isomer may well contain both chair forms in substantial amounts at equilibrium.

The presence of chair conformations has been shown by X-ray crystallographic measurements in a substantial number of pyranoid sugars;[36] in all cases the conformations are those predicted by Reeves.

d. Flexible and Half-Chair Conformations. In 1957 Reeves and Blouin[37] observed that some, but not all, methyl glycosides suffer a substantial change of optical rotation in alkaline solution. They suggested that ionization of axial hydroxyl groups causes a change from chair to boat forms, owing to increased repulsions, particularly when the boat conformation allows bulky substituents to become "equatorial." However, the reason for the rotational changes in alkaline solution is not clear; it may be due to a distortion of the chair forms, or to the rotational contribution of negatively charged oxygen atoms being different from that of hydroxyl groups.

For a similar reason Reeves had previously suggested[38] that amylose contains α-D-glucopyranose units in a boat form which allows the 1- and 4-substituents to take up "equatorial" positions. A study[39] of the amylose-iodine complex, however, appears to show that amylose consists

[36] (a) Sucrose: C. A. Beevers, T. R. R. McDonald, J. H. Robertson, and F. Stern, *Acta Cryst.*, **5**, 689 (1952); (b) α-D-glucosamine hydrobromide; E. G. Cox and G. A. Jeffrey, *Nature*, **143**, 894 (1939); α-D-glucose: T. R. R. McDonald and C. A. Beevers, *Acta Cryst.*, **5**, 654 (1952); α-L-rhamnose: H. McD. McGeachin and C. A. Beevers, *Acta Cryst.*, **10**, 227 (1957); β-D-arabinose: A. Hordvik, *Acta Chem. Scand.*, **15**, 16 (1961); β-cellobiose: R. A. Jacobson, J. A. Wunderlich, and W. N. Lipscomb, *Acta Cryst.*, **14**, 598 (1961).
[37] R. E. Reeves and F. A. Blouin, *J. Am. Chem. Soc.*, **79**, 2261 (1957).
[38] R. E. Reeves, *J. Am. Chem. Soc.*, **76**, 4595 (1954).
[39] C. T. Greenwood and H. Rossotti, *J. Polymer Sci.*, **27**, 481 (1958).

of a helix of glucopyranose units in the normal chair form. The latter view is now accepted by most workers.[39a]

Reeves, in his earlier papers, did not take the possible presence of skew forms into consideration. In 1958, however, he put the boat conformation into proper perspective as the energy maximum in the cycle of flexible conformations. The energy minimum being the skew-boat form, Reeves suggested,[4] after inspection of models, that this conformation appears attractive for the lyxose and altrose configurations. However, quantitative considerations (see end of this section) do not support this suggestion, and NMR spectroscopy has recently provided[40] strong evidence for the chair conformation of lyxose; nevertheless skew and half-chair forms have recently acquired an unjustified popularity.

Bentley's work[32] on the methyl β-D-idopyranosides may serve as an example. Reeves had found methyl 3-O-methyl-β-D-idopyranoside in the normal chair form (diaxial complex), whereas methyl 4,6-O-benzylidene-β-D-idopyranoside, which showed a dextrorotatory shift with cuprammonium—though to a lesser extent than the α-isomer—appeared to be at least partly in the alternative form. As additional examples, Bentley[32] investigated methyl β-D-idopyranoside (complex of low rotational shift), methyl 2-O-methyl-β-D-idopyranoside (levo complex), and methyl 2,3-di-O-methyl-β-D-idopyranoside (no complex). It must be clearly realized that *complex formation*, like other chemical reactions, *may shift the position of the conformational equilibrium*. Methyl 3-O-methyl-β-D-idopyranoside can complex *only* in the normal chair form[31] (see Fig. 14); hence a complex of this form will be observed even if the normal chair form is not the prevalent conformation (as long as the free energy change on complex formation is sufficient to overcome the free energy difference between the chair conformations). The 2-methyl and the 4,6-benzylidene derivatives can complex *only* in the alternative chair form (cf. Fig. 14), and complexes of that form are, indeed, observed. Methyl β-D-idopyranoside can complex in either conformation; the low rotational shift suggests the normal form (diaxial complex) but does not exclude the alternative form (levo and dextro complex).

Bentley assumed that each derivative of methyl β-D-idoside must form a complex in the *same* conformation. He found one half-chair form capable of fulfilling this condition, and assumed, unjustifiably, that this was the predominant conformation of methyl β-D-idopyranoside. In fact, in this case, where the free energy difference between the two conformations is small, the data on complex formation give no information, at all, on the conformation of the uncomplexed molecule.

[39a] For example V. S. R. Rao and J. E. Foster, *Biopolymers*, **1**, 527 (1963).

[40] L. D. Hall, L. Hough, K. A. McLauchlan, and K. Pachler, *Chem. Ind.* (*London*), **1962**, 1465.

Table 3. The Preferred Conformation of Aldopyranoses

Aldose	Conformations		Total Calculated Interaction Energies,[c] kcal./mole	
	Found[a]	Predicted[b]	N	A
α-Allose		N	4.2	6.65
β-Allose		N	2.85	6.45
α-Altrose	A,N	A,N	4.4	4.3
β-Altrose	A,N[d]	A,N	3.85	4.9
α-Galactose	N	N	3.2	5.9
β-Galactose	N	N	2.85	6.7
α-Glucose	N	N	2.3	7.1
β-Glucose	N	N	1.95	7.9
α-Gulose	N	A,N	4.75	5.1
β-Gulose		N	3.4	4.9
α-Idose	A	A	5.95	3.75
β-Idose		A,N	5.4	4.35
α-Mannose	N	N	2.85	5.1
β-Mannose	N	A,N	3.3	6.7
α-Talose		A,N	4.75	4.9
β-Talose		A,N	5.2	6.5
α-Arabinose	A	A	4.05	2.5
β-Arabinose	A	A	3.5	2.85
α-Lyxose	A,N	A,N	2.5	3.05
β-Lyxose	A,N	A,N	2.95	4.4
α-Ribose	N	N	3.85	4.85
β-Ribose	N	N	2.5	4.4
α-Xylose	N	N	1.95	5.05
β-Xylose	N	N	1.6	5.6

[a] Found by the cuprammonium method[31b] for the methyl glycopyranosides (not the free sugars).

[b] Predicted by Reeves.[31b] Both conformations are given when they differ

The skew-boat form of cyclohexane has a free energy content about 3.8 kcal./mole higher than the chair form (Sec. 2-1). If the interaction of substituents is also taken into account, it becomes clear that at least one of the chair conformations of every sugar pyranoside is more stable than any skew form. The half-chair conformation is *not* part of the flexible cycle but involves considerable angle strain; it is less stable than the skew form. Skew-boat and half-chair forms have never been found to be the predominant conformations of *monocyclic* pyranosides; the skew-boat form, however, takes over in some fused-ring systems (Sec. 6-2f), and the half-chair is the stable conformation of cyclohexene oxides (Sec. 6-2e). The boat, of course, is the conformation of 1,4-bridged six-membered rings, such as those found in 2,6-anhydropyranoses.

6–4. Free Energy and Conformation

a. Conformational Instability Factors of the Pyranose Ring. In order to explain which chair form of each aldopyranose is more stable, Reeves[31] defined the effects which increase the free energy of a chair form. He listed three of these "instability factors": (1) any atom or group (other than hydrogen) in an axial position; (2) the simultaneous presence of an axial hydroxyl group on C-1 and of an axial carbon atom on C-5 (this interaction had already been recognized by Hassel and Ottar[30] as being of decisive importance); (3) an axial oxygen atom on C-2 when the $C_{(2)}$—O bond bisects the angle between the two $C_{(1)}$—O bonds, i.e., when the oxygen atom on the anomeric carbon atom is equatorial. Reeves coined the term "$\Delta 2$" to describe this steric arrangement (Fig. 15).

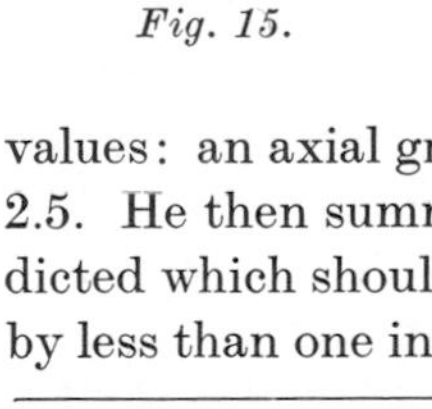

Reeves assumed that these instability factors were additive and assigned to them the following arbitrary values: an axial group, 1; the Hassel-Ottar effect, 0.5; the $\Delta 2$ condition, 2.5. He then summed the instability factors for each chair form and predicted which should be the more stable one; when the two forms differed by less than one instability unit, he assumed that both chair forms would

Fig. 15.

---

by not more than one instability unit, or when the more stable conformation has as many as 2.5 instability units.

c Including the anomeric effect and the $\Delta 2$ effect. The calculations are discussed in Sec. 6-4e.

d Reeves classifies the complex as "dextro-levo"; the rotational shift of −922°, however, indicates the preponderance of a levo complex, given only by the *N* conformation.

be detectable in equilibrium. Reeves's predicted conformations are listed in Table 3, in which N and A are used instead of the original Cl and 1C to describe them. The actual conformations found by the use of cuprammonium complexes are in excellent agreement with the predictions, except for methyl β-D-mannopyranoside. The values for the instability units were, of course, selected to give good agreement. One may notice, however, that the interaction between syn-axial hydroxyl groups was neglected, as was also that between vicinal hydroxyl groups; and there appears to be no reason for the high value ascribed to the $\Delta2$ condition. The use of this high value made the N form of methyl β-mannopyranoside appear less stable than it was found to be.

In order to explain the relationship between conformational stability and the refractive index of methylated pyranosides, Kelly[41] modified Reeves's instability factors as follows: the $\Delta2$ condition, 2.5; axial oxygen, 1; axial hydroxymethyl group, 2 (the latter is increased to 2.5 when an axial hydroxyl group is also present at C-1). This modification improves the agreement between the calculated and the experimentally determined conformations.

Rapid advances in conformational analysis since about 1950 make it now possible to calculate the free energy differences between pyranose conformations, with some approximations, as described in the following sections.

b. Interaction Energies.   Interaction energies are known for a number of groups and atoms attached to chair rings (Sec. 2-2b, 2-3a). The interactions of hydroxyl groups may vary with the nature of the solvent owing to solvation, dipole interaction, and hydrogen bonding; it is therefore advisable to use for the sugars the values shown in Table 1 which were determined on polyols in aqueous solutions.

These values, valid for cyclitols, can be applied to pyranoses with one limitation: because the C—O bonds are shorter than the C—C bonds, axial substituents on the two carbon atoms adjacent to the ring oxygen atom will interact more strongly than in a cyclohexane ring. This is particularly true for the interaction between an axial hydroxymethyl group on C-5 and the syn-axial anomeric β-hydroxyl group in aldopyranoses. Available data yield no quantitative value for this increased interaction which, fortunately, is of little importance, since it occurs only in the alternative conformations of the β-pyranosides, which are less stable than the normal conformations. The only exception is β-idopyranose (cf. Fig. 14) which appears to exist mainly in the alternative chair form despite the interaction between axial carbon and axial oxygen. It appears therefore justified,

[41] R. B. Kelly, *Can. J. Chem.*, **35,** 149 (1957).

as an approximation, to use 1.6 kcal./mole, determined for cyclitols, in the case of sugars too as the value of the interaction energy between an axial carbon and axial oxygen atom.

Another question arises: What is the magnitude of the interaction between an axial substituent and the ring oxygen atom? On the oxygen atom, an axial hydrogen atom is missing (compared to a cyclohexane ring). It is known, however, that the interaction is at least in part with the electron pair forming the H—O bond; an electron pair, though non-bonding, is also present on the ring oxygen atom in the same orbital. Estimates concerning the interaction with "lone" pairs vary. Aroney and Le Fèvre[42] claim that the volume requirement of a lone pair on a nitrogen atom exceeds that of a covalently bonded hydrogen atom (Sec. 3-8); Barton[43] makes a similar claim for a lone electron pair in a carbanion. More recent data (Sec. 3-8), however, indicate that interaction with an electron pair is less than with a hydrogen atom. There appear to be no reliable data concerning lone electron pairs on an oxygen atom. The only relevant figure available in carbohydrate chemistry is the free energy difference between α-D-glucose 1-phosphate and α-D-galactose 1-phosphate which was found to be 0.7 kcal./mole.[44] These two compounds differ only in the position of the hydroxyl group at C-4, which is axial in galactose; if this hydroxyl group interacts only with the axial hydrogen atom at C-2, the value of the interaction energy should be only 0.45 kcal./mole. Interaction with the electron pair on the ring oxygen therefore contributes to the free energy.

Existence of interaction between an axial group and the electron pair on a ring oxygen atom is also shown by the equilibration[45] of the 5-benzyloxy-2-phenyl-1,3-dioxanes (Fig. 16) in which the *trans* isomer predominates in the ratio of approximately 2.1:1 (at 50°). If there is no interaction with the electron pairs, the two isomers should be of equal stability. Calculation (using $\Delta G^\circ_{Ph} = 3.1$ kcal./mole) shows that the interaction of the benzyloxy group with the two electron pairs of the oxygen atoms has the approximate value 0.5 kcal./mole.

The question can be approached in yet another way. Inspection of models will show that the geometrical relation of an axial group to the ring oxygen is the same as that to an oxygen atom attached equatorially to a vicinal carbon atom—both are *gauche* relationships (but the ring

[42] M. Aroney and R. J. W. Le Fèvre, *J. Chem. Soc.*, **1958**, 3002.

[43] D. H. R. Barton, "Stereochemistry," in A. R. Todd, ed., *Perspectives in Organic Chemistry*, Interscience Division, John Wiley and Sons, New York, 1956, p. 83; cf. G. Roberts and C. W. Shoppee, *J. Chem. Soc.*, **1954**, 3418.

[44] R. G. Hansen, E. M. Craine, and P. Gray, *J. Biol. Chem.*, **208**, 293 (1954).

[45] N. Baggett, J. S. Brimacombe, A. B. Foster, M. Stacey, and D. H. Whiffen, *J. Chem. Soc.*, **1960**, 2574.

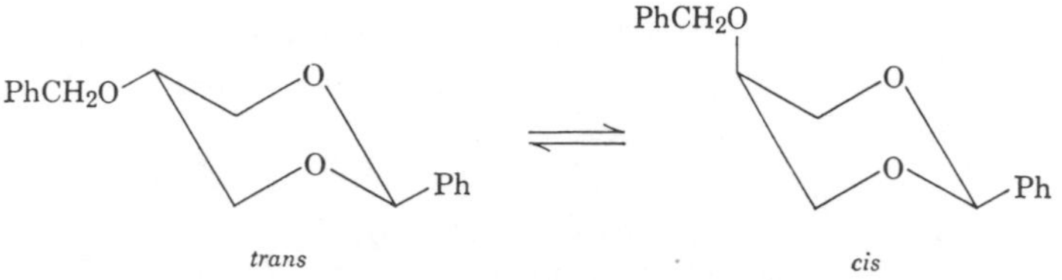

Fig. 16.

oxygen is fixed in orientation, whereas the equatorial oxygen has rotational freedom; this difference may affect the interactions). On the basis of this consideration, the interaction between the ring oxygen and an axial oxygen atom should be at least 0.35 kcal./mole (see Table 1).

From this discussion it is clear that the interaction of an axial group with the ring oxygen atom is somewhat, but not considerably, smaller than its interaction with an axial hydrogen atom. None of the available estimates for the magnitude of this interaction is very accurate, however. In the following calculations, for simplicity's sake, the interaction with the electrons of the ring oxygen will be taken as equal to that with an axial hydrogen atom.

At this point, consideration should also be given to the possible effect, on conformational stability, of an intramolecular hydrogen bond between an axial hydroxyl group and the ring oxygen atom. Such a bond has been shown to exist[45] in dilute carbon tetrachloride solution of *cis*-1,3-*O*-benzylideneglycerol (Fig. 17) (*cis*-5-hydroxy-2-phenyl-1,3-dioxane), and in other related compounds.[46] Even the *trans* isomer of 1,3-*O*-benzylideneglycerol exists,[47] to a considerable extent, in the hydrogen-bonded conformation (Fig. 17) under these conditions, although this conformer, having an axial phenyl and an axial hydroxyl group, is otherwise very

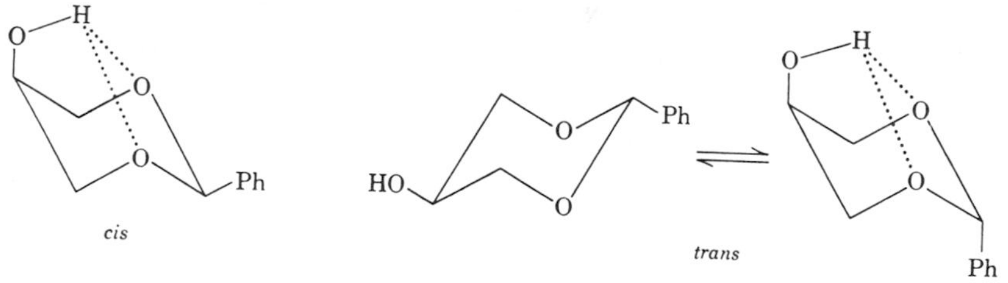

Fig. 17. 1,3-*O*-Benzylideneglycerols.

[46] S. A. Barker, J. S. Brimacombe, A. B. Foster, D. H. Whiffen, and G. Zweifel, *Tetrahedron*, **7**, 10 (1959).

[47] B. Dobinson and A. B. Foster, *J. Chem. Soc.*, **1961**, 2338; N. Baggett, M. A. Bukhari, A. B. Foster, J. Lehmann, and J. M. Webber, *ibid.*, **1963**, 4157.

unfavorable. The energy of the hydrogen bond can therefore have a considerable effect on conformational equilibria (see also Sec. 6-5b).

The conditions under which this hydrogen bonding is observed in the infrared spectrum are, however, designed to suppress *intermolecular* hydrogen bonding. When sugars are in the solid state, or are dissolved in hydroxylic solvents—the only ones which are good solvents for them—the opportunities for the formation of intermolecular hydrogen bonds are so manifold that the intramolecular bonds are probably relegated to a minor role.[48] Under these conditions intramolecular hydrogen bonds appear to have no influence on the conformations; in the following calculations they have been neglected. However, in the case of partially substituted sugar derivatives, which are soluble in aprotic solvents, the conformational equilibrium may change with the nature of the solvent, owing to hydrogen bonding. An example is methyl 4,6-O-benzylidene-α-D-idoside; in aqueous solution it complexes with cuprammonium ion in the alternative conformation,[31] but in dilute carbon tetrachloride solution the infrared spectrum shows hydrogen bonding between axial oxygen atoms,[49] which is possible only in the normal conformation (cf. Fig. 14). For other examples, see Secs. 6-5b and 6-5c.

Another set of values is available for interaction energies in a different solvent. Lemieux and Chü[50] have studied the anomerization of acetylated sugars in 1:1 acetic acid-acetic anhydride mixture in the presence of perchloric acid at 25°. They found that the free energy differences between the fully acetylated anomers of the four aldopentoses and of seven aldohexoses could be accounted for by the use of the following interaction energies: between an axial acetoxy group and an axial hydrogen atom, 0.18; between two *syn*-axial acetoxy groups, 2.08; between acetoxy groups on adjacent carbon atoms, 0.55 kcal./mole. The first of these values is much smaller than the corresponding value for a hydroxyl group in water (0.45 kcal./mole) and does not agree well with other estimates of this interaction.

c. The Anomeric Effect. Surprisingly, the 1-substituted derivatives of α-D-glucose—and many other sugars—are more stable than the corresponding β derivatives, although the former have the substituent in the axial position and the latter in the equatorial one. Thus equilibrium mixtures of methyl D-glucopyranosides,[51] penta-O-acetyl-D-glucopyranoses,[52]

[48] M. A. Kabayama and D. Patterson, *Can. J. Chem.*, **36**, 563 (1958); A. B. Foster, R. Harrison, J. Lehmann, and J. M. Webber, *J. Chem. Soc.*, **1963**, 4471.

[49] H. Spedding, *J. Chem. Soc.*, **1961**, 3617.

[50] R. U. Lemieux and N. J. Chü, Abstracts of Papers, *Am. Chem. Soc.*, **133**, 31N (1958).

[51] C. L. Jungius, *Z. Physik. Chem.*, **52**, 97 (1905).

[52] W. A. Bonner, *J. Am. Chem. Soc.*, **73**, 2659 (1951).

Fig. 18. Anomeric effect: direction of dipole moments.

and acetohalo-D-glucopyranoses[53,54] all contain more of the α than of the β form. There appears to exist an effect, named "anomeric effect" by Lemieux and Chü,[50] which makes an equatorial group on the anomeric carbon atom less stable, or an axial one more stable, than it would be in other positions of the ring. An explanation of this phenomenon was given by Edward[55] in terms of the dipole-dipole interaction between the carbon-oxygen bonds of the ring and the bond from the anomeric carbon to the substituent. These dipoles form a small angle when the substituent is equatorial, a large one when it is axial (Fig. 18).

In accordance with this explanation, the anomeric effect varies inversely with the dielectric constant of the solvent and is greatest when the substituent atom or group is highly electronegative. The acetylated glycosyl halides are stable only when the halogen atom occupies the axial position; the equatorial isomers, if they can be obtained at all, undergo ready and nearly complete anomerization.[54] Typical examples of this anomeric effect, taken from a field other than carbohydrate chemistry, are *trans*-2,5-dichloro-1,4-dioxane and *trans*-2,3-dibromo-1,4-dioxane, in which both halogen atoms have been shown to occupy axial positions.[56] The anomeric effect is closely related to the dipole interaction in 2-halo-cyclohexanones in which the axial position is favored by the halogen atom; this effect also increases with a decrease in the dielectric constant of the solvent (Sec. 7-3).

The anomeric effect on free hydroxyl groups in aqueous solution is small because of the high dielectric constant of water. In fact, in the equilibrium mixtures of glucose and of galactose in aqueous solution, the β-pyranose forms (with equatorial OH) predominate. It is easy to calculate the value of the anomeric effect in this case. The free energy of

[53] R. U. Lemieux, *Advan. Carbohydrate Chem.*, **9,** 1 (1954).

[54] L. J. Haynes and F. H. Newth, *Advan. Carbohydrate Chem.*, **10,** 207 (1955).

[55] J. T. Edward, *Chem. Ind. (London)*, **1955,** 1102.

[56] C. Altona, C. Romers, and E. Havinga, *Tetrahedron Letters*, No. 10, 16 (1959). See also Sec. 4-6c.

α-glucopyranose should be 0.9 kcal./mole higher than that of the β-anomer, owing to the interaction of one axial hydroxyl group with two *syn*-axial hydrogen atoms in the former. In fact, the equilibrium composition is 36% α : 64% β, corresponding to a free energy difference of only 0.35 kcal./mole. The difference between the calculated and observed value, 0.55 kcal./mole, is due to the anomeric effect. In calculating free energy differences by the use of interaction energies, this value has to be added whenever an anomeric hydroxyl group is equatorial. In solvents other than water, the anomeric effect of hydroxyl groups should be greater because the dielectric constant is smaller. In fact, in anhydrous methanol glucose has an equilibrium composition[57] of 50% α : 50% β, giving a calculated value of 0.9 kcal./mole for the anomeric effect. In pyridine, the equilibrium solution of mannose contains 85% of the α form,[58] compared to 67.5% in aqueous solution; the calculated anomeric effect is 1.15 kcal./mole.

From their data on the equilibria of pyranose acetates, Lemieux and Chü[50] calculated the magnitude of the anomeric effect as 1.3 kcal./mole for pentoses, 1.5 kcal./mole for hexoses; the difference between these values is presumably caused by the inductive effect of the oxygen atom on C-6. These values, which give excellent agreement with the experimental data, apply for acetoxy groups in a 1:1 mixture of acetic acid and acetic anhydride, and are understandably higher than the value for hydroxyl groups in water. The anomeric effect for halogen atoms in inert solvents should be even higher, but quantitative data are lacking.

The anomeric effect will be further discussed in connection with anomeric equilibria in Sec. 6-6b.

d. The $\Delta 2$ Effect. The theoretical explanation for the $\Delta 2$ effect is not clear, but it is probably due to dipole interaction. Molecules are known with the particular arrangement of three oxygen atoms, similar to the $\Delta 2$ condition, which show no signs of instability, e.g., diacetals of glyoxal. Reeves has probably overrated the importance of the $\Delta 2$ effect, but it does make some contribution to the free energy of certain sugar derivatives. For example, in contrast to glucose, mannose contains a higher proportion of the α form than the β form in equilibrium (68.8% α, 31.2% β). A calculation of free energy differences, using the interaction energies and the anomeric effect, gives the right figure only if a value of 0.45 kcal./mole is added for the $\Delta 2$ effect; this value is to be added to the interaction energies whenever the hydroxyl group on C-2 is axial and the one on C-1 is equatorial.

[57] H. H. Rowley and S. D. Bailey, *J. Am. Chem. Soc.*, **62,** 2562 (1940).
[58] P. A. Levene and D. W. Hill, *J. Biol. Chem.*, **102,** 563 (1933).

e. The Preferred Chair Conformation of Aldopyranosides. Using the values for the interaction energies, the anomeric effect, and the $\Delta 2$ condition given in the preceding sections, Angyal[59] calculated the relative free energy contents of both chair forms of the pentose and hexose pyranosides. These values are listed in Table 3 together with the conformations found by the cuprammonium method and those predicted by Reeves;[31a] his C1 and 1C notation has been transcribed into A and N. The conformation which has the lower free energy content will, of course, be the preferred one; in all cases where experimental data are available, they agree with the calculated predictions.

The free energy values listed in Table 3 may serve as a general guide to the behavior of pyranoid sugars, though the approximations and the uncertainties involved in the calculations should be kept in mind. The values, as pointed out previously, refer to aqueous solutions at room temperature. The application of these free energy values to carbohydrate equilibria will be discussed in Sec. 6-6.

A method completely different from the preceding one has been put forward by Barker and Shaw[60] to assess the relative stabilities of the conformations of pyranose rings. They calculated the total amount of atomic overlap of non-bonded atoms, overlap between each pair being calculated separately and added together. This calculation takes into account the exact shape of the pyranose ring and shows a greater overlap, for example, between an axial carbon at C-5 and an axial atom at C-1 than between the same carbon and an axial atom at C-3. In this respect the calculations are more exact than those discussed in the previous sections. However, they do not take into account the hydration of the hydroxyl groups, nor the anomeric effect; and the overlaps (expressed in angstrom units) are not readily translated into free energy values. In most cases the calculations lead to the same general conclusions as those of Reeves and of Angyal, but some results appear anomalous (e.g., that α-D-ribose prefers the alternative chair conformation).

f. The Furanose Forms. Carbohydrate chemists, reared mostly on common sugars like glucose, regard the furanose forms of sugars as inherently less stable than the pyranose forms; this view is, however, not necessarily always valid. It is true that six-membered carbon rings are more stable than five membered ones; the excess enthalpy of cyclopentane (over that of a polymethylene chain) is about 6.5 kcal./mole higher than that of cyclohexane (Sec. 4-3); but the difference may be less pronounced in the free energy contents because the flexible cyclopentane has greater entropy

[59] S. J. Angyal, unpublished results.
[60] G. R. Barker and D. F. Shaw, *J. Chem. Soc.*, **1959**, 584.

than the rigid cyclohexane. Replacement of one methylene group by an oxygen atom would lessen the difference between the stabilities of the five- and the six-membered ring systems by reducing the eclipsed interactions in the former; the difference in stability between tetrahydrofuran and tetrahydropyran has been estimated, from measurements of the heats of combustion, as 2.9 kcal./mole.[61]

A much smaller difference in stabilities is indicated by the data of Hurd and Saunders[62] who found that tetrahydropyran-2-ol and tetrahydrofuran-2-ol, in 75% aqueous dioxane, are in equilibrium with 6.1% and 11.4%, respectively, of the corresponding open chain hydroxyaldehydes* (see, e.g., Fig. 19). The difference in the free energy changes on cyclization between the two systems is therefore only 0.4 kcal./mole; this figure represents the difference in the stability between the two cyclic compounds. Here the anomeric effect is to be taken into account for the six-membered ring, but even then the difference is surprisingly small. The difference is, however, likely to be higher in sugar derivatives owing to the larger number of groups attached to the ring. The chair form of a six-membered ring, being staggered throughout, can accommodate substituents with less strain than the eclipsed five-membered ring.

Faced with a lack of reliable quantitative data, one can say only that the furanose ring, in itself, is less stable than the pyranose ring. The difference in stability is not large, however, and substituent groups, if suitably placed, may reverse this order of stability.

The conformations of furanose forms are similar to those of cyclopentane (Sec. 4-3); the ring is not planar, but one or two atoms are out of the plane containing the others (C_s and C_2, respectively). The non-planar nature of the ring has been shown in the crystalline state by X-ray crystallography (e.g., in the fructofuranose part of sucrose[36a]) and, more recently, in solution by proton magnetic resonance: depending on the nature and location of its substituents, the furanose ring was found in

Fig. 19. Reversible cyclization of a hydroxyaldehyde.

[61] R. C. Cass, S. E. Fletcher, C. T. Mortimer, H. D. Springall, and T. R. White, *J. Chem. Soc.*, **1958**, 1406.

[62] C. D. Hurd and W. H. Saunders, *J. Am. Chem. Soc.*, **74**, 5324 (1952).

* Calculated from the intensity of the ultraviolet absorption peak at 287 mµ. From the infrared spectrum[1] in carbon tetrachloride solution the approximate values of 3.0% and 4.5% can be obtained.

the C_s or in the C_2 conformation.[63-65] Bulky substituents are staggered as much as possible, and they take up positions of "equatorial" character at the most staggered carbon atoms. The oxygen atom can be expected to take up the least staggered section of the ring. Lemieux[66] has discussed in considerable detail the conformational behavior of furanoses based on the assumption that they take up C_s conformations.

In evaluating the importance of furanose forms, we are faced with two difficulties: (1) the flexible nature of the five-membered ring does not allow calculations of interaction energies similar to those which can be applied to the rigid chair form, and (2) experimental data are not available because in no case is the concentration of furanose forms in the equilibrium mixture of sugars known (except in the special case of D-fructose; see Sec. 6-6a); there is, in fact, no method known for determining the amount of furanose forms present in solution. Some quantitative data are, however, available on methyl pentofuranosides (see Sec. 6-6b).

In a purely qualitative way, one can state that the strongest non-bonded interaction in a furanose ring is the one between two *cis* substituents on adjacent carbon atoms, since all ring atoms are eclipsed to some extent, and some to a considerable extent. Consideration of these interactions can give a qualitative picture of the relative stabilities of furanose forms, as will be shown in Sec. 6-6a.

g. The *aldehydo* Forms. The equilibrium solution of pentoses and hexoses contains only very minor amounts of the open chain aldehyde forms[67] (0.0026% for glucose). In view of the considerable amount of aldehyde found[62] in the solutions of tetrahydrofuran-2-ol and tetrahydropyran-2-ol, this fact seems surprising; it is well known, however, that substituents favor cyclic versus open chain forms.[68] 4-Oxa-5α-cholestan-3α-ol (Fig. 20), a compound prepared[69] as a model for pyranoid sugars, shows no carbonyl peak in the ultraviolet or infrared absorption spectrum.

[63] R. U. Lemieux, *Can. J. Chem.*, **39**, 116 (1961).

[64] C. D. Jardetzky, *J. Am. Chem. Soc.*, **82**, 229 (1960); **83**, 2919 (1961); **84**, 62 (1962).

[65] R. J. Abraham, L. D. Hall, L. Hough, and K. A. McLauchlan, *J. Chem. Soc.*, **1962**, 3699.

[66] R. U. Lemieux, "Rearrangements and Isomerisations in Carbohydrate Chemistry," in P. de Mayo, ed., *Molecular Rearrangements*, Interscience Division, John Wiley and Sons, New York, 1963, p. 709.

[67] J. M. Los, L. B. Simpson, and K. Wiesner, *J. Am. Chem. Soc.*, **78**, 1564 (1956).

[68] G. S. Hammond, "Steric Effects on Equilibrated Systems," in M. S. Newman, ed., *Steric Effects in Organic Chemistry*, John Wiley and Sons, New York, 1956, p. 460; for an explanation, see N. L. Allinger and V. Zalkow, *J. Org. Chem.*, **25**, 701 (1960).

[69] J. T. Edward, P. F. Morand, and I. Puskas, *Can. J. Chem.*, **39**, 2069 (1961).

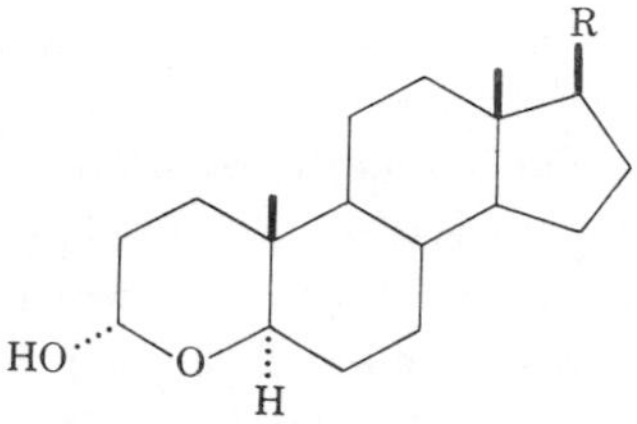

Fig. 20.

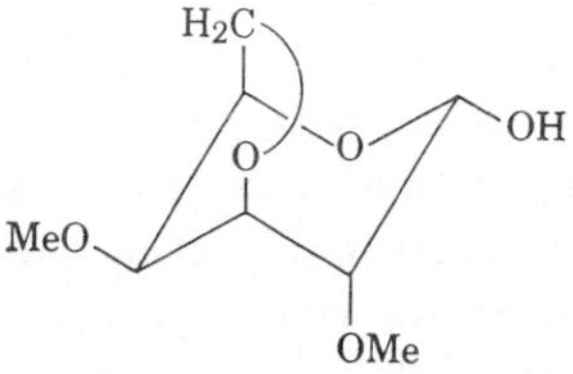

Fig. 21.

In considering the equilibria of sugars in solution the presence of the aldehyde forms will therefore be disregarded. It is possible, however, that the amounts of aldehyde are somewhat larger where all the conformations of the pyranose and the furanose forms are unfavorable. D-Idose, for example, has been reported[70] to give a positive Schiff reaction for aldehydes, unlike most other sugars. Similarly, 5-O-methyl glucose and 5,6-di-O-methyl glucose (which cannot form a pyranose ring) react with the Schiff reagent and with Fehling solution in the cold.[71]

Steric effects may stabilize the aldehyde form in special cases; thus 3,6-anhydro-2,4-di-O-methylgalactose[72](Fig. 21) and the corresponding glucose[73] and mannose derivatives[74] appear to be mainly in the aldehyde form. The hydroxyl group at C-5 is available for ring formation, but the pyranose form (shown in Fig. 21) would be considerably strained (cf. Sec. 6-6a). The methyl glycosides of these anhydro sugars are converted to the dimethyl acetals of the aldehyde form by methanolic hydrogen chloride.

3,6-Anhydro-D-galactose itself (the compound shown in Fig. 21 but with the two protecting methyl groups removed) also shows the properties of an aldehyde.[74] For this compound a furanose form would also be possible, since the hydroxyl group at C-4 is now free, but it would be even more unfavorable than the pyranose form because it would contain two *trans* fused five-membered rings.

6–5. Conformations and Physical Properties

a. Optical Rotation. The study of carbohydrates has provided an impressive accumulation of data on optical rotations of stereoisomers, yet

[70] L. von Vargha, *Chem. Ber.*, **87**, 1351 (1954).

[71] L. von Vargha, *Ber.*, **69**, 2098 (1936); M. R. Salmon and G. Powell, *J. Am. Chem. Soc.*, **61**, 3507 (1939).

[72] W. N. Haworth, J. Jackson, and F. Smith, *J. Chem. Soc.*, **1940**, 620.

[73] W. N. Haworth, L. N. Owen, and F. Smith, *J. Chem. Soc.*, **1941**, 88.

[74] A. B. Foster, W. G. Overend, M. Stacey, and G. Vaughan, *J. Chem. Soc.*, **1954**, 3367.

its contribution to the fundamental theory of optical rotatory power has been rather small compared with recent contributions from the steroid field. Most of the generalizations about relations between rotation and configuration of sugars are merely useful empirical rules, and their conformational implications have not been examined.

There are several reasons for this slow development in a field with a wealth of data. Firstly, the method of optical rotatory dispersion (Sec. 3-6) cannot easily be applied to carbohydrates. Cyclic sugar structures with carbonyl groups in the ring are quite rare, and rotatory dispersion curves generally will not be very informative for carbohydrates and their derivatives, as only plain curves may be expected in the currently accessible range of wavelengths. Nearly all recorded rotations are for the sodium D-line. On the other hand, the certainty that Cotton effects will only be encountered at short wavelengths does increase confidence in deductions already made on the basis of D-line rotations, since these rotations are often quite large. A number of thiocarbonates of carbohydrates are known,[75] and a study of their rotatory dispersions might be rewarding; xanthates have already been used to a limited extent for this purpose.[76] Other chromophoric groups have been introduced into carbohydrates, and some are mentioned below. However, it may be difficult to find suitable solvents for measurement of dispersion curves for all classes of carbohydrates.

Secondly, the stereochemical problem is complex. There is interconversion between acyclic forms and ring structures; the conformational picture is less sharply defined than for steroids, because nearly all well-studied derivatives of carbohydrates are monocyclic and therefore have fairly mobile conformations. Various rotational conformations are possible for polarizable substituents such as acyl groups, and some conformations may be stabilized by hydrogen bonding in unpredictable fashion.

In view of the purely empirical nature of many of the observations made about the optical rotations of carbohydrates, a discussion of all known correlations in detail is not justified. Attention will be confined mainly to those generalizations about cyclic carbohydrates that are closely related to the theoretical treatments developed by Brewster and Whiffen, and to some interesting correlations for acyclic polyhydroxy compounds. It should be noted that in most of the earlier discussions of the rotations of carbohydrates the molecular rotation is defined as the

[75] L. Hough, J. E. Priddle, and R. S. Theobald, *Advan. Carbohydrate Chem.*, **15**, 91 (1961).

[76] B. Sjöberg, D. J. Cram, L. Wolf, and C. Djerassi, *Acta Chem. Scand.*, **16**, 1079 (1962); Y. Tsuzuki, K. Tanaka, and K. Tanabe, *Bull. Chem. Soc. Japan*, **35**, 1614 (1962); Y. Tsuzuki, K. Tanabe, M. Akagi, S. Tejima, *ibid.*, **37**, 162 (1964).

product of the specific rotation and the molecular weight. In this discussion, in conformity with the conventions in steroid and terpene chemistry, all values have been calculated by the formula $[M]_D = [\alpha]_D \times$ (mol. wt./100).

Molecular Rotations of Cyclic Carbohydrates. The most widely used and theoretically interesting of the rules correlating rotation with configuration are Hudson's "rules of isorotation."[77] In their fully developed form they relate to derivatives of pyranose sugars (for lack of sufficient data on furanose forms): to the anomeric forms of methyl glycosides, of acetylated methyl glycosides, and of fully acetylated sugars. Hudson assumed that the principle of optical superposition will hold for such pairs of derivatives, and that the rotation of each anomeric form is the sum of two rotational contributions. One contribution (the A value) is characteristic of the substituent at the anomeric center, C-1, and independent of the configuration at the other ring atoms; anomerization causes the contribution from C-1 to be reversed in sign, but unaltered in magnitude. The other contribution (the B value) is characteristic of the configuration from C-2 onward, but independent of the nature of the substituent and configuration at C-1. In the D series of sugars the rules of isorotation lead to the following formulas for molecular rotations of anomeric derivatives:

$$[M]_D \text{ of } \alpha\text{-D form} = B + A$$
$$[M]_D \text{ of } \beta\text{-D form} = B - A$$

a change from the α- to the β-configuration being attended by a rotational shift of $+2A$.

The quantitative validity of the rules is surprisingly good for many aldohexose derivatives, and nearly constant differences, 2A, are found for corresponding anomeric derivatives of xylose, glucose, galactose, and gulose, and related heptoses. However, it was found that corresponding derivatives of mannose, rhamnose, altrose, and probably talose gave 2A values far outside the range for other sugars. This discrepancy has provoked much speculation; in conformational terms it seems that the constancy of 2A values fails to be maintained when the hydroxyl group at C-2 in a pyranose sugar is axial. The data for allose, idose, and talose are very incomplete.

Extensive data and discussions of the validity of the rules of isorotation have been given by Bates,[78] Isbell,[29] Pigman,[27] and Bose and Chatterjee.[79] The last-named authors sought, with fair success, to

---

[77] C. S. Hudson, *J. Am. Chem. Soc.*, **31**, 66 (1909).

[78] F. J. Bates and Associates, "Polarimetry, Saccharimetry and the Sugars," *Natl. Bur. Std. (U.S.) Circ.* C440 (1942).

[79] A. K. Bose and B. G. Chatterjee, *J. Org. Chem.*, **23**, 1425 (1958).

correlate empirically the sign of the rotational shift caused by epimerization of a secondary hydroxyl group with the stereochemical environment of the group for cyclic compounds generally, but they noted some anomalies. Korytnyk[80] described an interesting modification of the rules of isorotation that allows them to be applied to a wide range of derivatives containing highly polarizable substituents at the anomeric center; the rules in their original form failed with such compounds. To gain a better understanding of the reasons for the success and limitations of these empirical approaches to the correlation of rotational shifts, it is necessary to consider the conformational basis for the occurrence of optical activity.

Whiffen[81] described a method for calculating the molecular rotations of cyclic compounds, especially carbohydrates, by assigning rotational increments to conformational units in the structures. This approach has been developed by Brewster[82] to apply to a wide range of acyclic and cyclic compounds (see also Sec. 3-6). Optical activity is assumed to arise from a screw pattern of polarizability of the electrons around a center of optical activity. The screw pattern may arise from an asymmetric atom, but more usually the "center" of optical activity will be an asymmetric arrangement of bonds, and formally asymmetric atoms themselves will contribute very little to the rotation. It is further assumed that the total molecular rotation is obtained by simple summation of the partial contributions from all possible conformational units. All atoms are assumed to have a tetrahedral disposition of bonds.

If the sole source of asymmetry is the atom X in a compound XABCD, the net rotation of a randomly oriented assembly of molecules will be *dextro* for the configuration I (Fig. 22), if the polarizability of the attached atoms decreases in the order $A > B > C > D$. In the more general case of a chain of four atoms, A—C—C—A', the conformations shown in the

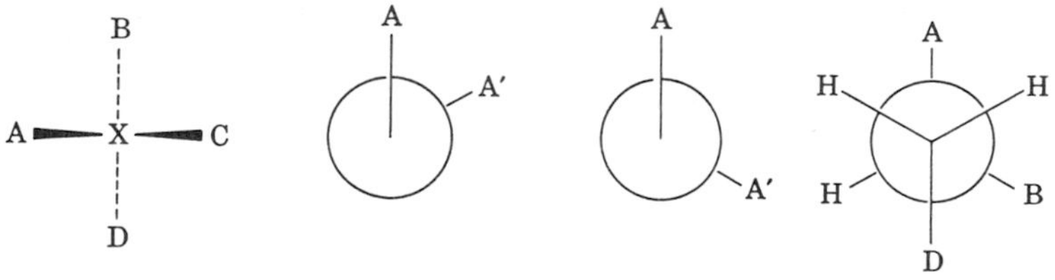

Fig. 22.

[80] W. Korytnyk, *J. Chem. Soc.*, **1959**, 650.
[81] D. H. Whiffen, *Chem. Ind. (London)*, **1956**, 964.
[82] J. H. Brewster, *J. Am. Chem. Soc.*, **81**, 5475, 5483 (1959).

Newman projections IIa and IIb (Fig. 22) have an asymmetric arrangement of the bonds A—C and A'—C, and the net effect of the asymmetry in each unit is a dextrorotatory increment for a randomly oriented assembly; it so happens that the value is the same for the staggered IIa and the eclipsed IIb conformation. If the atoms A—C—C—A' are coplanar, either eclipsed or *anti*, the asymmetry vanishes and the rotational contribution is zero. The total molecular rotation of the conformation III of a compound CHAB·CH$_2$D is the algebraic sum of six terms representing the contributions of the six asymmetric bond units, D—C—C—H, H—C—C—H, H—C—C—A, A—C—C—H, H—C—C—B, B—C—C—D. In practice, some terms may represent enantiomeric units, such as the third and fourth in this case, and cancel out. In dealing with acyclic compounds, Brewster imposed certain reasonable limitations to reduce the conformational analysis to manageable proportions. For cyclic compounds the conformational problem is simpler.

Although the fundamental theory permits the assignment of sign to the contributions from the different conformational units, it is necessary at present to select empirically the numerical magnitudes of the various units. To make the treatment as general as possible, Brewster derived the basic numerical values from the known configurations and rotations of simple molecules for which the conformational picture was clear. The magnitudes of the numerical values so assigned agree well with the polarizabilities of atoms and groups deduced from molecular refractions. The fundamental units derived for simple acyclic compounds can be used to calculate with fair accuracy the molecular rotations of carbocyclic compounds of known conformation, such as steroids. Many of the units are fairly large numerically; the value +45 is assigned to the important unit in carbohydrates in which A and A' in II (Fig. 22) are the oxygen atoms of hydroxyl groups. The source of the large shifts that normally result from epimerization at a single center in cyclic compounds (cf. ref. 79) is now apparent.

The calculation of molecular rotation is particularly simple in the case of cyclitols, where only one parameter is required. One looks along each C—C bond of the chair form in turn, with the hydroxyl group of the nearer carbon in the uppermost position. If a hydroxyl group on the rear carbon appears to the right, a value of +45° is taken; if to the left, −45° is used, the sum of these contributions being the predicted rotation. If the hydroxyl groups are *anti*, they do not contribute to the rotation of the molecule. The values thus calculated are in good agreement with the experimental data[81] as shown in Table 4. In two instances (those accompanied by superscripts d and e in the table), a successful prediction of the configuration was made on the basis of the molecular rotation.

Table 4. Calculated and Observed Molecular Rotations
of Polyhydroxycyclohexanes[a]

Compound	Molecular Rotation, °	
	Calculated[g]	Observed
Cyclohexane-		
(−)-1/2-diol	−45	−48
(−)-1,2/3-triol	−90	−92
(+)-1,2/3,4-tetrol[b]	+135, −90[c]	+106
(−)-1,2,3/4-tetrol	−45	−49
(−)-1,2,4/3-tetrol	−45	−57
(−)-1,2/3,5-tetrol[d]	−90	−90
(−)-1,2,5/3-tetrol[d]	+45	−12
(−)-1,3/2,4-tetrol	−45	−43
(+)-1,2,3/4,5-pentol[e]	+90	+100
(−)-1,2,3,5/4-pentol[f]	0	−9
(−)-1,2,4/3,5-pentol	−90	−91
(−)-1,2,5/3,4-pentol[e]	−135	−80
(+)-1,3,4/2,5-pentol	+45	+42
(−)-1,2,4/3,5,6-hexol	−135	−117

[a] For source of data see ref. 81 unless otherwise indicated.

[b] S. J. Angyal and P. T. Gilham, *J. Chem. Soc.*, **1958,** 375.

[c] Data calculated for both chair conformations since these are of nearly equal stability.

[d] D. H. Whiffen, personal communication; S. J. Angyal and P. A. J. Gorin, unpublished data.

[e] G. E. McCasland, S. Furuta, L. F. Johnson, and J. N. Shoolery, *J. Am. Chem. Soc.*, **83,** 2335 (1961).

[f] B. Magasanik, R. E. Franzl, and E. Chargaff, *J. Am. Chem. Soc.*, **74,** 2618 (1952).

[g] For a more elaborate method of calculation, which gives slightly better agreement with the observed results, see S. Yamana, *Bull. Chem. Soc. Japan*, **33,** 1741 (1960).

The values in the table were calculated for the predominant chair conformation of each compound. If the other chair is also taken into account, agreement with the observed values is improved; but the theoretical calculation of the rotations is not sufficiently accurate to allow the proportion of each chair form to be determined. This is shown particularly by the case of (+)-1,2/3,4-cyclohexanetetrol (Fig. 23) for which the two

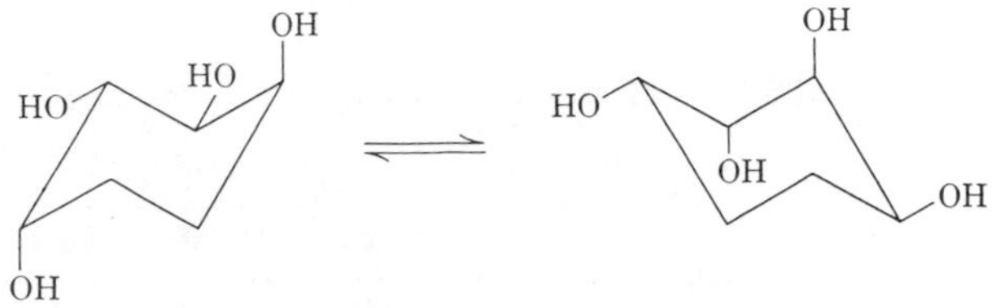

Fig. 23.

chair forms have nearly equal energy, yet optical rotation appears to indicate a great preponderance of form A.

Cyclic forms of carbohydrates include some additional conformational features for which allowance must be made. (1) The heterocyclic ring itself possesses an axis of polarizability difference (through the oxygen atom and C-3 in a pyranose ring), and the presence of dissimilar axial groups (usually H and OH) at C-1 and C-5 (Fig. 24, A) or at C-2 and C-4 (B) introduces a further axis of polarizability difference in each case, which forms a screw pattern of polarizability with the axis of the ring. The contribution of conformation (A) is assessed at +100, and of (B) at +60. (2) In hexopyranose rings the rotational asymmetry of the hydroxy-methyl group at C-5 necessitates a further empirical correction (cf. ref. 83). (3) The alkoxy group at C-1 in glycosides possesses rotational asymmetry; empirically, the large correction of 105 is found for methyl glycosides (dextro for an α-D form, levo for a β).

Brewster's numerical values[82] for the conformational parameters are used throughout this section, mainly because they have been developed as part of a picture embracing acyclic and alicyclic compounds as well as carbohydrates. Brewster himself has pointed out that the values assigned to a similar set of conformational factors by Whiffen are slightly more accurate when the sugars alone are considered, but the differences between the two sets are not important for the present purpose.

A simple calculation by Brewster's method shows that Hudson's 2A

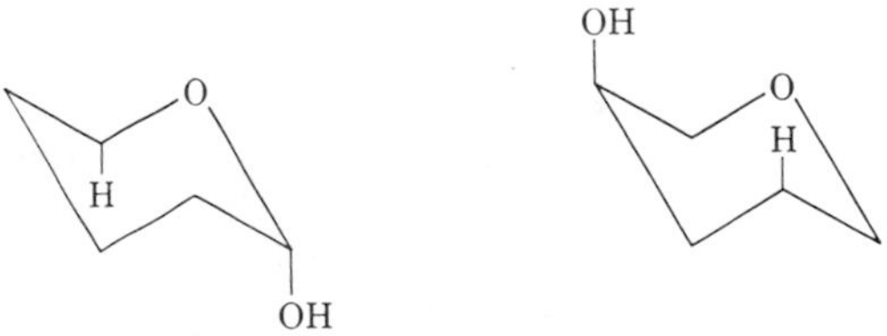

Fig. 24.

[83] S. Yamana, *Bull. Chem. Soc. Japan*, **31**, 558 (1958).

Table 5. Molecular Rotations of Methyl D-Glycopyranosides in Water

Parent Sugar	[M]$_D$ of methyl glycoside			Difference, 2A			Ref.[a]
	Obs.	N, calc.	A, calc.	Obs.	N, calc.	A, calc.	
α-D-glucose	+309	+325		375	400	265	
β-D-glucose	−66	−75					
α-D-galactose	+380	+375		380	400	265	
β-D-galactose	0	−25					
α-D-allose	—	+325		—	400	265	
β-D-allose	−103	−75					d
α-D-gulose	+232	+240		394	400	265	
β-D-gulose	−162	−160					
α-D-mannose	+154	+130		289	265	400	
β-D-mannose	−135	−135					
α-D-altrose	+239	+265		303	265	400	
β-D-altrose	−64	0					
α-D-idose	+190	+180		284	265	400	e
β-D-idose	−94	−85					f
α-D-talose	+204	+180		—	265	400	g
β-D-talose	—	−85					
α-D-xylose	+253	+250	+60	361	400	265	
β-D-xylose	−108	−150	−205				
α-D-arabinose	−28	+190	0	375	265	400	
β-D-arabinose	−403	−75	−400				
α-D-ribose	+170[b]	+250	+60	345[c]	400	265	h
β-D-ribose	−186	−150	−205				i
α-D-lyxose	+95	+55	+135	305	265	400	j
β-D-lyxose	−210	−210	−265				j

[a] When a reference is not specified, data are from ref. 81.
[b] In methanol.
[c] 2A for both anomers in methanol.
[d] B. Lindberg and O. Theander, *Acta Chem. Scand.*, **8**, 1870 (1954).
[e] E. Sorkin and T. Reichstein, *Helv. Chim. Acta*, **28**, 1 (1945).
[f] Ref. 32.
[g] P. A. J. Gorin, *Can. J. Chem.*, **38**, 641 (1960).
[h] G. R. Barker and D. C. C. Smith, *J. Chem. Soc.*, **1954**, 2151.
[i] J. Minsaas, *Ann.*, **512**, 286 (1934).
[j] H. S. Isbell and H. L. Frush, *J. Res. Natl. Bur. Std.*, **24**, 125 (1940).

value for anomeric methyl glycopyranosides should be $+400$ when the hydroxyl group at C-2 is equatorial in a normal chair ring, and $+265$ when this hydroxyl group is axial, and that configurational changes at C-3 and C-4 should have no further effect on the value of 2A. In Table 5 the molecular rotations of known methyl D-glycopyranosides are compared with the calculated values, in water as solvent.[84] The rotations of those members not discussed by Brewster have been calculated by his methods. Calculations for N and A conformations are included for the methyl pentopyranosides, but only calculations for the N form of hexopyranosides, as there are no authentic models from which to derive the numerical constants needed for the A forms of hexopyranosides.

For the methyl D-glycosides the agreement between observed and calculated rotations for the individual anomers, and between observed and calculated 2A values, is very good, considering that up to eight empirically derived numerical terms, each rounded to the nearest $5°$, are used in the calculation of each rotation. The agreement between calculated and observed rotations is equally good for the corresponding set of free sugars, 6-deoxy sugars, methyl 6-deoxy glycosides, methyl 2-deoxy glycosides, and 1-deoxypyranoses ($= 1,5$-anhydro hexitols). Of the pentosides, the methyl D-arabinosides clearly have the alternative conformation, in agreement with other evidence, and the D-xylosides the normal conformation, although the observed value for methyl β-D-xyloside does deviate more than usual from the calculated value.

Methyl β-D-altroside appears to be partially in the alternative conformation. On the other hand, agreement between the observed and calculated values for methyl α- and β-D-idoside does not necessarily prove that these sugars are in the normal conformation; the alternative chair forms may have similar rotations. Until molecular rotations can be calculated for the alternative conformations, the position cannot be fully clarified.

It is doubtful whether the study of optical rotations can decisively settle any conformational problems at present, even though Whiffen's and Brewster's investigations have clarified ideas of how conformational changes can effect rotations. At best, if the observed molecular rotation agrees with that calculated for a particular conformational model that fits other evidence, it may be regarded as a useful confirmation of the other deductions. On the other hand, if the observed rotation is very different from the calculated rotation, it probably means that the chosen conformational model is wrong or inexact, but the magnitude of the discrepancy can give little information about the true conformational equilibrium.

[84] A more extensive tabulation of sugar derivatives is given by W. Kauzmann, F. B. Clough, and I. Tobias, *Tetrahedron*, **13**, 57 (1961).

Perhaps the most significant contribution from these studies is the demonstration of the sensitivity of optical rotation to small conformational changes, such as will certainly occur at several centers in conformationally unstable pyranose rings undergoing conversion from one chair to the other through various flexible or half-chair forms.

Whiffen's and Brewster's approach to the problem of calculating optical rotations has been to assume that the principle of optical superposition holds for the various conformational elements of asymmetry, and that the individual elements do not have their partial rotations influenced significantly by neighboring elements. Brewster admitted that this is a simplification (cf. ref. 84), and it seems likely that a modified treatment will be necessary for compounds containing more highly polar substituents, such as phenyl glycosides and acylglycosyl halides—the groups that Korytnyk[80] found to require empirical modifications of Hudson's rules—because the influence of the polar group extends over several centers.

The fact that the calculations of the rotations of the pyranose conformations are applicable only to the unsubstituted sugars and glycosides in aqueous solution is a further limitation. Hudson's rules of isorotation seem to apply fairly well to acetylated sugars and glycosides, but the calculation of the necessary conformational parameters for these may be difficult, because of the rotational asymmetry of the acetoxy groups.

One other important empirical configurational correlation for cyclic carbohydrates that could probably be placed on a more nearly quantitative basis without difficulty is Hudson's "lactone rule."[85] In its original form it stated that the γ-lactone derived from an aldonic acid will be strongly dextrorotatory if the γ-hydroxy group that engages in lactone formation is on the right of the Fischer projection of the aldonic acid. Later the rule proved to be applicable to δ-lactones as well. The history of this rule, which led to the "generalized lactone rule" of Klyne,[86] has been summarized by Bose and Chatterjee.[79] If a suitable conformational model for lactones could be devised, it should be possible to calculate the conformational parameters for various configurations of hydroxyl groups within the ring, and so define the range of configurations for which this useful rule may safely be invoked.

The study of the optical rotations of cyclic carbohydrates has entered an important phase. The relatively simple treatment of conformational elements by Whiffen and Brewster has proved to be reasonably good for pyranose rings of well-defined conformations, and it should be applicable to many other classes of rings. There are many interesting groups of

[85] C. S. Hudson, *J. Am. Chem. Soc.*, **32**, 338 (1910).
[86] W. Klyne, *Chem. Ind. (London)*, **1954**, 1198.

cyclic derivatives of carbohydrates, of known conformation,[87] awaiting detailed correlation of optical rotation with configuration.

Acyclic Derivatives of Carbohydrates. It is generally assumed that the carbon chain of acyclic carbohydrates has the extended, planar, zig-zag conformation found in simple aliphatic compounds. This assumption accommodates most of the facts about the formation of cyclic acetals.[88] In such a chain, free rotation around the carbon-carbon bonds allows a balance of the interactions which cause optical rotations;[84] therefore, in agreement with expectations, the molecular rotations of the sugar alcohols and aldonic acids are uniformly very small in aqueous solutions.

The situation is different when the polyhydroxy chain is terminated by a highly polarizable group; the optical rotations of most derivatives of this type are far larger than is usual for acyclic compounds. An especially interesting class of compounds derivable from carbohydrates has an aromatic ring attached to a polyhydroxylated carbon chain. The best-known examples are the 2-(polyhydroxyalkyl)benzimidazoles, made by condensing aldonic acids with *o*-phenylenediamine,[89] and the "osotriazoles," made by oxidizing phenylosazones with cupric sulfate.[90] Some members are known of other series: polyhydroxyalkylbenzenes,[91] polyhydroxyalkylquinoxalines,[92] and polyhydroxyalkyl-1-pyrazole[3,4-*b*]quinoxalines ("sugar flavazoles").[93]

Some empirical correlations of rotation with configuration have been made in this group. Richtmyer and Hudson[94] noted that the sign of rotation of the benzimidazolyl derivatives was determined by the configuration of the secondary hydroxyl group adjacent to the aromatic nucleus. Their "benzimidazole rule" states that, if the acyclic polyhydroxy chain is drawn as a Fischer projection with the benzimidazole group at the top, all isomers in which the hydroxyl group at the uppermost asymmetric center lies on the right are dextrorotatory (for the D-line in dilute aqueous acid), and conversely those with the enantiomeric arrangement at this center are levorotatory. That is, all compounds with the partial configuration A (Fig. 25) are dextrorotatory, irrespective of the number and configuration of other asymmetric centers. Mills[95] has found

[87] J. A. Mills, *Advan. Carbohydrate Chem.*, **10,** 1 (1955).

[88] S. A. Barker, E. J. Bourne, and D. H. Whiffen, *J. Chem. Soc.*, **1952,** 3865.

[89] N. K. Richtmyer, *Advan. Carbohydrate Chem.*, **6,** 175 (1951).

[90] R. M. Hann and C. S. Hudson, *J. Am. Chem. Soc.*, **66,** 735 (1944); W. T. Haskins, R. M. Hann, and C. S. Hudson, *ibid.*, **70,** 2288 (1948).

[91] W. A. Bonner, *J. Am. Chem. Soc.*, **73,** 3126 (1951).

[92] H. Ohle and J. J. Kruyff, *Ber.*, **77,** 507 (1944).

[93] H. Ohle and G. A. Melkonian, *Ber.*, **74,** 279 (1941).

[94] N. K. Richtmyer and C. S. Hudson, *J. Am. Chem. Soc.*, **64,** 1612 (1942).

[95] (a) J. A. Mills, personal communication, 1962; (b) *Australian J. Chem.*, **17,** 277 (1964); H. El Khadem, *J. Org. Chem.*, **28,** 2478 (1963).

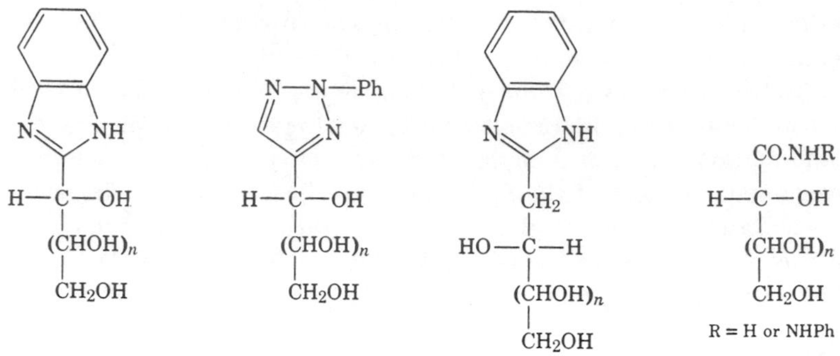

Fig. 25.

that a corresponding "phenylosotriazole rule" may be formulated: all isomers with the partial configuration B are dextrorotatory (in pyridine). No exception to either rule has been found. The configuration of the hydroxyl group in the α-position to the aromatic nucleus in A and B is that of the hydroxyl group at positions 2 and 3, respectively, in the sugar from which these derivatives were made, so the two rules together provide a means of assigning configurations to these positions in a reducing sugar.[95c]

The configuration of the hydroxyl groups at least as far as the γ-position in the chain has a discernible effect on the rotation. The benzimidazole derivatives of 2-deoxy sugars lack an asymmetric center at the α-position to the nucleus, and it seems that in these the configuration at the β-position determines the sign of rotation; all known compounds with the partial configuration C (Fig. 25) are dextrorotatory.[96] The phenyloso-triazoles show a striking relation between configuration and magnitude of rotation.[95a] Molecular rotations for eleven configurations are collected in Fig. 26; the values are calculated for the D-isomer in each case. Structures (1)–(4) with the *arabino* configuration at the first three asymmetric centers, have rotations near 230; structures (5)–(7), with the *xylo* configuration, near 130; and the remaining (incomplete) configurational possibilities have values of 80 or less.

A detailed study of the optical rotations of such compounds, in suitable solvents, would be rewarding in giving information on the partial rotations of secondary hydroxyl groups at varying distances from a chromophoric group and in differing conformational environments. These substances

---

[95c] For a generalized rotation rule, see H. El Khadem and Z. M. El-Shafei, *Tetrahedron Letters*, **1963**, 1887.

[96] A. J. Cleaver, A. B. Foster, and W. G. Overend, *J. Chem. Soc.*, **1957**, 3961.

would also be suitable subjects for the study of optical rotatory dispersion since they will have absorption bands at accessible wavelengths.

Aldonic acids and their simple salts have, like alditols, only small rotations, but the rotation is enhanced by conversion to amides or phenylhydrazides, with values of $[M]_D$ ranging up to 75. Here, again, the configuration of the hydroxyl group at the α-position determines the sign of the rotation. All such compounds with the α-hydroxyl group on the right in the Fischer projection (D, Fig. 25) are dextrorotatory. Hudson, arguing on the basis of the theory of optical superposition, showed[97] that the dominant rotational contribution from the α-position was modified by contributions from the β- and γ-positions. A more detailed study of

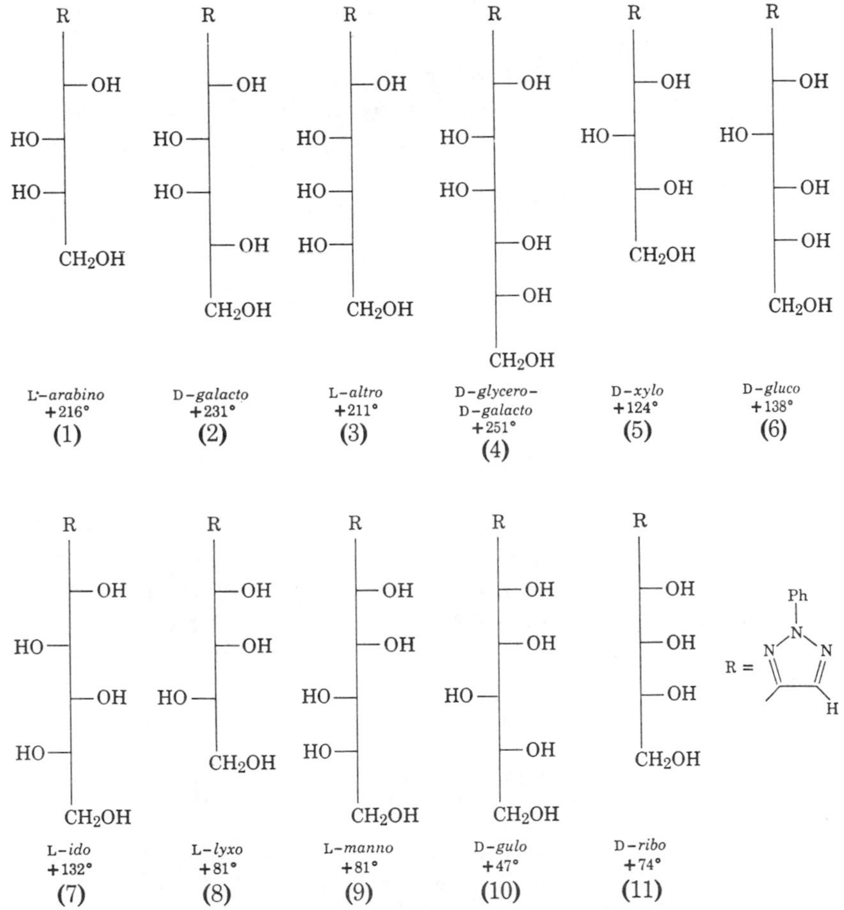

Fig. 26. Molecular rotations of phenylosotriazoles (in pyridine).

[97] C. S. Hudson, J. Am. Chem. Soc., **39**, 462 (1917); **40**, 813 (1918).

the optical rotations of such amides and phenylhydrazides, with due regard to the conformational fine structure, might be rewarding.

b. Infrared Absorption Spectra.[98] The infrared spectra of compounds such as carbohydrates, containing a plurality of similar groups, are always very complex, and the application of infrared spectroscopy to the study of carbohydrate conformations has naturally lagged behind its application to other classes of natural products. Nevertheless, considerable progess has been made since the first systematic studies began[99] about 1953. Partial success has been achieved in assigning bands to specific groups, particularly groups at the anomeric center, which are in an electronic environment different from that of substituents at other positions in the ring. This task was greatly assisted by the detailed study of the infrared spectrum of tetrahydropyran by Burket and Badger.[100] Recently, fairly complete assignments for simple analogs of carbohydrates have appeared;[101] they cover 2-hydroxymethyl- and 2-, 3-, and 4-hydroxytetrahydropyran and the corresponding derivatives of tetrahydrofuran.

Spectra have been recorded for carbohydrates and their derivatives as solids, dispersed as mulls or in alkali-halide discs, and dissolved (mainly as acetylated derivatives) in various solvents. The presumably fixed conformation in the solid state can thus be compared with the favored conformation in solution.

The Birmingham group carried out a detailed study of the infrared absorption of pyranose derivatives in the range 730–960 cm.$^{-1}$. They first examined[102] Nujol mulls of numerous derivatives of D-glucose (methylated sugars, methyl glucosides, free and methylated oligosaccharides, and polysaccharides) containing a pyranose ring of either α- or β-configuration. The spectra showed systematic differences that permitted the classification of any compound in these groups as either a member of the α or of the β series. Three types of absorption bands showing shifts diagnostic for α,β-anomerization were detected, and they were tentatively assigned to specific structural features as shown in Table 6.

The type 2 band at ca. 844 cm.$^{-1}$ (designated type 2a) was especially

[98] For a review, see W. B. Neely, *Advan. Carbohydrate Chem.*, **12**, 13 (1957).

[99] R. L. Whistler and L. R. House, *Anal. Chem.*, **25**, 1463 (1953); cf. L. P. Kuhn, *ibid.*, **22**, 276 (1950).

[100] S. C. Burket and R. M. Badger, *J. Am. Chem. Soc.*, **72**, 4397 (1950).

[101] N. Baggett, S. A. Barker, A. B. Foster, R. H. Moore, and D. H. Whiffen, *J. Chem. Soc.*, **1960**, 4565.

[102] S. A. Barker, E. J. Bourne, M. Stacey, and D. H. Whiffen, *J. Chem. Soc.*, **1954**, 171.

Table 6. Infrared Bands of Anomeric Glucopyranose Derivatives

Designation	Position of band, cm.$^{-1}$	Assignment
Type 1	α 917 $\pm$ 13; β 920 $\pm$ 5	Ring vibration
Type 2	α 844 $\pm$ 8; β 891 $\pm$ 7	$C_{(1)}$—H deformation
Type 3	α 768 $\pm$ 10; β 774 $\pm$ 9	Ring breathing

characteristic of the presence of α-D-glucopyranose units; the band at ca. 891 cm.$^{-1}$ (type 2b) was strongly marked for β-D-glucose units, but less certainly diagnostic, as type 1 bands of α-D-glucopyranose units could partly overlap its position. The occurrence of the type 2 bands in a region where C—H deformations may be expected to cause absorption, their sharpness, and the regularity of the shift caused by α-β isomerism, led to their allocation to the $C_{(1)}$—H bond, which is the only C—H bond in a distinctive electronic environment. The presence of type 2a absorption therefore indicates an equatorial hydrogen atom at the anomeric center, and of type 2b, an axial hydrogen at this site. The subsequent detection[103] of type 2a bands in the spectra of derivatives of α-D-galactopyranose and of α-D-mannopyranose, and of type 2b bands in spectra of the corresponding β-anomers, strengthened these assignments. The observation that type 2a bands occur in the spectra of derivatives of β-D- and β-L-arabinopyranose is in harmony with other proofs that derivatives of arabinose have the alternative chair conformation. Type 2a bands were not found for derivatives of xylopyranose of either α- or β-configuration, and their absence was tentatively explained by the markedly smaller change of dipole moment that would result from equatorial $C_{(1)}$—H deformation in the more symmetrical normal conformation of D-xylopyranose, which lacks both hydroxymethyl group and any axial group except that at C-1.

Almost all derivatives of D-galactopyranose and D-mannopyranose showed an extra type 2 absorption band, designated 2c, at ca. 875 cm.$^{-1}$, which was assigned to the equatorial C—H bond at positions 4 and 2, respectively, in these sugars. This band is absent for glucose derivatives which have only axial C—H bonds at positions other than C-1. This assignment was supported by the finding[1] of similar bands in the spectra of most cyclitols, in which one or more C—H bonds would be equatorial in the more stable conformation, whereas such bands were not found for *scyllo*-inositol and *scyllo*-quercitol (all hydroxyl groups equatorial).

Bands of types 1 and 3 occur in regions where C—C and C—O stretching modes are likely to be active, and by comparison with the assignments

[103] H. S. Isbell, F. A. Smith, E. C. Creitz, H. L. Frush, J. D. Moyer, and J. E. Stewart, *J. Res. Natl. Bur. Std.*, **59**, 41 (1957).

for tetrahydropyran[100] were assigned to ring vibration modes and ring breathing, respectively. Anomerization causes much smaller shifts in these bands than in type 2 bands; therefore they are less useful as a means of conformational assignment at the anomeric center.

Isbell and his colleagues[103] reported the results of an extended study of the infrared spectra of acetylated pyranose derivatives dissolved in various solvents. An important feature of the work is the inclusion of some heptopyranose derivatives; examples of *ido*, *gulo*, and *talo* configurations are readily accessible in the aldoheptose series. The conformationally interesting parts of the spectrum, which reflected conformational changes at the anomeric center, lay in the frequency range 1120–1220 cm.$^{-1}$. For fully acetylated pyranose sugars the significant absorption bands occurred at 1159 ± 6 and 1127 ± 8 cm.$^{-1}$. Axial anomers (i.e., α forms) of the acetates of conformationally stable sugars (*gluco*, *galacto*, and *manno* configurations) showed fairly strong absorption near 1159 and at most weak absorption near 1127 cm.$^{-1}$, whereas the converse was true of the equatorial anomers.

Molar absorbancy indices at 1159 and 1127 cm.$^{-1}$ and the differences between the indices were tabulated for a number of anomeric pairs of acetylated sugars. By comparing the differences in intensity of absorption, some information was obtained about the conformation at the anomeric center. The most interesting results were obtained with some derivatives of *gulo*, *talo*, and *ido* configurations. At least some axial component was present in hexa-*O*-acetyl-D-*glycero*-β-*ido*-heptopyranose and in hexa-*O*-acetyl-D-*glycero*-β-D-*gulo*-heptopyranose, and at least some equatorial component in hexa-*O*-acetyl-D-*glycero*-α-D-*gulo*-heptopyranose (for which the axial band at 1159 cm.$^{-1}$ was not detected) and in penta-*O*-acetyl-α-D-talose. It would seem from the spectra obtained that the alternative conformations might have predominated in solutions of these compounds, especially in the last two. Similar deductions could be made from the spectra of acetylated methyl glycopyranosides in solution.

More recently, Tipson and Isbell[104] reported a careful analysis of the spectra, over the widest possible range of frequencies, of acetylated and unacetylated methyl aldopyranosides pressed into discs of potassium chloride or iodide. The data should give information about preferred conformations in the solid state. The original papers must be consulted for details of the method of analysis of the spectra, but some of the conclusions are briefly as follows. Methyl α- and β-D-arabinopyranoside were found in the alternative conformation, and methyl β-D-gulopyranoside, methyl D-*glycero*-β-D-*gulo*-heptopyranoside, and cyclohexyl

[104] R. S. Tipson and H. S. Isbell, *J. Res. Natl. Bur. Std.*, **64A**, 239, 405 (1960).

D-*glycero*-β-D-*gulo*-heptopyranoside in the normal conformation, in agreement with expectations. A clear-cut result was not obtained for methyl D-glycero-α-D-*gulo*-heptopyranoside, although the axial arrangement of the methoxy group (normal conformation) was not excluded by the evidence; the group is not equatorial (i.e., not alternative conformation), but it could be in some conformational state not associated with a true chair ring. Indications of a possible mixture of conformations, or of a non-chair ring, were also found for methyl α-D-gulopyranoside and for methyl α- and β-D-lyxopyranosides.

The study of infrared spectra is therefore useful for determining the configuration and the conformation of anomeric carbon atoms; the same purpose, however, is now better served by NMR spectroscopy (Sec. 6-5c).

Information about intramolecular and intermolecular hydrogen bonding is conveniently obtained from infrared spectra. Polyhydroxy compounds are often unsuitable for this purpose, but simpler molecules, containing one or two hydroxyl groups, are available among cyclic acetals of carbohydrates, and the extent of hydrogen bonding in these can be related to the probable conformation.

Most applications of infrared spectroscopy to conformational problems involving hydrogen bonding depend on Kuhn's observation[105] that the hydroxyl stretching band near 3630 cm.$^{-1}$ is shifted to lower frequencies if the group engages in hydrogen bonding (Sec. 3-4a). Intramolecular hydrogen bonding is not observed if the hydrogen atom of the hydroxyl group is farther than about 3.3 Å from another oxygen atom to which bonding can occur. The effect of intermolecular hydrogen bonding is avoided by working with very dilute solutions in carbon tetrachloride. For more complex polyhydroxy compounds to which this method is not applicable, information about hydrogen bonding may be obtained by substituting deuterium for the hydrogen of the hydroxyl groups.[106]

In cyclic derivatives of carbohydrates, the presence of only one hydroxyl group is sufficient for hydrogen bonding to occur, if it is attached to a 1,3-dioxane ring[45] or a tetrahydropyran ring[46] and is suitably placed for bonding to an oxygen atom in the ring. This fact made possible assignments of configuration[45] to the *cis* and *trans* isomers of 5-hydroxy-2-phenyl-1,3-dioxane (1,3-*O*-benzylideneglycerol). One isomer showed strong absorption corresponding to a hydrogen-bonded hydroxyl group at 3593 cm.$^{-1}$; but absorption due to a free hydroxyl group could not be detected near 3630 cm.$^{-1}$, and therefore it was assigned the *cis* configuration, which permits complete hydrogen bonding (Fig. 17); whether the hydrogen

105 L. P. Kuhn, *J. Am. Chem. Soc.*, **74**, 2492 (1952); **76**, 4323 (1954).

106 S. A. Barker, E. J. Bourne, and D. H. Whiffen, *Methods Biochem. Anal.*, **3**, 213 (1956).

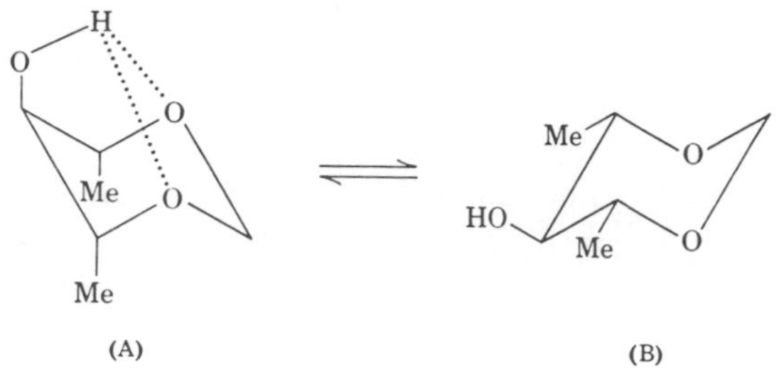

Fig. 27.

is bonded to both ring atoms simultaneously is open to question. The other isomer showed bands at both 3633 (free hydroxyl group) and 3593 cm.$^{-1}$ (hydrogen-bonded hydroxyl) and was assigned the *trans* configuration, with the explanation that the significant absorption at 3593 cm.$^{-1}$ was due to a proportion of the molecules existing in the alternative conformation, which, notwithstanding the unfavorable feature of an axial phenyl group, is stabilized by hydrogen bonding (Fig. 17).

A more extreme example of conformational stabilization resulting from intramolecular hydrogen bonding is that of 1,5-dideoxy-2,4-O-methylene-ribitol (Fig. 27). Its spectrum shows[107] a considerable contribution from an intramolecularly hydrogen-bonded species, which might be the chair form (A) with two axial methyl groups. However, the proportion of this conformation in equilibrium would be expected to be much less than that indicated by the spectral data, and it was suggested[107] that the hydrogen bonding is associated with conformations other than chair forms. The problem requires further study.

c. **Nuclear Magnetic Resonance Spectra** (cf. Sec. 3-4d). In the last few years the structures and configurations of several new sugars and sugar derivatives have been determined with the aid of nuclear magnetic resonance (NMR) spectra. Owing to the nature of this method, the conformations of the compounds were also established at the same time. NMR has thereby shown itself to be one of the most powerful techniques for the assignment of diastereomeric configurations, particularly in cyclic compounds; but so far relatively little use of it has been made in carbohydrate chemistry.

Nevertheless, early work of importance on the correlation of NMR spectra with conformations was carried out on carbohydrates as model compounds. Lemieux and his co-workers[2] studied the NMR spectra (at 40 Mc.p.s.) of acetylated cyclitols and sugars and found that, in general, equatorial

---

[107] S. A. Barker, A. B. Foster, A. H. Haines, J. Lehmann, J. M. Webber, and G. Zweifel, *J. Chem. Soc.*, **1963**, 4161.

hydrogen atoms produce signals at lower field (by about 25 c.p.s.) than similar but axial hydrogen atoms.[108] On the other hand, the hydrogen atoms of equatorial acetoxy groups produce their signal some 8 c.p.s. to higher field than those of axial acetoxy groups. This difference in chemical shifts between axial and equatorial groups* has been used to determine the configuration of deoxynitroinositols.[22,109] The important reservation has to be made that the presence of a *syn*-axial oxygen atom has a deshielding effect on an axial hydrogen atom; such a hydrogen atom may not show noticeable chemical shift from equatorial hydrogen atoms.[3,110]

More important still was the observation by Lemieux and his co-workers[2] that the spin-spin coupling constant for the hydrogens on neighboring carbon atoms is 2–3 times larger when the hydrogens are both axial (dihedral angle 180°) than when one or both of the hydrogens is equatorial (60°). This observation led to the development of the Karplus equation,[111] which gives the relation between dihedral angle and coupling constant (Sec. 3-4d).

The first application of coupling constants to the determination of configurations (and thereby of conformations) was in the dimethyl ethers of 1,2,3-cyclohexanetriols.[112] The configuration (and conformation) of mycarose,[113] a rare branched chain sugar, of actinamine,[114] an amino-cyclitol, and of several new quercitols[115] was recently established by this method.

Despite these successes, comparatively little work has yet been carried out on the conformations of the common sugars in aqueous solution. Such work could be easily done because deuterium oxide is a solvent suitable for determining NMR spectra. The possibilities are indicated in the work of Lenz and Heeschen[116] on glucose, mannose, and 2-deoxy-glucose. The proton on the anomeric carbon atom is deshielded by two oxygen atoms and gives rise to a signal at lower field than any of the other protons; it is therefore easily identified and studied. Here, again,

[108] See also E. L. Eliel, M. H. Gianni, T. H. Williams, and J. B. Stothers, *Tetrahedron Letters*, **1962**, 741.

[109] F. W. Lichtenthaler, *Chem. Ber.*, **94**, 3071 (1961).

[110] R. U. Lemieux, Lecture delivered at the I.U.P.A.C. Congress, Montreal, 1961.

[111] M. Karplus, *J. Chem. Phys.*, **30**, 11 (1959).

[112] R. U. Lemieux, R. K. Kullnig, and R. Y. Moir, *J. Am. Chem. Soc.*, **80**, 2237 (1958).

[113] W. Hofheinz, H. Grisebach, and H. Friebolin, *Tetrahedron*, **18**, 1265 (1962).

[114] G. Slomp and F. A. MacKellar, *Tetrahedron Letters*, **1962**, 521.

[115] G. E. McCasland, S. Furuta, L. F. Johnson, and J. N. Shoolery, *J. Am. Chem. Soc.*, **83**, 2335, 4243 (1961).

[116] R. W. Lenz and J. P. Heeschen, *J. Polymer Sci.*, **51**, 247 (1961).

* Tabulated data on the positions of the resonance of axial and equatorial hydrogens and acetoxy groups are given in refs. 2 and 109.

an equatorial proton appears at lower field (by about 35 c.p.s. at 60 Mc.p.s.) than an axial one: α- and β-glucose are readily distinguished and their amounts in mixtures can be determined. The method is therefore suitable for studying mutarotation and the composition of mutarotated mixtures, and it provides a good way to establish the configuration of anomeric centers (e.g., in thymidine,[63] in neomycin[117]).

By the use of the Karplus equation, which they modified by insertion of a proportionality factor, Lenz and Heeschen[116] derived values for the dihedral angles between C-1 and C-2 in several sugars. They obtained reasonable values for the dihedral angles in several cases but came to the unlikely conclusion that the chair of β-glucose is strongly distorted and that β-mannose is in a half-chair form.*

Lemieux, Stevens, and Fraser[118] pointed out that the coupling constant between hydrogens on adjacent carbon atoms depends not only on the dihedral angle but also on the electronegativity of substituents on the carbon atoms. They correct the Karplus equation not by a proportionality factor but by addition of a constant (i.e., by an upward displacement of the Karplus curve). Abraham and his co-workers[65] used a different modification of the Karplus equation in their studies on sugar acetals, whereas Jardetzky[64] used the unmodified equation to interpret the spectra of nucleosides.

It appears therefore that, while the Karplus equation gives the form of the angular dependence of the coupling constants, different systems require different parameters to yield reasonable results. Once these are known, it will be possible to obtain precise information on the conformations of carbohydrates in solution. A full analysis of the NMR spectra

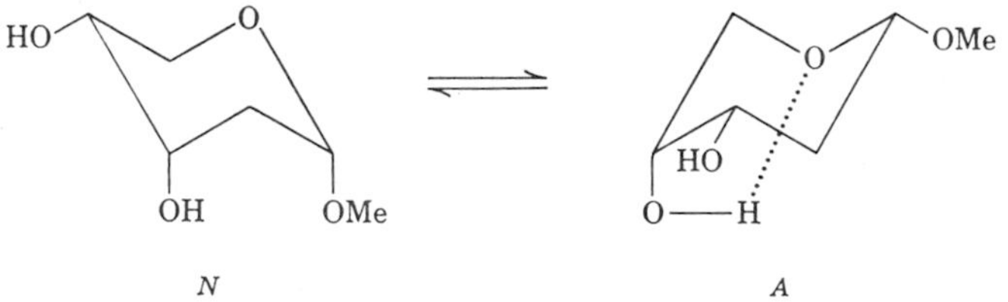

Fig. 28.

[117] K. L. Rinehart, Jr., W. S. Chilton, and M. Hichens, *J. Am. Chem. Soc.*, **84**, 3216 (1962).

[118] R. U. Lemieux, J. D. Stevens, and R. R. Fraser, *Can. J. Chem.*, **40**, 1955 (1962).

* The erroneous conclusion may be partly due to the fact that H-2 is not appreciably chemically shifted from H-3 and H-4. Under these conditions—as pointed out by F. A. L. Anet [*Can. J. Chem.*, **39**, 2262 (1961)] in discussing the fine structure of signals for methyl groups—the spacings of the peaks for H-1 do not give the true value of the coupling constant, J_{H1-H2}.

is desirable for this purpose; although the spectra of most carbohydrate derivatives are very complex, detailed assignment and interpretation have been achieved in several cases.[40],[64],[65]

Even the qualitative application of NMR yielded many interesting results. Thus Lemieux[110] found that methyl 2-deoxy-α-riboside changes its conformation as the solvent is changed; in water it occurs predominantly in the N, in chloroform in the A conformation (Fig. 28). Presumably the latter is stabilized, in the non-hydroxylic solvent, by an internal hydrogen bond (cf. the similar case of methyl 4,6-O-benzylidene-α-D-idoside, discussed in Sec. 6-4b).

In an ingenious way, Lemieux[110] used NMR to determine the equilibrium between the two chair forms of tetra-O-acetyl-β-L-arabinopyranose at room temperature. In both possible chair forms, there are two axial and two equatorial acetoxy groups, and the spectrum showed two signals of equal intensity. When the acetoxy group on C-1 was replaced by a fully deuterated acetoxy group, causing its signal to disappear from the spectrum, the relative intensity of the two signals had the value 1.46, from which it can be calculated that the alternate form is more stable by 0.75 kcal./mole.

NMR was also applied to conformational analysis in order to establish[119] the conformation of 1,3:2,4-di-O-methylene-L-threitol, the simplest example of an interesting group of bicyclic compounds. In this compound the carbon and oxygen atoms have the same arrangement as those in *cis*-decalin: two chair forms are possible but, unlike in *cis*-decalin, they are not equivalent. Mills discussed[87] structures of this type in detail and concluded—in order to rationalize the relative stabilities of cyclic acetals formed from sugar alcohols with aldehydes—that the so-called "O-inside" conformation (A) is more stable than the "H-inside" (B, Fig. 29). Analysis[119] of the NMR spectrum, as well as that of the dipole moment, proved that Mills's assumption was correct.*

NMR has clearly great potentialities in the study of carbohydrate conformations. The greatest advantage of the method is that it can be used on aqueous solutions of sugars, and therefore it yields information where it is most valuable. It is expected that in the near future NMR will yield quantitative data on the equilibria between α- and β-anomers,

[119] R. U. Lemieux and J. Howard, *Can. J. Chem.*, **41**, 393 (1963).

* Mills considered that the close approach of the unshared electrons on the four oxygen atoms in (A) is not likely to be so unfavorable as the close approach of the four axial hydrogen atoms in (B). Another factor may be more important, however: the unshared electrons in (A) suffer non-bonded interactions with oxygen atoms (of the other ring), whereas the axial hydrogen atoms in (B) are *syn*-axial with carbon atoms, causing greater non-bonded interaction.

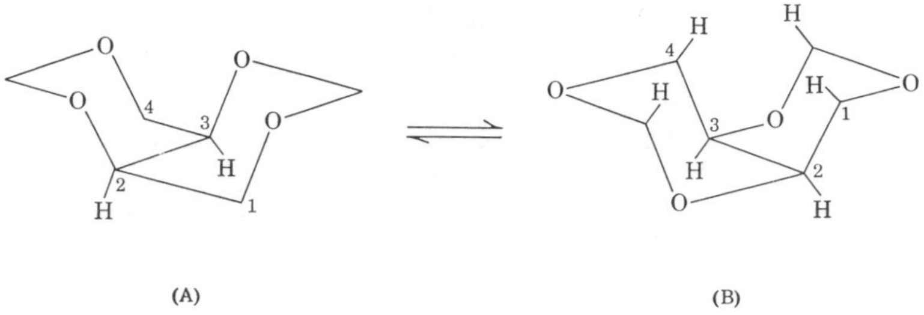

(A)   (B)

Fig. 29.

between pyranose and furanose forms, and between normal and alternative chair conformations.

6–6. Equilibria Dependent on Conformations

In this section several reactions will be discussed which lead to equilibrium. The position of the equilibria depends only on the free energy content of the components; in the examples discussed the free energy will be governed mainly by conformational factors. In this discussion it will be convenient to be able to refer to a collection of free energy values of sugars.

In Table 3 calculated free energy values were given for the two chair conformations of each sugar. In solution, each anomer of a sugar will consist of an equilibrium mixture of the two chair forms, and it will have a free energy lower than either has, owing to the entropy of mixing of the two. The free energy of each anomer of the aldohexoses and aldopentoses was calculated by the formula

$$G = G_N N_N + G_A N_A + RT(N_N \ln N_N + N_A \ln N_A)$$

that is, the weighted average of the free energy values was corrected for the entropy change on mixing (cf. Sec. 1-4). N_N and N_A, the mole fractions of the normal and the alternative conformers, respectively, were obtained from the equation $G_N - G_A = -RT \ln (N_N/N_A)$. The free energy values thus obtained are listed in the first two columns of Table 7.

When a sugar is in an equilibrated solution, i.e., as a mixture of its α- and β-anomers, its free energy will be lower than that of the individual anomers, owing to the entropy of mixing these anomers. The free energy of sugars in their equilibrium mixture has also been calculated, by this method, and it is given in the last column of Table 7. To be exact, the contribution of the furanose forms should also have been taken into

Table 7. Relative Free Energy Content of Pyranoses (kcal./mole) in Aqueous Solution

	α	β	Equil. Mixture
Allose	4.2	2.85	2.8
Altrose	3.95	3.75	3.45
Galactose	3.2	2.85	2.6
Glucose	2.3	1.95	1.7
Gulose	4.5	3.35	3.25
Idose	3.75	4.25	3.55
Mannose	2.85	3.3	2.65
Talose	4.4	5.15	4.25
Arabinose	2.45	2.7	2.15
Lyxose	2.3	2.9	2.1
Ribose	3.75	2.5	2.45
Xylose	1.95	1.6	1.35

account; the presence of considerable amounts of furanose forms would further lower the free energy of the sugar. However, since neither the amounts of furanoses in equilibrium, nor their free energies, are known, this correction—very small for most sugars—could not be applied.

The values in the table represent relative free energies, relative to an imaginary hexopyranose or pentopyranose in which there are no non-bonded interactions. Owing to the approximations involved in the calculation of the interaction energies, the figures in Table 7 are not to be taken as exact values; they serve only as a useful guide to the behavior of sugars in equilibria.

a. Pyranose-Furanose Interconversion. In aqueous solution, reducing sugars are present as complex mixtures of isomeric forms in equilibrium. When a crystalline sugar (which consists of one isomer only) is dissolved in water, its optical rotation changes until equilibrium is reached; this phenomenon is known as mutarotation. The fundamental problem in understanding the behavior of sugars in aqueous solution is to know the composition of the equilibrium mixture. In no case is this composition exactly known and, in particular, there are no methods for determining the amounts of the two furanose forms present in solution. The muta-rotation, however, can serve as a qualitative indication of the presence, or absence, of substantial amounts of furanose forms.

The pattern of mutarotation can fall into one of two types.[120] Some-times the mutarotation may be described by the equation for a first-order

[120] For a detailed discussion, see ref. 78, pp. 439 ff.

reversible reaction, whether approached from the α- or from the β-pyranose ("simple" mutarotation). Then the equilibrium mixture consists predominantly of the two pyranose forms. Some sugars which show this behavior are glucose, mannose, lyxose, and disaccharides in which these sugars are the reducing moieties, e.g., lactose.

Table 8. Occurrence of Furanose Forms in Equilibrium Solution of Sugars

Sugar	Configuration of Substituents in the Furanose Form	Occurrence of Fast Mutarotation
Lyxose	O-2, O-3, C-5 *cis*	—
Mannose		—
Gulose		—
Xylose	O-3 and C-5 *cis*	—
Glucose		—
D-*glycero*-D-*ido*-Heptose[a]		+
Ribose	O-2 and O-3 *cis*	+
Allose		b
Talose		+
Arabinose	all-*trans*	+
Altrose		+
Galactose		+

[a] Idose is not known in the crystalline state, hence mutarotation data are not available; the aldoheptose of the *ido* configuration is therefore listed in the table.

[b] The mutarotation of D-allose has not been determined with sufficient accuracy to allow a definite conclusion [F. P. Phelps and F. J. Bates, *J. Am. Chem. Soc.*, **56**, 1250 (1943)]; it appears to show fast mutarotation to a slight extent.

Other sugars, however, show "complex" mutarotation, consisting of an initial fast change, followed by a slow one, not necessarily in the same direction; more than two components are then involved in the equilibrium. The initial fast mutarotation is ascribed to a pyranose-furanose interconversion, and the slow one to the anomerization of pyranose forms. The occurrence of an initial fast mutarotation can be taken as an indication that substantial amounts of furanoses are present in equilibrium. Examples of such behavior are arabinose, ribose, galactose, and talose.

In comparing the experimental data with theoretical considerations, one has to bear in mind that the proportion of furanoses in equilibrium depends on the relative stabilities of the furanose and the pyranose forms. The stability of the furanose forms, in turn, depends on whether its substituents are *cis* or *trans* to each other. The data in Table 8 show that the presence of substantial amounts of furanose forms in some sugars, as indicated by the occurrence of fast mutarotation, is in accordance with theoretical considerations. When the furanose form has three substituents *cis* to each other (not counting the anomeric hydroxyl group which can always take up the configuration *trans* to its neighbor), the configuration is unfavorable, and fast mutarotation is not observed. When the three substituents are in the *trans-trans* arrangement, the furanose form is favorable and its presence is indicated by fast mutarotation. In the intermediate case of two *cis* substituents, substantial amounts of furanoses are present only if all the possible pyranose forms have high interaction energies; otherwise, as in xylose and glucose, the amount of furanoses present in equilibrium is small.

From the detailed mathematical analysis of the mutarotation data, on the assumption that only three components were present, it was calculated[121] that at 20° the equilibrium mixture of galactose contains 12% of an isomer different from the two pyranoses; presumably this is the β-furanose form, which is particularly favorable (all-*trans*). From a study of the optical rotation of freshly hydrolyzed sucrose it was concluded[122] that the equilibrium mixture of D-fructose at 25° contains 31.5% of the β-furanose form; in this case, again, the furanose is conformationally favorable (*trans-trans* at C-3, C-4, and C-5).

If there is considerable steric strain in the pyranoses, the furanose forms may be the more stable ones. Remarkable examples of such conditions are the methyl 3,6-anhydro-D-glucopyranosides[73] (Fig. 30) and some related compounds[74] which change spontaneously, in the presence of acids, to the corresponding furanosides. The anomeric configuration is

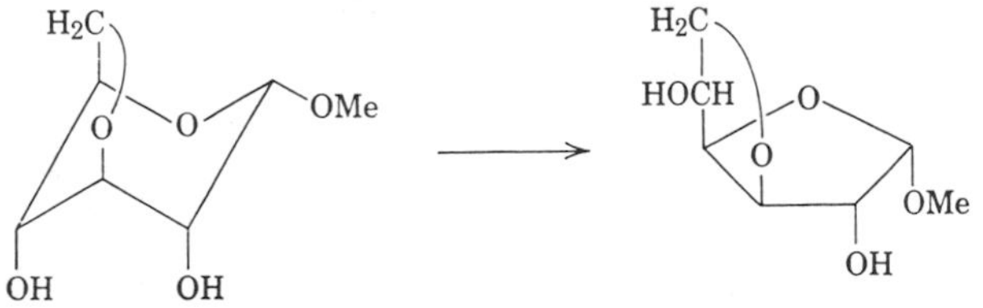

Fig. 30. Conversion of methyl 3,6-anhydro-α-D-glucopyranoside to the furanoside.

[121] G. F. Smith and T. M. Lowry, *J. Chem. Soc.*, **1928**, 666.
[122] B. Anderson and H. Degn, *Acta Chem. Scand.*, **16**, 215 (1962).

retained in this process (this retention is a remarkable feature which has not yet been explained). The anhydro bridge in the pyranoside requires considerable distortion of the chair form,* whereas the furanoside, with two *cis* fused five-membered rings, is free of strain. Presumably the corresponding free sugars, in equilibrium solution, also occur predominantly in furanose forms.

An equilibrium similar to that between furanoses and pyranoses exists between the γ- and the δ-lactones of glyconic acids; in this case, however, the presence of the carbonyl group increases the comparative stability of the five-membered ring. The position of the equilibrium varies from sugar to sugar, but there are no reliable data for conformational analysis; the problem may repay an investigation by modern methods. Some data are, however, available on the extent of the hydrolysis of fully methylated aldonolactones;[123] in most cases the δ-lactones are hydrolyzed to a greater extent in equilibrium than the γ-lactones, but the reverse is true for the xylonolactones (where the six-membered ring is particularly favorable, the five-membered one rather unfavorable). For a discussion, see ref. 66.

b. Anomerization (α-β Equilibrium)

Free Sugars. The proportion of α- and β-pyranose forms in equilibrium solution has been determined by two methods. If the mutarotation is not complicated by the presence of furanose isomers, and if both pyranose forms are available in the crystalline state and their rotations are therefore known, the proportion of the anomers can be calculated from the value of the equilibrium rotation. If the mutarotation is complex, it is still possible to calculate the contribution that the slow mutarotation (anomerization of the pyranoses) makes to the change in rotation. Provided, again, that

---

* The strained nature of this bridged chair form is shown by the difficulty with which 2,4-di-O-methyl-3,6-anhydro-D-mannonic acid is lactonized.[72] The acid can be sublimed unchanged *in vacuo* and is only partially converted to the lactone (A) by prolonged heating to 150°. The analogous gluconic acid derivative does not yield[73] the lactone which would have the conformation B, containing an additional interaction between two *syn*-axial methoxyl groups.

(A)   (B)

[123] S. R. Carter, W. N. Haworth, and R. A. Robinson, *J. Chem. Soc.*, **1930**, 2125.

both pyranose forms are known in the crystalline state, the extent of the slow mutarotation will give an approximate measure of the α- to β-pyranose ratio in equilibrium. The amount or the proportion of the two furanose forms cannot be determined in this way, however, because the rotations of the furanose isomers are not known.

The position of the equilibrium can also be studied by chemical methods, of which the oxidation of sugars by bromine has been most extensively investigated.[124] The α- and β-anomers differ widely in the rate of their reaction (see Sec. 6-7c). By comparing the rate of oxidation of an equilibrium solution of a sugar with the rates of the pure α and β modifications, as determined separately, it is possible to calculate the proportion of each modification in the equilibrium solution. This method yields useful information even when only one of the anomers is known in the crystalline state. The results agree well with those obtained from mutarotation data where the proportion of furanoses is low; otherwise the agreement is only moderately good.

By these methods the proportion of the α- and β-pyranose forms in equilibrium has been determined—fairly accurately for a few sugars, less accurately for a few others; for many sugars this proportion is not yet known. The available data can be used to check the validity of the quantitative conformational approach. The proportion of the α form of pyranoses in equilibrium has been calculated from the theoretical free energy values in Table 7 by the use of the equation $\Delta G = -RT \ln K$. The calculated values and the available experimental data are listed in Table 9; the agreement between the two sets of values is, in most cases, satisfactory.

These conformational considerations can be employed to determine the anomeric configuration of crystalline D-ribose. Only one form is known and, because mutarotation is complex, Hudson's rules cannot be used to decide whether it is the α or the β form. Isbell and Pigman[125] suggested that crystalline ribose is the β-isomer, but definite proof was lacking. Table 9 indicates that the β form should predominate considerably in the equilibrium mixture of ribose. Isbell and Pigman showed that about 90% of the equilibrium mixture is oxidized by bromine at the same rate as the pure crystalline isomer; and Tipson and Isbell[126] found recently that the infrared spectrum of crystalline ribose and that of the equilibrium mixture are practically identical. These facts prove that the crystalline compound is β-D-ribose. The complex mutarotation of ribose has caused it to be credited with very unusual behavior; in fact, however, the maximum

[124] Ref. 78, p. 454.
[125] H. S. Isbell and W. W. Pigman, *J. Res. Natl. Bur. Std.*, **18,** 141 (1937).
[126] R. S. Tipson and H. S. Isbell, *J. Res. Natl. Bur. Std.*, **66A,** 31 (1962).

variation in rotation during mutarotation is only 5°, indicating that there is little change in composition.

Although values of the α- to β-pyranose ratio are not known for the ketohexoses, similar considerations can be applied to them qualitatively. In these compounds there are two substituents on the anomeric carbon atom, C-2: a hydroxymethyl and a hydroxyl group. Since the hydroxymethyl group has a stronger interaction (by 0.9 kcal./mole) with axial

Table 9. Proportion of α-Pyranose Form in Equilibrium

	Calculated, %	Found from Rotation,[a] %	Found by Bromine Oxidation,[a] %
Glucose	36	36.2	37.4
Mannose	68	68.8	68.9
Galactose	36	29.6	31.4
Gulose	12.5	—	18.5
Talose	77.5	—	65.9
D-glycero-D-ido-Heptose	70	—	77.5
Lyxose	73	76	79.7
Xylose	36	34.8	32.1
Arabinose	61	73.5	67.6
Ribose	11	—	—

[a] Data from H. S. Isbell, J. Res. Natl. Bur. Std., **66A,** 233 (1962).

hydrogen atoms than the hydroxyl group, and an axial hydroxyl group is also favored by the anomeric effect (0.55 kcal./mole), the anomeric form of ketohexopyranoses which has an equatorial hydroxymethyl group will be more stable (by 1.45 kcal./mole) than the other anomer. In sorbose and tagatose, where the normal conformation is the more stable one (Fig. 31), the axial hydroxyl corresponds to the α configuration. Since the furanose forms are energetically not favorable, one would expect the α-pyranose form to predominate in the equilibrium mixture of these two sugars. In fact, crystalline sorbose and tagatose are both α-pyranoses and show very little mutarotation,[127] indicating that the equilibrium solution is composed mostly of that isomer which occurs in the crystalline state. Tipson and Isbell[126] found that the infrared spectra of the equilibrium mixtures were practically identical with those of the pure sugars.

On the other hand, the alternative conformation is more stable than the normal one (Fig. 31) for the pyranose forms of fructose, whether α or β.

[127] J. V. Karabinos, Advan. Carbohydrate Chem., **7,** 112 (1952).

In the alternative chair form it is the β-anomer which has the hydroxymethyl group in the equatorial position and is therefore the more stable isomer. In fact, crystalline D-fructose is the β-pyranose form; but, since the furanose forms are favorable (*trans-trans* at C-2, C-3, and C-4), there is considerable anomalous mutarotation.[128]

The fourth ketohexose, psicose, has not been obtained in crystalline form and, consequently, nothing is known about its equilibrium in solution.

Similar arguments apply to the ketoheptoses (heptuloses). Conformational analysis suggests that the known crystalline compounds of this series exist in the normal chair form, with their anomeric hydroxyl group axial, and are therefore all α-isomers. If this chair form is not unfavorable, and the furanose forms are not unduly favorable, there is little or no mutarotation: the α-pyranose form is overwhelmingly present in the equilibrium solution. In fact, *gluco*-,[129] *manno*-,[130] and *allo*-heptulose[131] show no mutarotation; there is some mutarotation with *galacto*-heptulose[132] (which is a homomorph of fructose) since the furanose form of the *galacto* configuration is very favorable; and *talo*-heptulose, with a very unfavorable pyranose form, shows strong mutarotation.[131] The other heptuloses are not known in the crystalline state.

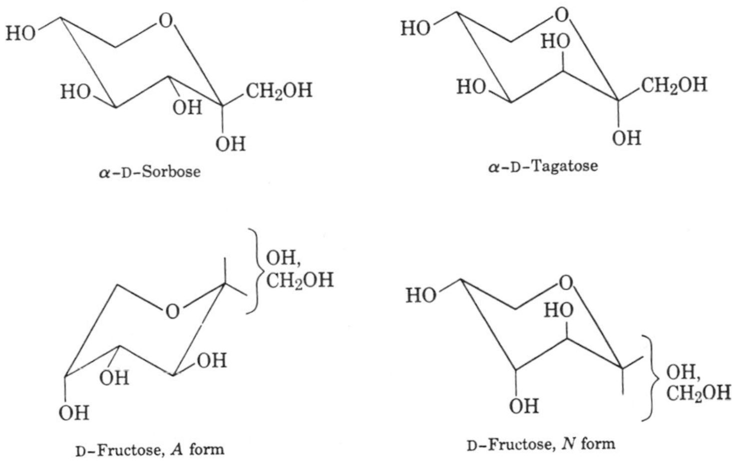

α-D-Sorbose

α-D-Tagatose

D-Fructose, A form

D-Fructose, N form

Fig. 31. Conformations of the ketohexoses.

[128] C. P. Barry and J. Honeyman, *Advan. Carbohydrate Chem.*, **7**, 55 (1952).

[129] W. C. Austin, *J. Am. Chem. Soc.*, **52**, 2106 (1930).

[130] F. B. LaForge, *J. Biol. Chem.*, **28**, 511, 517 (1917).

[131] J. W. Pratt and N. K. Richtmyer, *J. Am. Chem. Soc.*, **77**, 6326 (1955).

[132] M. L. Wolfrom, R. L. Brown, and E. F. Evans, *J. Am. Chem. Soc.*, **65**, 1021 (1943).

Table 10. Compositions[a] of Methyl Glycoside Mixtures at Equilibrium[133b] (By permission of the Editor of Canadian Journal of Chemistry)

Sugar	Methyl Glycoside, %			
	α-Furanoside	β-Furanoside	α-Pyranoside	β-Pyranoside
D-Xylose	1.9	3.2	65.1	29.8
	α + β		α + β	
3-O-Methyl-D-xylose	9.0		91.0	
2-O-Methyl-D-xylose	12.8		87.2	
2,3-Di-O-methyl-D-xylose	16.4		83.6	
D-Arabinose	21.5	6.8	24.5	47.2
3-O-Methyl-D-arabinose	50.7		49.3	
2-O-Methyl-D-arabinose	66.7		33.3	
2,3-Di-O-methyl-D-arabinose	75.4		24.6	
D-Lyxose	1.4	not detected	88.3	10.3
D-Ribose	5.2	17.4	11.6	65.8

[a] Sugar (2%) in 1% methanolic hydrogen chloride at 35°C.

Glycosides. In the presence of an alcohol and catalytic amounts of an acid, sugars give an equilibrium mixture of alkyl α- and β-furanosides and -pyranosides. Since the composition of this mixture remains unchanged once the acid has been removed, this system is particularly suitable for the study of the position of the equilibrium. It is regrettable therefore that reliable quantitative data have not been available until very recently, owing mainly to the difficulty of analyzing the four-component mixture.

As already mentioned in Sec. 6-4c, the methyl α-pyranoside predominates in the equilibrium mixture of the common sugars, and mannose gives methyl α-mannoside nearly exclusively. It is stated[51] that the equilibrium mixture of the methyl glucosides at 24° contains 77%, and that of the methyl galactosides 62%, of the α-isomer. (The latter figure, obtained from optical rotations, may not be accurate, owing to the presence of furanosides.) The ratio of α- to β-pyranoses in equilibrium is shifted therefore, compared to that of the free sugars in aqueous solution, in favor of the α-form. This must be due mainly to an increase in the anomeric effect when water is replaced by methanol as a solvent; a minor cause may also be a decrease of the interactions of the axial oxygen atom which may be solvated to a lesser extent in methanol than in water. The difference in interaction energies between an axial hydroxyl and an axial methoxyl group appears not to be significant (see Table 1, p. 44).

The first complete analyses of glycoside equilibrium mixtures were recently published in two important papers, by Bishop and Cooper,[133] on pentosides and some of their methylated derivatives. Gas-liquid partition chromatography was used to obtain the figures which are reproduced in Table 10. It is hoped that similar information relevant to other sugars will be obtained by the same method.*

Many important points emerge on inspection of Table 10 and on comparing its data with those in Table 9. First, the proportion of the α-pyranoside of xylose and lyxose, and of the β-pyranoside of arabinose, is considerably higher than that of the corresponding free sugars in aqueous equilibrium solution. The anomer with an axial methoxyl group predominates. This fact indicates that the anomeric effect on a methoxyl group in methanol is higher, on the average by 0.8 kcal./mole, than that

---

[133] C. T. Bishop and F. P. Cooper, (a) *Can. J. Chem.*, **40**, 224 (1962); (b) **41**, 2743 (1963).

[134] D. F. Mowery, Jr., *J. Org. Chem.*, **26**, 3484 (1961).

* After some initial failures, column chromatography has also been successfully used for the analysis of glycoside equilibrium mixtures.[134] The figures obtained for the methyl L-arabinosides agree well with those shown in the table; for the methyl D-mannosides, the results are 89% α- and 7% β-pyranoside, and about 2% of each furanoside.

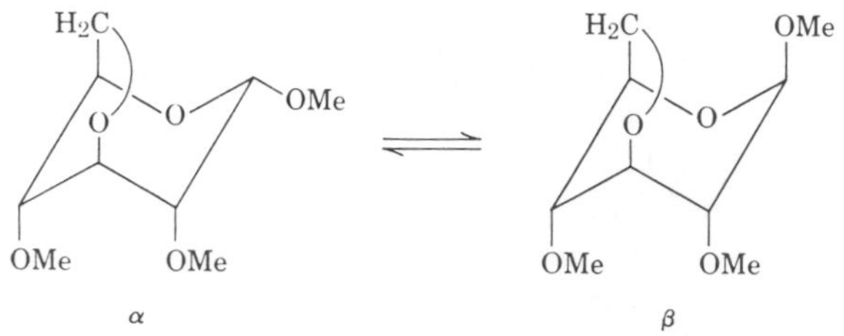

Fig. 32.

on a hydroxyl group in water. The anomeric effect for the methyl glycosides in methanol is therefore about 1.4 kcal./mole.*

Second, the proportion of furanosides, though small, is by no means negligible and is quite substantial in the case of the α-arabofuranoside which has the most favorable all-*trans* conformation. It is also interesting to note that methylation of xylose and of arabinose increases the proportion of furanosides: since the *trans* substituents at C-2 and C-3 are closer to each other in the six- than in the five-membered ring, substitution at these positions should decrease the stability of the pyranosides more than that of the furanosides. The methylated arabinosides represent the unusual case in which the furanosides are more stable than the pyranosides. Finally, one notes that furanosides in which O-1 is *trans* to O-2—β for xylose and ribose, α for arabinose and lyxose—are considerably more stable than their anomers.

An interesting problem is presented by methyl 3,6-anhydro-2,4-di-*O*-methyl-α-D-glucoside (Fig. 32) which anomerizes[73] very rapidly, in the presence of acids, to a mixture in which the β-anomer predominates (78% β, 22% α).† The corresponding α-galactoside undergoes the same change,[72] but the position of the equilibrium is not known; under the conditions of the experiment the anomerization is nearly complete because the syrupy α-anomer changes into the crystalline β-derivative. Other examples for which the position of the equilibrium has been determined are: methyl 3,6-anhydro-3-deoxy-4-*O*-methyl-D-glucoside, 45% β; 3,6-anhydro-2-deoxy-D-galactoside and its 4-methyl ether, about 30% β.[74]

These results are, at first sight, surprising because the β-glycosides

---

* J. T. Edward and I. Puskas [*Can. J. Chem.*, **40**, 711 (1962)] prepared 3-alkyloxy-4-oxa-5α-cholestanes as rigid models of hydroxyaldehydes in the pyranoside form. For a variety of alkyl groups (methyl, isopropyl, benzyl, cyclohexyl) the α-anomer represented 69–73% of the equilibrium mixture. From this result the anomeric effect can be calculated to be 1.6 kcal./mole.

† If the hydroxyl group on C-4 is not protected by methylation, the change is from pyranoside to furanoside (Sec. 6-6a).

(e.g., the glucoside, Fig. 32) have three axial substituents on the same side of the pyranose ring and might have been expected to be very unstable. However, a calculation of the free energy differences between the α- and β-anomers (Sec. 6-4e) gives results, for the three cases where the equilibrium is known, within 0.5 kcal./mole of the experimental data. In these calculations the anomeric effect was taken as 1.5 kcal./mole, and it was assumed that the interaction of an axial group with the —CH_2O— bridge is the same as would be its total interaction with an axial hydroxyl and an axial methyl group. This assumption is necessary because the values of interactions with the bridge are not known; actually, the interactions are probably smaller, and the agreement then is even better. The relative stability of the β-anomers is due to the large anomeric effect and to the $\Delta 2$ effect. An attempt[135] was made to explain the results by the assumption that the β-glycosides are predominantly in the boat form; the calculations (which were not published in full), however, do not appear to take the higher energy of the boat conformation into consideration.

The anomerization of sugar acetates has been discussed by Lemieux.[66]

c. The Lobry de Bruyn-Alberda van Ekenstein Transformation.

In the presence of bases, an aldose is partially converted to the epimeric aldose and to the corresponding ketose.[136] Since the reaction is reversible, it should be possible to predict the position of the equilibrium between epimeric aldoses by the use of the free energy values in Table 7. Unfortunately it is extremely difficult to determine experimentally the position of the equilibrium. Larger amounts of bases cause a progressive degradation of the sugar, and equilibrium of the two aldoses and the ketose is never reached. When the amount of base is reduced, to avoid the side reactions, the composition of the mixture becomes constant after a while; but by then the base has been neutralized by acids formed in the degradation, and what was described as equilibrium[137,138] is only an arrested reaction. However, from the careful measurements of Sowden and Schaffer[139] on a reaction starting from fructose, one can deduce that the ratio of glucose to mannose in equilibrium is approximately 6, in fair agreement with prediction.

When the hydroxyl group on C-2 is replaced by an acetamido group, formation of the ketose and subsequent side reactions are prevented, and equilibrium can be reached in the epimerization. Two such instances

135 A. B. Foster, W. G. Overend, and G. Vaughan, *J. Chem. Soc.*, **1954**, 3625.
136 J. C. Speck, Jr., *Advan. Carbohydrate Chem.*, **13**, 63 (1958).
137 M. L. Wolfrom and W. L. Lewis, *J. Am. Chem. Soc.*, **50**, 837 (1928).
138 C. E. Gross and W. L. Lewis, *J. Am. Chem. Soc.*, **53**, 2772 (1931).
139 J. C. Sowden and R. Schaffer, *J. Am. Chem. Soc.*, **74**, 499 (1952).

have been reported: the constant of the equilibrium between 2-acetamido-2-deoxy-D-glucose and 2-acetamido-2-deoxy-D-mannose[140] is 4–5, and that between corresponding arabinose and ribose derivatives[141] is approximately 2. The values calculated from the interaction energies for the parent sugars are 5.0 and 1.6, respectively.

Methylation of the hydroxyl group on C-2 also prevents ketose formation and side reactions; equilibria have been reached in the epimerization reactions of methylated aldoses but the results are unexpected. *Equal* amounts of 2,3,4,6-tetra-*O*-methyl-D-glucose and 2,3,4,6-tetra-*O*-methyl-D-mannose were found in equilibrium;[137] similarly, 2,4,6-tri-*O*-methyl-D-glucose and the analogous mannose derivative are formed in *equal* amounts.[142] Approximately 26% of 2,3,4-tri-*O*-methyl-D-xylose and 70% of 2,3,4-tri-*O*-methyl-D-lyxose were found after equilibration,[138] a reversal of the usual order of stability. These results are not easily explained; one can only speculate that accumulation of methoxyl groups in the equatorial belt favors axial groups, and favors particularly the arrangement of two axial groups on adjacent carbon atoms, there being no interaction between these groups. Such arrangement occurs in the α forms of mannose and lyxose. A better understanding of this problem would be gained by a study (apparently not yet accomplished) of the equilibrium between 2-*O*-methylglucose and 2-*O*-methylmannose.

Ketoses are the major products in the Lobry de Bruyn-Alberda van Ekenstein transformation, and the reaction has been employed almost exclusively for the synthesis of ketoses. One recent review[136] stated that "for some unknown reason, the best yields (50%) have consistently been obtained in the formation of D-*gluco*-heptulose from D-*glycero*-D-*gulo*-heptose." The reason is obvious: the product has the best possible (*gluco*) conformation whereas the starting material, with the *gulo* conformation, has some unfavorable interactions; the epimeric *ido*-heptose, which would also be formed, is even less favorable. One would expect a similar yield in the conversion of gulose to sorbose; the yield of this reaction under the same conditions (lime water) has, however, not been reported. With pyridine at a high temperature[143] the yield was 30%. Idose, which has an unfavorable conformation, is converted with particular ease to the corresponding ketose, sorbose.[70]

Here again, on account of the side reactions, it is very difficult to

[140] R. Kuhn and G. Baschang, *Ann.*, **636**, 164 (1960); P. M. Carroll and J. W. Cornforth, *Biochim. Biophys. Acta*, **39**, 161 (1960); C. T. Spivak and S. Roseman, *J. Am. Chem. Soc.*, **81**, 2403 (1959).

[141] B. Coxon and L. Hough, *J. Chem. Soc.*, **1961**, 1577.

[142] N. Prentice, L. S. Cuendet, and F. Smith, *J. Am. Chem. Soc.*, **78**, 4439 (1956).

[143] K. Gätzi and T. Reichstein, *Helv. Chim. Acta*, **21**, 456 (1938).

evaluate equilibrium constants. The interconversion of aldoses and ketoses can, however, be specifically catalyzed in some cases by enzymes, without the occurrence of other reactions. Equilibrium constants for several such interconversions have been determined by Palleroni and Doudoroff.[144] Only in the mannose-fructose interconversion did they find the equilibrium in favor of the ketose (29:71); in all the other cases studied, the aldoses predominated. These aldoses, however, were all pentoses or methylpentoses; the corresponding ketoses cannot form pyranose rings, and therefore these are equilibria between aldopyranoses and ketofuranoses. The predominance of the former is to be expected.

d. 1,6-Anhydrohexoses and Related Compounds. The 1,6-anhydrohexoses offer a particularly clear-cut example of an equilibrium controlled by conformational factors. Some sugars, like altrose and idose, are converted, in dilute acidic aqueous solution, to a considerable extent to their 1,6-anhydrides. Most of the common sugars, like glucose or mannose, however, do not appear to form an anhydride under these conditions; although such anhydrides can be prepared by more drastic treatment, they are rapidly hydrolyzed by acids in aqueous solution. It has only recently been realized that even in these cases there is an equilibrium between the sugar and its anhydride, but the equilibrium is much in favor of the free sugar.[145]

The 1,6-anhydrohexopyranoses (Fig. 33) are internal glycosides. If one of the hydroxyl groups in a sugar molecule is in a sterically favorable position, such glycoside formation should occur readily. In an aldohexopyranose, only the extra-annular hydroxyl group on C-6 is favorably situated and it will take part in anhydride formation; the conformations of the remaining hydroxyl groups will control the position of the equilibrium. The two extreme examples are illustrated in Fig. 33: the anhydride of idose, with all hydroxyl groups equatorial, has the most favorable, the

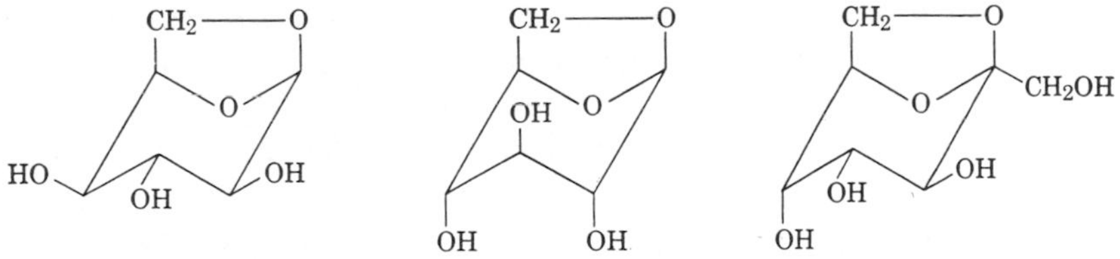

| 1,6–Anhydro–β–D–idopyranose | 1,6–Anhydro–β–D–glucopyranose | 2,7–Anhydro–β–D–altro–heptulopyranose |

Fig. 33.

[144] M. J. Palleroni and M. Doudoroff, *J. Biol. Chem.*, **218**, 535 (1956).

[145] For a summary, see L. C. Stewart, E. Zissis, and N. K. Richtmyer, *Chem. Ber.*, **89**, 535 (1956).

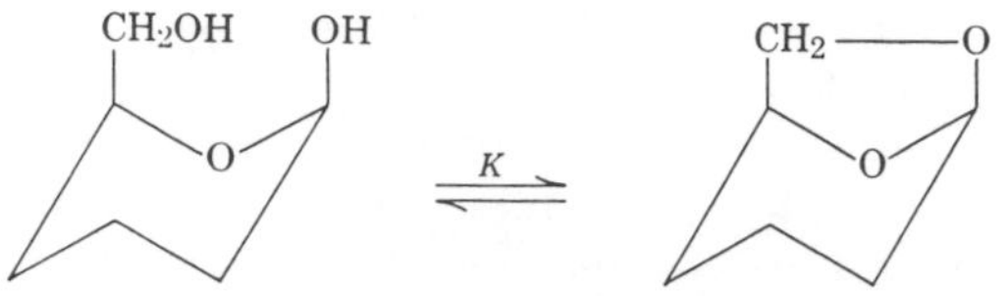

Fig. 34.

one of glucose, with all hydroxyl groups axial, the least favorable conformation. Accordingly, idose is converted to the greatest, glucose to the least, extent to the 1,6-anhydride.

Similarly, heptuloses are converted to 2,7-anhydroheptulopyranoses; in fact, the reaction was discovered[146] during the study of sedoheptulose (D-*altro*-heptulose) which is converted by warm dilute acids to the anhydride (Fig. 33) to the extent of 80%. The 2,7-anhydroheptuloses are analogous to the 1,6-anhydrohexoses and differ only in the presence of an additional hydroxymethyl group on the anomeric carbon atom.

In attempting to apply quantitative considerations to this equilibrium, several difficulties are encountered. First, the values of the equilibrium constants are, in most cases, not accurately known; often only the preparative yield of the anhydride has been recorded. Most of the data apply to an equilibrium temperature of 100° at which the values are considerably different from those measured at room temperature.[147] When it comes to calculations, several new interactions would have to be introduced, namely, those between the carbon-oxygen bridge and the other atoms. There is also a considerable distortion of the chair conformation when the bridge is formed.

Instead of calculating interaction energies, another approach can be used, at least qualitatively. The 1,6-anhydride is formed directly from the alternative β conformer of the sugars (Fig. 34); it is assumed that this formation involves the same free energy change for each sugar, (i.e., K is constant), and therefore the amount of anhydride in equilibrium is proportional to the amount of the alternative β form in equilibrium with the other forms of the sugar.

The amount of the latter will depend on the difference in free energy between the alternative β form (Table 3) and the equilibrium mixture (Table 7) of the sugar. These free energy differences are listed in Table 11, together with the available data on the proportion of 1,6-anhydrohexopyranoses and 2,7-anhydroheptulopyranoses in equilibrium. It can be seen that the proportion increases as the free energy difference decreases.

[146] F. B. La Forge and C. S. Hudson, *J. Biol. Chem.*, **30**, 61 (1917).

[147] N. K. Richtmyer and J. W. Pratt, *J. Am. Chem. Soc.*, **78**, 4717 (1956); E. Zissis, L. C. Stewart, and N. K. Richtmyer, *ibid.*, **79**, 2593 (1957).

Table II. Proportion of 1,6-Anhydrohexopyranoses and
2,7-Anhydroheptulopyranoses in Equilibrium

$G_{A\beta} - G_{equil.}$, kcal./mole	Anhydrohexoses	Anhydroheptuloses
	In Equilibrium, %	
gluco	6.2	0.07	0.7
manno	4.05	0.25	2.8
galacto	4.1	0.2	
allo	3.65	9	50
talo	2.25	7.2	27.5
gulo	1.65	41	87
altro	1.45	58.5	91
ido	0.8	80.5	90

The sugars of the *gluco* and *allo* configurations have a higher proportion of anhydride in equilibrium than one would expect from consideration of the free energy differences. In these anhydrides there are two *syn*-axial hydroxyl groups on C-2 and C-4; an inspection of models shows that formation of the anhydro-bridge, which requires "pulling together" O-1 and C-6, results in a simultaneous increase (from 2.5 to about 2.85 Å) in the distance between the axial hydroxyl groups on C-2 and C-4 (cf. the "reflex effect," Sec. 2-7). The interaction between these groups is thereby reduced, and the stability of the anhydro compounds is correspondingly increased.

A more detailed quantitative treatment of the equilibrium is possible but is not warranted until more accurate experimental data become available.

6–7. Reactions Whose Rates Depend on Conformations

a. Formation and Hydrolysis of Glycosides. The formation and hydrolysis of glycosides is a reversible reaction. As carried out in practice, however, it is forced, as far as possible, in one direction by the use of a large excess of an alcohol or water. The position of the equilibrium is of little importance, and not much is known about it; the rate of the reaction is of interest, however, and is dependent on conformations. Hence the reaction is best discussed in this section.

The formation of glycosides from free sugars is, kinetically, a very complicated reaction because the two furanosides and the two pyranosides are all produced, probably simultaneously. In a pioneering paper, Levene, Raymond, and Dillon[148] published data on the formation of

[148] P. A. Levene, A. L. Raymond, and R. T. Dillon, *J. Biol. Chem.*, **95**, 699 (1932).

methyl furanosides and pyranosides from nine sugars. These data are, as the authors emphasized, only approximate because the analytical methods then available did not clearly distinguish between furanosides and pyranosides. In all these reactions furanosides are formed faster than pyranosides; in some, like those of xylose and ribose, so much faster that practically all the pyranosides are formed from the furanosides and not directly from the sugars. This faster formation is in accord with the generally faster closure of five-membered, as compared to six-membered, rings.[149] It is also notable that formation of the furanoside is slowest in the case of mannose among the hexoses, and of lyxose among the pentoses, i.e., the sugars which have the least favorable (*cis-cis*) arrangement of substituents on C-2, C-3, and C-4 of the furanose ring. Apart from this observation, the results disclose no obvious conformational correlations. Accurate data are beginning to appear now,[133,134] but there are still insufficient examples to allow correlations to be drawn.

On the other hand, the hydrolysis of a single glycoside can readily be measured, and a considerable number of such measurements have been carried out. Much attention has been focused on this subject as a means for correlating the conformation of sugars with chemical reactivity. An adequate review[150] and a discussion[151] have recently appeared, and the reader is referred to them for full details.

Furanosides are hydrolyzed much faster than pyranosides (and so are the septanosides, the rare seven-membered glycosides). Among the methyl pyranosides, those with an equatorial glycosidic methoxyl group are hydrolyzed faster, i.e., the β-anomers of glucose, mannose, galactose, xylose, lyxose and the α-anomer of arabinose (and of gulose, which is an exception). The relative rates of hydrolysis of the methyl hexopyranosides parallel the free energy content of the parent sugars (Table 7); for the α-anomers they increase in the order glucose, mannose, galactose, altrose, idose, gulose, the relative rates at 60° being 1, 2.8, 5.0, 17.7, ~35, and ~58. Similarly, the relative rates of hydrolysis of the β-anomers of methyl xyloside, riboside, arabinoside, and lyxoside are 1, 1.5, 1.6, and ~5.6, respectively.[150,151] The pentosides are hydrolyzed faster than the hexosides of the same configuration; this point will be discussed later.

Complete conformational interpretation of these data is not easy. Acid-catalyzed hydrolysis of the glycosides can occur[152] by two different

[149] E. L. Eliel, *Stereochemistry of Carbon Compounds*, McGraw-Hill Book Co., New York, 1962, p. 198; J. A. Mills, *Advan. Carbohydrate Chem.*, **10**, 23 (1955).

[150] B. Capon and W. G. Overend, *Advan. Carbohydrate Chem.*, **15**, 33 (1960).

[151] W. G. Overend, C. W. Rees, and J. S. Sequeira, *J. Chem. Soc.*, **1962**, 3429.

[152] C. A. Bunton, T. A. Lewis, D. R. Llewellyn, and C. A. Vernon, *J. Chem. Soc.*, **1955**, 4419.

routes (Fig. 35): either through a cyclic (A) or through an acyclic (B) carbonium ion, both stabilized by resonance. The energy required to form these carbonium ions will determine the rate of the reaction. Route B leads initially to a hemiacetal which is then rapidly hydrolyzed further. It is not known to what extent each route is followed in the hydrolysis of any particular glycoside, nor is it easy to devise an experiment capable of distinguishing between them.

Shafizadeh[153] explained the dependence of hydrolysis rates on conformations in terms of route B. Since the steric strains in the ring are relieved by the formation of the acyclic ion, greater strain in the ground state should cause faster ring opening; less activation energy is required when the initial free energy is higher. Therefore gulosides are hydrolyzed faster than glucosides, lyxosides faster than xylosides. Similarly, furanosides, being less stable than pyranosides, are hydrolyzed faster. In the hydrolysis of fructose, where the stability of the furanose and pyranose forms does not differ to a great extent, the difference in the rate of hydrolysis between the furanosides and the pyranosides is relatively small.[154] For the 3,6-anhydro sugars, where the furanosides are more stable than the pyranosides (Sec. 6-6a), there is a reversal of the usual order of hydrolysis rates; in particular, methyl 3,6-anhydro-2,4-di-O-methyl-β-D-glucopyranoside is hydrolyzed faster than are the furanosides usually.[74] The reader will recall that this compound is considerably strained. It is very likely that hydrolysis of these anhydropyranosides occurs by route B, owing to the strain in the ring which would not be relieved by the formation of the cyclic ion (route A).

It is not quite clear how the faster hydrolysis of the equatorial, compared to the axial, methyl glycosides is to be explained in terms of route B. It is possible that here, again, the less stable anomer is always hydrolyzed faster, but it is not known with certainty which anomer of, for example,

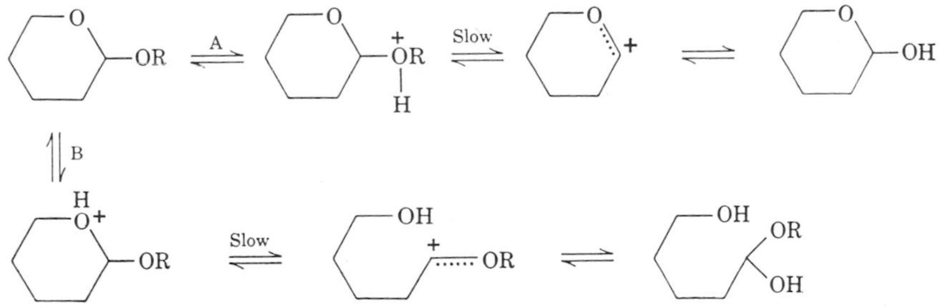

Fig. 35.

[153] F. Shafizadeh, *Advan. Carbohydrate Chem.*, **13**, 9 (1958).
[154] C. B. Purves and C. S. Hudson, *J. Am. Chem. Soc.*, **59**, 1170 (1937).

Fig. 36.

methyl glucoside is less stable in aqueous solution (β-glucose is more stable than α in water, but methyl α-glucoside is more stable than β in methanol).* It is a fact, however, that where the relative stability is known the less stable anomer reacts faster (β for mannose and lyxose, α for gulose†); and in the 3,6-anhydrohexose series the less stable anomer is hydrolyzed at a higher rate whether it has the α configuration (as for the glucoside) or the β configuration (as for the mannoside).[74] The rate of hydrolysis for a number of methyl aldofuranosides has been measured by Augestad and Berner;[155] in all cases the isomer with the substituents *cis* at C-1 and C-2, i.e., the less stable anomer, is hydrolyzed faster.

Although the variation in the hydrolysis rates of the glycosides can be explained by assuming that the reaction proceeds through an acyclic carbonium ion, there is considerable evidence for the importance of the cyclic ion in the reactions of glycosides.[156] Edward,[55] Huber,[157] and Overend[151] explain the variation in the rate of hydrolysis on the assumption that it occurs mainly by route A.

The faster hydrolysis of furanosides compared to that of pyranosides is readily explained because the introduction of trigonal atoms into a five-membered ring is energetically favored, whereas six-membered rings resist it (Sec. 4-1). The intermediate cyclic ion formed from pyranosides will, owing to its resonance, have a half-chair form like cyclohexene. Formation of this ion from the chair form involves rotation at C-2 and C-5 (Fig. 36); as a result of this change in the shape of the ring, one axial

[155] I. Augestad and E. Berner, *Acta Chem. Scand.*, **10,** 911 (1956).

[156] B. E. C. Banks, Y. Meinwald, A. J. Rhind-Tutt, I. Sheft, and C. A. Vernon, *J. Chem. Soc.*, **1961,** 3240.

[157] G. Huber, *Helv. Chim. Acta*, **38,** 1224 (1955).

* In any case the hydrolysis rates of the methyl α- and β-glycosides differ only by a factor of about 2, and the order of rates is reversed for the ethyl and phenyl glycosides. The rate differences result from differences in the entropy rather than in the energy of activation; this suggests that, for other than the smallest aglycons, the axial anomers are more highly oriented, owing to the restriction imposed on rotation of the aglycone, than are the equatorial anomers, and their increased reactivity is associated with greater loss of molecular order on passing to the transition state.[151]

† In the case of gulose, where the free energy difference between the anomers is particularly large, this effect seems to override all others, and the *axial* α-guloside is hydrolyzed faster than the β-anomer.

interaction of each axial group disappears and another one is weakened. The more axial hydroxyl groups a sugar has, the more easily will it achieve the conformation required by the cyclic ion. The order in the rates of hydrolysis, both for the hexopyranosides and for the pentopyranosides, is thereby explained.

The faster hydrolysis rates of the equatorial anomer of each glycoside may again be explained by assuming that it has higher free energy in its ground state[55]—since both anomers give the same cyclic ion—or by the following considerations. Equatorial oxygen atoms will be protonated to a greater extent than axial ones because, generally, an equatorial group is more basic than the same group in an axial position (Sec. 5-3a)* and, in the case of glycosides, because protonation destroys the polar interaction which causes the anomeric effect. The concentration of the oxonium ion, which gives rise to the cyclic carbonium ion by loss of methanol, will therefore be higher for the equatorial than for the axial anomer.

Glycosides of the deoxy sugars are hydrolyzed faster than those of the parent sugars;[153] among the causes for this difference, however, the steric factors, which were discussed by Edward,[55] appear to be small compared to the inductive effect. The glycosides of sugars are considerably stabilized by the inductive effect of hydroxyl groups. Removal of a hydroxyl group from C-2, the position adjacent to the anomeric carbon, in methyl α-D-glucoside causes a great increase in the rate of hydrolysis (about 2000-fold); the same effect is shown to a lesser extent by the 3-, 4-, and 6-deoxyglucosides. The glycosides of 2,3-dideoxy sugars are hydrolyzed very rapidly, and 2-methoxytetrahydropyran (which can be regarded as a 2,3,4-trideoxypentopyranoside) is very labile to dilute acids even at room temperature.

The pyanosides of pentoses are hydrolyzed more rapidly than those of hexoses of the same configuration. Here, again, removal of the inductive effect of the additional hydroxyl group,[157a] and the steric effect of the hydroxymethyl group,[157b] is responsible.

b. Mutarotation and Anomerization. Different sugars mutarotate at different rates. Although mutarotation rates have been measured, mainly by Isbell and Pigman,[125] for many sugars, their dependence on conformations has not yet been studied.

The "slow" mutarotation (Sec. 6-6a) is due to the anomerization of the pyranoses, whereas the "fast" mutarotation is generally believed to

[157a] L. K. Semke, N. S. Thompson, and D. G. Williams, *J. Org. Chem.* **29**, 1041 (1964).

[157b] T. E. Timell, *Chem. Ind. (London)*, **1964**, 503.

* As an example from carbohydrate chemistry, 2-amino-2-deoxy-D-glucose is a stronger base than 2-amino-2-deoxy-D-mannose (ref. 161; the explanation there offered involves, however, an impossible hydrogen bond).

indicate the velocity of the pyranose-furanose interconversion. The relative mutarotation coefficients, at 20°, of the "slow" mutarotation are:[125] glucose 1, galactose 1.3, mannose 2.75, gulose 3.0, talose 4.1, altrose[158] 12.5; xylose 3.2, arabinose 4.75, ribose 7.8, and lyxose 9.2. In both the hexose and the pentose series the variation of the mutarotation rates parallels that of the free energies,[159] except for altrose and lyxose which mutarotate faster than would be expected if such a correlation were generally valid. The mutarotation of 2-deoxy sugars is much faster.

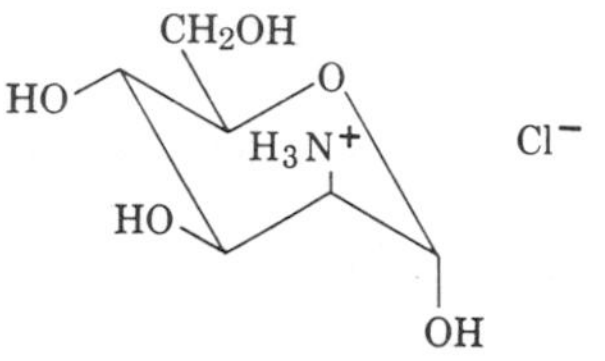

The anomerization of sugars is, of course, closely related to the formation and hydrolysis of glycosides, and it could also proceed either by a "cyclic" or by an "acyclic" mechanism; in this case, however, it is generally accepted that an open chain intermediate is involved.[160]

CH₂OH ... H₃N⁺ ... Cl⁻ ... HO ... HO ... OH

Fig. 37.

It is interesting, and apparently yet unexplained, that 2-amino-2-deoxy-D-mannose hydrochloride (Fig. 37) (and the corresponding altrose derivative) does not mutarotate.[161] The free base shows mutarotation, and so do the hydrochlorides of 2-amino-2-deoxy-D-glucose and -galactose. Maybe an axial (but not an equatorial) ammonium ion reduces the rate of protonation of the ring oxygen atom to such an extent that mutarotation becomes too slow to be observed.*

The anomerization of glycosides is a closely related reaction for which, however, very few quantitative data are yet available. It is known[133,162] that the xylo- and glucofuranosides are anomerized much faster than the corresponding pyranosides; and this is to be expected on the basis of either the "cyclic" or the "acyclic" mechanism. Moreover, anomerization becomes faster as the number of hydroxyl groups in the molecule is reduced[69] in the same way, and for the same reasons, as does the hydrolysis of glycosides.

The anomerization of sugar acetates has already been mentioned in Sec. 6-4b.

[158] N. K. Richtmyer and C. S. Hudson, *J. Am. Chem. Soc.*, **65**, 740 (1943).

[159] R. Bentley and D. S. Bhate, *J. Biol. Chem.*, **235**, 1225 (1960).

[160] For references, see ref. 153, p. 40.

[161] M. J. Carlo, A. Cosmatos, and H. K. Zimmerman, Jr., *Ann.*, **650**, 187 (1961).

[162] B. Capon, G. W. Loveday, and W. G. Overend, *Chem. Ind. (London)*, **1962**, 1537.

* The behavior of 2-amino-2-deoxy-D-mannose is also anomalous inasmuch as it is not oxidized to the corresponding -onic acid by bromine water or by mercuric oxide. This seems to be a conformational effect, since the analogous glucose and galactose derivatives behave normally, but an explanation is not apparent.

c. **Oxidation of Aldoses by Halogens.** The oxidation of aldoses by bromine water in the presence of barium carbonate has been extensively studied by Isbell and Pigman.[125,163] The reaction yields δ-lactones without opening of the ring. The α- and β-anomers of the sugars differ widely in their rates of reaction: the oxidation always occurs more rapidly if the anomeric carbon atom carries an equatorial rather than an axial hydroxyl group.[164] This is true irrespective of which anomer is the more stable one.

Isbell and Pigman[125] and Bentley[164] attached great significance to the ratio of the rate constants for the two anomers, which varies considerably from sugar to sugar, but it is now known that the measurements give a very inaccurate estimate of this ratio. Since mutarotation cannot be prevented during the oxidation, the less reactive anomer is gradually converted to the more reactive one, and the rate measured for the former is actually the sum of the rate of oxidation and the rate of mutarotation. Isbell assumed that oxidation is rapid in comparison with the rate at which one modification of the sugar changes into another. This assumption, however, proved to be wrong, at least for the less readily oxidizable anomers. Barker, Overend, and Rees[165] have recently shown that, in a buffer at pH 5.0, the rate constant of the oxidation of α-D-glucose was nearly identical with that of its mutarotation and varied with the temperature in the same way. It appears therefore that the α-anomer is oxidized predominantly after rate-determining anomerization to the β form. The same conclusion was made probable for several other aldoses. The actual rate of oxidation of the axial anomers appears to be very small.

On the other hand, the apparent oxidation rate of the equatorial anomers is lower than the actual one, owing to the concurrent anomerization which converts them to the axial anomers. The rate of oxidation for most equatorial anomers, however, is considerably faster than the mutarotation, and the effect of the latter is negligible except where mutarotation is particularly rapid (lyxose, altrose).

Several mechanisms have been proposed for the oxidation; the latest one,[165] which incorporates an ingenious explanation of the stereoselectivity and also of the strong dependence of the rate on the pH of the solution, envisages the steps shown in Fig. 38. It was assumed, though there is as yet no experimental evidence to support this assumption, that the final elimination of hydrogen and bromide ions requires an *anti* arrangement of these atoms, just like the elimination from a carbon-carbon bond. This conformation is readily achieved when the oxygen

[163] H. S. Isbell, *J. Res. Natl. Bur. Std.*, **66A**, 233 (1962).

[164] R. Bentley, *J. Am. Chem. Soc.*, **79**, 1720 (1957); *Nature*, **176**, 870 (1955).

[165] I. R. L. Barker, W. G. Overend, and C. W. Rees, *Chem. Ind.* (*London*), **1960,** 1297, 1298.

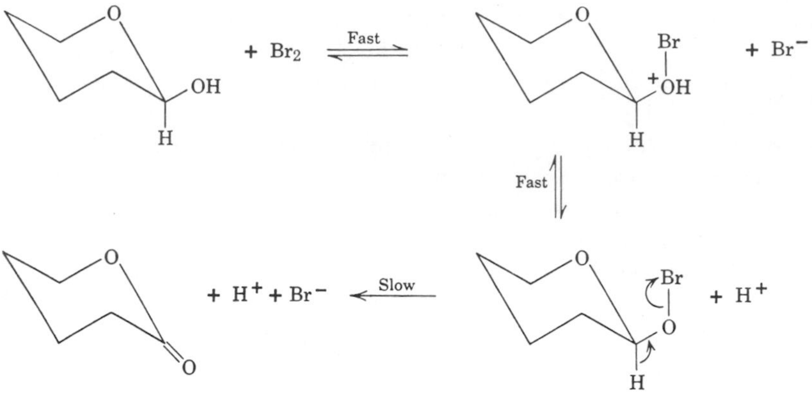

Fig. 38.

atom is equatorial (Fig. 38) but, in the case of an axial oxygen atom, establishment of the *anti* position by the bromine atom (Fig. 39) would involve very severe steric compression by interaction with the other two axial groups. This mechanism for the oxidation of *hemiacetals* differs from that postulated[166] for the oxidation of secondary *alcohols*; the latter is not specific for equatorial groups.

For the readily oxidized equatorial anomers the range of the reaction rates is not very wide. The free energy content of the sugars does *not* affect the rate; the most stable sugars, in fact, are oxidized at the fastest rate. Steric hindrance, however, has a strong effect: As shown in Fig. 40, an axial hydroxyl group on C-2 will hinder the bromine atom in taking up the *anti* position; an axial hydroxyl group at C-3, or an axial carbon atom on C-5, will hinder the approach of a nucleophilic reagent to the hydrogen atom which is to be eliminated.* The relative rate constants, as shown

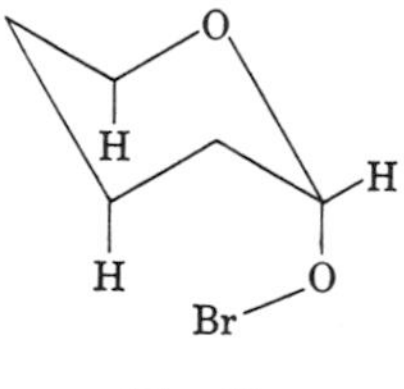

Fig. 39.

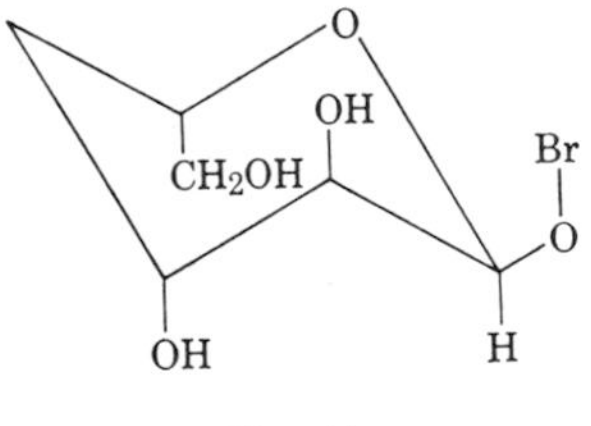

Fig. 40.

[166] I. R. L. Barker, W. G. Overend, and C. W. Rees, *Chem. Ind.* (*London*), **1961,** 558.

* Steric hindrance of an elimination reaction by *syn*-axial substituents has been suggested by E. L. Eliel and R. S. Ro, *Tetrahedron*, **2,** 353 (1958).

in Table 12, can be readily grouped according to the nature of the hindrance. The aldoheptoses are not listed in the table; they are all oxidized somewhat faster than the corresponding aldohexoses, but the rates fall into the same groups, according to the steric hindrance (with the exception of D-*glycero*-β-L-*manno*-heptose for which it is too fast).

Table 12. Relative Rates of Oxidation of Sugars by Bromine Water at 0°

Steric Hindrance	Sugar	Rel. Rate[163]
None	α-L-Arabinose	1.33
	β-D-Xylose	1.33
	β-D-Glucose	1.0
	β-D-Galactose	1.28
Axial OH on C-2	β-D-Lyxose	0.36[a]
	β-L-Rhamnose	0.61
	β-D-Mannose	0.61
	β-D-Talose	0.56
Axial OH on C-3	β-D-Ribose	0.16
	β-D-Gulose	0.32
	β-D-Allose	0.13
Axial OH's on C-2 and C-3	β-D-Altrose	0.14[b]
Axial C-6	α-D-Idose	0.17

[a] The true value is probably higher; the mutarotation of lyxoses being fast, some concurrent anomerization occurs to the less reactive α-lyxose.

[b] The mutarotation being very fast, this value probably refers to the equilibrium of the α and β forms.

Isbell[163] proposed an alternative explanation of the stereoselectivity involving the interaction energies, but his proposal does not appear as satisfactory as the one just described.

Similar considerations apply to the oxidation of methyl glycosides by chlorine.[164,167]

[167] A. Dyfverman, B. Lindberg, and D. Wood, *Acta, Chem. Scand.*, **5**, 253 (1951); A. Lindberg and D. Wood, *ibid.*, **6**, 791 (1952).

6–8. Miscellaneous Reactions

There are many other reactions in carbohydrate chemistry which have been studied or discussed from the conformational point of view. They will not be discussed here in detail because (1) for many of them the available data are insufficient for satisfactory interpretation, or (2) the interpretations are analogous to those already discussed in this chapter, or (3) recent reviews have adequately dealt with them. Only a short discussion will be given in this section of a few selected reactions in order to indicate further the scope of conformational analysis in carbohydrate chemistry, and to direct to suitable references those readers who are interested.

a. Cyclic Acetoxonium Ions. Cyclic acetoxonium ions (Fig. 41) are important intermediates in a number of carbohydrate reactions, such as formation of ortho esters and glycoside formation with retention of configuration from glycosyl halides.[168] Formation of these cyclic ions as a consequence of displacement of groups by neighboring acetoxy groups was actually postulated in carbohydrate chemistry by Isbell,[169] before Winstein's now classical work on neighboring group effects appeared.[170]

The five-membered ring of the cyclic acetoxonium ions will probably be much flatter, owing to considerable resonance, than that of neutral 1,3-dioxolanes. As a consequence the attachment of these rings to five- or six-membered rings is *always cis* (in contrast to that of ketals; Sec. 6-2f). When the cyclic ion is formed by nucleophilic displacement (Fig. 41), i.e., by inversion at the site of displacement, the steric requirement of the reaction is a *trans* disposition of the acetoxy group and the displaced group. It is not necessary, however (though it is advantageous),

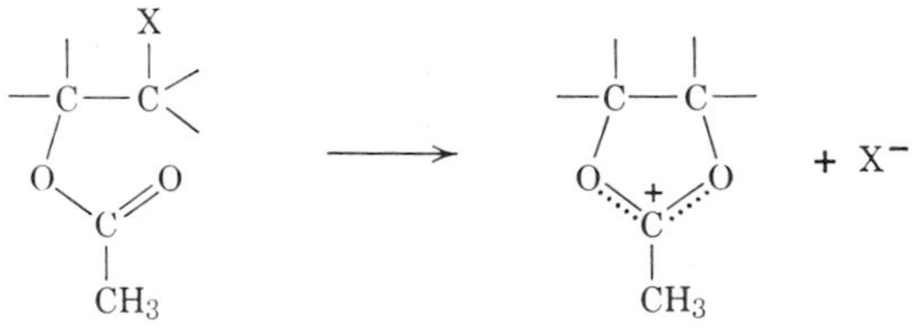

Fig. 41.

[168] R. U. Lemieux, *Advan. Carbohydrate Chem.*, **9**, 1 (1954).

[169] H. S. Isbell, *Ann. Rev. Biochem.*, **9**, 65 (1940); H. L. Frush and H. S. Isbell, *J. Res. Natl. Bur. Std.*, **27**, 413 (1941).

[170] S. Winstein and R. E. Buckles, *J. Am. Chem. Soc.*, **64**, 2780, 2787 (1942).

for these groups to be axial, as often postulated. Penta-O-acetyl-β-D-glucopyranose, for example, will form such a cyclic ion (see below), but it is unlikely that ring inversion to the all-axial alternative chair form would first occur. Korytnyk and Mills[171] have shown that glycosyl halides with a locked conformation (owing to fusion with another ring) and equatorial halogen atoms will undergo dissociation to the cyclic ion. It is possible that the reaction proceeds through a flexible form in which the two reacting groups are in *anti* relationship. It is not certain, however, that such *anti* relationship—necessary for the formation of three-membered rings, like that of epoxides—is actually required for the reaction leading to the five-membered cyclic ion.[150]

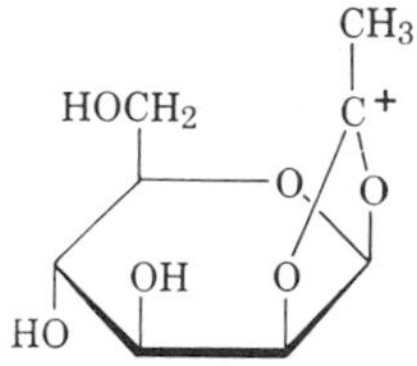

Light is thrown on the ease of formation of cyclic acetoxonium ions by Lemieux and Brice's study[172] of the exchange of anomeric acetoxy groups in the presence of C^{14}-labeled stannic trichloride acetate. This exchange is slow with 1,2-*cis*-acetates;[173] the rate is nearly the same as that of the anomerization,

Fig. 42.

and it appears that a carbonium ion is involved. In the 1,2-*trans*-acetates the rate of exchange is fast (owing to the anchimeric assistance of the neighboring group) and is much faster than the rate of anomerization: formation of the cyclic ion leads to exchange of the group at C-1 but not to inversion. The rate of exchange of the acetoxy groups can therefore be regarded as a measure of the rate of formation of acetoxonium ions.

The relative rates of the exchange of 1,2-*trans* sugar acetates with stannic trichloride acetate in chloroform at 40° are as follows:[172] pentoses: β-xylose 100, α-arabinose 100, β-ribose 11.9, α-lyxose 5.7; hexoses: α-altrose 21, β-glucose 11.9, β-galactose 10.6, β-allose 2.0, α-mannose 1.4. It is to be noted that the sugars with *cis*-2,3-acetoxy groups (ribose, lyxose, allose, mannose) react distinctly slower than those in which these two groups are *trans*; in the cyclic ions formed from these sugars (e.g., that from D-mannose shown in Fig. 42) the substituents on C-1, C-2, and C-3 are all *cis* to each other. The explanation of this difference leads to the recognition of an effect which is generally valid: that *cis* attachment of a five-membered to a six-membered ring is hindered by an adjacent *cis* substituent in the latter. The reason is that the distortion required by the attachment of the five-membered ring causes increased eclipsing with the neighboring substituent in the six-membered ring.

[171] W. Korytnyk and J. A. Mills, *J. Chem. Soc.*, **1959**, 636.

[172] R. U. Lemieux and C. Brice, *Can. J. Chem.*, **34**, 1006 (1956).

[173] R. U. Lemieux, C. Brice, and G. Huber, *Can. J. Chem.*, **33**, 134 (1955).

This effect, which was probably first recognized by Angyal and McHugh,[6] applies to a variety of five-membered rings. One example is the sluggish formation of ketals of myo-inositol, compared to those of other cyclitols. The occurrence of this effect in the formation of borate complexes and in periodate oxidations will be mentioned below.

A detailed discussion of acetoxonium ions is given in a review by Lemieux.[168]

b. Complex Formation with Borate. Historically, the work of Böeseken on complex formation of sugars with boric acid[174] is important, since it established the anomeric configuration of α- and β-D-glucose. In retrospect, however, the work loses much of its importance. Boric acid reacts to a considerable extent only with glycols in which the dihedral angle between the oxygen atoms is much less than 60°. Glycols of six-membered rings in the chair form, like sugars, complex very weakly, and the changes in conductivity observed by Böeseken were largely obscured by the effect of trace impurities in the sugars (at a time when ion exchange resins and chromatography were not yet used). Although Böeseken derived the correct configuration for α-D-glucose,[175] he came to the wrong conclusion concerning β-D-mannose.[176] Later workers[177] obtained different values for the conductivity of mannose in boric acid.

Borate ions at a pH of about 10, however, complex readily with glycols, the extent of complexing being dependent on the dihedral angle. Electrophoresis in a borate buffer is an easy (though not accurate) method for determining the extent of complex formation, since the electrophoretic mobility (measured as the M_G value) is dependent on the position of the equilibria, which is governed largely by the stereochemistry of the glycol group.[178] Favorably disposed hydroxyl groups form stable borate complexes and the M_G values are high. The strongest complexes are formed with cis-1,2-diols in five-membered rings.

The reaction of free sugars with borate ions is complex, pyranoses, furanoses, and open chain forms being involved.[179] A clearer picture is obtained by the study of the methyl glycosides (in which isomerizations are not possible); some of the results[180] are shown in Table 13.

In contrast to cuprammonium or periodate, borate ions react only with

[174] J. Böeseken, Advan. Carbohydrate Chem., **4**, 189 (1949).

[175] J. Böeseken and A. van Rossem, Rec. Trav. Chim., **30**, 392 (1911); J. Böeseken, Ber., **46**, 2612 (1913).

[176] J. Böeseken and H. Couvert, Rec. Trav. Chim., **40**, 354 (1921).

[177] H. T. Macpherson and E. G. V. Percival, J. Chem. Soc., **1937**, 1920.

[178] For a review, see A. B. Foster, Advan. Carbohydrate Chem., **12**, 81 (1957); also J. L. Frahn and J. A. Mills, Australian J. Chem., **12**, 65 (1959).

[179] A. B. Foster, J. Chem. Soc., **1953**, 982.

[180] A. B. Foster, J. Chem. Soc., **1957**, 4214.

Table 13. M_G Values of Some Methyl Glycopyranosides[180] (in a Borate Buffer)

Methyl Glycopyranoside	M_G Value	
	α-Anomer	β-Anomer
Xyloside	0.00	0.00
Arabinoside	0.38	0.38
Lyxoside	0.45	0.27
Mannoside	0.42	0.31
Guloside	0.59	0.72
Galactoside	0.38	0.38
Glucoside	0.11	0.19

cis-diol groups in cyclic systems, owing to the shortness of the boron-oxygen bond. Thus compounds lacking cis-1,2-hydroxyl groups, like the xylosides, have no electrophoretic mobility. In those glycosides in which the cis pair is on C-2 and C-3, the two anomers have different M_G values, owing to the operation of the effect mentioned in Sec. 6-8a: a cis-methoxy group on C-1 hinders formation of the five-membered borate ring. When the cis pair of hydroxyl groups is on C-3 and C-4, as in the arabinosides and the galactosides, the anomeric configuration does not effect the M_G values. In the glucosides there are no cis-1,2-hydroxyl groups; the mobilities are attributed to complex formation across the C-4 and C-6 hydroxyl groups. It is not clear, however, why the α-glucoside should migrate slower than the β-glucoside.

c. Reactions with Periodate. Glycol fission with periodate (and also with lead tetraacetate) involves cyclic intermediates, generally regarded as five-membered, and the rate of the reaction is therefore strongly dependent on the dihedral angle between the two hydroxyl groups. The action of periodate on carbohydrates[181] often requires conformational analysis for full interpretation of its rate and direction, and it can furnish information on the conformations of sugars. Carbohydrate chemistry provided the first examples—I,6-anhydro-β-D-glucofuranose (Fig. 43) and the corresponding galactose derivative[182]—of 1,2-diols which are resistant to periodate oxidation because the dihedral angle between the two oxygen atoms is rigidly held at a large angle (120° in this case). In glycol fission, also, a neighboring cis substituent has a retarding effect on the reaction of a cis-hydroxyl pair.[6]

---

[181] For a review, see J. M. Bobbitt, Advan. Carbohydrate Chem., 11, 1 (1956).

[182] R. J. Dimler, H. A. Davis, and G. E. Hilbert, J. Am. Chem. Soc., 68, 1377 (1946); B. H. Alexander, R. J. Dimler, and C. L. Mehltretter, ibid., 73, 4658 (1951).

Fig. 43.

Fig. 44.

During a study of the periodate oxidation of D-ribose, Barker and Shaw[60] found that consumption of several equivalents of the reagent was rapid in unbuffered solution, but at pH 7 one equivalent of periodate disappeared rapidly and the subsequent reaction was slow. A substantial proportion of the ribose could be recovered after the initial reaction. Barker and Shaw concluded that a complex is formed reversibly from equimolecular proportions of ribose and periodate. When the study was extended to other compounds it was found that complex formation required three contiguous hydroxyl groups in *ax-eq-ax* conformation. Accordingly, β-D-talose, α-D-gulose, β-D-lyxose, 1,6-anhydro-D-allose, and *cis-cis*-cyclohexane-1,2,3-triol all form such complexes. The complex is probably a tridentate ester, represented by the structure in Fig. 44 or by a hydrated form.

neo-Inositol, in which the required arrangment of hydroxyl groups occurs twice, forms a complex with two equivalents of periodate.[183] However, *myo*-inositol, although it has three contiguous *cis*-hydroxyl groups forms no complex; evidently the energy required (about 6 kcal./mole) to invert the stable chair form into the less stable one—which alone has the required *ax-eq-ax* sequence—is too high to be provided by the free energy of complex formation. On the other hand, β-D-allose, which forms a complex with periodate, can also react only in the less stable alternative conformation (Fig. 45); the energy required for ring inversion,

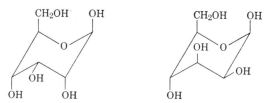

Fig. 45. β-D-Mannose and β-D-allose in their alternative conformations.

[183] G. R. Barker, *J. Chem. Soc.*, **1960**, 624.

3.6 kcal./mole (from Table 3), is apparently not prohibitive. It is somewhat puzzling that β-D-mannose does not form a complex: the energy required for inversion to the alternative form (Fig. 45), which alone has the required geometry, is about the same as that for β-D-allose. The two axial hydroxyl groups, however, are hindered by the axial hydroxymethyl group, and this hindrance may explain why complexing does not occur.

Complex formation with periodate is a method potentially useful for determining the configuration and the conformation of cyclic polyols. Complexing with molybdate ion, which is readily detected by paper ionophoresis, seems to show the same conformational requirement as complexing with periodate.[184]

d. Epoxides. The formation, opening, and migration of epoxides in carbohydrate chemistry offers a rich field for conformational analysis which has been the subject of numerous papers. The general principles have already been discussed in Sec. 6-2e; for details the reader is referred to the recent review by Newth.[185]

e. Formation of Acetals and Ketals. The large number of hydroxyl groups and the variation in ring forms of sugars allow the formation of numerous acetals and ketals, incorporating fused five- and six-membered rings, with substituents in selected axial or equatorial conformations. The sugar alcohols provide many more examples. A fascinating field of conformational analysis is hereby opened up, a field which, however, is too large to be discussed in this book. The reader who wishes to study these compounds, or perhaps investigate some of the still unexplored combinations, should first read the excellent and stimulating review of Mills[87] and then follow up with the review on the acetals and ketals of sugar alcohols by Barker and Bourne[186] and with the more recent review by Ferrier and Overend (cited under General References).

6–9. Natural Occurrence

In the carbohydrate field, at least, Nature appears to prefer the conformationally more stable isomers.[187] Among the inositols, for example, the most stable *myo* derivative is the most widely spread in Nature; the optically active inositols and *scyllo*-inositol, which have only slightly

[184] E. J. Bourne, D. H. Hutson, and H. Weigel, *J. Chem. Soc.*, **1960**, 4252.

[185] F. H. Newth, *Quart. Rev. (London)*, **13**, 30 (1959); see also G. Huber and O. Schier, *Helv. Chim. Acta*, **43**, 129 (1960), and M. Černý, I. Buben, and J. Pacák, *Collection Czech. Chem. Commun.*, **28**, 1569 (1963).

[186] S. A. Barker and E. J. Bourne, *Advan. Carbohydrate Chem.*, **7**, 137 (1952).

[187] S. J. Angyal and J. A. Mills, *Rev. Pure Appl. Chem.*, **2**, 185 (1952).

higher free energy (Sec. 6-2c), are also of common occurrence. On the other hand, inositols with two *syn*-axial hydroxyl groups have not been found in Nature (with the exception of one reported occurrence of *muco*-inositol[188]). The naturally occuring quercitols are also devoid of *syn*-axial hydroxyl groups. It is interesting that in all the ten naturally occurring inositol methyl ethers the methoxy groups are equatorial.[12]

Among the hexoses, the three most stable ones, glucose, mannose, and galactose, are very widely spread. (It is curious, however, that allose, of only slightly higher free energy, seems to be unknown in Nature.) Those with higher free energies occur rarely, if ever. Among the pentoses, ribose has the highest free energy in its pyranose form; although its natural occurrence is important, it always seems to be present in the furanose form. It is particularly striking that the polysaccharides, long chains of rings in chair conformations, never carry two *syn*-axial substituents on any ring.

This preference for the more stable configuration is not to be regarded as a definite rule. Nature produces some of the most unusual and unexpected chemical structures and configurations; these, however, are mostly freaks, to be encountered only rarely. For its everyday processes, Nature prefers the conformationally more stable compounds. It may serve as an encouragement to the reader who is struggling through this book that the principles he has just learned have been known to, and applied by, Nature for innumerable years.

General References

Capon, B., and W. G. Overend, "Constitution and Physicochemical Properties of Carbohydrates," *Advan. Carbohydrate Chem.*, **15**, 11–51 (1960).

Ferrier, R. J., and W. G. Overend, "Newer Aspects of the Stereochemistry of Carbohydrates," *Quart. Rev.* (London), **13**, 265–286 (1959).

Lemieux, R. U., "Rearrangements and Isomerizations in Carbohydrate Chemistry," in P. de Mayo, ed., *Molecular Rearrangements*, Interscience Division, John Wiley and Sons, New York, 1963, pp. 709–769

Michel, G., "Stéréochimie des Oses et de leurs Dérivés," *Bull. Soc. Chim. France*, **1960**, 2173–2187.

Reeves, R. E., "Cuprammonium-Glycoside Complexes," *Advan. Carbohydrate Chem.*, **6**, 107–134 (1951).

Shafizadeh, F., "Formation and Cleavage of the Oxygen Ring in Sugars," *Advan. Carbohydrate Chem.*, **13**, 9–61 (1958).

[188] S. K. Adhikari, R. A. Bell, and W. E. Harvey, *J. Chem. Soc.*, **1962**, 2829.

Chapter 7

Tabulation and a priori *Calculations* of *Conformational Energies*

This chapter deals with some matters which are of importance but which, for one reason or another, have not been covered in other parts of the book. Section 1 contains a documentation of all available conformational free energy differences in substituted cyclohexanes; it is thus an elaboration of Table 1 in Chap. 2. Section 2 deals with *a priori* calculations of conformational energies, a topic not taken up elsewhere in the book. Sections 3 and 4 deal with α-haloketones and boat (or twist) forms, respectively. These topics might have been taken up in Chap. 2 or 3 and, in fact, have been referred to in those chapters and in Chap. 5. However, because physical and chemical evidence as well as *a priori* calculations all bear extensively on the haloketone and boat problems, it was decided to deal with them separately in this chapter.

7–1. Tabulation of Conformational Energies in Cyclohexane

Conformational energy has been previously defined as the excess energy of a given conformation over that of the conformation of minimum energy of the same molecule. Conformational free energy in a substituted cyclohexane is thus the negative of the standard free energy change $(-\Delta G_X^\circ)$ for the equilibrium in Fig. 1.

All the values for free energy changes of this type which we have been able to find in the literature are given in Table 1, along with the conditions of temperature and solvent and the pertinent reference. Table 1 in

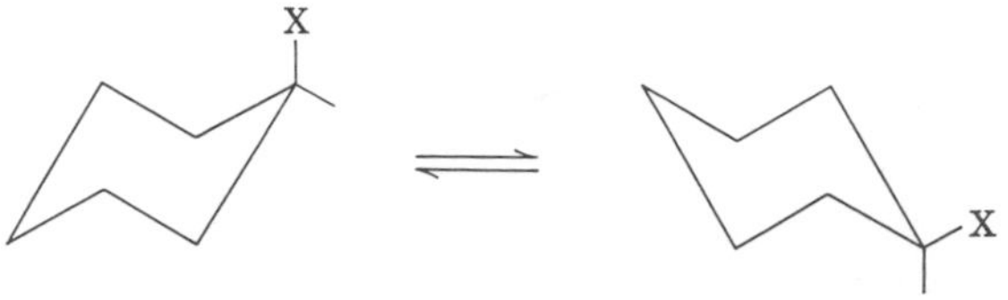

Fig. 1.

Chap. 2 is a summary of what we consider the likely ranges and best values of $-\Delta G_X^\circ$ based on the literature values listed in Table 1 here.

A few comments on the table are in order. Only experimental values are included; estimates made occasionally in the literature, even if reasonable, are not tabulated. (*A priori* calculation of $-\Delta G_X^\circ$ will be taken up in Sec. 7-2.) Values which manifestly rest on an erroneous theoretical foundation, such as values based on intensity ratios of equatorial and axial C—X stretching bands in the infrared[1] neglecting the fact that the extinction coefficients are unequal, are generally disregarded. Data based on differences in potential barrier between axial and equatorial isomers as determined by ultrasonic measurements (Sec. 3-10) are also excluded because of experimental unreliability. A number of good values in the literature have no doubt been missed.

Values of $-\Delta H_X^\circ$ have been included when the given reference gives only $-\Delta H_X^\circ$ but not $-\Delta G_X^\circ$; or, if the experimental values of $-\Delta G_X^\circ$ are at temperatures very different from 25°C., such values are denoted by a footnote reference "a."

Many of the methods used in obtaining the values in Table 1, such as the kinetic method (Secs. 2-2c and 3-13), analogous methods based on equilibrium constants (e.g., pK: Sec. 3-11), direct equilibrium methods (Sec. 2-4b,c) and various indirect equilibrium methods (Sec. 2-4d), have already been discussed in the text, as has the trick of determining the conformational equilibrium of *cis*-4-methylcyclohexyl-X instead of cyclohexyl-X and then determining $-\Delta G_X^\circ$ by means of the equation $-\Delta G_X^\circ = -\Delta G_{Me}^\circ - \Delta G'$ (p. 78). A few of the physical methods will, however, be summarized here. In the nuclear magnetic resonance method denoted "NMR-1" the equilibrium constant K is derived from the fact that the position of a salient proton peak in cyclohexyl-X is given as $\nu = \nu_e N_e + \nu_a N_a$, where N_e and N_a are the mole fractions of the contributing equatorial and axial conformers, respectively, and ν_e and ν_a

---

[1] C. G. Le Fèvre, R. J. W. Le Fèvre, R. Roper, and R. K. Pierens, *Proc. Chem. Soc.*, **1960**, 117.

are the chemical shifts corresponding to ν in these pure conformers.[2] Since $N_e + N_a = 1$, both can be evaluated if ν, ν_e, and ν_a are known; then $K = N_e/N_a$. In method "NMR-1a" ν_e and ν_a are obtained from 4-t-butyl-substituted cyclohexyl-X, whereas in method "NMR-1b" they are derived from low temperature NMR measurements on cyclohexyl-X itself, i.e., at a temperature where interconversion of the axial and equatorial conformers is sufficiently slowed down that one can observe the spectrum of each one individually, rather than an average spectrum. Under these circumstances it is also possible to obtain the proportion of the equatorial and axial conformers directly from the measured ratio of the intensity of the appropriate signals (method "NMR-2"). Infrared measurement may lead to $-\Delta H_X^\circ$ through measurement of appropriate band intensities as a function of temperature ("IR-1"),† or it may lead to $-\Delta G_X^\circ$ through comparison of appropriate bands in cyclohexyl-X with corresponding bands in cis- and $trans$-4-t-butylcyclohexyl-X ("IR-2"). Methods based on electron diffraction patterns, Kerr constants (Sec. 3-8), and dipole moments (Sec. 3-5; especially p. 159) are essentially based on a comparison of the observed patterns or values with appropriately calculated values, assuming various proportions of the two pure conformational isomers.

7–2. Calculation of Conformational Energies

Conformational analysis has now progressed to the point where it is possible to make fairly good predictions regarding equilibria between reasonably simple conformers for which proper analogies are available. It seems feasible to attempt to reproduce experimental data on conformational equilibria by *a priori* calculations, both for purposes of understanding the factors involved, and also for practical application in cases where the experimental data do not yet exist. A few such calculations have been reported, and, although the agreement with experiment has not been spectacular, it has been sufficient to encourage the development of better *a priori* methods.

In principle, of course, all molecular problems should be attacked by quantum mechanical methods. In practice one usually seeks an alternative approach. The Schrödinger equation for the hydrogen atom was

[2] E. L. Eliel, *Chem. Ind.* (*London*), **1959**, 568.

† This method is discussed in detail on p. 147.

Table I

Atom or Group	$-\Delta G_X^\circ$, kcal./mole	Method	Temperature, °C.	Solvent or State	Reference
F	0.17	E.D.[h]	R.T.[w] ?	Gas phase	3
	0.24	NMR-1b[i]	R.T.	3M in CS_2	4
	0.25	NMR-2[j]	−93 ±1	3M in CS_2	4
Cl	<0.26	E.D.	?	Gas phase	5
	0.3 to 0.4[a]	IR-1[k]	?	Pure liquid	6
	0.34	IR-1	117 to 200	Gas phase	7
	0.33	IR-1	25	CS_2	7
	0.44	IR-1	25	CH_3COOCH_3	7
	0.41	NMR-2	−93 to −80	CS_2	8
	0.48	NMR-1b	R.T.	3M in CS_2	4
	0.47	Kerr[n]	R.T.	CCl_4 (infinite dilution)	1
	0.51	NMR-2	−81 ±1	3M in CS_2	4
Br	0.2[a]	IR-1	?	Pure liquid	6
	0.24	NMR-1a[i]	25	Pure liquid	2
	0.26	Kerr	R.T.	CCl_4 (infinite dilution)	1
	0.44	NMR-1b	R.T.	3M in CS_2	4
	0.46[b]	IR-2[l]	25?	Various solvents	9
	0.48	NMR-2	−81 ±1	3M in CS_2	4
	0.51	NMR-2	−104 to −86	CS_2	8
	0.6	Kin.[o]	25	87% EtOH	10
	0.61	IR-2	30	CS_2	9
	0.63[c]	IR-2	25?	Various solvents	9
	0.7	[p]	?	Gas phase (?)	12
I	0.73	Kin.	25.1 ±0.1	87% EtOH	13
	0.94	NMR-1a	25	$CHCl_3$	2
	0.30	Kerr	R.T.	CCl_4 (infinite dilution)	1
	0.41	NMR-1b	R.T.	CS_2	4
	0.43	NMR-2	−81 ±1	CS_2	4

	Value	Method	Temp.	Solvent	Ref.
OH	0.29 to 0.37	IR-2	R.T. ?	CH_2Cl_2, $CHCl_3$, CCl_4, CS_2, $Br(CH_2)_2Br$, $(CH_3)_2CO$, xylene, toluene, C_6H_6, n-hexane	14
	0.31 to 0.41[a]	IR-1	?	Pure liquid	6
	0.33	Thermo.[q]	25	H_2O (pH = 5.5)	15, 16
	0.33 to 0.35	IR-2	R.T.	$CHCl_3$, CH_2Cl_2, CS_2, CCl_4, $(CH_3)_2CO$	17
	0.33 to 0.37	IR-2	R.T. ?	Xylene, toluene, C_6H_6, n-hexane	14
	0.36 to 0.41	IR-2	R.T. ?	Pure liquid, CH_2Cl_2, $CHCl_3$, CCl_4, $Br(CH_2)_2Br$, CS_2, $(CH_3)_2CO$	14
	0.38	IR-1	-30 to $+50$(?)	Pure liquid	11
	0.4	IR-2	20	CS_2	18
	0.47 ± 0.04	D. Eqb.[r]	110 to 170	C_6H_{12}	19
	0.52 ± 0.04	D. Eqb.	90	Abs. EtOH	19
	0.52 to 0.55	Kin.	25	C_5H_5N	20
	0.6	NMR-1a	R.T.	ca. 10% CCl_4	21
	0.64 to 0.66	Raman	30-68	Pure liquid	22
	0.73 ± 0.02[m]	NMR-1a	30	CCl_4	23
	0.8[f]	Kin.	40	75% HOAc	24
	0.82	NMR-1a	36	CCl_4	24a
	0.83	NMR-1a	36	C_5H_5N	24a
	0.84	NMR-1a	36	Isoöctane	24a
	0.88 ± 0.02	NMR-1a	30	2-D-2-PrOH	23
	0.9	p	22	H_2O	25
	0.96	D. Eqb.	89	i-PrOH	26
	0.97	NMR-1a	36	C_6H_5Br	24a
	0.98 to 1.12	D. Eqb.	Reflux temp.	C_6H_6, toluene, xylene	19
	1.0	D. Eqb.	89	i-PrOH	19
	1.0	NMR-1a[z]	28	CCl_4	27
	1.05	NMR-1a	36	t-BuOH	24a

Table I (Continued)

Atom or Group	$-\Delta G^\circ_X$, kcal./mole	Method	Temperature, °C.	Solvent or State	Reference
OH (contd.)	1.07	NMR-1a	R.T.	10% CCl$_4$	28
OCH$_3$	1.25	NMR-1a^z	28	D$_2$O	27
	0.6	NMR-1a	R.T.	CCl$_4$	21
	0.73	I. Eqb.^{s}	140	H$_2$O	29
	0.74	I. Eqb.^{s}	100	H$_2$O	29
OC$_2$H$_5$	0.89	I. Eqb.^{s}	140	H$_2$O	29
	0.98	I. Eqb.^{s}	100	H$_2$O	29
OCHO	0.27	NMR-1b	25	CS$_2$	30
OCOCF$_3$	0.68	NMR-1b	25	CS$_2$	30
OCOCH$_3$	0.36	D. Eqb.	25	HOAc:Ac$_2$O (1:1)	31
	0.66	NMR-1b^z	28	CS$_2$	27
	0.68	NMR-1b	25	CS$_2$	30
	0.7	NMR-1a	R.T.	CCl$_4$	21
	ca. 0.76	NMR-2	−110	CS$_2$	27
o-pNB^d	1.5	Kin.	40	H$_2$O-dioxane (1:1)	32
	1.6	Kin.	15 to 63	H$_2$O-dioxane (1:1, 1:3)	33
	0.98	Kin.	25	80% aq. (CH$_3$)$_2$CO	34
OTs^e	0.6	NMR-1a	R.T.	CCl$_4$	21
	0.7	Kin.	25	87% EtOH	35
	0.7	Kin.	25.1	90% EtOH	36
	1.7^f	Kin.	25, 50, 75	EtOH, AcOH, HCOOH	24
ONO$_2$	0.59	NMR-1b	25	CS$_2$	30
SH	−0.41 ± 0.04^a	IR-1	−50 to +25	?	37
	0.6	pK^t	25	DMF^x	38
	0.9	NMR-1a	R.T.	CCl$_4$	39

Substituent	Value	Method	Temp.	Solvent	Ref.
S⁻	1.3	pK	25	DMF[x]	38
SCH₂⁻	0.4	I. Eqb.[u]	35	Ether	40
	0.7	I. Eqb.[u]	35	Ether	41
	−0.4	I. Eqb.[u]	80	C_6H_6	42
SCH₃	0.7	NMR-1a	R.T.	CCl_4	39
SC₆H₅	0.8	NMR-1a	R.T.	CCl_4	21
SOC₆H₅	1.9	Kin.	0	90% 2-PrOH	41
SO₂C₆H₅	2.5	NMR-1a	R.T.	CCl_4	41
NH₂	1.1 to 1.2	NMR-1a	R.T.	C_6H_{12}, CCl_4, CH_3CN, C_5H_5N	43
	1.4 to 1.5	NMR-1a	R.T.	$CDCl_3$, 95% EtOH, 90% t-BuOH, t-BuOH	43
NH₃⁺	1.7	pK	20	80% 2-Methoxyethanol	44
	1.8	Kin.	25.1 or 56.6	98% EtOH	43
	1.55 to 1.86	NMR-1a	R.T.	CF_3COOH, HOAc, CF_3COOH:HOAc (1:8), CF_3COOH:t-BuOH (1:8)	43
N(CH₃)₂	2.0	pK	20	80% 2-Methoxyethanol	44
	2.1	pK	20	80% 2-Methoxyethanol	44
	2.4	pK	20	80% 2-Methoxyethanol	44
NH(CH₃)₂⁺	1.5 ± 0.1	I. Eqb.[u]	35	Anhyd. ether	47
	1.53 to 1.70	I. Eqb.[u]	35	Anhyd. ether	48
CH₃	1.54 to 1.72[a]	Thermo.	25	Liquid phase	49
	1.54 to 2.47	I. Eqb.[s]	100	H_2O	29
	1.60 ± 0.06	NMR-1a	30	2-D-2-PrOH	23
	1.67	I. Eqb.[u]	25	2-PrOH	49a
	1.68 ± 0.06[m]	NMR-1a	30	CCl_4	23
	1.69 to 2.18	I. Eqb.[s]	140	H_2O	29

Table I (Continued)

Atom or Group	$-\Delta G_X^\circ$, kcal./mole	Method	Temperature, °C.	Solvent or State	Reference
CH₃ (contd.)	1.70	NMR-Ia	32	CCl₄	24a
	1.75 ± 0.10	D. Eqb.	200 to 300	Liquid phase	50
	1.87	I. Eqb.ᵘ	25	EtOH	51
	1.87 to 1.96ᵃ	Thermo.	25	Gas phase	46, 49
	1.9 to 2.1	D. Eqb.	25	99.8% H₂SO₄	45
	1.91 to 1.96ᵃ	Thermo.	25	Gas phase	52
	1.97 ± 0.30ᵃ	D. Eqb.	257 to 327	Liquid phase	53
	1.97	IR-2ᵖ	30	CS₂	9
	2.0 to 2.1ᵃ	D. Eqb.	25 to 300	Gas phase	54
C₂H₅	1.62 to 1.82	I. Eqb.ᵘ	35	Anhyd. ether	48
	1.68 ± 0.06ᵐ	NMR-Ia	30	CCl₄	23
	1.77 ± 0.07	NMR-Ia	30	2-D-2-PrOH	23
	1.80	I. Eqb.ᵘ	25	EtOH	51
	1.8 ± 0.3ᵃ	D. Eqb.	258 to 320	Liquid phase	55
	1.86	I. Eqb.ᵘ	25	2-PrOH	49a
	2.0	pKᵖ	25	50% EtOH-H₂O	56
	2.09	I. Eqb.ˢ	143	H₂O	29
	2.27	I. Eqb.ˢ	100	H₂O	29
CH(CH₃)₂	1.84 to 2.20	I. Eqb.ᵘ	35	Anhyd. ether	48
	1.91 ± 0.01	D. Eqb.	287	Liquid phase	53
	1.91	D. Eqb.	287	?	51
	2.1 ± 0.1	Kin.ᵖ	100 to 110	AcOH	57
	2.11	I. Eqb.ᵘ	25	EtOH	51
	2.22 ± 0.08ᵐ	NMR-Iaᵖ	30	CCl₄	23
	2.25 ± 0.08	NMR-Iaᵖ	30	2-D-2-PrOH	23

Group	Value	Method	Temperature	Solvent	Ref.
—C(CH₃)₃	2.43 to 2.62	I. Eqb.ᵘ	25	MeOH, 2-PrOH	49a
	2.5	pKᴾ	25	50% EtOH-H₂O	56
	2.63	I. Eqb.ˢ	139	H₂O	29
	3.55	I. Eqb.ˢ	100	H₂O	29
	>3.6g	I. Eqb.ˢ	140	H₂O	29
	>4.2	D. Eqb.	220	Neat	58
	>4.43g	I. Eqb.ˢ	100	H₂O	29
CH₂C(CH₃)₃	2.0	pKᴾ	25	50% EtOH-H₂O	56
C₆H₅	2.0	Dipoleᵛ	25	C₆H₆	59
	ca. 2.6	I. Eqb.ᵘ	35	Anhyd. ether	47
	3.1	D. Eqb.	39, 56.5, 80, 118, 155.5	Dry (CH₃)₂SO	60
C₆H₁₁	2.15	NMR-1a	32	CCl₄	24a
CH₂OTs	1.7 to 1.8	Kin.	100 to 110	AcOH	61
CO₂H	0.7	Kin.	30	Abs. EtOH	65
	1.15	pK	25	80% 2-Methoxyethanol	64
	1.2	D. Eqb.	230, 286	C₁₂H₂₆	66
	1.2	I. Eqb.ᵘ	78	MeOH-H₂O	49a
	1.5 to 1.6	pK	25	50% EtOH-H₂O	56
	1.58 to 1.86	pK	25 (?)	66% DMF: 34% H₂O	62
	1.6 ± 0.3	pK	25 ± 0.05	80% 2-Methoxyethanol	63
CO₂⁻	1.9	p	25	66% DMF: 34% H₂O	49a
	2.0 to 2.8	pK	25 (?)	66% DMF: 34% H₂O	62
	2.1 to 2.3	pK	25	50% EtOH-H₂O	56
	2.2 ± 0.3	pK	25 ± 0.05	80% 2-Methoxyethanol	63
CO₂CH₃	1.05	D. Eqb.	66	Abs. MeOH	67
	1.15	D. Eqb.	25	Abs. MeOH	49a

Table I (Continued)

Atom or Group	$-\Delta G°_X$, kcal./mole	Method	Temperature, °C.	Solvent or State	Reference
—$CO_2C_2H_5$	1.0 to 1.2	Kin.	25.2	70% EtOH	68
	1.1	NMR-1a	25	CCl_4	21
	1.12 ± 0.14	D. Eqb.	80	Abs. EtOH	67
	1.2	D. Eqb.	80	EtOH	68
—$CO_2CH(CH_3)_2$	0.96	D. Eqb.	25	2-PrOH	49a
COCl	1.2	D. Eqb.	130 to 140	Neat liquid	69
CN	0.15	D. Eqb.	25	t-BuOH	70
	0.25	D. Eqb.	66	THF[y]	71
HgBr	0	D. Eqb.	95	C_5H_5N	72
HgCl	0.3	Opt. Rotation[p]	25 (?)	EtOH	73

[a] ΔH value.   [b] Recalculated from ref. 6 using the ratio of extinction coefficients.
[c] Recalculated from ref. 1 using the ratio of extinction coefficients.   [d] pNB = p-nitrobenzoate.
[e] Ts = p-toluenesulfonate.   [f] This is an overall average value reported by the authors.
[g] By the nature of the method used (see footnote s) these values are only minima of no very great significance.
[h] Electron diffraction.
[i] NMR-1a: nuclear magnetic resonance method: $K = (v_a - v)/(v - v_e)$, where v_a and v_e are the chemical shifts of cis- and trans-4-t-butylcyclo-C_6H_{10}X, respectively. Calculated parameters are used for v_a and v_e in ref. 28. NMR-1b: Similar to NMR-1a, except that v_a and v_e are obtained from the cyclohexyl-X spectrum at low temperature.
[j] NMR-2: percentages of the e and the a conformers are obtained by measuring the area under the corresponding peak at low temperature.
[k] IR-1: Infrared method. ΔH value obtained by application of the van't Hoff equation (intensity ratio vs. temperature).
[l] IR-2: K (equilibrium constant) from molar extinction coefficient.
[m] Recalculated value.   [n] Molar Kerr constant.   [o] Kinetic method.   [p] Indirect argument.
[q] Thermochemical method.   [r] Equilibrium method.   [s] Lactone-hydroxy acid equilibrium method.
[t] pK measurement—dissociation constant.   [u] Indirect equilibrium method.   [v] Dipole measurement.
[w] Room temperature.   [x] Dimethyl formamide.   [y] Tetrahydrofuran.   [z] Using coupling constants.

[3] P. Andersen, *Acta Chem. Scand.*, **16**, 2337 (1962).

[4] A. J. Berlin and F. R. Jensen, *Chem. Ind. (London)*, **1960**, 998.

[5] V. A. Atkinson, *Acta Chem. Scand.*, **15**, 599 (1961).

[6] G. Chiurdoglu, L. Kleiner, W. Masschelein, and J. Reisse, *Bull. Soc. Chim. Belges*, **69**, 143 (1960).

[7] K. Kozima and K. Sakashita, *Bull. Chem. Soc. Japan*, **31**, 796 (1958).

[8] L. W. Reeves and K. O. Strømme, *Can. J. Chem.*, **38**, 1241 (1960).

[9] F. R. Jensen and L. H. Gale, *J. Org. Chem.*, **25**, 2075 (1960).

[10] E. L. Eliel and R. G. Haber, *Chem. Ind. (London)*, **1958**, 264.

[11] W. Masschelein, *J. Mol. Spectr.*, **10**, 161 (1963).

[12] E. J. Corey, *J. Am. Chem. Soc.*, **75**, 2301 (1953).

[13] E. L. Eliel and R. G. Haber, *J. Am. Chem. Soc.*, **81**, 1249 (1959).

[14] G. Chiurdoglu and W. Masschelein, *Bull. Soc. Chim. Belges*, **69**, 154 (1960).

[15] M. A. Kabayama and D. Patterson, *Can. J. Chem.*, **36**, 563 (1958).

[16] M. A. Kabayama, D. Patterson, and L. Piche, *Can. J. Chem.*, **36**, 557 (1958).

[17] G. Chiurdoglu and W. Masschelein, *Bull. Soc. Chim. Belges*, **70**, 29 (1961).

[18] R. A. Pickering and C. C. Price, *J. Am. Chem. Soc.*, **80**, 4931 (1958).

[19] G. Chiurdoglu and W. Masschelein, *Bull. Soc. Chim. Belges*, **70**, 767 (1961).

[20] E. L. Eliel and C. A. Lukach, *J. Am. Chem. Soc.*, **79**, 5986 (1957).

[21] E. L. Eliel and M. H. Gianni, *Tetrahedron Letters*, **1962**, 97.

[22] P. Neelakantan, *Proc. Indian Acad. Sci.*, **57**, A94 (1963).

[23] A. H. Lewin and S. Winstein, *J. Am. Chem. Soc.*, **84**, 2464 (1962).

[24] S. Winstein and N. J. Holness, *J. Am. Chem. Soc.*, **77**, 5562 (1955).

[24a] J. Reisse, J. C. Celotti, D. Zimmermann, and G. Chiurdoglu, *Tetrahedron Letters*, **1964**, 2145.

[25] S. J. Angyal and D. J. McHugh, *Chem. Ind. (London)*, **1956**, 1147.

[26] E. L. Eliel and R. S. Ro, *J. Am. Chem. Soc.*, **79**, 5992 (1957).

[27] F. A. L. Anet, *J. Am. Chem. Soc.*, **84**, 1053 (1962).

[28] E. L. Eliel, M. H. Gianni, T. H. Williams, and J. B. Stothers, *Tetrahedron Letters*, **1962**, 741.

[29] D. S. Noyce and L. J. Dolby, *J. Org. Chem.*, **26**, 3619 (1961).

[30] E. A. Allan, E. Premuzic, and L. W. Reeves, *Can. J. Chem.*, **41**, 204 (1963).

[31] R. U. Lemieux and P. Chu, Abstracts, San Francisco Meeting, Am. Chem. Soc., 31N (1958).

[32] N. B. Chapman, R. E. Parker, and P. J. A. Smith, *J. Chem. Soc.*, **1960**, 3634.

[33] E. A. S. Cavell, N. B. Chapman, and M. D. Johnson, *J. Chem. Soc.*, **1960**, 1413.

[34] G. F. Hennion and F. X. O'Shea, *J. Am. Chem. Soc.*, **80**, 614 (1958).

[35] E. L. Eliel and R. S. Ro, *J. Am. Chem. Soc.*, **79**, 5995 (1957).

[36] E. L. Eliel and R. S. Ro, *Chem. Ind. (London)*, **1956**, 251.

[37] G. Chiurdoglu, J. Reisse, and M. Vander Stichelen Rogier, *Chem. Ind. (London)*, **1961**, 1874.

[38] B. P. Thill, Ph.D. Dissertation, University of Notre Dame, Notre Dame, Indiana, 1964.

[39] E. L. Eliel and B. P. Thill, *Chem. Ind. (London)*, **1963**, 88.

[40] E. L. Eliel and L. A. Pilato, *Tetrahedron Letters*, **1962**, 103.

[41] E. L. Eliel, E. W. Della, and M. Rogić, *J. Org. Chem.*, **30**, 855 (1965).

[42] M. P. Mertes, *J. Org. Chem.*, **28**, 2320 (1963).

[43] E. L. Eliel, E. W. Della, and T. H. Williams, *Tetrahedron Letters*, **1963**, 831.

[44] J. Sicher, J. Jonáš, and M. Tichý, *Tetrahedron Letters*, **1963**, 825.

accurately solved as long ago as 1926,[74] and the next simplest atom, helium, has had its ground state energy calculated to eight significant figures.[75] While this may not be exact, it is better by an order of magnitude than experimental measurement. The hydrogen molecule has also been accurately treated.[76] The computational difficulties increase rapidly,

[45] A. K. Roebuck and B. L. Evering, *J. Am. Chem. Soc.*, **75**, 1631 (1953).

[46] J. E. Kilpatrick, H. G. Werner, C. W. Beckett, K. S. Pitzer, and F. D. Rossini, *J. Res. Natl. Bur. Standards*, **39**, 523 (1947).

[47] E. L. Eliel and M. N. Rerick, *J. Am. Chem. Soc.*, **82**, 1367 (1960).

[48] E. L. Eliel and T. J. Brett, *J. Am. Chem. Soc.*, **87**, 5039 (1965).

[49] E. J. Prosen, W. H. Johnson, and F. D. Rossini, *J. Res. Natl. Bur. Std.*, **39**, 173 (1947).

[49a] B. J. Armitage, G. W. Kenner, and M. J. T. Robinson, *Tetrahedron*, **20**, 747 (1964).

[50] W. Szkrybalo and F. A. VanCatledge, unpublished results.

[51] N. L. Allinger, L. A. Freiberg, and S.-E. Hu, *J. Am. Chem. Soc.*, **84**, 2836 (1962).

[52] C. W. Beckett, K. S. Pitzer, and R. Spitzer, *J. Am. Chem. Soc.*, **69**, 2488 (1947).

[53] N. L. Allinger and S.-E. Hu, *J. Org. Chem.*, **27**, 3417 (1962).

[54] C. J. Egan and W. C. Buss, *J. Phys. Chem.*, **63**, 1887 (1959).

[55] N. L. Allinger and S.-E. Hu, *J. Am. Chem. Soc.*, **84**, 370 (1962).

[56] H. van Bekkum, P. E. Verkade, and B. M. Wepster, *Koninkl. Ned. Akad. Wetenschap. Proc.*, **64B**, 161 (1961).

[57] N. Mori and F. Suda, *Bull. Chem. Soc. Japan*, **36**, 227 (1963).

[58] N. L. Allinger and L. A. Freiberg, *J. Am. Chem. Soc.*, **82**, 2393 (1960).

[59] N. L. Allinger, J. Allinger, M. A. DaRooge, and S. Greenberg, *J. Org. Chem.*, **27**, 4603 (1962).

[60] E. W. Garbisch, Jr. and D. B. Patterson, *J. Am. Chem. Soc.*, **85**, 3228 (1963).

[61] N. Mori, *Bull. Chem. Soc. Japan*, **34**, 1299 (1961); **35**, 1755 (1962).

[62] R. D. Stolow, *J. Am. Chem. Soc.*, **81**, 5806 (1959).

[63] M. Tichý, J. Jonáš, and J. Sicher, *Collection Czech. Chem. Comm.*, **24**, 3434 (1959).

[64] J. Sicher, private communication.

[65] N. B. Chapman, J. Shorter, and K. J. Toyne, *J. Chem. Soc.*, **1964**, 1077.

[66] E. L. Eliel and M. Reese, unpublished results.

[67] N. L. Allinger and R. J. Curby, *J. Org. Chem.*, **26**, 933 (1961).

[68] E. L. Eliel, H. Haubenstock, and R. V. Acharya, *J. Am. Chem. Soc.*, **83**, 2351 (1961).

[69] E. L. Eliel and R. Gerber, unpublished results.

[70] B. Rickborn and F. R. Jensen, *J. Org. Chem.*, **27**, 4606 (1962).

[71] N. L. Allinger and W. Szkrybalo, *J. Org. Chem.*, **27**, 4601 (1962).

[72] F. R. Jensen and L. H. Gale, *J. Am. Chem. Soc.*, **81**, 6337 (1959).

[73] V. I. Sokolov and O. A. Reutov, *Dokl. Akad. Nauk SSSR*, **148**, 867 (1963); English transl., p. 123.

[74] E. Schrödinger, *Phys. Rev.*, **28**, 1049 (1926).

[75] C. L. Pekeris, *Phys. Rev.*, **112**, 1649 (1958); **115**, 1216 (1959).

[76] W. Kolos and C. C. J. Roothaan, *Rev. Mod. Phys.*, **32**, 219 (1960).

however, as the number of particles (electrons plus nuclei) in the molecule increases, and the calculated ground state energy of lithium hydride, after the expenditure of a prodigious amount of labor, is still off by about 0.5% from the experimental value.[77] Although this error is quite small percentagewise, it does in fact amount to some 18 kcal./mole. A very detailed quantum mechanical attack on the ethane molecule was reported[78] in 1963. The calculations indicated that the staggered conformation was of lower energy than the eclipsed conformation by 3.3 kcal./mole (cf. Sec. 1-2). The absolute energies of these conformations, though accurate to about 1%, differed from the experimental value by some 500 kcal./mole. There is reason to believe that the errors in the ground state energies will cancel out when the two conformations are compared, however, and the 3.3 kcal. value appears to be accurate.

There has been no shortage of ideas as to how the rotational barrier in ethane might be explained.[79] At the present time, however, we simply do not know with any certainty what the barrier is due to. It may well be that there is no simple mechanical or electrostatic model by means of which the barrier can be explained. If one solves the Schrödinger equation for ethane, the barrier is contained in the solution, and that is really all that one can expect.

The simple examples quoted above do not offer much promise for a quantum mechanical approach to conformational analysis. A few attempts have been made to apply this approach to more complicated systems,[80] necessarily in more approximate form. Until the early 1960's the prospects for such calculations could only be described as dismal. Progress is being made in this area, however, the availability of really fast electronic computers is most helpful, and more effort is now going into semi-empirical approaches.[81,82] In such an approach one does not try to calculate everything a priori but makes use of available experimental data in the calculation, and, especially important, one tries to calculate the differences between similar states (such as conformations) rather than the absolute values.

At this writing, it is fair to state that the quantum mechanical approach to conformational problems has not as yet been very useful. It is the

[77] D. D. Ebbing, J. Chem. Phys., **36**, 1361 (1962).

[78] R. M. Pitzer and W. N. Lipscomb, J. Chem. Phys., **39**, 1995 (1963).

[79] E. B. Wilson, Jr., Advan. Chem. Phys., **2**, 367 (1959). Subsequent papers of interest are: M. Karplus and R. G. Parr, J. Chem. Phys., **38**, 1547 (1963); and ref. 78. Also see pp. 7–8.

[80] K. E. Howlett, J. Chem. Soc., **1957**, 4353; R. Pauncz and D. Ginsburg, Tetrahedron, **9**, 40 (1960).

[81] R. Hoffmann, J. Chem. Phys., **39**, 1397 (1963).

[82] J. B. Pedley, Trans. Faraday Soc., **57**, 1492 (1961).

fundamentally correct approach, however, and it will probably pay off in the long run. In the meantime, a more simple and more accurate, if fundamentally less sound, approach to *a priori* calculations must necessarily be used. In this approach† the molecular system is treated as a classical mechanical system, upon which one can superimpose such quantum mechanical restrictions as seem desirable.[82,83]

In the classical approach, if the relative energies of two conformations are to be calculated, one needs only to calculate those quantities which change in going from one conformation to another. The conformational energy of a single conformation for a hydrocarbon‡ is given by

$$E = E_F + E_T + E_\Theta + E_V \tag{1}$$

where E_F is the sum of the compression and stretching energies required to deform the bond lengths of the system from the normal to the observed values; E_T is the sum of the torsional energies due to any "Pitzer strain" or deviation from torsional arrangements of minimum energy; E_Θ is the sum of the bending energies required to deform the bond angles in the system from the normal to the actual values; and E_V is the sum of the van der Waals energies (attractive or repulsive) between every pair of atoms in the system other than those pairs which are covalently bound to one another.

To calculate the difference in energy between conformations, it is in principle only necessary to evaluate E in equation (1) for each conformation. There are a number of practical problems which are met when such calculations are attempted, but before considering these it will be convenient to discuss how each type of term in equation 1 is evaluated.

The E_F term is the sum of the longitudinal deformation energies of all the bonds in the molecule. This term will be significant if there are in the molecule one or more bonds which are forced by their environment to have unnatural bond lengths. Thus in hexaphenylethane, for example, the central C—C bond might be expected to be stretched as the two

[83] F. H. Westheimer, in M. S. Newman, ed., *Steric Effects in Organic Chemistry*, John Wiley and Sons, New York, 1956, p. 523.

[84] M. M. Kreevoy and E. A. Mason, *J. Am. Chem. Soc.*, **79**, 4851 (1957); E. J. Corey, *ibid.*, **75**, 2301 (1953); J. Allinger and N. L. Allinger, *Tetrahedron*, **2**, 64 (1958).

† An excellent review of this approach is given in ref. 83.

‡ A molecule other than a hydrocarbon would be treated in the same way if it contained only one polar group. If two or more polar groups were present, the electrostatic interaction between the dipoles would have to be included as an additional term in the equation. See, for example, ref. 84.

bulky end groups endeavor to move out of one another's way.† The distortion can be assumed to obey Hooke's law, so that the energy E_f is given by

$$E_f = (k/2)(\Delta l)^2 \qquad (2)$$

where Δl is the difference between the actual and natural bond length. The stretching constant k can be evaluated from vibrational spectra. The value of k is *approximately* constant for a given type of bond, regardless of its environment. Some typical values are listed in Table 2. If

Table 2. Force Constants for Bond Distortions[a]

Bond	$k/2$ Stretching, kcal./mole Å^2	Bond Angle	$k/2$ Bending, kcal./mole radian^2
C—H	346	C—C—H	39.5
C—C	324	C=C—H	48.9
C=O	870	H—C—H	23.0
C=C	690	Ar—H	61.8
C—Br	224	C—C—C	57.5
(Ar) C—H	360		

[a] From ref. 83. These "constants" may vary by $\pm 10\%$.

there are several bonds of unnatural length in the molecule, $E_F = \Sigma\, E_f$. For a given amount of nuclear motion, it costs much less energy to bend a bond than to stretch or compress it, and consequently the E_F term is usually negligible in comparison with other terms in equation (1). (For exceptions, see refs. 86 and 87.) Usually, therefore, one can neglect the E_F term in conformational calculations.

The next term in equation (1), E_T, is the sum of the torsional strains about individual bonds: $E_T = \Sigma\, E_t$. For an open chain or for simple cyclohexane rings, E_T is usually close to zero, and an almost perfectly staggered geometry exists. The small and medium rings are structures which necessarily exhibit torsional strain. That present in a single C—C bond (three-fold barrier) is given by

$$E_t = (V_0/2)(1 + \cos 3\omega) \qquad (3)$$

[85] L. S. Bartell, *J. Am. Chem. Soc.*, **83**, 3567 (1961).

[86] F. H. Westheimer and J. E. Mayer, *J. Chem. Phys.*, **14**, 733 (1946); F. H. Westheimer, *ibid.*, **15**, 252 (1947); M. Rieger and F. H. Westheimer, *J. Am. Chem. Soc.*, **72**, 19 (1950).

[87] N. L. Allinger, W. Szkrybalo, and M. A. DaRooge, *J. Org. Chem.*, **28**, 3007 (1963).

† Actually, since the atoms in molecules vibrate there is some ambiguity about the "position" of an atom. The calculations discussed here are not yet of sufficient accuracy for this ambiguity to cause concern, but it will ultimately have to be considered. See ref. 85.

where ω is the dihedral angle measured from $\omega = 0°$ corresponding to an eclipsed conformation. The ethane value, 2.8 kcal./mole, is used for V_0. If the barrier is not three-fold, or if it involves elements other than carbon, appropriate adjustments must be made; thus, in the dioxane calculation (Sec. 4-6b), the value of V_0 for the C—O bond is 1.1 kcal./mole, and for piperidine the C—N value is 1.9 kcal./mole. Some examples of the calculation of E_T will be given later in this section.

The next term in equation (1), E_Θ, is the sum of the energies required to bend the various bonds in the molecule from their normal angles to those observed. If an atom has a symmetrical tetrahedral arrangement (methane, neopentane), it will have bond angles of 109.5°. If the atom has different groups attached to it, as in propane, the angles are not exactly tetrahedral. The differences may result from van der Waals repulsions between the substituents, from bonding (hybridization) effects, or from both. The simplest approach is to say that the 112° C—C—C angle in a hydrocarbon corresponds to $E_\theta = 0$, and to measure deviations from that value.† The energy of deforming a single bond angle is given by

$$E_\theta = (k/2)(\Delta\theta)^2 \qquad (4)$$

where $\Delta\theta$ is the angular deformation from the normal value and k is the bending force constant (Table 2). If one angle at an atom is distorted, the other angles at that atom will adjust to minimize the total bending strain. Thus, if the C—C—C angle in propane is widened, the H—C—H angle will contract slightly. All of these angle changes have to be taken into account. As a good approximation[88] for angle distortions of less than 10° a "composite" bending constant may be used and applied to the angle of major distortion. In propane, then, if the C—C—C angle is distorted, a value of $k/2$ of 64 kcal./mole radian² can be used for that angle alone, instead of accounting separately for the H—C—H and C—C—H angles. If the substituents attached to the propane were carbons instead of hydrogens the constant would be 70, and if there were one hydrogen and one carbon the constant would be 67. It can be calculated from these data that a 5° deviation of a C—C—C angle leads to an energy increase of 0.5 kcal./mole, while a 10° deviation corresponds to 1.9 kcal./mole. These energies are small enough that molecules suffering from any sizable amount of van der Waals compression when their bond angles are normal can usually lower their energies very greatly by small or moderate bond angle deformations. Since the increase in E_θ is

[88] J. B. Hendrickson, *J. Am. Chem. Soc.*, **83**, 4537 (1961).

† The normal C—C—C angles in paraffins are reported to have various values in the range 112 ± 1°, depending on the compound (p. 133).

proportional to the square of the bending, distorting each of two angles by 5° costs only half as much as distorting one angle by 10°, and hence strain tends to distribute itself throughout a system, rather than being concentrated in one place.

The final term in equation (1), the van der Waals term, is usually the most troublesome. An interaction of the van der Waals type is considered to exist between any two atoms which are near one another and not bound together, whether they are part of the same molecule or not.[89] The van der Waals interactions in molecules are almost always sizable, and often the conformation of a molecule can be predicted simply by a minimization of the van der Waals forces. The *anti* conformation of butane, for example, is more stable than the *gauche* because of the less favorable van der Waals interactions in the latter. For this molecule, the other terms in equation (1) are essentially identical in the two conformations.

In view of the importance of E_V, it is desirable to have data on van der Waals functions between atoms in the first two rows of the Periodic Table—hydrogen, the halogens, and a few other atoms. Unfortunately, however, these data are not available by any simple experimental method. Accurate data are available for the rare gases and for a few other special cases. Consequently the usual approach is to utilize data of the latter type and adapt them to the desired purpose. Different adaptations have been used,† and one of them will be discussed here.

A van der Waals function is in theory composed of two parts. At longer distances (r) there is an attractive force between two atoms which is proportional to $1/r^6$. This force is due to electron correlation and is sometimes called the London or dispersion force.[90] At shorter distances, as the atoms come sufficiently close that their van der Waals radii interpenetrate, the sum of the repulsions between the nuclei and between the electrons outweighs the attractive force. The repulsion energy is usually approximated by $\exp(-ar)$ or by $1/r^{12}$. Since the repulsive force becomes tremendously important quite abruptly, whereas the attractive one increases in importance only more gradually, the combination of these forces leads to the familiar van der Waals curve (Fig. 2).

Having now the general shape of a semitheoretical curve, we wish to fit empirical parameters to it in such a way as to describe the behavior of various covalently bound atoms. Hill has shown that such a curve can reproduce the experimental data for a number of atoms by using

---

[89] R. McWeeny, *Proc. Roy. Soc.*, **A253**, 242 (1959); O. Sínanoğlu, *J. Chem. Phys.*, **37**, 191 (1962).

[90] K. S. Pitzer, *Advan. Chem. Phys.*, **2**, 59 (1959).

† A number of alternatives are discussed by Hendrickson in ref. 88.

only two parameters.[91] One parameter (α) is a function of the distance coordinate alone, and the other (ϵ) is a function of the energy coordinate. A more general way of making a van der Waals plot is to express the distance separating the two atomic centers in question in units of α, the

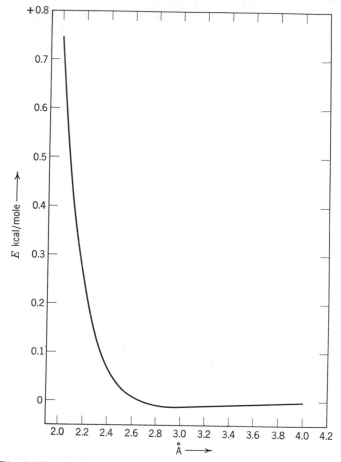

Fig. 2. The energy of two helium atoms as a function of distance.[91]

sum of the van der Waals radii of the two atoms (i.e., $\alpha = r/(r_1^* + r_2^*)$, where r_1^* and r_2^* are van der Waals radii of atoms 1 and 2). This means simply that the curve shown is either stretched out or compressed along the horizontal axis depending on whether one is dealing with large atoms or small atoms. Likewise, along the vertical axis one can plot E_v/ϵ, where ϵ is a parameter which varies with the size of the atom, since clearly it is more difficult to push two large atoms, like iodine for example,

[91] T. L. Hill, *J. Chem. Phys.*, **16**, 399 (1948).

together to some small value of α than it is to push two hydrogen atoms together to the same value of α. To obtain E_v in this way for a given pair of atoms, the following equation, which is the van der Waals curve in analytical form, can be used.

$$E_v/\epsilon = -2.25\alpha^{-6} + 8.28 \times 10^5 \exp\,(-\alpha/0.0736) \qquad (5)$$

(In practice it is convenient to prepare a graph of E/ϵ against α.)

In his original work, Hill evaluated ϵ for a number of atoms.[91] It is proportional to the critical temperature of the element if the element is monatomic in the liquid and gas phases; it is also proportional (but the proportionality constant is different) if the substance is diatomic. No simple relationship has been devised for substances like carbon which are polyatomic, but roundabout methods have been used to evaluate ϵ for such atoms. In accordance with Hill's data, ϵ has been evaluated for a number of atoms as listed in Table 3. Thus ϵ has the value 0.042 kcal./mole for the interaction of two hydrogens, and 0.067 kcal./mole for the interaction between a carbon and a hydrogen, and so on.

Calculations of this type have been applied to a number of monosubstituted cyclohexanes in an attempt to determine the energy differences between conformations. Since entropy considerations have been neglected, and since the calculations do not include solvation effects, the best that can be hoped is that the calculated and observed energies of the groups will agree to about ± 0.5 kcal./mole, as they usually do.

It should be emphasized that, in this kind of treatment, whole atoms are considered, not bonds or lone pairs, and the atoms are assumed to be spherical. Fortunately, when conformations are compared, most of the van der Waals interactions cancel and, hopefully, most of the errors cancel too. It seems particularly advisable to omit from the calculation those terms between any two atoms bound to the same atom, because differences in those terms are at least in part compensated for in the E_θ terms.

The use of equation (1) can now be illustrated by application to a group of examples, and a few have been chosen which exhibit various features of interest. Suppose we wish to calculate the axial $\rightleftharpoons$ equatorial equilibrium between the chair forms of chlorocyclohexane. This problem is relatively easy, since the molecule is nearly strain-free in either conformation, and a normal undistorted geometry may be assumed.† The

---

† Electron diffraction studies [V. A. Atkinson and O. Hassel, *Acta Chem. Scand.*, **13**, 1737 (1959)] indicate that the axial chlorine is bent out from the main axis of the ring by 6.3°. An axial hydrogen is bent out 4.1° (*vide infra*), and substitution of the axial hydrogen by a halogen might be expected to bend the latter 1–2° toward the equatorial hydrogen [T. Ukaji and R. A. Bonham, *J. Am. Chem. Soc.*, **84**, 3627 (1962)]. No significant additional bending therefore occurs.

Table 3. Numerical Values for ε and r*

ϵ^a	H	C	N	O	F	S	Cl	Br	I	r^{*b}	Ref.
H	0.042								0.162	1.20	91
C	0.067	0.107							0.258	1.70	92
N	0.063	0.100	0.095						0.243	1.50	—
O	0.069	0.111	0.105	0.116					0.268	1.40	93
F	0.068	0.108	0.102	0.112	0.109				0.259	1.35	94
S	0.115	0.183	0.172	0.190	0.185	0.314			0.442	1.85	—
Cl	0.115	0.183	0.172	0.190	0.185	0.314	0.314		0.442	1.80	91
Br	0.136	0.215	0.203	0.224	0.217	0.369	0.369	0.434	0.522	1.95	91
I	0.162	0.258	0.243	0.268	0.259	0.442	0.442	0.522	0.623	2.15	—

[a] In kilocalories per mole.
[b] In angstroms.

[92] N. L. Allinger and W. Szkrybalo, *J. Org. Chem.*, **27,** 4601 (1962).
[93] N. L. Allinger, M. A. DaRooge, and R. B. Hermann, *J. Org. Chem.*, **26,** 3626 (1961).
[94] N. L. Allinger, M. A. DaRooge, and C. L. Neumann, *J. Org. Chem.*, **27,** 1082 (1962).

E_F term is therefore zero in both conformations, and so is the E_T term since the system is perfectly staggered. We can begin by calculating E_V for each conformation. There are eighteen atoms in the molecule, and hence for each conformation there are 153 van der Waals interactions between pairs of atoms! Properly, each of these should be evaluated. If the repulsions in the axial form were significant, as is commonly supposed, then we might expect that the chlorine would bend out away from the ring, thus introducing some E_θ terms. We shall then have to recalculate E_V for the new geometry, and we shall have to keep repeating the calculation of E_θ and E_V until we find the geometry where the total conformational energy is a minimum for the axial form, and similarly for the equatorial form. Calculations of this type *as such* are only practical with the aid of an electronic computer, and they have been carried through for a few molecules. The conformational energy of a methyl group,[95] for example, was calculated to be 1.0 kcal./mole, in fair agreement with the experimental value.

This brings us to the second type of difficulty, which is that the basic quantities required for such a calculation, namely, the bending and stretching constants and especially the van der Waals functions, are not accurately known, and indeed they are not even accurately constant from one molecule to another. Nevertheless, such calculations have yielded surprisingly good results in a number of cases, the calculations of the racemization rates of orthosubstituted biphenyls by Westheimer[86] and of the preferred conformation of the cycloheptane ring by Hendrickson[88] (Sec. 4-4a) being classic examples. Such calculations will undoubtedly become more commonplace in the future. The major bottleneck now is the lack of truly adequate, computer programs for handling the calculations in any convenient way in the general case. No really fundamental problems are involved in such programming, but the labor required has so far been a sizable deterrent. Thus, at present it is expedient to seek simplified solutions to conformational calculations. Sometimes approximations can be made which will permit equation (1) to be solved with a slide rule to the necessary accuracy, and a discussion of these approximations is now in order.[96]

When calculations of this type were first applied in conformational analysis, it was hoped that in a given conformation one or two interactions

[95] J. B. Hendrickson, *J. Am. Chem. Soc.*, **84**, 3355 (1962).

[96] An alternative approach to this simplification through the use of empirical data is given by A. I. Kitaygorodsky, *Tetrahedron*, **9**, 183 (1960); **14**, 230 (1960). It is not widely applicable at present, however, as the necessary data are not often available. It is to be noted that the method accurately predicts the observed bond angles in alkanes as being deformed from their tetrahedral values to 112.5° and 108° for C—C—C and H—C—H, respectively.

Table 4

Atom	Coordinates				Substituent Coordinates			H-Substituent Coordinates		
	X	Y	Z		X	Y	Z	X	Y	Z
Coordinates of Atoms in Cyclohexane										
a	−1.325	0	−0.651	ax.	−1.325 + 0.223d	0	−0.651 − 0.975d	−1.079	0	−1.727
				eq.	−1.325 − 0.997d	0	−0.651 + 0.077d	−2.426	0	−0.565
b	−0.764	−1.263	0	ax.	−0.764 − 0.333d	−1.263 − 0.067d	0.940d	−1.132	−1.337	+1.038
				eq.	−0.764 − 0.333d	−1.263 − 0.804d	−0.492d	−1.132	−2.151	−0.543
c	0.764	−1.263	0	ax.	0.764 + 0.333d	−1.263 − 0.067d	−0.940d	1.132	−1.337	−1.038
				eq.	0.764 + 0.333d	−1.263 − 0.804d	0.492d	1.132	−2.151	0.543
d	1.325	0	0.651	ax.	1.325 − 0.223d	0	0.651 + 0.975d	1.079	0	1.727
				eq.	1.325 + 0.997d	0	0.651 − 0.774d	2.426	0	0.565
e	0.764	1.263	0	ax.	0.764 + 0.333d	1.263 + 0.067d	−0.940d	1.132	1.337	−1.038
				eq.	0.764 + 0.333d	1.263 + 0.804d	0.492d	1.132	2.151	0.543
f	−0.764	1.263	0	ax.	−0.764 − 0.333d	1.263 + 0.067d	0.940d	−1.132	1.337	1.038
				eq.	−0.764 − 0.333d	1.263 + 0.804d	−0.492d	−1.132	2.151	−0.543
Coordinates of Atoms in Cyclohexanone										
a	0	0	0	ax.						
				eq.						
b	0.795	1.272	0	ax.	0.795 + 0.433d	1.272 + 0.123d	−0.893d	1.271	1.407	−0.982
				eq.	0.795 − 0.614d	1.272 + 0.777d	0.141d	0.120	2.126	0.156
c	1.868	1.265	1.090	ax.	1.868 − 0.436d	1.265 + 0.063d	1.090 + 0.898d	1.389	1.335	2.078
				eq.	1.868 + 0.585d	1.265 + 0.804d	1.090 − 0.103d	2.512	2.150	0.977
d	2.726	0	1.028	ax.	2.726 + 0.533d	0	1.028 − 0.846d	3.313	0	0.097
				eq.	2.726 + 0.649d	0	1.028 + 0.761d	3.440	0	1.865
e	1.868	−1.265	1.090	ax.	1.868 − 0.436d	−1.265 − 0.063d	1.090 + 0.898d	1.389	−1.335	2.078
				eq.	1.868 + 0.585d	−1.265 − 0.804d	1.090 − 0.103d	2.512	−2.150	0.977
f	0.795	−1.272	0	ax.	0.795 + 0.433d	−1.272 − 0.123d	−0.893d	1.271	−1.407	−0.982
				eq.	0.795 − 0.614d	−1.272 − 0.777d	0.141d	0.120	−2.126	0.156

would outweigh all the others, and that an estimate of conformational energy might be thus made rather simply. Unfortunately, this has not generally proved to be the case, although some success has been attained in special instances.

Let us return now to chlorocyclohexane. Examination of models shows that with such a pair of conformers most of the interactions are identical, or at least very similar in each conformation, and may therefore be disregarded. Consequently, instead of comparing 153 interactions for each conformation, we may try to consider only 8, which makes the problem arithmetically reasonable. Moreover, because of the approximate nature of the van der Waals potential curves and force constants used, the energy differences obtained by such a simplified treatment are probably comparable in accuracy to those that would be obtainable by a more elaborate calculation at the present time anyway.

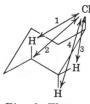

Fig. 3. The van der Waals interactions considered for the axial form in the conformational calculation.

The interactions considered for the axial form are shown in Fig. 3. It is convenient for calculations of this type to have available the relative coordinates of all the atoms of a cyclohexane ring.† Such coordinates are available from the literature for the molecule with tetrahedral bond angles.[97] It is now known that the C—C—C angles are somewhat larger than tetrahedral, and the H—C—H angles are smaller.[98] A set of improved coordinates is therefore furnished in Table 4, both for cyclohexane and for cyclohexanone with any substituents X,‡ the axes being arranged as in Fig. 4.

Bond Angles		Bond Lengths	
C—C—C	111.6°	C—C($sp^3 - sp^3$)	1.53§
H—C—H	107.3°	C—C($sp^3 - sp^2$)	1.50
H—C—C	109.5°	C=O	1.22
C—C—O	122.0°	C—H	1.10
		C—X	d

From the data in Table 4, the interaction (1) involves a hydrogen and a chlorine at a distance of 2.80 Å; therefore $\alpha = (2.80/3.00) = 0.934$.

[97] E. J. Corey and R. A. Sneen, *J. Am. Chem. Soc.*, **77**, 2505 (1955).

[98] M. Davis and O. Hassel, *Acta Chem. Scand.*, **17**, 1181 (1963).

† The coordinates can be obtained from measurements on Dreiding models, but the error of measurement is usually large enough to introduce still further uncertainty into E_V.

‡ We are indebted to Dr. J. Tai for the calculation of these coordinates.

§ This value was used for cyclohexanone. The value 1.528 was used for cyclohexane.

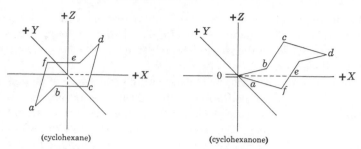

(cyclohexane)   (cyclohexanone)

Fig. 4. The coordinate systems used for cyclohexane and cyclohexanone.

Referring to equation (5), it is seen that $E_v/\epsilon = -0.85$, and, since from Table 3 $\epsilon = 0.115$, there is an attraction between these atoms of 0.10 kcal./mole. Similarly, the interaction (2) between carbon and chlorine gives an attraction of 0.06 kcal./mole, while the interactions (3) and (4) give attractions of 0.11 and 0.07 kcal./mole, respectively. The total of these van der Waals interactions (each of which occurs twice) gives a stabilization of 0.34 kcal./mole. The calculation is then carried out for the equatorial chlorine, considering the same pairs of atoms, and the stabilization calculated is 0.40 kcal./mole. Thus by difference one predicts that the equatorial halogen will be more stable than the axial by 0.12 kcal./mole, which may be compared with the experimental values of 0.3–0.5 kcal./mole (Table 1). It may be noted that *repulsions* between the *syn*-axial hydrogens and the chlorine do not cause the instability of the axial halogen, as is often assumed. Rather, there are actually attractions between the axial chlorine and the other atoms, but the sum of the attractions is greater when the chlorine is equatorial.

A similar calculation (with normal bond angles and lengths) gave a value of 0.96 kcal./mole for the conformational energy of iodine. In this case there are small repulsions between the iodine and the β-methylenes, but the attractions still dominate the situation. This calculated value is definitely too high. Because the C—I bond is quite long, a small bending of the C—C—I angle can move the iodine quite a distance from the β-methylenes. Consequently the actual molecule can lower its van der Waals repulsion energy considerably by expending a small amount of energy in bond bending, and a few degrees of bending would bring the calculated value down to the experimental range.†

† If the axial iodine were to move toward the equatorial hydrogen attached to the same carbon by bending the C—I bond 5°, E_θ would be 0.30 kcal./mole, E_v would be reduced by 0.83 kcal./mole, and the conformational energy would be 0.43 kcal./mole, in good agreement with the experimental value (0.2–0.4 kcal./mole, Table 1).

It appears to be established both from electron diffraction measurements (Sec. 3-2b) and by nuclear magnetic resonance spectra (Sec. 3-4d) that the equatorial position for the chlorine in cyclohexyl chloride is more favorable by about 0.4 kcal./mole. It seems equally well established from the electron diffraction studies on propyl chloride[99] that the *gauche* form is of lower enthalpy by some 0.3 kcal./mole, and this conclusion is supported by studies on *n*-butyl chloride[100] (p. 17). The reason for the apparent discrepancy is not clear, but a likely possibility is that, while the axial chlorine of cyclohexyl chloride is unfavorably constrained, a moderate torsional deformation about the C_{1-2} bond in propyl chloride might be more than compensated for by a favorable change in the van der Waals interactions of the system. The torsional angles in cases of the latter type seem indeed to be larger than 60°, but the experimental errors are too large to allow definite conclusions. The all-*anti* form of *n*-butyl chloride has an enthalpy lower by 0.6 kcal./mole than the form which has a *gauche* arrangement at the methyl end.[100] This number may be smaller than one-half the conformational energy of the methyl in cyclohexane, but again the situation is obscured by the experimental error (p. 43).

The results of this type of calculation for methylcyclohexane were mentioned on p. 453. We might next inquire as to the modifications necessary to carry out a similar calculation for ethylcyclohexane. The molecule being staggered and undeformed to a first approximation, the E_T, E_θ, and E_F terms may all be taken as zero. There are a number of conformations which have to be considered for each of the axial and equatorial isomers (previously discussed on p. 61), which have different enthalpies and hence different mole fractions. The entropies of mixing of these conformations must also be taken into account (p. 24). The conformational free energy of the ethyl group is thus calculated to be only 0.06 kcal./mole greater than that of the methyl group, and this is in good agreement with experiment.[101]

There are two general conclusions that can be drawn from these *a priori* calculations regarding conformational energies of cyclohexane derivatives. The first is that a single atom attached to a cyclohexane ring has a rather small conformational energy regardless of the size of the atom (experimentally it is less than 1 kcal./mole in all known cases). The main reason for this situation is that, as the atoms considered become larger, their bond lengths become longer. Thus, although iodine is much larger than chlorine, and might therefore be expected to interfere more

[99] Y. Morino and K. Kuchitsu, *J. Chem. Phys.*, **28**, 175 (1958).
[100] T. Ukaji and R. A. Bonham, *J. Am. Chem. Soc.*, **84**, 3631 (1962).
[101] N. L. Allinger and S. E. Hu, *J. Am. Chem. Soc.*, **84**, 370 (1962).

seriously with the *syn*-axial hydrogens, its center is in turn much farther from these hydrogens; hence the interference is not so bad as might have been expected. In addition, the longer bond length allows a greater motion of the atom for a given angular distortion, and the result is that all the halogens have small and similar conformational energies (Table 1).

The second conclusion is that the presence of substituents on the atom attached to the ring does not cause any large change in the conformational free energy of the group unless the number and arrangement of the groups is such that they cannot avoid severe interaction with the ring. Thus, methyl, ethyl, and isopropyl groups all have conformational energies in the range 1.8–2.1 kcal./mole (Table 1), because, even with axial isopropyl, no severe interaction need exist (Fig. 5). The group loses some of its freedom of rotation, but this has only a very minor effect on the free energy, and there may be a small bond angle change. The *t*-butyl group, on the other hand, has no comfortable arrangement if it is axial, and its conformational energy is quite high. Similarly it seems quite certain that the methylamino or dimethylamino group will have essentially the same conformational energy as amino; —S—CH$_3$ will be the same as —S—H, or —S—CH$_2\phi$, etc., as long as solvation effects do not play an important role. Some of the predictions are, in fact, confirmed by the data shown in Table 1 (see also p. 44).

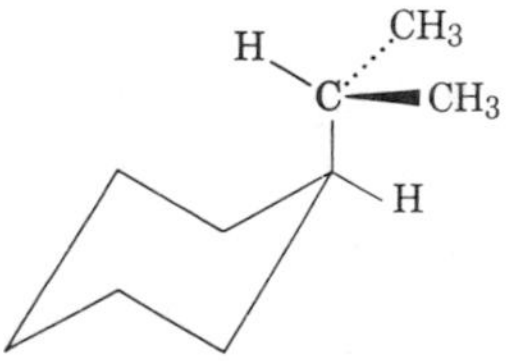

Fig. 5. The favorable conformation of axial isopropyl.

Calculations of the preferred geometries of the medium and large rings have met with some success. In the seven- and eight-membered rings the torsional angles—the E_T term in equation (1)—tend to be of dominant importance. The earliest calculations (without the aid of machine computation and hence rather simplified) predicted a skewed chair conformation for cycloheptane and a twisted crown (D_2) for cyclo-öctane and showed that the known heats of combustion could be adequately accounted for by such conformations.[102] More recently these molecules have been considered in greater detail.[88,103] When the cycloheptane molecule was treated about as rigorously as the classical method will allow (with machine computation),[88] the skewed chair was still found to be favored, and there seems to be no reason to doubt that this is the actual conformation.

The more recent calculations on cycloöctane[103] have not been quite so complete as those on cycloheptane, as the additional freedom in the

[102] N. L. Allinger, *J. Am. Chem. Soc.*, **81**, 5727 (1959).
[103] J. B. Hendrickson, *J. Am. Chem. Soc.*, **86**, 4854 (1964).

C_s $\qquad\qquad\qquad\qquad\qquad$ D_2

Fig. 6. The C_s and D_2 forms of cycloöctane.

larger molecule complicates the calculations very considerably. These calculations suggest that the twisted crown (D_2) and a boat-like form with a center of symmetry (C_s) (Fig. 6) are about equal in energy. The experimental evidence is not clear-cut, but it has all been previously interpreted in terms of basically a crown form of some sort, and it is consistent with such an interpretation (Sec. 4-4b).

Cyclononane was calculated to have two conformations which are more stable than the other possibilities, but similar in energy to one another.[103] The conformation observed (Sec. 4-4c) for the amine hydrochloride in the crystal is actually the one calculated to be the second most stable. The calculated difference in energy between these two forms is small enough, however, that the presence of substituents, or the lattice energy of the crystal, could well invert their stability order.

Cyclodecane is geometrically spectacular (Sec. 4-4c). The bond angles are very much distorted. If the molecule were a symmetrical crown with normal bond angles, the excess heat of combustion would be 27.5 kcal./mole because of the unfavorable torsion.[102] The simplified calculations, from equation (1) setting $E_F = 0$, indicate that, by adopting the conformation which it does, E_T is reduced to 3.9 kcal./mole, at the expense of adding 5.2 kcal./mole for E_θ. The molecule is sufficiently flexible that E_T could be further improved without changing E_θ, but only at the expense of increasing E_V. The repulsions giving rise to the latter are already appreciable, however, so apparently the molecule cannot further improve its energy in this way. There are six pairs of hydrogens with a separation of 1.8 Å in the known cyclodecane conformation, corresponding to a total E_V of 4.7 kcal./mole. There is no simple way to allow for the *gauche* interactions, but, if they are taken to be the same as in cyclohexane, the total heat of combustion of cyclodecane can be calculated to be 1588.3 kcal./mole, in good agreement with the experimental value of 1586.5 kcal./mole. It is easy enough to calculate the (approximate) heat of combustion once it is known what the conformation is, but it did not prove possible to calculate the correct conformation in advance. Subsequently, in his very elegant study, Hendrickson[103] calculated the energies of all the likely conformations of cyclodecane, and found that the observed conformation is in fact very much more stable than any other of the possibilities and is in agreement with the observed heat of combustion.

The cyclododecane molecule turned out to have a geometry quite different from any which had been suggested for it (Sec. 4-4c). From the known geometry, the heat of combustion can be calculated from equation (1), setting the terms other than E_T and E_θ equal to zero. Some of the C—C—C bond angles are considerably larger than the normal values, and the torsional angles are not quite optimal. From the experimental values for these quantities, the strain energies due to E_T and E_θ were calculated, using equations (3) and (4), and were found to be 5.3 and 2.2 kcal./mole, respectively, for a total strain energy of 7.5 kcal./mole. The heat of combustion of cyclohexane is 944.70 kcal./mole, or 157.45 kcal./methylene group. Since this molecule contains six *gauche* interactions, a completely strainless rigid ring might be expected to have a heat of combustion of 156.55 kcal./methylene group. Cyclododecane contains eight *gauche* interactions, so the heat of combustion calculated for the whole molecule is $156.55 \times 12 + 8 \times 0.9 + 7.5 = 1893.3$ kcal./mole, in good agreement with the experimental value of 1892.8 ± 1.2 kcal./mole.

In summary, although not a great deal has so far been done in the way of *a priori* calculations by the methods of classical mechanics, the methods are well understood and can be applied with only some minor problems. They appear to give at least qualitative (and sometimes better) agreement with experiment, and the approximations required thus appear to be pretty good. While the quantum mechanical approach will probably yield better results in the long run, the classical approach appears to offer more promise for the immediate future.

7–3. α-Haloketones

The α-haloketones represent a class of compounds which are amenable to study by a large variety of physical, chemical, and theoretical methods.†
Let us consider first as an example 2-bromocyclohexanone, which is a mixture of two conformational isomers, I*e* with equatorial bromine and I*a* with axial bromine (Fig. 7). The position of equilibrium between I*a* and I*e* is determined by effects which may be conveniently divided into polar and steric. The steric effects are in principle the same effects as are found in the hydrocarbons (Sec. 7-2), but the polar effects add a new feature. "Polar" effects would automatically be allowed for if the molecule were treated by quantum mechanical methods. Because of the

† For a review see C. Djerassi, *Optical Rotatory Dispersion*, McGraw-Hill Book Co., New York, 1960, p. 115.

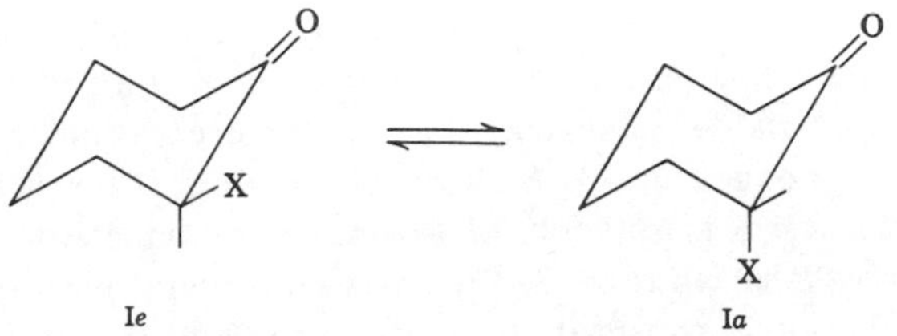

Fig. 7. The conformational equilibrium in 2-halocyclohexanone.

difficulty of this approach, however, only rather approximate quantum mechanical calculations on such molecules have been carried out;[104] in fact, the system has been treated in much more detail on a classical basis,[105–110] which requires separate consideration of the electrostatic interactions between the dipoles. Since the positive ends of the dipoles are located at the same distance from one another in Ie as in Ia, while the negative ends are much closer in Ie, there will be an electrostatic repulsion tending to convert Ie to Ia. This interaction is fairly strong, but it decreases as the polarity of the solvent in which the substance is dissolved increases (Sec. 3-5). Quantitatively, the classical electrostatic interaction energy of the two dipoles is given by

$$E_\mu = \frac{\mu_1\mu_2}{Dr^3}(\cos\chi - 3\cos\alpha_1\cos\alpha_2) \tag{6}$$

where μ_1 and μ_2 are the magnitudes of the dipole moments, r is the distance between them, D is the dielectric constant of the medium, and χ, α_1, and α_2 are angular terms defining their relative orientation.† This

[104] J. Tai and M. A. Miller, unpublished results.

[105] W. D. Kumler and A. C. Huitric, *J. Am. Chem. Soc.*, **78**, 3369 (1956).

[106] J. Allinger and N. L. Allinger, *Tetrahedron*, **2**, 64 (1958).

[107] N. L. Allinger and J. Allinger, *J. Am. Chem. Soc.*, **80**, 5476 (1958).

[108] N. L. Allinger, J. Allinger, and N. A. LeBel, *J. Am. Chem. Soc.*, **82**, 2926 (1960).

[109] J. M. Lehn and G. Ourisson, *Bull. Soc. Chim. France*, **1963**, 1113.

[110] N. L. Allinger, J. G. D. Carpenter and M. A. DaRooge, *J. Org. Chem.*, **30**, 1423 (1965).

[111] J. H. Jeans, *Mathematical Theory of Electricity and Magnetism*, 5th ed., Cambridge University Press, 1933, p. 377.

[112] C. P. Smyth, R. W. Dornte, and E. B. Wilson, Jr., *J. Am. Chem. Soc.*, **53**, 4242 (1931).

† The theoretical relationship (equation 6) derived by Jeans[111] was applied by Smyth and co-workers[112] to the present type of system. One of the angles used by Smyth and co-workers was in fact the complement of the angle used by Jeans. Most subsequent calculations have used the Smyth formulation, and the values calculated are unreliable. The correct formulation is given by J. M. Lehn and G. Ourisson.[109] By using the corrected equation, a repetition of the calculation in ref. 106 gives a prediction which is in almost exact agreement with experiment.

equation is not expected to hold accurately when the dipoles are as close together as they are in the α-haloketones, but it is the best approximation available. Because the dielectric constant between the dipoles enters into equation (6), E_μ becomes small if D is very large, but it is a significant term in solvents of low dielectric constant. The value of D for the solvent which is needed to solve equation (6) is unfortunately not the bulk dielectric constant, but rather the effective dielectric constant in the vicinity of the dipoles. This quantity is not easily determined by experiment, nor can it be very readily calculated.[113] Only a rough agreement between the calculated conformational equilibrium for α-bromocyclohexanone and the experimental value is therefore to be expected, and such is found.[105-110]

Analogous calculations have also been made for α-chloro-[114] and α-fluorocyclohexanone[115,116] (I, X = Cl, F). The agreement with experiment is somewhat poorer here, but it is as good as can be expected from such a classical treatment. The calculated values for the chloro, bromo, and fluoro compounds are all essentially the same, while experimentally they differ quite appreciably.

There are numerous physical methods through which equilibria of this type may be measured. The infrared method was the first to be applied. It was originally observed in the steroid field[117] that substitution of a bromine atom at the axial position adjacent to a carbonyl group left the carbonyl stretching frequency essentially unchanged, while substitution of a bromine atom at the corresponding equatorial position raised the stretching frequency by some 20 cm.$^{-1}$. This effect can be understood by recognizing that the carbonyl group has two principal resonance forms (a and b in Fig. 8). The presence of the C—Br dipole adjacent to the carbonyl tends to decrease the importance of form (b) and hence makes the C=O bond stronger (i.e., enhances its double bond character), which in turn causes an increase in the C=O stretching frequency. Thus one can usually tell the conformation of a conformationally pure isomer from the position of the carbonyl stretching frequency compared with that of the parent compound. Corey subsequently showed that the relationship found in steroids also holds in simple cyclohexane derivatives;[12,118] and

[113] J. G. Kirkwood and F. H. Westheimer, *J. Chem. Phys.*, **6**, 506, 513 (1938).

[114] N. L. Allinger, J. Allinger, L. A. Freiberg, R. F. Czaja, and N. A. LeBel, *J. Am. Chem. Soc.*, **82**, 5876 (1960).

[115] A. S. Kende, *Tetrahedron Letters*, No. 14, 13 (1959).

[116] N. L. Allinger and H. M. Blatter, *J. Org. Chem.*, **27**, 1523 (1962).

[117] R. N. Jones, D. A. Ramsay, F. Herling, and K. Dobriner, *J. Am. Chem. Soc.*, **74**, 2828 (1952).

[118] E. J. Corey and H. J. Burke, *J. Am. Chem. Soc.*, **77**, 5418 (1955).

sometimes, depending on the resolution of the two carbonyl bands, the composition of a conformational mixture could thus be estimated.[12,106-108]

The C=O stretching frequency is increased by 15 to 22 cm.$^{-1}$ by the presence of an equatorial bromine, and by −3 to +3 cm.$^{-1}$ by an axial bromine.[119] In α-bromocamphor[120] where the dihedral angle is between the extremes for equatorial and axial halogen, the shift is also intermediate (14 cm.$^{-1}$). For equatorial and axial chlorocyclohexanones, the C=O stretching frequency increases are 18–31 cm.$^{-1}$ and 2–10 cm.$^{-1}$, respectively.[119,121] The corresponding fluoro compounds[116] show shifts of about 27 and 18 cm.$^{-1}$, while the limited data available suggest shifts of about 8 and −4 cm.$^{-1}$ for the iodo derivatives.[119]

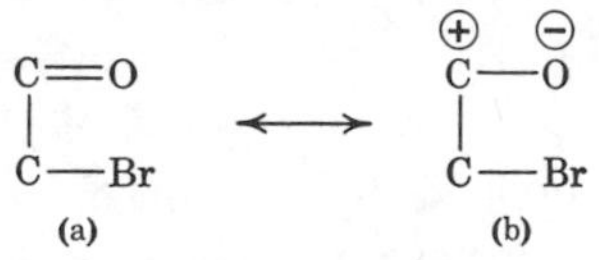

$$\underset{\text{(a)}}{\begin{array}{c} \text{C} = \text{O} \\ | \\ \text{C} - \text{Br} \end{array}} \quad \longleftrightarrow \quad \underset{\text{(b)}}{\begin{array}{c} \overset{\oplus}{\text{C}} - \overset{\ominus}{\text{O}} \\ | \\ \text{C} - \text{Br} \end{array}}$$

Fig. 8. The principal resonance forms of an α-bromoketone.

It is not necessary to examine the carbonyl band specifically to determine the conformation of the halogen, and with 2-chlorocyclohexanone the C—Cl stretching frequencies actually proved to be more useful.[114,122] This substance shows two strong bands in the region where the C—Cl stretching frequency is expected. To use these bands for conformational analysis it was necessary to establish which of them was due to the axial halogen, and which was due to the equatorial. The assignment of the higher frequency was made to the latter (Sec. 3-4a) and was confirmed by the fact that the high frequency band increased in intensity relative to the low frequency band as the solvent polarity was increased. From a study of the temperature dependence of the band ratio, the enthalpy difference between conformations was found (equation 5, p. 147) to be 1.1 kcal./mole in the pure liquid,[114,122] and 0.8 kcal./mole in carbon disulfide,[122] the equatorial conformer being the more stable in each case. These values may be compared with values of 0.4–0.6 kcal./mole found in a relatively polar solvent (dioxane), and values of −0.7 kcal. to −0.2 kcal./mole found in non-polar solvents from dipole moments, ultraviolet spectra, and by direct equilibration[105,114] (all of these methods are discussed below). The energy difference found for the conformers of the pure liquid (which constitutes its own, highly polar solvent) is in agreement with that

[119] For a summary of available data see E. G. Cummins and J. E. Page, *J. Chem. Soc.*, **1957**, 3847.

[120] F. V. Brutcher, Jr., T. Roberts, S. J. Barr, and N. Pearson, *J. Am. Chem. Soc.*, **81**, 4915 (1959).

[121] An earlier suggestion (M. L. Josien and C. Castinel, *Bull. Soc. Chim. France*, **1958**, 801) that the spectral shifts are in the opposite direction in the simple systems from those in the steroids has been shown to be incorrect (ref. 114).

[122] (a) K. Kozima and E. Hirano, *J. Am. Chem. Soc.*, **83**, 4300 (1961); (b) K. Kozima and Y. Yamanouchi, *J. Am. Chem. Soc.*, **81**, 4159 (1959).

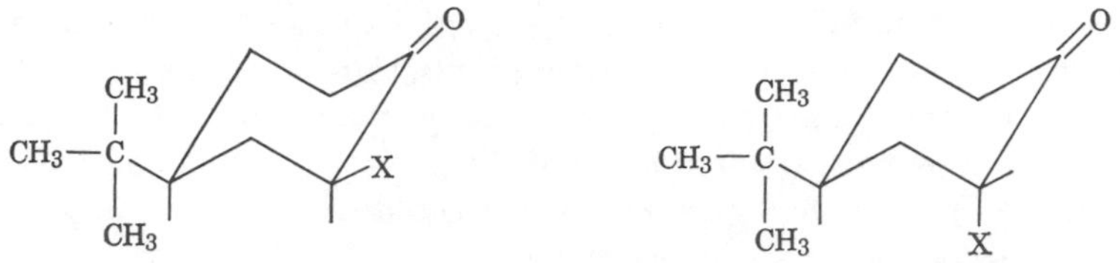

Fig. 9. The epimers of the 2-halo-4-*t*-butylcyclohexanones.

obtained by the other methods. The value favoring the equatorial conformer in carbon disulfide is qualitatively opposite that found by all other methods and is probably in error. The reason for the discrepancy is not clear, but it may be due to the fact that the compound is still associated at quite low concentrations, as shown by cryoscopic measurements.

The dipole moment method has been most useful for determining the position of the conformational equilibria of α-haloketones. The axial and equatorial conformers (I*a* and I*e*, Fig. 7) have the carbonyl and halogen dipoles oriented quite differently, and hence they are expected to exhibit quite different resultant moments. From the known cyclohexanone coordinates (relative positions of all the atoms)† and the moments of the individual dipoles, the resultant moment of each conformational isomer was calculated. The magnitude of the equatorial C—X moment in 2-halocyclohexanone is smaller than that in the corresponding halocyclohexane, because the strong C=O dipole causes an induced opposing moment in the polarizable C—X bond. The magnitude of this induced moment is about 0.3 D in an equatorial C—Br bond; thus, compared to a C—X moment of 1.5 D, it is not negligible and must be taken into account, at least approximately.[12,105,106] Similarly, the C—X moment induces a moment in the C—O bond when X is equatorial. The magnitudes of these induced moments are expected to be much less if the halogen is axial.

An independent and better method for determining the dipole moments of the conformers made use of model compounds. Winstein and Holness introduced the use of the *t*-butyl group as an anchor for holding conformation.[24] Utilizing this idea, one can study separately the isolable geometric isomers of II (Fig. 9) and assign certain of the observable properties of these isomers to the conformers of I.

† The calculations[12,106] were actually done at a time when the only available coordinates for the molecule were somewhat less accurate than those given in Table 4, but the use of better coordinates would lead to only small changes in the calculated values.

There is one possible source of error in taking the dipole moments or other properties of the corresponding isomers of I and II as synonomous; it is that a conformational distortion may occur in the vicinity of the t-butyl group (Sec. 2-7), and this distortion may be transmitted to the other end of the molecule, which may be distorted in turn (evidence for such a conformational transmission in steroids[123] was discussed in Sec. 5-9). The assumption is ordinarily made that as a first approximation the distortion introduced by non-polar substituents which involve no

Table 5. Percent of Axial Conformer in the 2-Halocyclohex-anoncs from Dipole Moment Measurements

| X | Solvent | | | Ref. |
	Heptane, %	Benzene, %	Dioxane, %	
F	48	23	15	114, 115
Cl	76	56	37	113
Br	85	76	62	12, 105, 106
Calculated[a]				
Br	84	78	79[b]	

[a] From equation (6) using the correct formulation (ref. 109), placing the dipoles midway between the bound atoms, using the bulk dielectric constants for D, and otherwise following ref. 106. The values calculated for chlorine and fluorine are essentially the same as for bromine.

[b] Dioxane is more polar than its bulk dielectric constant would imply; see, for example, J. B. Hyne, J. Am. Chem. Soc., **85**, 304 (1963).

serious steric interactions can be neglected. Since the values of the dipole moments of the two conformations of the α-haloketones which are calculated from group moments are almost the same as those found for the 4-t-butyl-derivatives (IIe and IIa with X = Cl,[114] for example, have dipole moments of 4.29 and 3.17 D, respectively, while the values calculated for Ie and Ia from bond moments are 4.28 and 3.07 D), it appears that the moments are correctly determined either way. From the moments of the two conformers of II, the composition of the equilibrium mixture of I was determined (equation 12 on p. 159) in different solvents. The equilibrium compositions of the halocyclohexanones (I) thus found are given in Table 5. There is substantial variation with solvent, as expected. The position of equilibrium also depends to a marked extent on which halogen is present.

[123] D. H. R. Barton, F. McCapra, P. J. May, and F. Thudium, J. Chem. Soc., **1960**, 1297.

In principle, the variation of the dipole moment with temperature could be used to determine the enthalpy difference between the conformers (equation 5 on p. 147), but the moment of α-chlorocyclohexanone was found not to vary significantly over the conveniently accessible temperature range.†

The equilibration method (Sec. 2-4) has also been applied to a study of the conformational equilibria I$e \rightleftharpoons$ Ia indirectly by studying the epimerization II$e \rightleftharpoons$ IIa in the presence of the corresponding halogen acid. The equilibrated epimeric mixture of IIa and IIe was then analyzed using the fingerprint region in the infrared. The equilibrium constants found varied with changing solvent,[107,114] for X = Br, the percentage of IIa was 63 in dioxane and 78 in carbon tetrachloride. These values were in satisfactory agreement with those determined by the dipole moment measurements, which were 51 % in dioxane and 68 % in heptane.

Ultraviolet spectra have been used widely in determining the axial or equatorial nature of a halogen adjacent to a carbonyl (Sec. 3-4c). If the halogen is equatorial it exerts a small or negligible effect, but if it is axial the $n \rightarrow \pi^*$ band is shifted to longer wavelength by about 20–30 mμ. Equatorial halogen or axial fluorine exerts no effect on the extinction coefficient of the band, but for the other halogens in the axial position, ϵ is increased three- to ten-fold. Thus, if a conformational mixture is present, since the bands cannot be resolved the minor one is submerged in the larger one, and even a comparatively small amount of the axial conformation (except for fluorine) yields a wavelength shift similar to that found for the pure axial isomer, but with a reduced value for ϵ.[107]

The rotatory dispersion curve of a carbonyl compound is closely related to its ultraviolet spectrum, and the point at which the Cotton effect curve crosses the axis of zero rotation is approximately the wavelength of maximum extinction in the ultraviolet. The axial haloketones therefore have their Cotton effect curves shifted to longer wavelength relative to the parent ketones by 15–30 mμ. The pronounced change in the amplitude of the Cotton effect curve that results from the introduction of a polarizable halogen (not fluorine) into the axial position adjacent to a carbonyl has been discussed earlier (Sec. 3-6).

The applications of several additional available physical methods to α-haloketones were discussed in Chap. 3: nuclear magnetic resonance in Sec. 3-4d, rotatory dispersion in Sec. 3-6, and polarography in Sec. 3-9; and further discussion of some applications of these methods will be given in the next section.

---

[124] P. Mauret and J. Petrissans, *Compt. Rend.*, **254**, 3662 (1962).

† An earlier report to the contrary [S. Yaroslavsky and E. D. Bergmann, *J. Chem. Phys.*, **33**, 635 (1960)] appears to be based on erroneous data (see ref. 124).

A few additional simple bromocyclohexanones have also been studied. 2,6-Dibromocyclohexanone was equilibrated in acetic acid in the presence of hydrobromic acid and yielded a mixture of *cis* and *trans* isomers in which the latter predominated to the extent of about 85%.[125] The *cis* isomer is itself a mixture of diaxial and diequatorial conformers in which the former was reported to be present to the extent of at most 2%. It would appear that the major source of instability in the diaxial form is the steric repulsion between the bromines.† If the bond angles and bond lengths in the molecule remained normal, the distance between the bromine atoms (from the coordinates in Table 4) would be 3.02 Å, and their van der Waals repulsion would be 4.8 kcal./mole (Sec. 7-2). Bond angle bending will lower this energy somewhat, but the interaction will clearly be appreciable.

Since the methyl group of a 2-methylcyclohexanone shows a considerable tendency to remain equatorial (about 1.6 kcal./mole)[126,127] while the bromine atom of a 2-bromocyclohexanone shows a small tendency to be axial,[106–108] it was predicted and found[128] that 2-bromo-2-methylcyclohexanone existed very largely in conformation I, (Fig. 10) rather than II. Similarly, the molecules III to VII (Fig. 10) appear to exist mainly in the conformations indicated, rather than in the alternatives available to them.[128,129]

The conformational equilibrium constants were determined for V and VI by the dipole moment method;[129] they gave an energy for the *syn*-axial methyl-bromine interaction of 2.2 kcal./mole in these ketones. Similar studies with VII showed that the corresponding methyl-methyl interaction energy was larger than this.[129]

An unusual conformational situation was observed by Sandris and Ourisson,[130] and a quantitative study has been reported by Waegell and Ourisson.[131] They have shown that, although *trans*-2,6-dibromocyclohexanone predominates over the *cis* form at equilibrium, the corresponding 3,3,5,5-tetramethylated compound (VIII) exists preferentially in the *cis* form with diequatorial bromines; and the difference was attributed to the diaxial methyls pushing one another apart with the simultaneous

[125] E. J. Corey, *J. Am. Chem. Soc.*, **75**, 3297 (1953).

[126] N. L. Allinger and H. M. Blatter, *J. Am. Chem. Soc.*, **83**, 994 (1961).

[127] B. Rickborn, *J. Am. Chem. Soc.*, **84**, 2414 (1962).

[128] E. J. Corey, T. H. Topie, and W. A. Wozniak, *J. Am. Chem. Soc.*, **77**, 5415 (1955).

[129] G. L. Wang and H. F. Chow, unpublished results.

[130] C. Sandris and G. Ourisson, *Bull. Soc. Chim. France*, **1958**, 1524.

[131] B. Waegell and G. Ourisson, *Bull. Soc. Chim. France*, **1963**, 495, 496.

† There is also an electrostatic repulsion between the *syn*-axial bromine atoms, but it is smaller than the electrostatic repulsions of the diequatorial conformer.

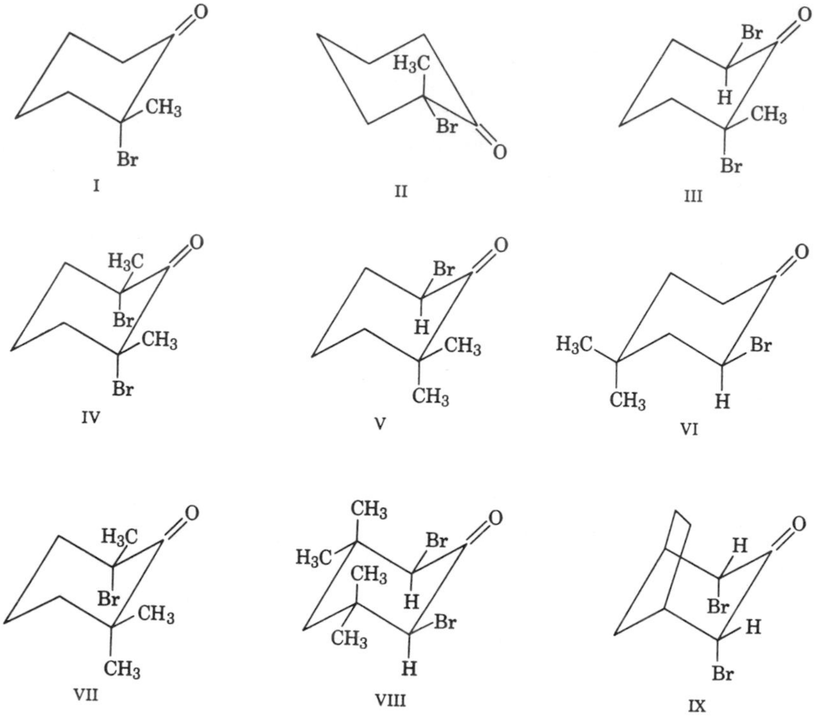

Fig. 10. Conformations of some cyclohexanones.

squeezing together of the 2 and 6 diaxial substituents (reflex effect, Sec. 2-7). While a bromine on an undeformed cyclohexane or cyclohexanone ring is sterically rather comfortable in the axial position, it would suffer serious van der Waals interactions with the ring if twisted inward by this reflex effect, and hence in VIII both bromines prefer to be equatorial. This case may be contrasted with that of IX, where, instead of methyls pushing one another apart, the axial carbon atoms are bonded together and they twist the adjacent ring carbons in the reverse direction.[132] Here the diaxial conformation was the product of the bromination, and it is probably the thermodynamically stable isomer. If so, this would represent an example of the "reverse reflex effect" (p. 247).

As outlined above, the principles affecting the stability of a given α-halocyclohexanone are pretty well understood from both experimental and theoretical points of view, as long as the ring is in a chair from. The thermodynamic product or products resulting from the bromination of

[132] C. W. Jefford and B. Waegell, *Tetrahedron Letters*, **1963**, 1981.

such a cyclohexanone can now be predicted quite successfully. The kinetic product of bromination and the properties of cyclohexanones in which the rings are not in chair forms are distinct problems which will be discussed in the following section.

Acyclic α-haloketones,[132a] α-halocyclobutanones,[132b] -pentanones,[105,132c] the medium-ring α-haloketones,[132d] as well as α-haloesters,[132a,132e] and related compounds[132a,133] have been studied in considerable detail by a number of physical methods, mainly spectroscopic. The principles involved are the same as for the α-halocyclohexanones discussed earlier, but with the other rings the conformational situation is more complicated and quantitative conclusions are more difficult to come by.

7-4. Boat Forms†

Sachse first postulated the possibility of the existence of the flexible or boat form of cyclohexane in 1890.[136] Fifty-five years later, however, Pitzer[137] was able to predict that such a form was not to be found in any appreciable concentration in simple cyclohexane derivatives. This prediction was based on a consideration of the existence of two eclipsed ethane linkages in the boat form, leading to an estimate of the enthalpy of this form as 5.6 kcal./mole above that of the stable chair. The thermodynamic properties of cyclohexane at elevated temperatures proved to be consistent with this estimate.[138] Hazebroek and Oosterhoff[139] studied the boat form by the methods of statistical mechanics and emphasized the

[132a] L. J. Bellamy and R. L. Williams, *J. Chem. Soc.*, **1957**, 4294.

[132b] J.-M. Conia and J.-L. Ripoll, *Bull. Soc. Chim. France*, **1963**, 768.

[132c] F. V. Brutcher, Jr., T. Roberts, S. J. Barr, and N. Pearson, *J. Am. Chem. Soc.*, **81**, 4915 (1959); J. Fishman and T. Nambara, *Chem. Ind. (London)*, **1961**, 79; J. Fishman, *J. Org. Chem.*, **27**, 1745 (1962).

[132d] N. J. Leonard and F. H. Owens, *J. Am. Chem. Soc.*, **80**, 6039 (1958).

[132e] H. O. House and J. W. Blaker, *J. Am. Chem. Soc.*, **80**, 6389 (1958).

[133] L. J. Bellamy and R. L. Williams, *J. Chem. Soc.*, **1958**, 3465.

[134] J. Levisalles *Bull. Soc. Chim. France*, **1960**, 551; M. Balasubramanian, *Chem. Rev.*, **62**, 591 (1962).

[135] D. H. R. Barton and G. A. Morrison, *Progress Chem. Org. Natural Products*, **19**, 165 (1961).

[136] H. Sachse, *Ber.*, **23**, 1363 (1890); *Z. Physik. Chem. (Leipzig)*, **10**, 203 (1892).

[137] K. S. Pitzer, *Science*, **101**, 672 (1945).

[138] C. W. Beckett, K. S. Pitzer, and R. Spitzer, *J. Am. Chem. Soc.*, **69**, 2488 (1947).

[139] P. Hazebroek and L. J. Oosterhoff, *Discussions Faraday Soc.*, **10**, 87 (1951).

† For reviews see refs. 134 and 135. An introduction to the subject of boat forms is given in Sec. 2-1.

fact that, while the chair is rigid and incapable of any relative motion of the ring carbons without distortion of bond angles, the boat form is really a flexible form in which a considerable amount of such motion is possible. This pseudorotation leads from a classical boat through a stretched or twist form (cf. p. 38) back to another classical boat where different carbons are now the prow and the stern. This motion may be continued, and each carbon will in turn become the prow of the boat. No bond angle distortion is required for this pseudorotation to occur. The variation in energy of the boat form as it undergoes this motion has been calculated[88,139] to be about 1.6 kcal./mole. Thus the boat form is really quite flexible, and the word "boat" is used here and in general to mean the physical arrangement as it actually exists.† Because of the flexibility and the relatively low symmetry of the boat form, it has a somewhat higher entropy than does the chair form, and the difference (which is a function of temperature) is calculated statistically to be about 2 e.u. While the amount of boat form in equilibrium with the chair is small at room temperature because of the large enthalpy difference, it increases with increasing temperature (as $T \Delta S$ becomes larger) and it should amount to about 8 % of the equilibrium mixture at 600°K.[139]

The energy differences between the boat and chair forms of various specific cyclohexane derivatives have been measured in two independent ways, by heats of combustion and by equilibration methods, and found to amount to 4–6 kcal./mole. The experimental values indicate that the original value of 5.6 kcal./mole[137] is still quite acceptable for the conformational enthalpy of the boat form of cyclohexane.

The enthalpy of the boat form of a cyclohexane ring (specifically excluding unsaturated rings) is so much greater than that of the chair that molecules which are free to assume either form have been demonstrated to exist in the boat form in only a few rare instances.‡ trans-1,3-Di-t-butylcyclohexane was discussed in Sec. 2-1, and related compounds behave similarly.[140] The only other type of saturated ring for which a boat form has been convincingly demonstrated at this writing is one in which there are either two or three syn-axial interactions in the chair form which are relieved in the boat form, and in addition a hydrogen bond in the boat which is not present in the chair. An example is the boat

[140] R. D. Stolow, J. Am. Chem. Soc., 83, 2592 (1961); R. D. Stolow and M. M. Bonaventura, 85, 3636 (1963).

† This is the original meaning[138,139] attached to the word "boat." The terms classical boat and stretched form will be used to refer to more specific geometries.

‡ Simple molecules rigidly held in the classical boat form (bicyclo[2.2.2]octane), and stretched or twisted boat [twistane: H. W. Whitlock, Jr., J. Am. Chem. Soc., 84, 3412 (1962)] as well as the chair form (ademantane), are, of course, well known.

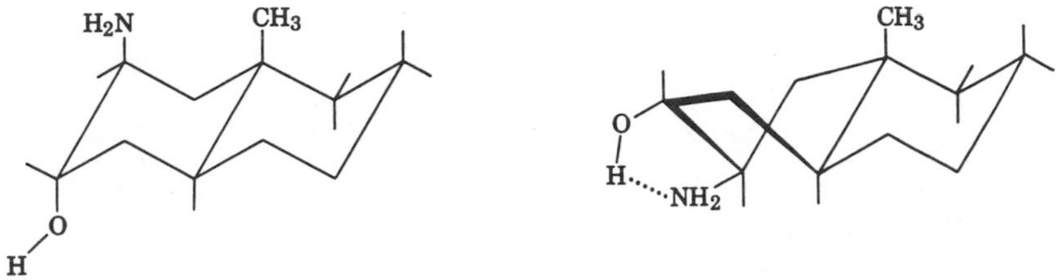

Fig. 11. Conformations of ring A in 2β-amino-5α-cholestan-3α-ol.

form of ring A in 2β-aminocholestan-3α-ol.[141]† As can be seen in Fig. 11, there is a syn-axial repulsion between the 19 methyl and the amino group in the chair form, while a hydrogen bond is possible only in the boat form. The infrared spectrum (Sec. 3-4a) shows the presence of a significant amount of intramolecularly hydrogen-bonded hydroxyl,[142] which requires a boat form for ring A. This form is present in the equilibrium mixture to the extent of perhaps 10%. The effects which combine to stabilize the boat include the repulsion between the 2β-amino group and the 19-methyl substituent, the energy of the hydrogen bond§ formed in the boat, and the fact that the bending of the 19-methyl group away from the amino group is opposed by the presence of the B and C rings.

A similar case is that of the 2β-bromo-3α-hydroxy-4,4-dimethyl structure (Fig. 12).[144] The weakness of the hydrogen bond is more than compensated by the severe syn-triaxial interaction, and such compounds exist with ring A largely or exclusively in the boat form.

In sharp contrast to the cyclohexane ring, which is to be found in a boat form only under very special circumstances, stands the cyclohexanone ring, a wide variety of derivatives of which have been shown

[141] M. Svoboda, M. Tichý, J. Fajkoš, and J. Sicher, *Tetrahedron Letters*, **1962**, 717.

[142] For a review see M. Tichý, *Chem. Listy*, **54**, 506 (1960).

[143] H. H. Freedman, *J. Am. Chem. Soc.*, **83**, 2900 (1961).

[144] (a) J. Klinot and A. Vystrčil, *Chem. Ind.* (*London*), **1963**, 738; (b) J. Levisalles, private communication.

† This work[141] leads to a number of interesting conclusions other than those stated by the authors. Since the necessary numerical data are not included in the paper, numerical values for the quantities below cannot be given here, but the qualitative conclusions are: (a) the amino group and the dimethylamino group have nearly the same conformational energy, and (b) the conformational energy of the amino group is only slightly larger than that of the hydroxyl group (in tetrachloroethylene solvent). Cf. the data on p. 439 and the *a priori* conclusions on p. 458.

§ The hydrogen bond from a hydroxyl to an amino nitrogen appears to be quite strong.[143] No numerical values for this bond energy are known to the authors, but from the data given by J. Pitha, J. Sicher, F. Šipoš, M. Tichý, and S. Vašíčková, *Proc. Chem. Soc.*, **1963**, 301, one can calculate that the enthalpy of hydrogen bond formation between diequatorial amino and hydroxyl is −2.9 kcal./mole.

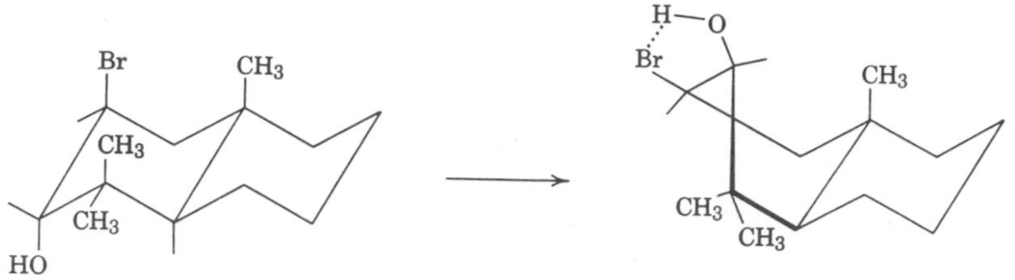

Fig. 12. Conformations of ring A in 2β-bromo-3α-hydroxy-19β,28-epoxy-18α-oleanane.

to exist in the boat form. It was predicted some time ago that the energy of the boat form of cyclohexanone would be considerably lower than that of cyclohexane,[102] and, although an experimental value for this energy difference has not been reported at this writing, there are many reasons for believing the original prediction was correct. To facilitate discussion of the boat form of cyclohexanone it is convenient to introduce a pseudo-rotational parameter θ.† If the cyclohexanone is imagined in the classical boat form with the carbonyl at the prow, the value of θ is taken as zero (Fig. 13). If the ring is allowed to pseudorotate until the carbon adjacent to the carbonyl becomes the prow, θ has a value of 60° and θ continues to increase as the pseudorotational motion continues. When $\theta = 180°$, the carbonyl has moved to the stern of the boat, and at $\theta = 360°$ it has returned to its initial position.

Returning to the energy of the cyclohexanone boat, if $\theta = 0°$, the repulsion which in cyclohexane exists between the prow and the stern hydrogens is eliminated. This geometry for cyclohexanone still leaves the molecule with two eclipsed ethane linkages, and therefore it should have an energy of about 5.6 kcal./mole above that of the chair. At $\theta = 60°$, on the other hand, there is only one eclipsed ethane linkage, the others all being staggered. The repulsions between the prow and stern methylenes are small, and, if θ is increased or decreased by a few degrees, they become negligible. To calculate the enthalpy of the $\theta = 60°$ form, we must also note that the carbonyl group, which in the chair form is nearly eclipsed by the two α equatorial hydrogens, is now perfectly staggered with respect to the hydrogens on one side, but approximately eclipsed on the other side. The data needed for the calculation are the barrier height for acetone, and its torsional arrangement at the minimum of energy. The barrier height is 0.8 kcal./mole (p. 140), but the conformation of minimum energy is unknown. One might suppose that it would be similar to that in acetaldehyde and analogous compounds, where the barriers are of the

† An exact definition of θ is quite complicated (ref. 139), and only a qualitative use of the quantity will be made here.

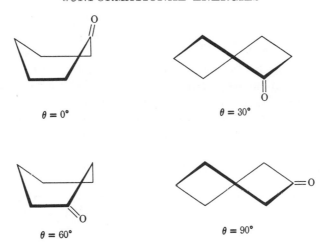

$\theta = 0°$

$\theta = 30°$

$\theta = 60°$

$\theta = 90°$

Fig. 13. The pseudorotational arrangements of the boat conformation of cyclo-hexanone.

order of 1 kcal./mole (p. 140), and the conformation of minimum energy is that in which the oxygen is eclipsed by a methyl hydrogen. Calorimetric data,[145] on the other hand, indicate that the cyclohexane → cyclohexanone change is less favorable than the propane → acetone change by 2.0 kcal./mole, suggesting that the eclipsing of the carbonyl by an α-hydrogen is unfavorable by 1.0 kcal./mole. The apparently conflicting data have been discussed at length,[102,146–150] but the preferred torsional arrangement about the bond in question remains uncertain. All the evidence is consistent with the barrier being small, however, no more than about 1 kcal./mole. Thus it seems safe to say that the staggered arrangement of the methyl group of a ketone with respect to the carbonyl is better than the eclipsed arrangement by 0 ± 1 kcal./mole. The conformational energy of the $\theta = 60°$ form thus becomes 2.8 ± 1 kcal./mole. The torsional energy of the $\theta = 90°$ form is about the same as for $\theta = 60°$, and hence the boat form pseudorotates in an energy well from $\theta = 60°$ to $\theta = 120°$ with almost no energy change. If there are constraints on the cyclo-hexanone ring in the chair form, which can be relieved in the boat form,

[145] Unpublished data of E. J. Prosen, quoted in ref. 146.

[146] H. C. Brown, J. H. Brewster, and H. Shechter, *J. Am. Chem. Soc.*, **76**, 467 (1954).

[147] W. D. Cotterill and M. J. T. Robinson, *Tetrahedron Letters*, **1963**, 1833.

[148] N. L. Allinger, J. Allinger, and M. A. DaRooge, *J. Am. Chem. Soc.*, **86**, 4061 (1964).

[149] J. D. Cox, *Tetrahedron*, **19**, 1175 (1963).

[150] K. S. Pitzer and W. E. Donath, *J. Am. Chem. Soc.*, **81**, 3213 (1959).

the amount of energy required is much smaller if the cyclohexanone can assume a value of θ in the range 60–120° than if it is constrained to a value near 0°.

The height of the barrier to the reaction chair → boat cyclohexane has long been of interest, since the fact that the two forms were not separately isolable was for many years taken as evidence against the Sachse theory. Mohr suggested that an equilibrium between the forms was possible,[151] and Shoppee made a rather accurate estimate of the barrier height as early as 1946.[152] Very careful calculations[88] have now predicted a ΔH of 12.7 kcal./mole for this barrier. Experimental values available from NMR[153] (Sec. 3-4d) and ultrasonic measurements (Sec. 3-10) for cyclohexane and a number of derivatives have been discussed in Sec. 2-1, and they are in satisfactory agreement with the calculated values.

Attempts have been made to measure the corresponding barriers in cyclohexanones from low temperature NMR spectra of cyclohexanone and some of its simple alkyl derivatives down to −75°.[154] Although it has not yet been possible to reach sufficiently low temperatures to freeze the conformational inversion, the implication is that the inversion barrier may be substantially lower than in cyclohexane itself. This conclusion also follows from ultrasonic measurements on cyclohexanone (Sec. 3-10).

The boat form of cyclohexane-1,4-dione (or the 1,3 isomer) should, by reasoning along the above lines, have an enthalpy relative to the chair form of 0 ± 2 kcal./mole. The dipole moment of cyclohexane-1,4-dione has long been known to be small but non-zero,[155] and this fact was originally interpreted to be indicative of an equilibrium between boat and chair forms.[155–157] If the torsional energy of the boat form is calculated for different pseudorotational arrangements,[158] it is found that a minimum of energy is obtained for the form having $\theta = 90°$ (analogous to Fig. 13) but that a fairly wide oscillatory motion is possible, during the course of which an arrangement approaching that with $\theta = 60°$ is reached without much energy increase. The dipole moment calculated for the oscillating boat agreed well with the experimental value.[158]

[151] E. Mohr, J. Prakt. Chem., [2] 98, 315 (1918).

[152] C. W. Shoppee, J. Chem. Soc., 1946, 1138.

[153] F. A. L. Anet, M. Ahmad, and L. D. Hall. Proc. Chem. Soc., 1964, 145; F. A. Bovey, F. P. Hood, E. W. Anderson, and R. L. Kornegay, ibid., 146.

[154] J. B. Stothers, personal communication.

[155] C. G. Le Fèvre and R. J. W. Le Fèvre, J. Chem. Soc., 1935, 1696; O. Hassel and E. Naeshagen, Tidsskr. Kjemi, Bergresen Met., 10, 81 (1930).

[156] C. G. Le Fèvre and R. J. W. Le Fèvre, Rev. Pure Appl. Chem., 5, 261 (1955).

[157] M. T. Rogers and J. M. Canon, J. Phys. Chem., 65, 1417 (1961).

[158] N. L. Allinger and L. A. Freiberg, J. Am. Chem. Soc., 83, 5028 (1961).

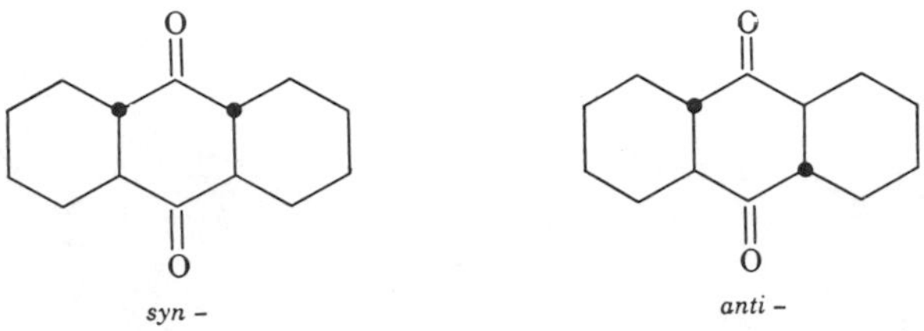

syn –   anti –

Fig. 14. The *trans-trans* isomers of 9,10-perhydroanthraquinone.

Since the infrared spectrum of cyclohexane-1,4-dione showed no bands in solution not present in a mull, an equilibrium between conformations seemed unlikely, and, since there were a number of coincidences in the Raman and infrared (p. 150), a boat conformation was assigned to the compound.[158] The boat form must have a somewhat higher entropy than would the chair form, but the fact that the molecule seems to be exclusively in this form at room temperature must be largely due to the lack of unfavorable torsional arrangements in the boat compared to those in the chair. X-ray studies[159,160] on the crystalline compound have confirmed the boat structure and indicate that in the crystal no oscillation of the boat is apparent, and that the molecules exist in a form having a value of θ intermediate between 60° and 90°, with an angle of 152–156° between the dipoles.

When these studies are extended to a larger system, it is found that there are five possible diastereomers of 9,10-perhydroanthraquinone, of which three are known.[161] The three give, respectively, *cis*, *trans*, and a mixture of *cis* and *trans* hexahydrophthalic acids on oxidation and they must therefore possess *cis-cis*, *trans-trans*, and *cis-trans* stereochemistry respectively. One ambiguity lies in the *trans-trans* isomer, of which *syn* and *anti* diastereomers are possible (Fig. 14). Only one of these compounds is known experimentally, the thermodynamically stable one. Models show that the *trans-syn-trans* isomer has the center ring constrained to a chair form, while the *trans-anti-trans* isomer has a corresponding boat form. The latter is (from models) somewhat less flexible than cyclohexane-1,4-dione itself, and its dipole moment was predicted to be a little smaller. The observed moment[162] of the known isomer was 1.09 D, and the *trans-anti-trans* stereochemistry was therefore assigned to the compound.[162] The *anti* isomer would be expected to be much more stable than the *syn*, in view of the fact that the amount of chair form in equilibrium with the boat form in the parent cyclohexane-1,4-dione was

[159] P. Groth and O. Hassel, *Proc. Chem. Soc.*, **1963**, 218.

[160] A. Mossel, C. Romers, and E. Havinga, *Tetrahedron Letters*, **19**, 1247 (1963).

[161] K. Alder and G. Stein, *Ann.*, **501**, 247 (1933).

[162] L. A. Freiberg and W. Szkrybalo, unpublished results.

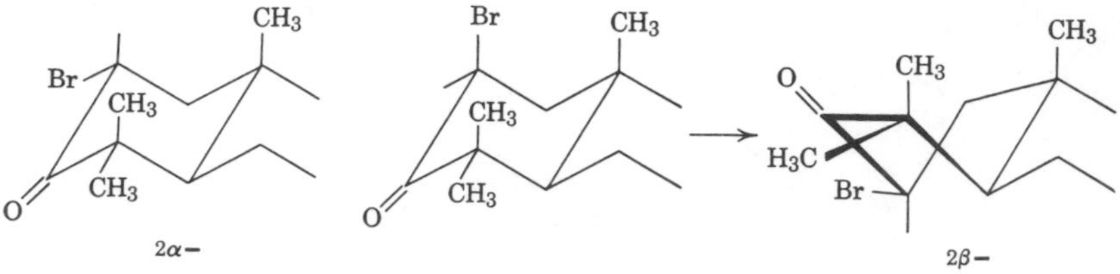

Fig. 15. Ring A conformations in the 2-bromolanost-8-en-3-ones.

too small to detect. Evidence has also been presented which indicates that both cis- and trans-2,5-di-t-butylcyclohexane-1,4-dione exist preferentially in boat forms.[163]

The amount of boat form present in cyclohexanone itself is not yet known, but if the chair form is of lower entropy than the boat by 2 e.u.,[†] while the latter is of higher enthalpy by 2.8 ± 1.0 kcal./mole, the amount of boat form in equilibrium with chair form may be estimated to be in the range 1–12% at room temperature. Thus, if there is any sizable steric interaction in the chair conformation of a substituted cyclohexanone, which can be alleviated by the adoption of a boat conformation, the latter may be a significant or major contributor to the actual ground state of the molecule.

The first compound for which it was experimentally demonstrated that a compound free to adopt either a boat or a chair conformation actually preferred the former was 2β-bromolanost-8-en-3-one, studied by Barton and co-workers.[164] According to the infrared and ultraviolet spectra (Sec. 7-3), this compound had an equatorial bromine atom. By the same criteria, the bromine atom in the 2α epimer was also equatorial. The interpretation made was that ring A exists in the ordinary chair form when the bromine is α, but when it is β the simultaneous interaction of the three syn-axial β substituents is sufficient to push ring A over into a boat conformation (Fig. 15), with θ probably near 30°. Chemical methods may be used to show in which compound the bromine is α and in which it is β. The usual chemical method for determining the configuration of the carbon to which the halogen is attached in a compound such as 2-bromo-lanost-8-en-3-one is to reduce the compound to a bromohydrin, and to determine whether this is cis or trans by the type of product obtained on treatment with base (ketone or epoxide, respectively; cf. p. 107). Reductive removal of the bromine then leads to an alcohol of known configuration, thereby establishing the configuration at C-2.

[163] R. D. Stolow and C. B. Boyce, J. Am. Chem. Soc., 83, 3722 (1961).

[164] D. H. R. Barton, D. A. Lewis, and J. F. McGhie, J. Chem. Soc., 1957, 2907.

† Estimated from the data in Ref. 139 for cyclohexane, allowing for the changes in symmetry.

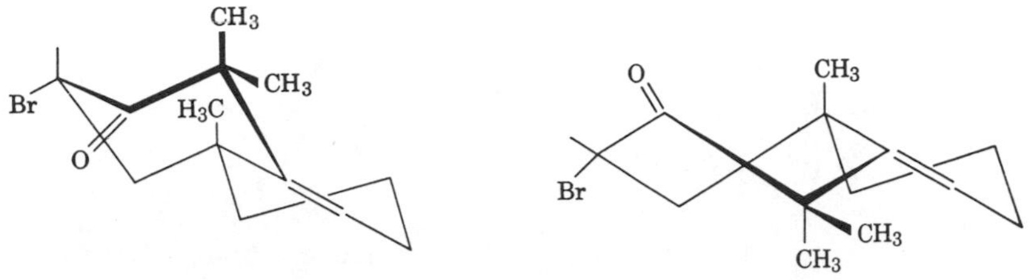

Fig. 16. 2α-Bromo-Δ⁵-4,4-dimethylcholesten-3-one.

An interesting case is the bromination product of Δ^5-4,4-dimethylcholestenone (Fig. 16) in which the halogen is at C-2 and which must, from the infrared spectrum, have the halogen axial. The rotatory dispersion curve showed a negative shift relative to the parent, while the 2β-bromo derivative with ring A in a chair form would be expected to show a positive shift. The only acceptable qualitative interpretation was that ring A was in a boat form,[165] although there is no immediately obvious reason why the molecule prefers this form. Chemical interrelation and NMR spectra later confirmed the α-configuration at C-2.[166] The quantitative effect of the bromine on the rotatory dispersion curve was suspiciously small, however, and moreover it changed with changing solvent.[167] An examination of the dipole moment of the compound was then carried out, and it was concluded that a mixture of boat and chair forms was present, the exact ratio being dependent on solvent, but the amounts were roughly comparable.[167] The reason for the stability of the boat form here appears to a combination of two factors, the tendency of the bromine to want to be axial, and the reduced unfavorable torsion in the boat form at the C 4-5 bond when the double bond is present at C 5-6.

Quite a number of additional cases of the occurrence of boat forms have been reported in steroidal systems.[134,135,168–170] These forms usually occur in cyclohexanones because of a severe 1,3-diaxial repulsion which can be relieved by going over to a boat form. There remain, however, a

[165] D. T. Cropp, B. B. Dewhurst, and J. S. E. Holker, *Chem. Ind.* (*London*), 209 (1961); G. R. Chaudhry, T. G. Halsall, and E. R. H. Jones, *J. Chem. Soc.*, **1961**, 2725.

[166] B. B. Dewhurst, J. S. E. Holker, A. Lablache-Combier, and J. Levisalles, *Chem. Ind.* (*London*), **1961**, 1667; R. J. Abraham and J. S. E. Holker, *J. Chem. Soc.*, **1963**, 806.

[167] M. A. DaRooge, unpublished work.

[168] C. W. Shoppee, M. E. H. Howden, R. W. Killick, and G. H. R. Summers, *J. Chem. Soc.*, **1959**, 630. See also Sec. 5-5a.

[169] F. Sondheimer, Y. Klibansky, Y. M. Y. Haddad, G. H. R. Summers, and W. Klyne, *J. Chem. Soc.*, **1961**, 767.

[169a] C. Djerassi and W. Klyne, *Proc. Natl. Acad. Sci. U.S.*, **48**, 1093 (1962).

[170] J. W. Huffman, *J. Org. Chem.*, **24**, 447 (1959).

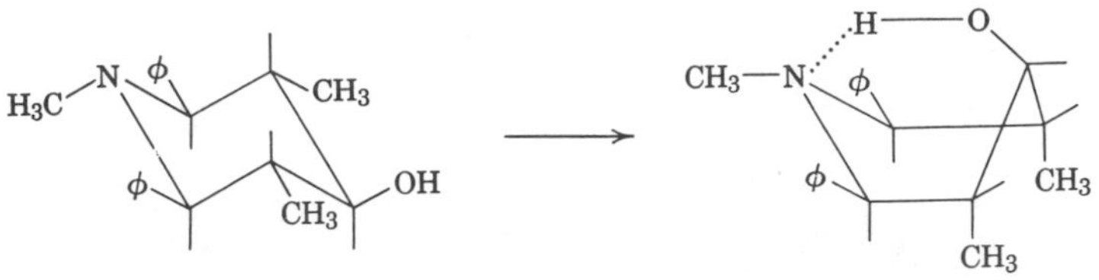

Fig. 17. The suggested conformation of α-1,3,5-trimethyl-2,6-diphenyl-4-piperidinol.

number of cases which are not understood as yet. Sondheimer and coworkers have presented evidence that, although 2α-methylcholestan-3-one is more stable than the corresponding 2β epimer, this stability order is reversed by the introduction of a double bond in the 6-position.[169] The greater stability of the 2β epimer is consistent only with its existence in a boat form. No explanation has been offered for the stability of the boat form in this system.

Huffman[170] has suggested that, judging from the infrared spectrum, lanost-8-en-3α-ol has ring A in a boat form. The weight of the evidence[171] leads to the conclusion that Huffman's compound was in fact the slightly impure 3β epimer, and neither the authentic 3α nor the 3β compound shows any evidence for a boat conformation.

It has been reported[172] that the α-form of 1,3,5-trimethyl-2,6-diphenyl-4-piperidinol shows roughly comparable amounts of free and intramolecularly hydrogen-bonded hydroxyl in the O—H stretching region of the infrared spectrum, which was attributed to a chair → boat equilibrium (Fig. 17). Although the hydrogen bond tends to stabilize the boat, and the *gauche* interactions of the hydroxyl with the methyls tend to destabilize the chair, these energies would not be expected to be sufficient to overcome the 1,3-interaction between the methyls in the boat form in addition to the energy of the boat itself, and this case merits further study.

A more unusual case which has led to considerable discussion involves the 4,4-dimethyl-3-keto steroids and related compounds. Before the relatively low energy of the boat form of a cyclohexanone was recognized, it was assumed that ring A in such compounds would be in an ordinary chair form.[173] After the magnitude of the 1,3-diaxial-dimethyl interaction was determined,[174] it seemed likely that ring A would exist in a boat form, or at least that a boat ⇌ chair equilibrium would exist, and experimental

[171] I. L. Allsop, A. R. H. Cole, D. E. White, and R. L. S. Willix, *J. Chem. Soc.*, **1956,** 4868.

[172] M. Balasubramanian and N. Padma, *Tetrahedron Letters,* **1963,** 49.

[173] For example, C. Djerassi, O. Halpern, V. Halpern, and B. Riniker, *J. Am. Chem. Soc.*, **80,** 4001 (1958).

[174] N. L. Allinger and M. A. Miller, *J. Am. Chem. Soc.*, **83,** 2145 (1961).

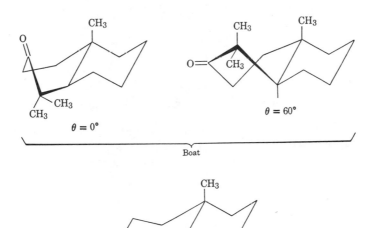

Fig. 18. Possible conformations of 4,4-dimethyl-5α-cholestan-3-one.

evidence was obtained which was interpreted in terms of such an equilibrium.[175] An alternative suggestion,[176] which after considerable discussion[175–179] has now become accepted, is that ring A exists in a flattened chair form, neither a boat nor an ordinary chair, but between the two (Fig. 18). The key experiment in the solution of this problem was the determination of the dipole moment of 4,4-dimethylandrostan-3,17-dione.[176] The observed moment was too small to correspond to an

[175] J. Lehn, J. Levisalles, and G. Ourisson, *Tetrahedron Letters,* **1961,** 682.

[176] N. L. Allinger and M. A. DaRooge, *Tetrahedron Letters,* **1961,** 676; *J. Am. Chem. Soc.,* **84,** 4561 (1962).

[177] J. S. E. Holker and W. B. Whalley, *Proc. Chem. Soc.,* **1961,** 464.

[178] A. K. Bose, M. S. Manhas, and E. R. Malinowski, *J. Am. Chem. Soc.,* **85,** 2795 (1963).

[179] J. Lehn, J. Levisalles, and G. Ourisson, *Bull. Soc. Chim. France,* **1963,** 1096.

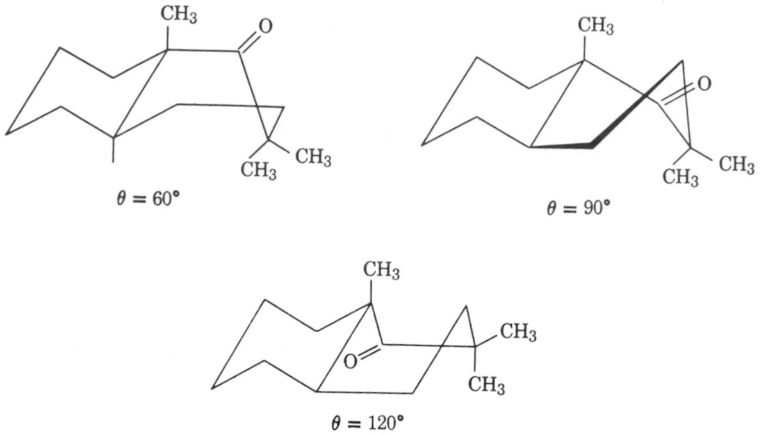

$\theta = 60°$ $\theta = 90°$

$\theta = 120°$

Fig. 19. Boat conformations of the 17,17-dimethyl-17a-keto-D-homosteroids.

ordinary chair or to any kind of boat form, and required the two dipoles to be approximately co-planar. This type of compound is a special case, because there is present a severe *syn*-diaxial dimethyl interaction in the chair form, which is not relieved in the favorable boat ($\theta = 60°$). The alternative boat ($\theta = 0°$) is not utilizing the favorable type of torsion possible for a cyclohexanone, and any conformation having θ between $0°$ and $60°$ has some of the bad features of both. The molecule finds a way out of the dilemma by adopting the flattened chair form, in which C-3 moves up toward (but not quite to) the plane defined by C-1, C-2, C-4, and C-5. Quantitative calculations[148] (with some fairly crude approximations) show that this behavior is reasonable, and at the same time predict that the 17,17-dimethyl-17a-keto-D-homosteroids, which at first sight appear to be analogous to the 4,4-dimethylcholestanones, should prefer boat to flattened chair forms. The experimental data are consistent with the prediction.[148,169a] The fundamentally different conformation of the D-homo compounds results from their having boat forms which are constrained by the fusion of the C ring to values of θ from $60°$ to $120°$ (Fig. 19). At the former limit there is a serious repulsion between the 17α-methyl and the 14α-hydrogen, while at $\theta = 120°$ the 18-methyl is too close to the 16β-hydrogen. At $\theta = 90°$ neither of these interactions is very serious, and the ring energy is a minimum for the flexible form. Again, quantitative calculations[148] are consistent with experiment in predicting that, contrary to the ring A cases discussed above (in which a value of $\theta = 90°$ is precluded by the ring fusion), here the boat form with $\theta = 90°$ should be the energetically favored conformation.

A comparison of the ultraviolet spectra of Δ^5-3-keto steroids and their 4,4-dimethyl derivatives shows that the two groups of compounds have rather different conformations,[180] but the conformational details cannot be deduced from this information alone.

Djerassi and co-workers have shown that the axial haloketone rule could be used in conjunction with rotatory dispersion measurements to show quickly and unambiguously the conformation of ring A of the epimeric 2-bromo-2-methylcholestanones (p. 169).[181] Thus it was concluded that 2β-bromo-2α-methylcholestanone had ring A in the chair form, while the C-2 epimer had ring A in the boat form. Subsequently, *a priori* calculations indicated that, while the former compound could be predicted[182] to show a strong preference for a chair conformation, the latter should have been a mixture of a boat and a chair in similar amounts. The dipole moment was then determined for the 2α-bromo-2β-methyl compound,[182] and the data accurately confirmed the predictions made by *a priori* calculation. Additional confirmation came from the variation of the circular dichroism curve of the compound with temperature.[183] Interestingly, the rather different behavior of the two epimers appears to be due, not to the difference between the methyl-methyl and the methyl-bromine repulsions, but rather to the inherent tendency of the bromine to prefer an axial position when adjacent to a carbonyl group (Sec. 7-3).

In most cases the thermodynamically stable epimer of an α-haloketone of the type discussed above can now be predicted by *a priori* calculation[148,182] (Sec. 7-2). The kinetically controlled product of the halogenation of a ketone (or ketone derivative) can sometimes be obtained under the proper reaction conditions, and for prediction purposes Corey set forth the generalization that, for stereoelectronic reasons, the halogen would tend to prefer to attack from an axial position, and the kinetic reaction product could be predicted on that basis (Sec. 5-5a).[184] It will be advantageous to review briefly the basis for such a generalization, and to try to decide what circumstances might lead to a breakdown of Corey's rule.

The mechanism of the halogenation of ketones has been extensively studied and is moderately well understood.[185,186] The kinetics of the

[180] M. Gorodetsky and Y. Mazur, *Tetrahedron Letters*, **1964**, 227.

[181] R. Villotti, H. J. Ringold, and C. Djerassi, *J. Am. Chem. Soc.*, **82**, 5693 (1960).

[182] N. L. Allinger, J. G. D. Carpenter, and M. A. DaRooge, *J. Org. Chem.*, **30**, 1423 (1965).

[183] C. Djerassi, personal communication.

[184] E. J. Corey, *Experientia*, **9**, 329 (1953).

[185] E. J. Corey, *J. Am. Chem. Soc.*, **75**, 4832 (1953).

[186] C. K. Ingold, *Structure and Mechanism in Organic Chemistry*, Cornell University Press, Ithaca, New York, 1953, p. 569.

Fig. 20. Intermediates in the bromination of a carbonyl compound.

reaction under acidic conditions makes it clear that enolization is the rate-determining step. Less is known experimentally about the details of the subsequent reaction steps, although the general course of events is clear. By analogy with the addition of bromine to ordinary olefins, one may write the intermediate bromonium ion (I in Fig. 20) which in principle could react as does that derived from an ordinary olefin (Sec. 5-4b). In practice it would be expected that, because of the loss of translational entropy during such a bimolecular process, displacement of the bromine from the carbonyl carbon by an electron pair from oxygen (leading to II in Fig. 20) would be more rapid, and some evidence for this path has been obtained.[181] If this mechanism is accepted, predictions can be made about the stereochemistry of the kinetically controlled product.[187] Its stereochemistry is that resulting from the path whereby I (Fig. 20) is formed and reacts to give II at the fastest over-all rate. When positively charged bromine or its equivalent adds to cyclohexene, a cyclic bromonium ion is formed which is geometrically analogous to an epoxide, and the bromine is bound to one equatorial and one axial bond of the ring. The cyclic bromonium ion, in the simple case, opens upon axial attack of a bromide ion to yield a *trans*-diaxial bromide, again quite analogous to the behavior of an epoxide (the Fürst-Plattner rule, p. 102).[188–189]

A qualitative consideration of a simple addition-elimination reaction from the quantum mechanical-conformational point of view can very easily rationalize the order of preference of elimination of (or addition to yield); *anti*-coplanar > *syn*-coplanar > non-coplanar groups.[190] The Fürst-Plattner rule,[189] the preference of *trans*-diaxial elimination, and the like are special cases of the generalization above. If we take time as our reaction coordinate and consider the energy of the system where X_2 is added to an olefin (Fig. 21), in the general case, according to LCAO-MO theory, an approximate unnormalized wave function for the system is

[187] For an excellent review on the stereochemistry of the addition of bromine to cyclohexene and related reactions, see J. Valls and E. Toromanoff, *Bull. Soc. Chim. France*, **1961**, 758.

[188] A. Fürst and P. A. Plattner, *Helv. Chim. Acta*, **32**, 275 (1949).

[189] (a) S. J. Angyal, *Chem. Ind. (London)*, **1954**, 1230; (b) R. E. Parker and N. S. Isaacs, *Chem. Rev.*, **59**, 737 (1959).

[190] C. H. DePuy, R. D. Thurn, and G. F. Morris, *J. Am. Chem. Soc.*, **84**, 1314 (1962).

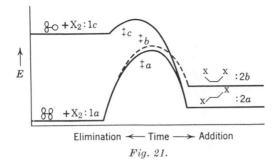

Fig. 21.

$\psi = \lambda(t)\psi_{1a} + \psi_{2a}$, where ψ_{1a} is the wave function of the olefin $+ X_2$ and ψ_{2a} is the wave function of the adduct. The coefficient λ is a function of time which has values ranging from infinity (olefin $+ X_2$) down to zero (adduct). In state $1c$, the olefin is twisted and of higher energy. To the extent that the corresponding function (ψ_{1c}) contributes to the transition state function ($\psi_{\ddagger c}$), the energy of the transition state $\ddagger_c$ will be raised above that of $\ddagger_a$. Hence a bimolecular E_2 elimination is slower when the X groups are not coplanar by an amount which depends on the coefficient λ, and which, since $1c$ is higher in energy than $1a$ by some 60 kcal./mole, may be very large. Instead of an elimination proceeding with an activation energy of anything like that value, the mechanism changes to the E_1 or E_{2cb} carbanion mechanism, which are really unimolecular mechanisms. In a unimolecular elimination reaction, ψ_1 will ordinarily contribute little to the transition state, and thus a very unfavorable transition state can be avoided by a change in mechanism. Other facts are apparent from Fig. 21. For instance, additions to a non-planar olefin such as *trans*-cyclooctene should be very fast[191] and quite exothermic[192] (approximately from $1c$ to $2a$), and this is what is found experimentally.

The reason for the usual preference of *anti* over *syn* elimination is a little more difficult to see, and this does not appear to have previously interpreted theoretically. The *syn* adduct ($2b$) would be of higher energy than the *anti* one ($2a$), and part of this higher energy is due to a repulsion between the X's, which would be relieved in the transition state. Steric acceleration of *syn* elimination might therefore be expected if the adduct is rigidly held in a *syn* conformation throughout the course of the reaction. If $2b$ and $2a$ can interconvert, then the rate of *syn* elimination depends only on the difference in free energy between $2a$ and $\ddagger_b$, and it will be slower than the rate of *anti* elimination, which depends on the free energy difference between $2a$ and $\ddagger_a$ (see p. 28). On this basis the energy

[191] K. Ziegler and H. Wilms, *Ann.*, **567**, 1 (1950).
[192] R. B. Turner and W. R. Meador, *J. Am. Chem. Soc.*, **79**, 4133 (1957).

difference between $\ddagger_a$ and $\ddagger_b$ would have to be pretty small, perhaps 1 or 2 kcal./mole in typical cases, and this is hardly enough to explain the observed rate differences. A molecular orbital picture of the different transition states is quite suggestive. No detailed calculations have been carried out for such a reaction, but qualitatively it is easy to see the direction of the effect. The atomic orbitals concerned can be sketched as in Fig. 22. In the Hückel approxima-

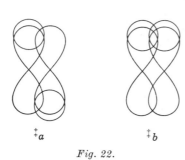

$\ddagger_a$ $\ddagger_b$

Fig. 22.

tion the resonance integral between the X's has the value zero in $\ddagger_a$ and one in $\ddagger_b$. The orbital energies of $\ddagger_a$ and $\ddagger_b$ are then $\pm1.6\beta$, $\pm0.6\beta$, and $\pm2\beta$, 0, 0, respectively, and the total energy of $\ddagger_a$ is 4.4β, while that of $\ddagger_b$ is only 4.0β. The orbital energies will behave qualitatively in this way, irrespective of the size of X or the exact distances involved.† Thus there is a good quantum mechanical

reason why the X's want to keep apart in the transition state. (In view of the known facts, it seems rather certain that a more elaborate quantum mechanical calculation would confirm the conclusions given here and might give some useful numerical values. While a proper calculation by the ASMO method would be difficult, it is feasible at the present time.)

Considerations such as given above indicate that "equatorial attack," which appears to necessitate formation of either a bent C—X bond, or a twisted olefin (like $\ddagger_c$) in the transition state, would be a very slow reaction. It seems improbable that such a reaction path will be followed as long as an alternative one is available. In the bromination of cyclohexanone itself, there are two reaction paths available which allow for transition states of the type $\ddagger_a$, as indicated in Fig. 23. The cation is said to be adding by parallel or antiparallel attack, depending on whether it approaches from the same or opposite side of the double bond as the axial hydrogen on the adjacent carbon, and the initial reaction products are a boat and a chair, repectively.[187] Since these products differ in energy by only about 2.8 kcal./mole, the transition states must differ in energy by even less. Hence one expects a substantial predominance of the product (*a*), but not a 100% predominance, and the energy difference between $\ddagger_a$ and $\ddagger_b$ is sufficiently small that it may on some occasions be outweighed by other factors.

Boat forms of cyclohexenes usually do not differ much from chair forms in energy (Sec. 2-6a), and this fact has led to a consideration of the

† A similar formulation predicts that an S_N2 reaction prefers to proceed with inversion of configuration while an S_E2 reaction prefers retention.

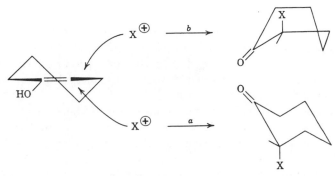

Fig. 23. Parallel (*b*) and antiparallel (*a*) addition to cyclohexanone enol.

bromination of the boat form of the enol of cyclohexanone. The boat form of a cyclohexene does not, however, correspond to a minimum of conformational energy,[193] and it seems to be better considered as one extreme in the vibration of the chair form of the molecule. It therefore does not appear to be meaningful to discuss reactions or properties of the boat form of a cyclohexene in a simple molecule. If sufficient constraint is present in the molecule, it is possible that the energy minimum will not correspond to the ordinary chair (or half-chair) but will in fact be more of a boat. A case in point happens not to be a cyclohexene, but the rather analogous epoxide, 2β,3β-epoxylanostane (Fig. 24). The

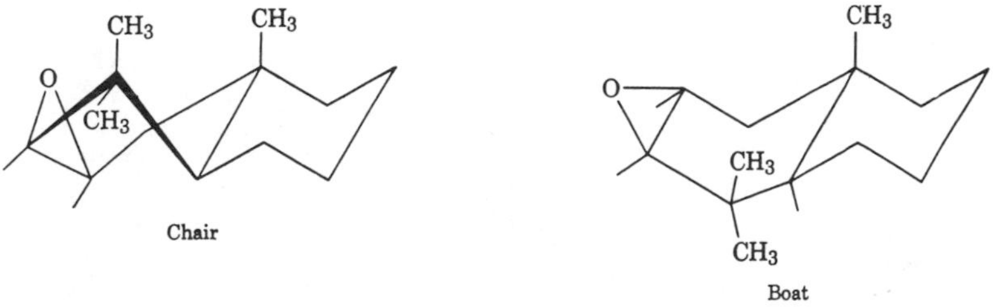

CH₃ ... CH₃ ... CH₃ ... O ... CH₃ ... Chair ... CH₃ ... O ... CH₃ ... CH₃ ... Boat

Fig. 24. Conformations of 2β,3β-epoxylanostane.

ordinary chair form is unfavorable because of the interaction of the 19-methyl and 4β-methyl groups, while this interaction is relieved in the boat form, and some conformational compromise can be anticipated (although it has not yet been demonstrated in the ground state). When the epoxide is opened with hydrogen bromide, the Fürst-Plattner rule suggests that the bromide ion would attack from the α-side and give the diaxial 3α-bromo-2β-hydroxy derivative. In fact the 2α-bromo-3β-hydroxy compound is obtained.[164] The transition state leading to the latter can

[193] R. Bucourt and D. Hainaut, *Compt. Rend.*, **258**, 3305 (1964).

have the desired *anti* arrangement of groups and still avoid the serious interaction between the methyl groups if it has ring A in a boat form. The eclipsing which develops in the ring in the transition state leading to the boat is more serious than that in the corresponding transition state leading to the chair, but evidently the methyl-methyl interaction is the determining factor.

Returning to the case of the bromination of the enol acetate of 2-methylcholestanone, the kinetic product is 2α-bromo-2β-methylcholestanone, which has the halogen axial and ring A is in the boat form (p. 309). In this case, the reaction must go via a path like that of *b* in Fig. 23. Equatorial attack on the chair form is quite improbable, and axial attack on the boat form is not really a meaningful concept. The bromination of 2-chlorocholestanone and the chlorination of 2-bromocholestanone lead to epimeric kinetic products in which the addition has in each case taken place from the α-side of the molecule, analogous to the bromination of 2-methylcyclohexanone.[194]

Deformation in cyclohexane systems seems destined to receive more study and become increasingly important in the development of the theoretical basis of conformational analysis. The peculiarities of the 4,4-dimethyl-3-keto steroids, the reflex effect, conformational transmission[195] (Sec. 5-9), and many related phenomena show that, even in the relatively simple cases of chair forms in cyclohexane systems, our understanding has only reached the level of the first approximation.†

[194] E. W. Warnhoff, *J. Org. Chem.*, **28,** 887 (1963).
[195] R. Bucourt, *Bull. Soc. Chim. France*, **1962,** 1983; **1963,** 1262.
† Short reviews of all these topics are to be found in ref. 135.

Author Index

487

Subject Index

509